HIGHER RECURSION THEORY AND SET THEORY

LECTURE NOTES SERIES
Institute for Mathematical Sciences, National University of Singapore

Series Editors: Zuowei Shen and Chen-Bo Zhu
Institute for Mathematical Sciences
National University of Singapore

ISSN: 1793-0758

Published

Vol. 44 *Higher Recursion Theory and Set Theory*
edited by James Cummings, Andrew Marks, Yue Yang & Liang Yu

Vol. 43 *On the Langlands Program: Endoscopy and Beyond*
edited by Wee Teck Gan, Dihua Jiang, Lei Zhang & Chen-Bo Zhu

Vol. 42 *Aspects of Computation and Automata Theory with Applications*
edited by Noam Greenberg, Sanjay Jain, Keng Meng Ng, Sven Schewe, Frank Stephan, Guohua Wu & Yue Yang

Vol. 41 *Density Functionals for Many-Particle Systems: Mathematical Theory and Physical Applications of Effective Equations*
edited by Berthold-Georg Englert, Heinz Siedentop & Martin-Isbjörn Trappe

Vol. 40 *Modeling and Simulation for Collective Dynamics*
edited by Weizhu Bao, Peter A Markowich, Benoit Perthame & Eitan Tadmor

Vol. 39 *Models and Methods for Quantum Condensation and Fluids*
edited by Weizhu Bao, Yongyong Cai, Ionut Danaila & Peter A Markowich

Vol. 38 *Genealogies of Interacting Particle Systems*
edited by Matthias Birkner, Rongfeng Sun & Jan M Swart

Vol. 37 *Mathematics of Shapes and Applications*
edited by Sergey Kushnarev, Anqi Qiu & Laurent Younes

Vol. 36 *The Geometry, Topology and Physics of Moduli Spaces of Higgs Bundles*
edited by Richard Wentworth & Graeme Wilkin

Vol. 35 *Combinatorial and Toric Homotopy: Introductory Lectures*
edited by Alastair Darby, Jelena Grbić, Zhi Lü & Jie Wu

Vol. 34 *White Noise Analysis and Quantum Information*
edited by Luigi Accardi, Louis H Y Chen, Takeyuki Hida, Masanori Ohya, Si Si & Noboru Watanabe

Vol. 33 *Sets and Computations*
edited by Sy-David Friedman, Dilip Raghavan & Yue Yang

For the complete list of titles in this series, please go to
https://www.worldscientific.com/series/LNIMSNUS

Lecture Notes Series, Institute for Mathematical Sciences,
National University of Singapore
Vol. 44

HIGHER RECURSION THEORY AND SET THEORY

Editors

James Cummings
Carnegie Mellon University, USA

Andrew Marks
University of California, Berkeley, USA

Yue Yang
National University of Singapore, Singapore

Liang Yu
Nanjing University, China

World Scientific

NEW JERSEY • LONDON • SINGAPORE • BEIJING • SHANGHAI • HONG KONG • TAIPEI • CHENNAI

Published by

World Scientific Publishing Co. Pte. Ltd.
5 Toh Tuck Link, Singapore 596224
USA office: 27 Warren Street, Suite 401-402, Hackensack, NJ 07601
UK office: 57 Shelton Street, Covent Garden, London WC2H 9HE

Library of Congress Control Number: 2025000010

British Library Cataloguing-in-Publication Data
A catalogue record for this book is available from the British Library.

Lecture Notes Series, Institute for Mathematical Sciences,
National University of Singapore — Vol. 44
HIGHER RECURSION THEORY AND SET THEORY

ISBN 978-981-98-0657-7 (hardcover)
ISBN 978-981-98-0658-4 (ebook for institutions)
ISBN 978-981-98-0851-9 (ebook for individuals)

For any available supplementary material, please visit
https://www.worldscientific.com/worldscibooks/10.1142/14141#t=suppl

https://doi.org/10.1142/9789819806584_fmatter

Contents

https://doi.org/10.1142/9789819806584_fmatter

Foreword

The Institute for Mathematical Sciences (IMS) organizes thematic programs of longer duration as well as shorter workshops and public lectures. The themes are selected from among areas at the forefront of current research in the mathematical sciences and their applications. Each volume of the *IMS Lecture Notes Series* is a compendium of papers based on lectures and tutorials delivered at the IMS. The aim is to make original papers and expository articles on a subject of current interest available to the international research community. These volumes also serve as a record of activities that took place at the IMS. We hope that through regular publication of these Lecture Notes the Institute will achieve, in part, its objective of reaching out to the community of scholars in the promotion of research in the mathematical sciences.

Zuowei Shen
Chen-Bo Zhu
Series Editors

https://doi.org/10.1142/9789819806584_fmatter

Preface

This volume arises from the IMS programme "Higher Recursion Theory and Set Theory — in celebration of the research work of Professors Theodore A. Slaman and W. Hugh Woodin (20 May–14 June 2019), held at the Institute for Mathematical Sciences at the National University of Singapore. The programme was dedicated to celebrating the research achievements of Professors Slaman and Woodin on the occasion of their 65th birthdays in 2019.

The workshop focused on two primary research areas and their interactions: higher recursion theory and set theory, where Professors Slaman and Woodin have made significant contributions. By "higher recursion theory", we refer to aspects of recursion theory that either study subsets of the natural numbers beyond arithmetical sets or investigate computability or definability on domains extending beyond natural numbers. This includes topics such as hyperarithmetic sets, descriptive set theory, α-recursion theory and E-recursion theory, along with areas like the fine structure theory of the constructible universe L and Q theory. Although "set theory" was not specified in the title, the programme focused on areas closely linked to definability, such as Woodin's program on ultimate L, the HOD conjecture, and descriptive inner model theory.

Higher recursion theory and the parts of set theory mentioned above have a long history of interaction. In the late 1960s and 1970s when Sacks and his students systematically explored α-recursion theory, they applied many ideas from fine structure theory of Ronald Jensen to study computability on initial segments of L. On the other hand, Jensen's fine structure theory in some sense can be viewed as recursion theory on L. The interaction remains ongoing, producing notable outcomes such as Slaman and Woodin's work on the global structure theory of the Turing degrees, and definability of the jump operator in the 1980s and 1990s, and Montalban and Shore's recent calibration of determinacy in reverse mathematics.

The chapters in this volume address a wide range of topics related to higher recursion theory, set theory and their interactions. Below, we provide an outline of the essential background to these themes.

One of the foundational results in recursion theory is Turing's theorem, which demonstrates that there is no algorithm for solving the halting problem — a theorem that has had profound effects across mathematics. It introduced the fundamental tools used to study the dividing lines between computability and incomputability. In the internal development of recursion theory, Turing's result also plays an indispensable role. The relativized version of Turing's proof shows that for any oracle X, the halting problem relative to X is incomputable from X. This operator, known as the Turing jump, offers a canonical way of increasing computational complexity and has been central to the study of the global structure of the Turing degrees, including in Slaman and Woodin's work on automorphisms of the Turing degrees. Two chapters in this volume study local degree structures: one by Cholak, Downey, Greenberg and Turetsky, and another on pseudojump inversion by Jananthan and Simpson.

In higher recursion theory, hyperarithmetical sets or Borel sets played a central role, and closely tied to the classification problems in descriptive set theory. The field of Borel reducibility of equivalence relations has provided a framework for analyzing the complexity of classification problems in mathematics, a major research topic in descriptive set theory over the past 25 years. Hyperarithmetical tools have recently been applied to the study of mathematical structures. Seven chapters in this volume reflect this theme, contributions from Bezhenov and Mustafa; Greenberg, Richter, Shelah and Turetsky; Jananthan and Simpson on higher Posner-Robinson Theorem; Monin and Yu; Montalban; Shore; and Westrick. Additionally, two chapters on generalized computation and functional interpretation by Kjos-Hansson and Towsner, respectively, complement these discussions.

In modern set theory, one of the central themes is the development of inner model theory. The prototypical inner model, Gödel's constructible universe L, is the smallest transitive model of ZFC set theory which contains the class of ordinals. Inner model theory has since involved the discovery of larger L-like models which are canonical in a similar way, and which can be used to illuminate the structure of the full set-theoretic universe V. Inner models have a rich internal structure, which can be explored using the fine structure theory initiated by Jensen. For example, inner models typically satisfy a strong form of the Generalised Continuum Hypothesis known as diamond, and a powerful construction principle known as square. Inner models also interact with the external universe V. Inspired by

Silver's work on the Singular Cardinal Hypothesis, Jensen showed that either zero-sharp exists or L covers V in the sense that every uncountable set of ordinals in V is covered by a set in L of the same V-cardinality. This result, known as the Covering Lemma, is best viewed as a very sharp dichotomy concerning the relationship between L and V: a) If zero-sharp exists then V is "far from L", for example, every uncountable regular V-cardinal is weakly compact in L. b) If zero-sharp does no exist then V is "close to L", for example, L computes the successor of every V-singular cardinal correctly. The Covering Lemma and its generalisations to larger inner models have played an important role in analysing V, and in proving lower bounds in consistency strength for combinatorial hypotheses such as the Singular Cardinal Hypothesis.

One of the characteristic properties of inner models is that they have a definable global well-ordering. There is a largest model with such a well-ordering, namely the universe HOD of hereditarily ordinal definable sets. Woodin proved a striking covering result for HOD known as the HOD Dichotomy: assuming that there is an extendible cardinal delta then either a) Every V-singular cardinal greater than delta is singular in HOD, and HOD computes its successor correctly or b) Every regular cardinal greater than delta is measurable in HOD. Woodin's HOD Conjecture asserts roughly that alternative b) does not hold. It is actually formulated as a conjecture about HOD in ZFC, which strongly negates b) when delta is extendible. If it holds the HOD Conjecture would have striking consequences for inner model theory and choiceless set theory. For example, it would imply that HOD is quite rigid (in particular there is no analogue of zero-sharp for HOD) and would radically limit the possibilities for choiceless large cardinals.

The prototypical inner model L can only contain rather small large cardinals, in particular there is no measurable cardinal in L. The central problem in inner model theory is to construct L-like models which can contain large cardinals, and then to explore their internal structure and their external interactions with V. An extender is a directed system of measures which can be used to generate an elementary embedding from V into an inner model by taking a limit ultrapower. Large cardinal hypotheses up to the level of a superstrong cardinal can be expressed using short extenders (those whose measures concentrate on sets whose size is equal to the critical point), while stronger hypotheses require the use of long extenders. Classical inner models are of the form $L[E]$ where E is some sequence of short extenders, obeying a list of technical conditions. The two main ingredients in analysing such models are fine structure and the comparison of mice via iteration. Very roughly speaking, mice are set approximations to class models of the form $L[E]$, and comparison is a process in which two mice M and N are iterated side by side to produce models M' and N where M' is an initial segment of N' or vice versa. In the early stages of the inner model program, mice were compared via linear iterations. This gave models of the form $L[E]$ which could contain strong cardinals and which had a Δ^1_3 well-ordering of the reals (so could not be models of PD). Martin and Steel introduced iteration trees, a form of non-linear iteration: this development made it possible to produce inner models of the form $L[E]$ which contain Woodin cardinals, and was also crucial in proving PD and $\mathrm{AD}^{L(\mathbb{R})}$ from the optimal hypotheses. In general building iteration trees will require iteration strategies, which dictate how the tree should be built.

More recently, inner model theory has been infused with ideas from descriptive set theory, giving rise to descriptive inner model theory. Work by Woodin played a central role here:

- The isolation of AD-plus, a strengthened form of AD which holds in natural models of AD, and the study of its models.
- The proof (building on work of Steel) that if $V = L(\mathbb{R})$ and AD holds then HOD has the form $L[E, \Sigma]$ where E is a sequence of extenders and Σ is a sequence of iteration strategies.
- The discovery of the core model induction, which fuses determinacy methods with inner model theory to give a powerful tool for proving lower bounds in consistency strength.

Descriptive inner model theory involves classical mice built from extenders, hybrid mice built from extenders and iteration strategies, and models of AD-plus. Woodin's Mouse Set Conjecture says roughly that the methods of descriptive inner model theory are adequate for all large cardinals up to the level of a superstrong cardinal.

In this volume, four chapters contribute to these themes, authored by Gitik; Larson; Shi and Steel.

In addition to Professors Slaman and Woodin's remarkable research work, their long-standing association with NUS and IMS has helped establish Singapore become one of the centers for mathematical logic. They have been visiting Singapore every year for the last 20 years. It is fitting that this workshop was held in their honor to mark their 65th birthdays.

This volume contains 15 contributions related to the topics presented during the workshop, reflecting the current state of art in these fields. The editors wish to express their gratitude to the Institute of Mathematical Sciences, its staff, and advisory board for funding, organizing and providing an excellent support for the programme. We are also grateful to World Scientific Publisher for their assistance and patience, and to authors for their valuable contributions to this volume.

The organizers of the programme were:
James Cummings (Carnegie Mellon University)
Andrew Marks (UC Berkeley)
YANG Yue (NUS)
YU Liang (Nanjing University)

November 2024 *Volume Editors*

https://doi.org/10.1142/9789819806584_0001

Elementary Theories of Rogers Semilattices in the Analytical Hierarchy

Nikolay Bazhenov[a]
Sobolev Institute of Mathematics,
4 Acad. Koptyug Ave., Novosibirsk, 630090, Russia
bazhenov@math.nsc.ru

Manat Mustafa[b]
Department of Mathematics, School of Sciences and Humanities,
Nazarbayev University,
53 Qabanbaybatyr Ave., Astana, 010000, Kazakhstan
manat.mustafa@nu.edu.kz

Investigations of elementary theories for Rogers semilattices constitute one of the core directions in the theory of numberings. In recent years, the studies of significative differences in isomorphism types of these semilattices, witnessed by their algebraic and first-order properties, flourished (especially, in the area of numberings belonging to the levels of various recursion-theoretic hierarchies). Nevertheless, there are not many known results on the algorithmic complexity of elementary theories for Rogers semilattices.

In this chapter, we investigate initial segments of Rogers semilattices in the analytical hierarchy, and we study decidability for fragments of the corresponding first-order theories. Let n be a non-zero natural number. For an arbitrary non-trivial Rogers semilattice R induced by a Σ^1_n-computable family of sets, we obtain the following results. The Π_4-fragment of the theory $Th(R)$ in the signature of partial orders is hereditarily undecidable. The Σ_1-fragment of $Th(R)$ in the signature of

[a]The work of Bazhenov was carried out within the framework of the state contract of the Sobolev Institute of Mathematics (project no. FWNF-2022-0011).
[b]The work was supported by Nazarbayev University Faculty Development Competitive Research Grants 201223FD8823. Mustafa was also partially supported by the Science Committee of the Ministry of Education and Science of the Republic of Kazakhstan (Grant No. AP19676989).

upper semilattices is decidable. Similar results hold for families in the arithmetical hierarchy, starting with the Σ^0_4 level.

1. Introduction

Consider a family $\mathcal{S}$ of mathematical objects. The theory of numberings provides a fruitful formal framework for studying uniform computations of $\mathcal{S}$. The roots of the theory can be traced back to the incompleteness theorems of Gödel [1], and the monograph of Kleene [2] (e.g., his construction of a universal partial computable function). Kolmogorov and Uspenskii [3, 4] and, independently, Rogers [5] established the theory of numberings as a separate field of study.

Let $\mathcal{S}$ be a countable set. A *numbering* of $\mathcal{S}$ is a surjective map ν from the set of natural numbers ω onto $\mathcal{S}$. The *reducibility* between numberings provides a standard method for gauging their algorithmic complexity. A numbering ν is *reducible* to another numbering μ, denoted by $\nu \leq \mu$, if there is total computable function $f(x)$, such that $\nu(x) = \mu(f(x))$ for all $x \in \omega$. In other words, there is an effective procedure which transforms a ν-index of an object $a \in \mathcal{S}$ into a μ-index for a.

This chapter considers only families $\mathcal{S}$ containing subsets of ω, i.e., we always assume that $\mathcal{S} \subseteq P(\omega)$ and $\mathcal{S}$ is countable.

From the 1960s to 1990s, research in the theory of numberings has been mainly focused in the area of *computable numberings* and the corresponding *Rogers semilattices* (to be explained below). Goncharov and Sorbi [6] started developing the theory of generalized computable numberings. One of their approaches to generalized computations can be summarized as follows.

Let Γ be a complexity class (e.g., Σ^0_1, d-Σ^0_1, Π^0_n, or Σ^1_n). A numbering ν of a family $\mathcal{S}$ is Γ-*computable* if the set

$$G_\nu = \{\langle k, x\rangle : x \in \nu(k)\}$$

belongs to the class Γ. We emphasize that in the literature, they typically use the term "computable numbering" as a synonym for a "Σ^0_1-computable numbering". We say that a family $\mathcal{S}$ is Γ-*computable* if it has a Γ-computable numbering.

In a standard recursion-theoretic way, the notion of reducibility between numberings gives rise to the corresponding upper semilattice: Assume that we fixed a class Γ. For a Γ-computable family $\mathcal{S}$, the *Rogers semilattice* of $\mathcal{S}$ contains the degrees of all Γ-computable numberings of $\mathcal{S}$. As usual, here two numberings have the same degree if they are reducible to each other.

Informally speaking, the supremum of two numberings is provided by their join, see Sec. 2.1 for a more formal exposition.

We refer the reader to, e.g., [7, 8, 9, 10] for the results and bibliographic references on Γ-computable numberings (and the corresponding Rogers semilattices). One of important problems in the theory of Rogers semilattices can be formulated as follows:

Problem 1.1: *Given a class Γ, study the elementary theories of Rogers semilattices for Γ-computable families.*

For the sake of brevity, we say that an upper semilattice $\mathcal{U}$ is a *Rogers Γ-semilattice* if $\mathcal{U}$ is isomorphic to the Rogers semilattice of some Γ-computable family. A semilattice $\mathcal{U}$ is *non-trivial* if it contains more than one element.

V'yugin [11] proved that there are infinitely many pairwise elementarily non-equivalent Rogers Σ^0_1-semilattices. Badaev, Goncharov and Sorbi [12] proved that for any natural number $n \geq 2$, there are infinitely many pairwise elementarily non-equivalent Rogers Σ^0_n-semilattices. In the paper [13], authors obtained a partial result for the analytical hierarchy: For any $n \geq 1$, there are at least three pairwise elementarily non-equivalent, non-trivial Rogers Σ^1_n-semilattices.

Recall that $\mathcal{E}$ is the lattice of c.e. sets under set-theoretic inclusion, and $\mathcal{E}^*$ is the quotient of $\mathcal{E}$ modulo finite sets. Podzorov (Theorem 1 of [14]) proved that for any $n \geq 2$, every non-trivial Rogers Σ^0_n-semilattice contains an ideal, which is isomorphic to one of the following:

(a) $\mathcal{E}^*$, or
(b) $\mathcal{E}^* \setminus \{\bot\}$, i.e. $\mathcal{E}^*$ with the least element omitted.

Recall that the elementary theory $Th(\mathcal{E}^*)$ is hereditarily undecidable [15]. Therefore, it is not difficult to show that for a non-trivial Rogers Σ^0_n-semilattice $\mathcal{U}$, where $n \geq 2$, the theory $Th(\mathcal{U})$ is also hereditarily undecidable (see Corollary 2.2.2 of [16]). We note that (to the best of our knowledge) it is still open whether this result can be extended to Rogers Σ^0_1-semilattices (see, e.g., Question 7 in [16]).

By proceeding in a similar way to [14], Dorzhieva [17] established that for any $m \geq 1$ and any non-trivial Rogers Σ^1_m-semilattice $\mathcal{R}$, the theory $Th(\mathcal{R})$ is hereditarily undecidable.

In this chapter, we continue the studies of Rogers semilattices in the analytical hierarchy developed in [10, 13]. We consider the following problem:

Problem 1.2: *Let $m \geq 1$, and $\mathcal{R}$ be a non-trivial Rogers Σ^1_m-semilattice. What can one say about undecidability of the* fragments *of $Th(\mathcal{R})$?*

The structure of this chapter is as follows. Section 2 contains necessary preliminaries. In Sec. 3, we prove that for any $\mathcal{R}$ described above, the Π_4-fragment of the theory $Th(\mathcal{R})$ in the signature $\{\leq\}$ is hereditarily undecidable (Theorem 3.1). We note that this theorem cannot be easily obtained by employing an ideal isomorphic to $\mathcal{E}^*$ (recall the results of Podzorov and Dorzhieva): It is known that the Π_6-fragment of $Th(\mathcal{E}^*)$ in the signature $\{\vee, \wedge\}$ is hereditarily undecidable (Nies, Theorem 4.6.1 of [18]), but it is still open whether Π_5–$Th(\mathcal{E}^*)$ is undecidable.

Section 4 proves that the Σ_1-fragment of $Th(\mathcal{R})$ in the signature $\{\leq, \vee\}$ is decidable (Theorem 4.1). Section 5 contains further discussion.

2. Preliminaries

Upper semilattices are usually treated as structures in the signature $L_{usl} = \{\leq, \vee\}$. Given an upper semilattice $\mathcal{U}$, we will work with two kinds of theories:

(a) $Th_{\leq,\vee}(\mathcal{U})$ is the first-order theory of $\mathcal{U}$ in the signature L_{usl};
(b) $Th_{\leq}(\mathcal{U})$ is the elementary theory of $\mathcal{U}$ in the signature of partial orders $L_{po} = \{\leq\}$.

If $\mathcal{A}$ is an upper semilattice and $a \leq_{\mathcal{A}} b$ are elements from $\mathcal{A}$, then the *interval* $[a; b]_{\mathcal{A}}$ is the semilattice

$$[a; b]_{\mathcal{A}} := (\{c \in \mathcal{A} : a \leq_{\mathcal{A}} c \leq_{\mathcal{A}} b\}; \leq_{\mathcal{A}}, \vee_{\mathcal{A}}).$$

2.1. *Numberings*

Suppose that ν is a numbering of a family $\mathcal{S}_0$, and μ is a numbering of a family $\mathcal{S}_1$. Note that the condition $\nu \leq \mu$ implies that $\mathcal{S}_0 \subseteq \mathcal{S}_1$.

Numberings ν and μ are *equivalent* (denoted by $\nu \equiv \mu$) if $\nu \leq \mu$ and $\mu \leq \nu$. The numbering $\nu \oplus \mu$ of the family $\mathcal{S}_0 \cup \mathcal{S}_1$ is defined as follows:

$$(\nu \oplus \mu)(2x) = \nu(x), \quad (\nu \oplus \mu)(2x+1) = \mu(x).$$

The following fact is well-known (see, e.g., p. 36 in [7], or Lemma 6 of [19]): If ξ is a numbering of a family $\mathcal{T}$, then

$$(\nu \leq \xi \,\&\, \mu \leq \xi) \;\Leftrightarrow\; (\nu \oplus \mu \leq \xi).$$

For further background on numberings, the reader is referred to, e.g., [7, 8, 19, 20, 21, 22].

Let Γ be a complexity class with the following properties:

(a) If ν is a Γ-computable numbering and μ is a numbering such that $\mu \leq \nu$, then μ is also Γ-computable.
(b) If numberings ν and μ are both Γ-computable, then the numbering $\nu \oplus \mu$ is also Γ-computable.

It is not difficult to show that for any non-zero natural number n, each of the classes Σ^0_n, Σ^{-1}_n and Σ^1_n has these properties.

Consider a Γ-computable family $\mathcal{S}$. By $Com_\Gamma(\mathcal{S})$, we denote the set of all Γ-computable numberings of $\mathcal{S}$. Since the relation $\equiv$ is a congruence on the structure $(Com_\Gamma(\mathcal{S}); \leq, \oplus)$, we use the same symbols $\leq$ and $\oplus$ on numberings of $\mathcal{S}$ and on $\equiv$-equivalence classes of these numberings.

The quotient structure $\mathcal{R}_\Gamma(\mathcal{S}) := (Com_\Gamma(\mathcal{S})/{\equiv}; \leq, \oplus)$ is an upper semilattice. We say that $\mathcal{R}_\Gamma(\mathcal{S})$ is the *Rogers semilattice* of the Γ-computable family $\mathcal{S}$. For the sake of convenience, we use the following notation:

$$\mathcal{R}^1_n(\mathcal{S}) := \mathcal{R}_{\Sigma^1_n}(\mathcal{S}).$$

We emphasize that the isomorphism types of Rogers Π^1_n-semilattices are *essentially the same* as those of Rogers Σ^1_n-semilattices, due to Duality Lemma below. Thus, we do not consider semilattices $\mathcal{R}_{\Pi^1_n}(\mathcal{S})$ in this chapter.

Let $\Gamma \in \{\Sigma^1_n, \Pi^1_n : n \geq 1\}$. Then $\breve{\Gamma}$ denotes the dual class:

$$\breve{\Gamma} = \begin{cases} \Pi^1_n, & \text{if } \Gamma = \Sigma^1_n, \\ \Sigma^1_n, & \text{if } \Gamma = \Pi^1_n. \end{cases}$$

Suppose that $\mathcal{S}$ is a countable family of subsets of ω. By $Dual(\mathcal{S})$, we denote the following family:

$$Dual(\mathcal{S}) := \{A : \ \omega \setminus A \in \mathcal{S}\}.$$

It is not hard to prove the following fact (a proof sketch is given in, e.g., Lemma 3.1 of [13]).

Lemma 2.1: (Duality Lemma). *Suppose that* $\Gamma \in \{\Sigma^1_n, \Pi^1_n\}$. *Then the operator* Dual: $\mathcal{S} \mapsto Dual(\mathcal{S})$ *is a bijection from*

$$\{\mathcal{S} \subset P(\omega) : \mathcal{S} \textit{ is } \Gamma\textit{-computable}\} \textit{ onto } \{\mathcal{T} \subset P(\omega) : \mathcal{T} \textit{ is } \breve{\Gamma}\textit{-computable}\}.$$

Furthermore, the semilattices $\mathcal{R}_\Gamma(\mathcal{S})$ *and* $\mathcal{R}_{\breve{\Gamma}}(Dual(\mathcal{S}))$ *are isomorphic.*

2.2. *Hereditary undecidability*

We follow [23, 24, 25] and discuss the necessary background on elementary definability and hereditary undecidability.

For a computable signature L, we use the following notations: K_L is the class of all L-structures, Sen_L is the set of all first-order L-sentences, and Val_L is the set of all valid L-sentences. If n is a non-zero natural number, $\mathrm{C} \in \{\Sigma_n, \Pi_n\}$, and $\Theta \subseteq Sen_L$, then

$$\mathrm{C}\text{–}\Theta = \{\psi \in \Theta : \psi \text{ is a C-sentence}\}.$$

For simplicity, here we consider only finite relational signatures, and we assume that the equality relation $=$ can be freely used in constructing first-order formulas. If K is a class of L-structures, then $Th(K)$ denotes the first-order theory of K.

Suppose that L_1 and L_2 are signatures. A *Σ_k-L_2-scheme S in L_1* is a list of Σ_k-L_2-formulas which consists of:

- $\psi_U(x, \bar{y})$,
- $\psi_R(x_1, \ldots, x_n, \bar{y})$, $\psi_{\neg R}(x_1, \ldots, x_n, \bar{y})$ for each n-ary relation R from L_1, and
- $\psi_=(x_1, x_2, \bar{y})$.

A class $K_1 \subseteq K_{L_1}$ is *Σ_k-elementary definable with parameters* (or *Σ_k-e.d.p.* for short) in a class $K_2 \subseteq K_{L_2}$ if there is a Σ_k-L_2-scheme S in L_1 with the following property: For any $\mathcal{A} \in K_1$, there are a structure $\mathcal{B} \in K_2$ and a tuple $\bar{p}$ from $\mathcal{B}$ such that

(1) the set $U = \{b \in \mathcal{B} : \mathcal{B} \models \psi_U(b, \bar{p})\}$ is not empty;
(2) for every R from L_1, the sets $X_R = \{\bar{b} \in U^n : \mathcal{B} \models \psi_R(\bar{b}, \bar{p})\}$ and $X_{\neg R} = \{\bar{b} \in U^n : \mathcal{B} \models \psi_{\neg R}(\bar{b}, \bar{p})\}$ are disjoint, and $X_R \cup X_{\neg R} = U^n$;
(3) the set $E = \{(b_0, b_1) \in U^2 : \mathcal{B} \models \psi_=(b_0, b_1, \bar{p})\}$ is a congruence relation on the L_1-structure $\mathcal{B}'$, where the domain of $\mathcal{B}'$ is equal to U and for each R from L_1, we define

$$\mathcal{B}' \models R(\bar{b}) \text{ iff } \mathcal{B} \models \psi_R(\bar{b}, \bar{p});$$

(4) the quotient structure $\mathcal{B}'/E$ is isomorphic to $\mathcal{A}$.

If the tuple $\bar{y}$ in the scheme S is empty, then we say that K_1 is *Σ_k-elementary definable* (or *Σ_k-e.d.* for short) in K_2.

Recall that two disjoint sets $A, B \subseteq \omega$ are *computably inseparable* if there is no computable set C with $A \subseteq C \subseteq \overline{B}$. A set Θ of L-sentences is *hereditarily undecidable* if for any $\Xi \subseteq Sen_L$, we have

$$Val_L \cap \Theta \subseteq \Xi \subseteq \Theta \;\Rightarrow\; \Xi \text{ is undecidable.}$$

In other words, Θ is hereditarily undecidable iff the sets $(Val_L \cap \Theta)$ and $(Sen_L \setminus \Theta)$ are computably inseparable.

Proposition 2.2: (Nies, Lemma 3.1 of [24]). *Suppose that* $2 \leq r < \omega$, $1 \leq k < \omega$, $K_1 \subseteq K_{L_1}$, *and* $K_2 \subseteq K_{L_2}$.

(1) Assume that K_1 *is* Σ_k*-e.d. in* K_2*. If the set* Σ_r–$Th(K_1)$ *is hereditarily undecidable, then* Σ_{r+k-1}–$Th(K_2)$ *is also hereditarily undecidable.*
(2) Assume that K_1 *is* Σ_k*-e.d.p. in* K_2*. If* Π_{r+1}–$Th(K_1)$ *is hereditarily undecidable, then* Π_{r+k}–$Th(K_2)$ *is also hereditarily undecidable.*

3. Undecidable Fragments

Theorem 3.1: *Let* $n \geq 1$, *and* $\mathcal{U}$ *be a non-trivial Rogers* Σ^1_n*-semilattice. Then the fragment* Π_4–$Th_{\leq}(\mathcal{U})$ *is hereditarily undecidable.*

In order to obtain the theorem, it is sufficient to establish the following fact:

Proposition 3.2: *Let* $\mathcal{D}$ *be a finite distributive lattice. Then there are elements* a *and* b *from* $\mathcal{U}$*, such that the interval* $[a;b]_{\mathcal{U}}$ *is isomorphic to* $\mathcal{D}$.

Indeed, assume that Proposition 3.2 is already proved. Let FDL be the class of all finite distributive lattices in the signature $\{\vee, \wedge\}$. Nies (Theorem 4.8 of [24]) proved that the fragment Π_3–$Th(FDL)$ is hereditarily undecidable.

On the other hand, Proposition 3.2 provides a Σ_2-scheme, which gives an elementary definition of FDL inside the *partial order* $\mathcal{U}$: Informally speaking, if one wants to define a structure $\mathcal{D} \in FDL$ inside $\mathcal{U}$, then one just chooses appropriate elements a and b from $\mathcal{U}$, and says that the interval $[a;b]_{\mathcal{U}}$ is a copy of $\mathcal{D}$. More formally, the following scheme suffices for our purposes:

- $\psi_U(x;a,b) := (a \leq x) \,\&\, (x \leq b)$;
- $\psi_\vee(x_1,x_2,z;a,b) := (x_1 \leq z) \,\&\, (x_2 \leq z) \,\&\, \forall w[(x_1 \leq w) \,\&\, (x_2 \leq w) \to z \leq w]$ and $\psi_{\neg\vee}(x_1,x_2,z;a,b) := \neg\psi_\vee$;

- $\psi_\wedge(x_1, x_2, z; a, b) := (z \leq x_1) \,\&\, (z \leq x_2) \,\&\, \forall w[(w \leq x_1) \,\&\, (w \leq x_2) \rightarrow w \leq z]$ and $\psi_{\neg\wedge}(x_1, x_2, z; a, b) := \neg\psi_\wedge$;
- $\psi_=(x_1, x_2; a, b) := (x_1 = x_2)$.

In this scheme, a and b are parameters; the formulas $\psi_\vee$ and $\psi_\wedge$ define the graphs of the lattice functions $\vee$ and $\wedge$, respectively.

Therefore, we apply the second item of Proposition 2.2 (for $k = 2$ and $r = 2$), and deduce that Π_4–$Th_\leq(\mathcal{U})$ is hereditarily undecidable.

3.1. *Proof of Proposition 3.2*

The main tool for obtaining "good" intervals inside $\mathcal{U}$ is provided by the following result from [10].

Let $\mathbf{D}_m$ denote the semilattice of all m-degrees. As usual, we assume that $\mathbf{D}_m$ does not contain the degrees $\deg_m(\emptyset)$ and $\deg_m(\omega)$. For a non-empty set $A \subset \omega$, let $\mathbf{D}_m(\leq A)$ be the principal ideal $\{\mathbf{d} : \mathbf{d} \leq \deg_m(A)\}$ in the semilattice $\mathbf{D}_m$.

Suppose that ν and μ are numberings of a family $\mathcal{S}$. We say that ν is Δ^1_n*-reducible* to μ, denoted by $\nu \leq_{\Delta^1_n} \mu$, if there is a total Δ^1_n function $f(x)$ such that $\nu(x) = \mu(f(x))$, for all $x \in \omega$. The numberings ν and μ are Δ^1_n*-equivalent*, denoted by $\nu \equiv_{\Delta^1_n} \mu$, if $\nu \leq_{\Delta^1_n} \mu$ and $\mu \leq_{\Delta^1_n} \nu$.

Proposition 3.3: (Essentially, Proposition 3.2 of [10]) *Let $\mathcal{S}$ be a Σ^1_n-computable family, such that $\mathcal{S}$ contains at least two elements. Suppose that a set $U \subset \omega$ is immune and Δ^1_n. Then for any Σ^1_n-computable numbering α of the family $\mathcal{S}$, there is a Σ^1_n-computable numbering β of $\mathcal{S}$ with the following properties:*

(a) $\beta \equiv_{\Delta^1_n} \alpha$.
(b) If $\mathcal{S}$ is finite, then the principal ideal induced by β inside $\mathcal{R}^1_n(\mathcal{S})$ (this ideal is denoted by $[\widehat{\beta}]^1_n$) is isomorphic to the structure $\mathbf{D}_m(\leq U)$.
(c) If $\mathcal{S}$ is infinite, then the ideal $[\widehat{\beta}]^1_n$ is isomorphic to the structure $\mathbf{D}_m(\leq U) \setminus \{\mathbf{0}\}$, i.e. to the ideal $\mathbf{D}_m(\leq U)$ with the least element omitted.

Before continuing the proof, we give two comments about Proposition 3.3. First, note that the original proposition of [10] is formulated for Π^1_n*-computable* numberings. Nevertheless, this is not an obstacle for us, since we can transfer from $\mathcal{R}_{\Sigma^1_n}(\mathcal{S})$ to $\mathcal{R}_{\Pi^1_n}(Dual(\mathcal{S}))$, and vice versa, by Duality Lemma. Second, the proof of Proposition 3.3 is entirely based on Theorem 2 of Podzorov [14].

In order to successfully apply Proposition 3.3, we need to work with m-degrees of immune sets:

Proposition 3.4: *Let $\mathcal{D}$ be a finite distributive lattice. Then there exists a set $A \subseteq \omega$ with the following properties:*

(1) the principal ideal $\mathbf{D}_m(\leq A)$ is isomorphic to $\mathcal{D}$;
(2) A is immune, and A is Σ^0_3.

Proof: We give a sketch of the proof. The desired result is obtained via a simple combination of known constructions. Lachlan [26] proved that the topped finite initial segments of $\mathbf{D}_m$ are precisely the finite distributive lattices. Thus, one can follow, e.g., the exposition of Odifreddi (Proposition VI.2.3 in [27]), and get a construction, which builds a set A, such that $\mathbf{D}_m(\leq A)$ is isomorphic to the given lattice $\mathcal{D}$. We note that the construction also builds a set $B \subset \omega$ with $A \cap B = \emptyset$.

Our additional worry is that we need to ensure the immunity of the constructed set A. This is guaranteed by the following straightforward modification. We split the stages of our construction into even and odd. Even stages make the same actions as originally intended by Lachlan. At an odd stage $2e+1$, we ask the oracle $\mathbf{0}^{(2)}$, whether the c.e. set W_e is infinite. If W_e is infinite, then we choose a fresh big element $w \in W_e$ and enumerate the whole (finite) box containing w into B (see [27] for the details on boxes). If W_e is finite, then just proceed to the next stage.

Since the sets A and B do not intersect, it is clear that A is immune. An analysis of the construction shows that both A and B are c.e. in $\mathbf{0}^{(2)}$.

Further technical details can be recovered from [27] and Proposition 3.3 of [10] — this proposition proves a similar fact: Given a countable linear order $\mathcal{L}$ with the least and greatest elements, one can build an immune set A, such that $\mathbf{D}_m(\leq A)$ is isomorphic to $\mathcal{L}$, and A is c.e. in $\mathbf{0}^{(2)} \oplus$ (the atomic diagram of $\mathcal{L}$). □

Now we are ready to finish the proof of Proposition 3.2. Fix a Σ^1_n-computable family $\mathcal{S}$, such that $\mathcal{R}^1_n(\mathcal{S})$ is isomorphic to our structure $\mathcal{U}$. If $\mathcal{S}$ contains only one element (i.e. $\mathcal{S} = \{Y\}$ for some Σ^1_n set Y), then clearly, $\mathcal{S}$ has *only one* numbering $\nu(k) := Y$, $k \in \omega$. Therefore, the non-triviality of $\mathcal{U}$ implies that $\mathcal{S}$ contains at least two elements.

If $\mathcal{S}$ is finite, then we combine Propositions 3.4 and 3.3, and obtain that any finite distributive lattice $\mathcal{D}$ is isomorphic to some initial segment of the semilattice $\mathcal{U}$.

If $\mathcal{S}$ is infinite, then we have to be a little bit more careful. Let $\mathcal{D}$ be a finite distributive lattice. We define a new lattice $\mathcal{D}^{\perp}$ by appending an external least element $\perp$ to $\mathcal{D}$. It is not hard to show that the lattice $\mathcal{D}^{\perp}$ is also distributive. By Proposition 3.4, we get an immune set A, such that $\mathbf{D}_m(\leq A)$ is isomorphic to $\mathcal{D}^{\perp}$. Then Proposition 3.3 gives an interval inside $\mathcal{U}$, which is isomorphic to $\mathcal{D} = \mathcal{D}^{\perp} \setminus \{\perp\}$.

This concludes the proof of Proposition 3.2 and Theorem 3.1.

4. Decidable Fragments

Theorem 4.1: *Let $n \geq 1$, and $\mathcal{U}$ be a non-trivial Rogers Σ^1_n-semilattice. Then the fragment Σ_1–$Th_{\leq,\vee}(\mathcal{U})$ is decidable.*

First, we note that in the theory of upper semilattices T_{usl}, the formulas $(x \leq y)$ and $(x \vee y = y)$ are logically equivalent. Hence, without loss of generality, this section will work only with first-order formulas in the signature $\{\vee\}$.

A typical example of a Σ_1-sentence ψ is the following:

$$\psi = \exists x \exists y \exists z \exists w[(x, y, z, w \text{ are pairwise distinct}) \,\&\, x \vee y = z \,\&\, x \vee w = z].$$

We note that the quantifier-free part of ψ gives us only a *partial information* about the atomic diagram on the domain $\{x, y, z, w\}$: in particular, even by appealing to T_{usl}, we cannot deduce what is the value of $y \vee w$.

Thus, to prove Theorem 4.1, we have to consider the algebraic properties of *partial* upper semilattices.

4.1. *Partial upper semilattices*

The exposition here mainly follows Sec. 5.4 of Ch. 1 in [28], where partial lattices are discussed.

Definition 4.2: Let A be a non-empty set, and $\vee$ be a partial binary function on A. We say that the structure $(A; \vee)$ is a *weak partial upper semilattice* (or *weak partial usl*, for short) if it satisfies the following conditions: for all $a, b, c \in A$,

(i) $(a \vee a)\downarrow = a$, i.e. the value $a \vee a$ is defined and is equal to a.
(ii) If $(a \vee b)\downarrow$, then $(b \vee a)\downarrow = a \vee b$.
(iii) If the values $(a \vee b)$, $(a \vee b) \vee c$ and $(b \vee c)$ are defined, then $a \vee (b \vee c)$ is also defined and $a \vee (b \vee c) = (a \vee b) \vee c$.

If $(A; \vee)$ is a weak partial usl, then the relation $\leq_A$ is defined as follows:

$$a \leq_A b \Leftrightarrow (a \vee b) \downarrow = b.$$

It is easy to show that $\leq_A$ is a partial order.

Definition 4.3: Let A be a non-empty set, and $\vee$ be a partial binary function on A. One can consider $(A; \vee)$ as a structure in a relational signature: as usual, the graph of $\vee$ is treated as a ternary predicate. We say that $(A; \vee)$ is a *partial upper semilattice* (or *partial usl*) if $(A; \vee)$ is a relational substructure of some upper semilattice (in relational signature).

It is clear that every partial usl is a weak partial usl. The main goal of this subsection is to answer the following question: In which cases the inverse is true, i.e. when a weak partial usl is a partial usl?

This question is very important to us, since roughly speaking, the quantifier-free part of a Σ_1-sentence ψ provides us only with a *weak partial usl* $(A; \vee)$, and it is crucial for us to see whether this $(A; \vee)$ can be extended to a "real" upper semilattice.

Definition 4.4: Let $(A; \vee)$ be a weak partial usl. A non-empty set $I \subseteq A$ is an *ideal* of $(A; \vee)$ if it satisfies the following:

(a) If $b \in I$ and $a \leq_A b$, then $a \in I$.
(b) If $a \in I$, $b \in I$ and $(a \vee b) \downarrow$, then $(a \vee b) \in I$.

Consider the set

$$\mathrm{Id}_0(A) := \{\emptyset\} \cup \{I : I \text{ is an ideal of } (A; \vee)\}.$$

One can prove that $\mathrm{Id}_0(A)$ forms a lattice with respect to set-theoretic inclusion. Let I and J be ideals, then:

(a) Their infimum $I \wedge J$ is equal to the intersection $I \cap J$.
(b) The supremum $I \vee J$ can be recovered via the following procedure:

$$\Psi_0(I,J) := I \cup J; \quad \Psi_{m+1}(I,J) := \Psi_m(I,J) \cup$$
$$\{x \in A : (\exists a, b \in \Psi_m(I,J))[(a \vee b) \downarrow \,\&\, x \leq_A a \vee b]\};$$
$$I \vee J = \bigcup_{m \in \omega} \Psi_m(I,J).$$

It is not hard to show that for any $a \in A$, the set

$$id(a) := \{b \in A : b \leq_A a\}$$

is an ideal of $(A; \vee)$.

Now we are ready to give a characterization of weak partial usls, which are also partial usls.

Proposition 4.5: (Funayama [29], see also Theorem 84 in [28]). *Let $(A; \vee)$ be a weak partial usl. Then the following conditions are equivalent:*

(1) $(A; \vee)$ is a partial usl.
(2) All $a, b, c \in A$ satisfy the following: if $id(a) \vee id(b) = id(c)$ in the lattice $\mathrm{Id}_0(A)$, then $a \vee b$ exists in A and is equal to c.

Proof: (1)⇒(2). Since $(A; \vee)$ is a partial usl, there is an upper semilattice $\mathcal{U}$, such that $(A; \vee)$ is a relational substructure of $\mathcal{U}$. Let d be the element of $\mathcal{U}$, such that $\mathcal{U} \models (a \vee b = d)$. Note that in general, d can be not a part of A.

For $x \in A$, by $id_A(x)$ we denote the ideal $id(x)$ from $\mathrm{Id}_0(A)$. Similarly, $id_{\mathcal{U}}(x)$ denotes the corresponding ideal from $\mathrm{Id}_0(\mathcal{U})$. Let a, b, c be elements of A, and suppose that inside $\mathrm{Id}_0(A)$, we have

$$id_A(a) \vee id_A(b) = id_A(c).$$

Since $a \leq_{\mathcal{U}} d$ and $b \leq_{\mathcal{U}} d$, we obtain $id_A(a) \vee id_A(b) \subseteq id_{\mathcal{U}}(d)$. In particular, this implies $c \leq_{\mathcal{U}} d$.

On the other hand, $a \leq_A c$ and $b \leq_A c$, i.e. c is an upper bound for both a and b. Since d is the supremum of a and b, we deduce $d \leq_{\mathcal{U}} c$. Therefore, c is equal to $d = a \vee_{\mathcal{U}} b$.

(2)⇒(1). Consider an injective map $\theta \colon A \to \mathrm{Id}_0(A)$, such that $\theta(a) = id(a)$. We will prove that θ isomorphically embeds $(A; \vee)$ into $(\mathrm{Id}_0(A); \vee)$ (as a relational substructure). Since $\mathrm{Id}_0(A)$ is a lattice, this fact implies that $(A; \vee)$ is a partial usl.

Let $a, b \in A$. First, assume that $(a \vee b)\downarrow = c$. Consider the ideal $J := id(a) \vee id(b)$ from $\mathrm{Id}_0(A)$. We need to show that $J = id(c)$.

Since J is an ideal and $a, b \in J$, we deduce that $c = a \vee b$ belongs to J. This implies $id(c) \subseteq J$.

On the other hand, note that $a \leq_A c$ and $b \leq_A c$. Indeed, since both values $(a \vee a)$ and $(a \vee b)$ are defined, the value $a \vee (a \vee b)$ is also defined, and we have:

$$a \vee c = a \vee (a \vee b) = (a \vee a) \vee b = a \vee b = c \Rightarrow a \leq_A c.$$

In a similar way, $b \leq_A c$. Hence, we obtain $id(a) \subseteq id(c)$ and $id(b) \subseteq id(c)$. Thus, $J = id(a) \vee id(b) \subseteq id(c)$.

We established the following: if $(a \vee b)\downarrow = c$, then $id(a) \vee id(b) = id(c)$. The inverse direction (i.e. if $id(a) \vee id(b) = id(c)$ for some c, then $(a \vee b)\downarrow$

$= c$) is precisely item (2) of the proposition. Therefore, we deduce that the map $\theta\colon a \mapsto id(a)$ is an isomorphic embedding of relational structures, and $(A; \vee)$ is a partial usl. Proposition 4.5 is proved. □

We will also employ the following auxiliary fact:

Lemma 4.6: *Let $(A; \vee)$ be a finite partial usl. Then $(A; \vee)$ can be isomorphically embedded into a finite distributive lattice.*

Proof: Since $(A; \vee)$ is a partial usl, there is a (total) upper semilattice $\mathcal{U}$ and an isomorphic embedding $\xi\colon (A; \vee) \to \mathcal{U}$. Consider a finite set

$$B := \left\{ \bigvee_{a \in F} \xi(a) : F \text{ is a non-empty subset of } A \right\}.$$

It is clear that the structure $\mathcal{B} := (B; \vee_{\mathcal{U}})$ is a finite subsemilattice of $\mathcal{U}$. Furthermore, the map ξ can be treated as an isomorphic embedding from $(A; \vee)$ into $\mathcal{B}$.

To prove the proposition, it is sufficient to embed the (total) semilattice $\mathcal{B}$ into a finite distributive lattice. Consider the lattice $P(B)$ of all subsets of B under set-theoretic inclusion. Define an injective function

$$\theta(a) := \{x \in B : a \not\leq_B x\}.$$

In order to show that θ is an isomorphic embedding of upper semilattices, we need to establish the following two facts:

(i) If $a \vee b = c$, then $\theta(c) = \theta(a) \cup \theta(b)$.
(ii) If $\theta(a) \cup \theta(b) = \theta(c)$, then $c = a \vee b$.

Note that the condition $\theta(a) \cup \theta(b) = \theta(c)$ is equivalent to $\overline{\theta(a)} \cap \overline{\theta(b)} = \overline{\theta(c)}$, where $\overline{X} := B \setminus X$ is the complement of X inside B.

(i) Assume that $a \vee b = c$. Then

$$x \in \overline{\theta(a)} \cap \overline{\theta(b)} \;\Leftrightarrow\; (a \leq_B x) \& (b \leq_B x) \;\Leftrightarrow\; (c \leq_B x) \;\Leftrightarrow\; x \in \overline{\theta(c)}.$$

(ii) Let $d := a \vee b$ (recall that the semilattice $\mathcal{B}$ is total), and assume that $\overline{\theta(a)} \cap \overline{\theta(b)} = \overline{\theta(c)}$. Thus, we deduce $(a \leq_B x) \& (b \leq_B x) \Leftrightarrow (c \leq_B x)$. This equivalence easily implies that c is equal to d.

Therefore, θ is an isomorphic embedding from $\mathcal{B}$ into $P(B)$. Lemma 4.6 is proved. □

4.2. *Proof of Theorem 4.1*

Consider a Σ_1-sentence ψ in the signature $\{\vee\}$:

$$\psi = \exists x_1 \exists x_2 \dots \exists x_n \varphi(x_1, x_2, \dots, x_n),$$

where φ is a quantifier-free formula. We describe an algorithm for checking whether the sentence ψ is true in the Rogers semilattice $\mathcal{U}$.

Recall that an atomic formula in the signature $\{\vee\}$ is *unnested* if it is a formula of one of the following two forms:

(1) $x = y$, where x and y are variables; or
(2) $x \vee y = z$, where x, y, z are variables.

A first-order formula is *unnested* if all its atomic subformulas are unnested.

It is known that given an existential formula $\theta(\bar{y})$, one can effectively find an unnested existential formula $\theta'(\bar{y})$, which is logically equivalent to θ (see, e.g., Corollary 2.6.2 of [30]). Therefore, without loss of generality, we assume that both our formulas ψ and φ are unnested.

In a standard way, we replace the formula $\varphi(\bar{x})$ with its disjunctive normal form:

$$\varphi(\bar{x}) \sim \bigvee_{1 \le i \le M} \varphi_i(\bar{x}),$$

where each φ_i is a conjunction of unnested atomic formulas and their negations. As usual, we may assume that for every j and k, such that $1 \le j < k \le n$, the formula φ_i states precisely one of the following facts: either $(x_j = x_k)$, or $(x_j \neq x_k)$.

Since the original Σ_1-sentence ψ is logically equivalent to

$$\bigvee_{1 \le i \le M} \exists \bar{x} \varphi_i(\bar{x}),$$

it is sufficient to work only with sentences of the form

$$\psi' := \exists \bar{x} \varphi_1(\bar{x}).$$

Without loss of generality, we will assume that:

- The formula $\varphi_1(\bar{x})$ says that all elements from $\bar{x}$ are pairwise distinct. If φ_1 states that, say, $x_1 = x_2$, then one can just replace all occurrences of x_2 with x_1.
- For every j and k with $1 \le j, k \le n$, precisely one of the following two cases holds:

(1) There is a unique p, such that φ_1 states $x_j \vee x_k = x_p$. For all $q \neq p$, φ_1 includes $x_j \vee x_k \neq x_q$.
(2) For all p, the formula φ_1 states $x_j \vee x_k \neq x_p$.

Note the following: If φ_1 includes, say, $(x_1 \vee x_2 = x_3)$ and $(x_1 \vee x_2 = x_4)$, then this produces a contradiction with our previous assumption $(x_3 \neq x_4)$. In this case, clearly, φ_1 would be unsatisfiable.

The first important stage of our algorithm checks the conditions given by the definition of a weak partial usl (Definition 4.2):

(i) If φ_1 states that $x_j \vee x_j \neq x_j$ for some j, then it is clear that both φ_1 and ψ' cannot be satisfied in an upper semilattice. In this case, one can immediately declare that $\mathcal{U} \not\models \psi'$.
(ii) If φ_1 states that $x_j \vee x_k = x_p$ and $x_k \vee x_j \neq x_p$ for some j, k, p, then again $\mathcal{U} \not\models \psi'$.
(iii) Assume that φ_1 says that $x_j \vee x_k = x_p$, $x_p \vee x_q = x_r$ and $x_k \vee x_q = x_t$. Then if φ_1 wants to preserve the associative property $(x_j \vee x_k) \vee x_q = x_j \vee (x_k \vee x_q)$, it *must* say "$x_j \vee x_t = x_r$". Otherwise, we also declare that $\mathcal{U} \not\models \psi'$.

If φ_1 successfully passed all three tests above, we proceed to the second stage: Define a weak partial usl $(A[\varphi_1]; \vee)$ as follows:

- The domain of $A[\varphi_1]$ is equal to $\{1, 2, \dots, n\}$.
- We set $j \vee k := p$ if and only if φ_1 says that $x_j \vee x_k = x_p$.

The last ingredient of the proof is provided by the following lemma:

Lemma 4.7: *Assume that we already defined the weak partial usl $(A[\varphi_1]; \vee)$. Then the following conditions are equivalent:*

(a) The sentence $\psi' = \exists \bar{x} \varphi_1(\bar{x})$ is true in the semilattice $\mathcal{U}$.
(b) The structure $(A[\varphi_1]; \vee)$ is a partial usl.

Proof: (a)⇒(b). Choose elements $a_1, \dots, a_n$ from $\mathcal{U}$, such that $\mathcal{U} \models \varphi_1(a_1, \dots, a_n)$. Then it is clear that the function $g\colon i \mapsto a_i$ is an isomorphic embedding (of relational structures) from $(A[\varphi_1]; \vee)$ into $\mathcal{U}$. Therefore, $(A[\varphi_1]; \vee)$ is a partial usl.

(b)⇒(a). Suppose that $(A[\varphi_1]; \vee)$ is a partial usl. By Lemma 4.6, the structure $(A[\varphi_1]; \vee)$ can be isomorphically embedded into a finite distributive lattice $\mathcal{D}$. By Proposition 3.2, one can choose an interval $[a; b]_{\mathcal{U}}$ inside

$\mathcal{U}$, which is isomorphic to $\mathcal{D}$. These facts together easily imply that there are elements $c_1, \ldots, c_n \in \mathcal{U}$, such that $\mathcal{U} \models \varphi_1(c_1, \ldots, c_n)$. □

Lemma 4.7 shows that the third stage of the algorithm can be arranged as follows: We check whether $(A[\varphi_1]; \vee)$ is a partial usl. By Proposition 4.5, this procedure is effective — one needs to verify some specific conditions for a finite number of finite sets (namely, the ideals of $(A[\varphi_1]; \vee)$).

Finally, the condition $\mathcal{U} \models \psi'$ holds if and only if $(A[\varphi_1]; \vee)$ is a partial usl. This concludes the proof of Theorem 4.1.

5. Further Discussion

First, we note that our methods can be transferred to the setting of the arithmetical hierarchy. We need to employ Theorem 2 of [14], and keep in mind that Proposition 3.4 produces an immune set that belongs to the Σ^0_3 level.

Corollary 5.1: *Let $n \geq 4$ be a natural number. Then any non-trivial Rogers Σ^0_n-semilattice $\mathcal{U}$ satisfies the following:*

(a) Π_4-$Th_{\leq}(\mathcal{U})$ is hereditarily undecidable, and
(b) Σ_1-$Th_{\leq,\vee}(\mathcal{U})$ is decidable.

We leave open the following question:

Problem 5.2: *What can one say about decidability for Π_2- and Π_3-fragments of first-order theories for Rogers Σ^1_n-semilattices?*

References

1. K. Gödel, Über formal unentscheidbare Sätze der Principia Mathematica und verwandter Systeme. I., *Monatsh. Math. Phys.* **38** (1931), 173–198. doi: 10.1007/BF01700692.
2. S. C. Kleene, *Introduction to metamathematics.* Van Nostrand, New York, 1952.
3. A. N. Kolmogorov and V. A. Uspenskii, On the definition of an algorithm, *Uspehi Mat. Nauk.* **13** (1958), 3–28. In Russian.
4. V. A. Uspenskii, Systems of denumerable sets and their enumeration, *Dokl. Akad. Nauk SSSR.* **105** (1958), 1155–1158. In Russian.
5. H. Rogers, Gödel numberings of partial recursive functions, *J. Symb. Logic.* **23** (1958), 331–341. doi: 10.2307/2964292.
6. S. S. Goncharov and A. Sorbi, Generalized computable numerations and non-trivial Rogers semilattices, *Algebra Logic.* **36** (1997), 359–369. doi: 10.1007/BF02671553.

7. Y. L. Ershov, *Theory of numberings*. Nauka, Moscow, 1977. In Russian.
8. S. A. Badaev and S. S. Goncharov, The theory of numberings: Open problems. In eds. P. Cholak, S. Lempp, M. Lerman and R. Shore, *Computability Theory and Its Applications*, vol. 257, *Contemp. Math.*, pp. 23–38. American Mathematical Society, Providence, 2000. doi: 10.1090/conm/257/04025.
9. S. Badaev and S. Goncharov, Computability and numberings. In eds. S. B. Cooper, B. Löwe and A. Sorbi, *New Computational Paradigms*, pp. 19–34. Springer, New York, 2008. doi: 10.1007/978-0-387-68546-5_2.
10. N. Bazhenov, S. Ospichev and M. Yamaleev, Isomorphism types of Rogers semilattices in the analytical hierarchy. In eds. N. Greenberg, S. Jain, K. M. Ng, S. Schewe, F. Stephan, G. Wu and Y. Yang, Aspects of Computation and Automata Theory with Applications, vol. 42, *Lect. Notes Ser. Inst. Math. Sci. Natl. Univ. Singapore*, pp. 97–114. World Scientific, Singapore, 2023. doi: 10.1142/9789811278631_0004.
11. V. V. V'yugin, On some examples of upper semilattices of computable enumerations, *Algebra Logic.* **12** (1973), 287–296. doi: 10.1007/BF02218587.
12. S. A. Badaev, S. S. Goncharov and A. Sorbi, Elementary theories for Rogers semilattices, *Algebra Logic.* **44** (2005), 143–147. doi: 10.1007/s10469-005-0016-x.
13. N. Bazhenov and M. Mustafa, Rogers semilattices in the analytical hierarchy: The case of finite families, 2019. Submitted. arXiv:2010.00830.
14. S. Y. Podzorov, Initial segments in Rogers semilattices of Σ_n^0-computable numberings, *Algebra Logic.* **42** (2003), 121–129. doi: 10.1023/A:1023354407888.
15. L. Harrington and A. Nies, Coding in the partial order of enumerable sets, *Adv. Math.* **133** (1998), 133–162. doi: 10.1006/aima.1997.1687.
16. S. Badaev, S. Goncharov, S. Podzorov and A. Sorbi, Algebraic properties of Rogers semilattices of arithmetical numberings. In eds. S. S. Goncharov and S. B. Cooper, *Computability and Models*, pp. 45–78. Springer, New York, 2003. doi: 10.1007/978-1-4615-0755-0_3.
17. M. V. Dorzhieva, Undecidability of elementary theory of Rogers semilattices in analytical hierarchy, *Sib. Elektron Mat. Izv.* **13** (2016), 148–153. doi: 10.17377/semi.2016.13.013. In Russian.
18. A. Nies, Coding methods in computability theory and complexity theory, 2013. arXiv:1308.6399.
19. J. L. Eršov, Theorie der Numerierungen I, *Z. Math. Logik Grundlagen Math.* **19** (1973), 289–388. doi: 10.1002/malq.19730191901.
20. J. L. Eršov, Theorie der Numerierungen II, *Z. Math. Logik Grundlagen Math.* **21** (1975), 473–584. doi: 10.1002/malq.19750210164.
21. J. L. Eršov, Theorie der Numerierungen III, *Z. Math. Logik Grundlagen Math.* **23** (1977), 289–371. doi: 10.1002/malq.19770231902.
22. Y. L. Ershov, Theory of numberings. In ed. E. R. Griffor, *Handbook of Computability Theory*, vol. 140, *Stud. Logic Found. Math.*, pp. 473–503. North-Holland, Amsterdam, 1999. doi: 10.1016/S0049-237X(99)80030-5.
23. Y. L. Ershov, I. A. Lavrov, A. D. Taimanov and M. A. Taitslin, Elementary theories, *Russ. Math. Surv.* **20** (1965), 35–105. doi: 10.1070/RM1965v020n04ABEH001188.

24. A. Nies, Undecidable fragments of elementary theories, *Algebra Univers.* **35** (1996), 8–33. doi: 10.1007/BF01190967.
25. S. O. Speranski, A note on hereditarily Π_1^0- and Σ_1^0-complete sets of sentences, *J. Log. Comput.* **26** (2016), 1729–1741. doi: 10.1093/logcom/exu066.
26. A. H. Lachlan, Initial segments of many-one degrees, *Can. J. Math.* **22** (1970), 75–85. doi: 10.4153/CJM-1970-010-6.
27. P. Odifreddi, *Classical recursion theory.* vol. 125, *Stud. Logic Found. Math.*, Elsevier, Amsterdam, 1992.
28. G. Grätzer, *Lattice theory: Foundation.* Springer, Basel, 2011.
29. N. Funayama, Notes on lattice theory. IV. On partial (semi) lattices, *Bull. Yamagata Univ. Natur. Sci.* **2** (1953), 171–184.
30. W. Hodges, *Model theory.* vol. 42, *Encyclopedia of Mathematics and Its Applications*, Cambridge University Press, Cambridge, 1993.

https://doi.org/10.1142/9789819806584_0002

Realizing Computably Enumerable Degrees in Separating Classes

Peter Cholak
Department of Mathematics, University of Notre Dame
South Bend, Indiana USA
cholak@nd.edu

Rod Downey†, Noam Greenberg‡ and Daniel Turetsky§
School of Mathematics and Statistics,
Victoria University of Wellington
PO Box 600, Wellington, New Zealand
†*rod.downey@vuw.ac.nz*, ‡*noam.greenberg@vuw.ac.nz*, §*dan.turetsky@vuw.ac.nz*

We investigate what collections of c.e. Turing degrees can be realised as the collection of elements of a separating Π^0_1 class of c.e. degree. We show that for every c.e. degree $\mathbf{c}$, the collection $\{\mathbf{c}, \mathbf{0}'\}$ can be thus realized. We also rule out several attempts at constructing separating classes realizing a unique c.e. degree. For example, we show that there is no *super-maximal* pair: disjoint c.e. sets A and B whose separating class is infinite, but every separator of c.e. degree is a finite variant of either A or $\overline{B}$.

1. Introduction

This chapter is concerned with (computably bounded) Π^0_1 classes. Of course we can consider these (up to Turing degree) as being a collection of infinite paths through a computable binary tree. They have deep connections with computability theory in general, as well as reverse mathematics, algorithmic randomness and many other areas. See, for example, Cenzer and Jockusch [1].

Rod Downey and Noam Greenberg are partially supported by Marsden Fund of New Zealand. Peter Cholak is partially supported by NSF-DMS-1854136, a FRG grant from the NSF. We dedicate this chapter to Hugh Woodin and Ted Slaman as part of their combined birthday celebrations.

The meta-question we attack concerns realizing c.e. degrees as members of Π^0_1 classes. In this regard, we follow some earlier studies of Csima, Downey and Ng [2] and Downey and Melnikov [8], but in this case, we will be looking at *separating classes*.

Recall that one of the fundamental theorems in this area is the Computably Enumerable Basis Theorem which says that each Π^0_1 class has a member of computably enumerable degree. Indeed, if α is the left- or right-most path of a Π^0_1 class P, then there is a c.e. set W_e, such that $W_e \equiv_T \alpha$.

Definition 1.1: ([2]). We will say that a c.e. degree $\mathbf{w}$ is *realized* in a Π^0_1 class P iff there exists some $\beta \in P$ with $\deg_T(\beta) = \mathbf{w}$.

The fundamental question attacked in [2] was "What sets of c.e. degrees can be realized in a Π^0_1 class?" In [2], Csima, Downey and Ng give a surprising characterization of the sets of c.e. degrees that can be realized. They showed that the question is related to one of representing index sets.

For a set $\mathcal{W}$ of c.e. degrees, let

$$I(\mathcal{W}) = \{e : W_e \in \mathcal{W}\}$$

be the set of indices of c.e. sets whose degrees are in $\mathcal{W}$. Letting $\mathcal{R}$ be the collection of all c.e. degrees, for a Π^0_1 class P, let

$$\mathcal{W}[P] = \{\mathbf{w} \in \mathcal{R} : \mathbf{w} \text{ is realized in } P\}.$$

A calculation shows:

Proposition 1.2: ([2]). *For any Π^0_1 class P, the index set $I(\mathcal{W}[P])$ is Σ^0_4.*

The first natural question is whether Proposition 1.2 reverses, that is, if every Σ^0_4 index set is realized as the collection of indices of c.e. sets whose degrees are realized in some Π^0_1 class. The answer is negative (see a discussion in [2]). To describe which sets can be realized, we observe that collections of degrees whose index sets are complicated can nonetheless have simple *representations*, in the following sense. For a set $S \subseteq \mathbb{N}$, let

$$\mathcal{D}(S) = \{\deg_T(W_e) : e \in S\}$$

be the collection of degrees of c.e. sets whose indices are in S. Since Turing equivalence is Σ^0_3, if S is Σ^0_4, then so is its closure under Turing equivalence, namely, $I(\mathcal{D}(S))$ (denoted by $G(S)$ in [2]), is also Σ^0_4. But it is possible for S to be very simple but $I(\mathcal{D}(S))$ to be complicated; for example, $I(\{\mathbf{0}'\})$ is Σ^0_4-complete, but of course $\{\mathbf{0}'\} = \mathcal{D}(S)$ where S is a singleton. For a complexity class Γ, we say that a set $\mathcal{W}$ of c.e. degrees is *Γ-representable* if $\mathcal{W} = \mathcal{D}(S)$ for some set S in the class Γ. This notion gives the following characterization:

Theorem 1.3: (Csima, Downey and Ng [2]). *The following are equivalent for a set $\mathcal{W}$ of c.e. degrees:*

(i) $\mathcal{W}$ is Σ^0_3-representable;
(ii) $\mathcal{W}$ is computably representable;
(iii) $\mathcal{W} = \mathcal{W}[P]$ for some Π^0_1 class P.

They also show that the class P in (iii) can be taken to be perfect. As mentioned, not every set of c.e. degrees whose index-set is Σ^0_4 is thus realized; in [2], the authors give an example of a Π^0_3-represented set of c.e. degrees which contains $\mathbf{0}$ but is not realized as the collection of c.e. degrees of elements of some Π^0_1 class.

Theorem 1.3 implies that, for example, the superlow c.e. degrees, the K-trivial c.e. degrees, and all upper cones are realizable. Lower cones were classified by Downey and Melnikov.

Theorem 1.4: (Downey and Melnikov [8]). *The lower cone $[\mathbf{0}, \mathbf{c}]$ is realizable iff $\mathbf{c}$ is* low_2 *or $\mathbf{0}'$.*

In this chapter, we will consider similar questions for what are arguably the most important Π^0_1 classes in terms of applications in Reverse Mathematics; the *separating classes.* Recall that P is a separating class if there exist c.e. disjoint sets A, B, such that $P = S(A, B) = \{Z \mid Z \supseteq A \wedge Z \cap B = \emptyset\}$. For example, if A represents the sentences provable in PA and B the sentences refutable, then $S(A, B)$ represents the complete extensions of PA. Also, as is well-known, if we desire to show that a theorem of second order arithmetic is as strong as WKL_0, then it suffices to code in separating classes, in spite of the fact that Weak König's Lemma says that infinite binary trees have paths. That is, we don't have to look at all infinite binary trees, only separating classes. For example, it is known that the theorem "Every countable commutative ring with identity has a prime ideal" is equivalent to WKL_0, and this is proven by Friedman, Simpson and Smith [10] by coding a given separating class as the prime ideal structure of a commutative ring with identity, but their earlier claim ([9]) that, given a Π^0_1 class P there is a (computable) commutative ring with identity whose prime ideals are in 1-1 correspondence with the members of the class remains open. We refer the reader to e.g. Downey and Hirschfeldt [6] or

Simpson [15]. The questions of what $\mathcal{W}[S(A,B)]$ can be for various A, B was asked in both [2] and [8].

1.1. *Our results*

Whilst an exact characterization is elusive, we do prove some results to show how different the situation is for separating classes.

We do know that strange "separation sprectra" can occur. For example, Jockusch and Soare [11] showed that there two pairs of pairwise disjoint c.e. sets A_1, B_1, A_2, B_2 such that every (not necessarily c.e.) separator of A_1 and B_1 is Turing incomparable with every separator of A_2 and B_2. This result was extended by Downey, Jockusch and Stob [7] who showed that four such sets can be below a c.e. degree $\mathbf{a}$ iff $\mathbf{a}$ is array noncomputable. Moreover, Jockusch and Soare [12] extended their earlier result to construct the sets so that every separator of A_1, B_1 forms a minimal pair with each separator of A_2, B_2. This result was shown to be realizable below each *promptly* array noncomputable c.e. degree by Downey and Greenberg [5].

The basic result used in the arguments in [2] is that any c.e. singleton can be realized as a spectrum. Then, the authors use some computability theory approximation arguments for Σ^0_3-representable sets to get one direction of the characterization.

We do know of two singletons which can be realized. One is the trivial one $\mathbf{0}$ if we had a computable A and considered $S(A,\overline{A})$, but in the non-trivial situation where $\mathbb{N} - (A \sqcup B)$ is infinite, the only singleton we know of is the PA-one, namely $\mathbf{0}'$. The following question is open:

Question 1.5: *Is any other singleton* $\mathbf{a}$ *realizable as* $\mathcal{W}(S(A,B))$*?*

We conjecture that the answer is "no".

We can show that the answer is no in the case that $A \equiv_{wtt} B$, where $\leq_{wtt}$ denotes weak truth table reducibility. In fact, in §2, we prove the following "upward closure" result.

Theorem 1.6: *Suppose* $A \equiv_{wtt} B$ *are c.e. sets, such that:*

- $A \cap B = \emptyset$; *and*
- $|\omega \setminus (A \cup B)| = \infty$.

Then for every $C \geq_T A$*, there is a separator of* A *and* B *of the same Turing degree as* C*.*

On the other hand, we really do need "$A \equiv_{wtt} B$" in the hypothesis, to force upward closure, as $A \leq_{wtt} B$ is not enough, as we see in the next result.

Theorem 1.7: *There are c.e. sets $A \geq_{wtt} B$, such that:*

- $A \cap B = \emptyset$;
- $|\omega \setminus (A \cup B)| = \infty$; *and*
- *No separator of A and B computes $\emptyset'$.*

One of the most natural ways to possibly get a singleton would be to have $A \equiv_T B$ and such that $S(A, B)$ was highly constrained in that members of c.e. degree X would be "close" to A and B. One place where this idea was used is where "c.e. degree" was replaced by "c.e." in, for example, Downey [3]. There (A, B) is called a *maximal pair* if whenever X is a c.e. set separating A and B then either $X - A$ or $X - B$ is finite (also see Muchnik [14]). Any simple c.e. set can be split into a maximal pair. They are quite useful in reverse mathematics as, for example, in [4]. Thus, it would be very nice if there was a stronger version of maximal pair with c.e. *set* replaced by set of c.e. *degree.*

Definition 1.8: Two c.e. sets A and B form a *super-maximal pair* if the following hold:

- $A \cap B = \emptyset$;
- $|\omega - (A \cup B)| = \infty$; and
- If X is of c.e. degree with $A \subseteq X$ and $B \subseteq \overline{X}$, then $X =^* A$ or $\overline{X} =^* B$.

Alas, no such pairs exist.

Theorem 1.9: *Super maximal pairs do not exist.*

We believe that this result is of independent interest aside from our interest in degrees of members of separating classes. The proof is surprisingly difficult and requires *three* levels of nonuniformity, in the same way that the Lachlan Nondiamond Theorem (Lachlan [13]) needs one level of nonuniformity. The only other example where exactly 3 levels are needed occurs in an unpublished manuscript of Slaman where he shows that there is a c.e. degree $\mathbf{a} \neq \mathbf{0}$ which is not the top of a diamond lattice in the Turing degrees. Thus, this proof is of some technical interest.

Giving up on *one* degree, we ask whether *two* degrees are possible. This time the answer is yes, provided that one is $\mathbf{0}'$. One easy way to see this is to take A a complete c.e. set with $\overline{A}$ introreducible. Then $S(A, \emptyset)$ has spectrum $\mathbf{0}', \mathbf{0}$. The next result shows that $\mathbf{0}$ can be replaced by any c.e. degree.

Theorem 1.10: *For every c.e. degree* $\mathbf{c}$*, the separating spectrum* $\{\mathbf{c}, \mathbf{0}'\}$ *is possible.*

We remark that we are unaware of any other definite spectrum which can be realized, even for the two degree case.

2. Wtt-results

In this section, we prove the comparability results about weak truth table reducibility for A and B. The first shows that we can have no upward closure whilst having comparability.

Theorem 2.1: *There are c.e. sets* $A \geq_{wtt} B$*, such that:*

- $A \cap B = \emptyset$;
- $|\omega \setminus (A \cup B)| = \infty$; *and*
- *No separator of* A *and* B *computes* $\emptyset'$.

Proof: We construct such sets. To achieve $A \geq_{wtt} B$, we promise to never enumerate an element into B unless we simultaneously enumerate a smaller element into A.

We build an auxiliary c.e. set D and meet the following requirements:

N_k: $(\exists x > k)[x \notin (A \cup B)]$.
R_e: For any separator $Z \in S(A, B)$, $\Phi_e^Z \neq D$.

Clearly, this will suffice.

Strategy for N_k*:* Wait for a stage $s > k$. By construction, $s \notin (A_s \cup B_s)$. Restrain $(A \cup B) \upharpoonright_{s+1}$.

Strategy for R_e*:* We fix a restraint r to be the stage at which the strategy was last initialized, such that the strategy will not be permitted to enumerate elements below r into $A \cup B$. The strategy repeats the following loop:

(1) Claim a large n not yet claimed by any strategy.
(2) At stage s, search for a $\sigma \in 2^s$, such that:
   - $\Phi_e^\sigma \upharpoonright_{n+1} [s] = D_s \upharpoonright_{n+1}$; and
   - For all $x < |\sigma|$, $x \in A_s \to \sigma(x) = 1$, and $x \in B_s \to \sigma(x) = 0$.

(3) Having found such a σ, if there is an $m < |\sigma|$, such that $m \geq r$, $\sigma(m) = 1$ and $m \notin A_s$, fix the least such. For all $x \in [m, |\sigma|)$, if $\sigma(x) = 1$, then enumerate x into A_{s+1}, and if $\sigma(x) = 0$, then enumerate x into B_{s+1}.
(4) Regardless of whether a desired m exists, enumerate n into D and return to Step (1).

Construction: Arrange the strategies into a priority ordering. At stage s, run the first s strategies, in order of priority. Whenever a strategy acts, initialize all lower priority strategies.

Verification: By construction, we never enumerate a number into B unless we simultaneously enumerate a smaller number into A, and so $B \leq_{wtt} A$. Also, we never enumerate a number into $(A \cup B)[s+1]$ unless that number is smaller than s.

Claim 2.2: *Suppose the R_e-strategy is only initialized finitely many times. Then it only enumerates finitely many numbers into D.*

Proof: Let r be the final restraint imposed on the strategy. Towards a contradiction, suppose there are infinitely many stages at which the strategy enumerates an element into D, and list those which occur after the final time the strategy was initialized as $s_0 < s_1 < \ldots$. Fix n_i the element enumerated at stage s_i, σ_i the witnessing σ, and let m_i be the selected m, if it exists, and otherwise let $m_i = s_i$. By well-ordering properties, there are infinitely many i, such that for all $j > i$, $m_i \leq m_j$. Fix such an i.

Fix a $j > i$. By construction, we have $n_i < n_j$. So it cannot be that σ_j extends σ_i, as that would give $\Phi_e^{\sigma_j}(n_i) = \Phi_e^{\sigma_i}(n_i) = 0$, but $n_i \in D_{s_i+1} \subseteq D_{s_j}$, contrary to our choice of σ_j. So there must be some $y < |\sigma_i|$ with $\sigma_i(y) \neq \sigma_j(y)$. By construction, $y < m_i$.

If $y \geq r$, we claim that $\sigma_i(y) = 0$. For if not, then $\sigma_{i+1}(y) = 0$ implies $y \notin A_{s_{i+1}} \supseteq A_{s_i}$, and so y contradicts our choice of m_i.

So if $y \geq r$, $y \notin A_{s_i}$. Higher priority strategies will never act after stage s_i, and lower priority strategies were initialized and so have restraints greater than y, so neither can enumerate y into A. By choice of i, $y < m_i \leq m_k$ for all $k \in [i, j]$, and so the action of this strategy cannot enumerate y into A before stage s_j. So $y \notin A_{s_j}$, and so if $y \geq r$, then y is a viable candidate for m_j. This contradicts $m_i < m_j$. It follows that $y < r$, and so $\sigma_i \restriction_r \neq \sigma_j \restriction_r$, for all $j > i$.

But there are only finitely many strings in 2^r, and so there can be only finitely many i, such that for all $j > i$, $m_i \leq m_j$, contrary to our earlier observation. The claim follows. □

It is now a simple induction to show that each strategy is only initialized finitely many times. Clearly, each N_k-strategy ensures its requirement. Also, each R_e-strategy must eventually wait forever at Step (2), and so there can be no separator computing D. This completes the proof. □

The second result shows that if there is a realizable singleton which is not $\mathbf{0}$ or $\mathbf{0}'$, then we cannot use non-adaptive reductions.

Theorem 2.3: *Suppose $A \equiv_{wtt} B$ are c.e. sets, such that:*

- $A \cap B = \emptyset$; *and*
- $|\omega \setminus (A \cup B)| = \infty$.

Then for every $C \geq_T A$, there is a separator of A and B of the same Turing degree as C.

Proof: Fix wtt-operators Γ and Δ with $\Gamma^A = B$ and $\Delta^B = A$, and let f be a computable function bounding the use of both Γ and Δ. We assume that f is monotonic and $f(x) > x$ for all x. We split the argument into two cases, depending on whether or not there are infinitely many x with $(x, f(x)] \subseteq A \cup B$.

Case 1. Suppose there are only finitely many x with $(x, f(x)] \subseteq A \cup B$.

Fix k such that there are no such $x \geq k$. Define the computable sequence: $m_0 = k$; $m_{i+1} = f(m_i)$. Then we define a separator Z such that if $n \notin C$, then Z agrees with A on $(m_n, m_{n+1}]$, and if $n \in C$, then Z agrees with $\overline{B}$ on $(m_n, m_{n+1}]$. As C computes A (and thus B), $Z \leq_T C$.

To establish $C \leq_T Z$, note that Z cannot agree with both A and $\overline{B}$ on any $(m_n, m_{n+1}]$, as $(m_n, m_{n+1}] \nsubseteq A \cup B$. Thus $n \in C$ iff Z agrees with $\overline{B}$ on $(m_n, m_{n+1}]$ iff Z differs from A on $(m_n, m_{n+1}]$. So Z can determine whether $n \in C$ by waiting for a stage s, such that it agrees with either A_s or $\overline{B}_s$ on $(m_n, m_{n+1}]$.

Case 2. Suppose there are infinitely many x with $(x, f(x)] \subseteq A \cup B$.

Define the following sequence:

- $m_0 = -1$;
- m_{n+1} is the least $x > m_n$, such that:
  - $(m_n, x] \nsubseteq A \cup B$; and
  - $(x, f(x)] \subseteq A \cup B$.

By assumption, m_n exists for all n. We define a separator Z such that if $n \notin C$, then Z agrees with A on $(m_n, m_{n+1}]$, and if $n \in C$, then Z agrees with $\overline{B}$ on $(m_n, m_{n+1}]$.

First, observe that since C computes A (and thus B), C computes $(m_n)_{n\in\omega}$. Thus, C computes Z.

Next, we argue that $(m_n)_{n\in\omega}$ is computable from Z. Suppose we have determined m_i for $i \leq n$. Then let s be a stage, such that the following hold:

- For every $i < n$, Z on $(m_i, m_{i+1}]$ agrees with either A_s or $\overline{B}_s$;
- For every $i < n$, $(m_i, f(m_i)] \subseteq A_s \cup B_s$; and
- There is some $x > m_n$, such that:
  - $\Gamma^{A_s} \restriction_{x+1} = B \restriction_{x+1}$ and $\Delta^{B_s} \restriction_{x+1} = A \restriction_{x+1}$;
  - Z on $(m_n, x]$ agrees with either A_s or $\overline{B}_s$;
  - $(m_n, x] \not\subseteq A_s \cup B_s$; and
  - $(x, f(x)] \subseteq A_s \cup B_s$.

Fix y the least such x.

Claim 2.4: $y = m_{n+1}$.

Proof: Suppose not. If $m_{n+1} > y$, then by minimality of m_{n+1}, it must be that $(m_n, y] \subseteq A \cup B$. By choice of y, this means that some element $z \leq y$ is enumerated into $A \cup B$ after stage s.

If $m_{n+1} < y$, then by minimality of y, it must be that $(m_{n+1}, f(m_{n+1})] \not\subseteq A_s \cup B_s$. By monotonicity, $f(m_{n+1}) \leq f(y)$, so there must be some element $z \leq f(y)$ enumerated into $A \cup B$ after stage s. By choice of y, $z \leq y$.

In either case, we see that there is an element $z \leq y$ enumerated into $A \cup B$ after stage s. By choice of s and correctness of Δ, if $z \in A$, then there must be a $w \leq f(z)$, such that w is enumerated into B after stage s, and by monotonicity and choice of s and y, $w \leq y$. If $z \in B$, then symmetric reasoning shows there is a $w \leq y$ enumerated into A after stage s. Without loss of generality, assume that z is the least element enumerated into $A \cup B$ after stage s

If there is an $i < n$ such that $z \in (m_i, m_{i+1}]$, then by choice of s and correctness of Γ and Δ, $w \in (m_i, m_{i+1}]$. But Z agrees with either A_s or $\overline{B}_s$ on $(m_i, m_{i+1}]$ by choice of s, so there cannot be such z and w: e.g. if Z agrees with A_s, then $z \in A \setminus A_s$ contradicts Z being a separator. So it must be that $z, w \in (m_n, x]$. But again, Z agrees with either A_s or $\overline{B}_s$ on $(m_n, x]$, so this is a contradiction. □

Having computed $(m_n)_{n\in\omega}$, Z can determine whether $n \in C$ by waiting for a stage s, such that Z on $(m_n, m_{n+1}]$ agrees with either A_s or $\overline{B}_s$ (in fact, the stage s used to determine m_{n+1} suffices). □

3. No Super-Maximal Pairs

Recall that super-maximal pairs are c.e. sets A, B where all X separating of c.e. degree must be finite variants of either A or B.

Theorem 3.1: *Super maximal pairs do not exist.*

Proof: Let A and B be two c.e. sets satisfying $A \cap B = \emptyset$ and $|\omega - (A \cup B)| = \infty$. We will build a separating set X of c.e. degree, such that $X \neq^* A$ and $\overline{X} \neq^* B$. This construction is necessarily non-uniform, however, so we may make up to three attempts. If the first attempt fails, it will be because our set has either $X =^* A$ or $\overline{X} =^* B$. If the second attempt fails, it will fail in the opposite manner. Then the third attempt will succeed.

The First Attempt. We build a computable sequence $(X_{1,s})_{s \in \omega}$ approximating X_1. We begin with $X_{1,0} = \emptyset$. We also define the auxiliary sets $Y^n_{1,0} = \emptyset$ for all n.

At stage $s+1$, we first define a sequence $x^1_{-1,s+1} < x^1_{0,s+1} < \cdots < x^1_{k,s+1} = s$:

- $x^1_{-1,s+1} = -1$.
- Given $x^1_{n,s+1} < s$, if $x^1_{n+1,s}$ is undefined or $x^1_{n+1,s} \leq x^1_{n,s+1}$, we let $x^1_{n,s+1} = s$.
- Given $x^1_{2i,s+1} < x^1_{2i+1,s} < s$, if there is a $y \in (x^1_{2i,s+1}, x^1_{2i+1,s}]$ satisfying:
  - $y \in X_{1,s} - (A_{s+1} \cup B_{s+1})$; or
  - $y \notin A_{s+1} \cup B_{s+1}$, and there is some $z < y$ with $z \in X_{1,s} \cap B_{s+1}$ or with $z \in \overline{X}_{1,s} \cap A_{s+1}$; or
  - $y \notin A_{s+1} \cup B_{s+1}$ and $y = s$,

  then we let $x^1_{2i+1,s+1} = x^1_{2i+1,s}$. Otherwise, we let $x^1_{2i+1,s+1} = s$.
- Given $x^1_{2i+1,s+1} < x^1_{2i+2,s} < s$, if there is a $y \in (x^1_{2i+1,s+1}, x^1_{2i+2,s}]$ satisfying:
  - $y \in \overline{X}_{1,s} - (A_{s+1} \cup B_{s+1})$; or
  - $y \notin A_{s+1} \cup B_{s+1}$, and there is some $z < y$ with $z \in X_{1,s} \cap B_{s+1}$ or with $z \in \overline{X}_{1,s} \cap A_{s+1}$; or
  - $y \notin A_{s+1} \cup B_{s+1}$ and $y = s$,

  then we let $x^1_{2i+2,s+1} = x^1_{2i+2,s}$. Otherwise, we let $x^1_{2i+2,s+1} = s$.

Note that the sequence we define is strictly increasing. It terminates when it achieves s.

On each interval $(x^1_{2i,s+1}, x^1_{2i+1,s+1}]$, we will attempt to ensure that $X \neq^* A$ by putting whatever elements we can into X_1. On each interval $(x^1_{2i+1,s+1}, x^1_{2i+2,s+2}]$, we will do the opposite. Of course, this is restricted by our need to make X_1 a separator of c.e. degree. Accordingly, we make the following definition

Definition 3.2: We say that y is *permitted at stage* $s+1$ (for X_1) if $y \notin A_{s+1} \cup B_{s+1}$, and $y = s$ or there is some $z < y$ with $z \in X_{1,s} \cap B_{s+1}$ or with $z \in \overline{X}_{1,s} \cap A_{s+1}$.

We now define $X_{1,s+1}$ as follows:

- If $y \in A_{s+1}$, then $y \in X_{1,s+1}$.
- If $y \in B_{s+1}$, then $y \notin X_{1,s+1}$.
- If $y \in (x^1_{2i,s+1}, x^1_{2i+1,s+1}]$ and is permitted at stage $s+1$, then $y \in X_{1,s+1}$.
- If $y \in (x^1_{2i+1,s+1}, x^1_{2i+2,s+1}]$ and is permitted at stage $s+1$, then $y \notin X_{1,s+1}$.
- If none of the above apply, then $x \in X_{1,s+1} \iff x \in X_{1,s}$.

This completes the first construction.

Claim 3.3: *$(X_{1,s})_{s\in\omega}$ converges (in a Δ^0_2 fashion) to a set X_1, and X_1 is a separator of A and B of c.e. degree.*

Proof: Observe first that if $x \in X_{1,s} \triangle X_{1,s+1}$, then one of the following must hold:

(1) $x \in X_{1,s} \cap B_{s+1}$;
(2) $x \in \overline{X}_{1,s} \cap A_{s+1}$;
(3) For some $z < x$, one of (1) or (2) holds; or
(4) $x = s$.

By induction on x, each of these can occur only finitely many times for each x, and so the limit X_1 exists.

Now, let $W = \{x : \exists s\, x \in X_{1,s} \cap B_{s+1} \vee x \in \overline{X}_{1,s} \cap A_{s+1}\}$ with the obvious c.e. approximation $(W_s)_{s\in\omega}$. If $x \in W_{s+1} - W_s$, then $x \in X_{1,s} \triangle X_{1,s+1}$. Conversely, if $x \in X_{1,s} \triangle X_{1,s+1}$ for some $s > x$, then there is a $z \le x$ with $z \in W_{s+1} - W_s$. Thus $X_1 \equiv_T W$, and so X_1 is of c.e. degree.

That $A \subseteq X$ and $X \cap B = \emptyset$ is immediate from our definition of $X_{1,s+1}$.

□

Now, for each n, we consider the sequence $(x^1_{n,s})_{s\in\omega}$. Note that not every $x^1_{n,s}$ will be defined. However, if $x^1_{n,s}$ is defined for almost every s, we can consider whether the sequence has a limit.

Claim 3.4: *Suppose $x^1_n = \lim_s x^1_{n,s}$ exists with $x^1_n < \infty$ (and, implicitly, $x^1_{n,s}$ is defined for almost every s). Then for every $j < n$, $x^1_j = \lim_s x^1_{j,s}$ exists, and $x^1_j < x^1_n$. Further, if $n > -1$, then there is a $y \in (x^1_{n-1}, x^1_n]$ with $y \notin A \cup B$, and $y \in X_1$ if and only if n is odd.*

Proof: By construction, for $j < n$ and every s at which $x^1_{n,s}$ is defined, $x^1_{j,s}$ is also defined. So $x^1_{j,s}$ is defined for almost every s. By construction, $x^1_{j,s}$ is nondecreasing in s and bounded by x^1_n, so $x^1_j < x^1_n$ exists.

For s with $x^1_{n,s+1} = x^1_n < s$, there is some y witnessing that $x^1_{n,s+1} \neq s$. This y is in $X_{1,s+1}$ if and only if n is odd by construction. So by pigeon hole, there is some y in $(x^1_{n-1}, x^1_n]$ with this property for infinitely many s, and thus for almost every s (as the approximation to X_1 converges). Thus, $y \in X_1$ if and only if n is odd. □

So if $x^1_n < \infty$ exists for every n, then X_1 is as desired. So instead assume ℓ_1 is the greatest n, such that $x^1_n < \infty$ exists. Let $k_1 = \ell_1 + 1$, and WLOG assume k_1 is odd. Observe that $x^1_{k_1,s+1}$ is defined for every $s > x^1_{\ell_1}$.

Claim 3.5: *For every s with $x^1_{\ell_1,s} = x^1_{\ell_1}$ and $x^1_{k_1,s}$ defined, and for every $y \in (x^1_{\ell_1}, x^1_{k_1,s}] \cap X_{1,s}$, $y \in A \cup B$.*

Proof: Fix $t \geq s$, such that $x^1_{k_1,t+1} \neq x^1_{k_1,t} = x^1_{k_1,s}$. Then, by definition of $x^1_{k_1,t+1}$, every such y must be in $A_{t+1} \cup B_{t+1} \cup \overline{X}_{1,t}$. But, by induction on $s' \in (s,t]$, if $y \notin B$, then $y \in X_{1,s'}$. □

It follows that for any $y > x^1_{\ell_1}$, $y \in X_1 \iff y \in A$.

The Second Attempt. We make a second attempt based on the knowledge of how the first attempt failed. First, we perform a speedup of the enumerations of A and B and of the previous construction, such that the following all hold:

- For each $n < k_1$ and every s, $x^1_{n,s} = x^1_n$.
- For every s, $x^1_{k_1,s} > s$ is defined.
- For every $y \in (x^1_{\ell_1}, s]$, $y \in A_s \cup \overline{X}_{1,s}$.

We now build X_2 in this new timeline. We simply repeat the construction of X_1, except that we refer to the sequence we build at each stage as $(x^2_{n,s+1})$, to avoid confusion, and we always begin with $x^2_{-1,s+1} = x^1_{\ell_1}$. Analogues of Claims 3.3 and 3.4 for X_2 proceed as the originals. Note that for every n and s with $x^2_{n,s}$ defined, $x^2_{n,s} < x^1_{k_1,s}$.

Again, if $x^2_n < \infty$ exists for every n, then X_2 is our desired separator. So instead assume ℓ_2 is the greatest n, such that $x^2_n < \infty$ exists, and let $k_2 = \ell_2 + 1$. Again, $x^2_{k_2,s+1}$ is defined for every $s > x^2_{\ell_2}$.

Claim 3.6: *k_2 is even.*

Proof: Suppose not. Fix an $s > x^2_{\ell_2} > x^1_{\ell_1}$ with $s \notin A \cup B$.

Fix t least with $x^2_{k_2,t+1} \geq s$. We claim that s is permitted for X_2 at stage $t+1$. This is immediate if $s = t$, so suppose not. Then $x^2_{k_2,t} < s < t$, so there is some $y \in (x^2_{\ell_2}, x^2_{k_2,t}] \cap X_{2,t}$ witnessing that $x^2_{k_2,t} < t-1$. As this y does not suffice for $t+1$, it must be that $y \in A_{t+1} \cup B_{t+1}$. In second case, y witnesses that s is permitted for X_2 at stage $t+1$. In the first case, since $y \notin A_t$, $y \in \overline{X}_{1,t}$. So y witnesses that s is permitted for X_1 at some point in the time period between stages t and $t+1$ of our speedup. But as $x^1_{\ell_1} < s < x^1_{k_1,s} \leq x^1_{k_1,t}$, a definition $s \in X_1$ would be made. This contradicts Claim 3.5.

So s is permitted for X_2 at stage $t+1$, and so $s \in X_{2,t+1}$. But then s will forever witness that $x^2_{k_2,s'}$ does not need to change, contrary to choose of k_2. □

Analogously to Claim 3.5, we obtain the following.

Claim 3.7: *For every s with $x^2_{\ell_2,s} = x^2_{\ell_2}$ and $x^2_{k_2,s}$ defined, and for every $y \in (x^2_{\ell_2}, x^2_{k_2,s}] \cap \overline{X}_{2,s}$, $y \in A \cup B$.*

It follows that for any $y > x^2_{\ell_2}$, $y \in X_2 \iff y \notin B$.

The Final Attempt. We make our final attempt based on the knowledge of how the first two attempts failed. Again, we begin with a speedup of our previous speedup, such that the following all hold:

- For each $n < k_2$ and every s, $x^2_{n,s} = x^2_n$.
- For every s, $x^2_{k_2,s} > s$ is defined.
- For every $y \in (x^2_{\ell_2}, s]$, $y \in X_{2,s} \cup B_s$.

Again, we always begin with $x^3_{-1,s+1} = x^2_{\ell_2}$. We again prove analogues of Claims 3.3 and 3.4. Define k_3 to be the least, such that $x^3_{k_3} < \infty$ does not exist. We show two versions of Claim 3.6, one showing that k_3 cannot be odd by arguing to a contradiction with Claim 3.5, and another showing that k_3 cannot be even by arguing to a contradiction with Claim 3.7. It follows that there is no such k_3, and so X_3 is as desired. □

4. Two Degrees

In this section, we examine degree spectra containing two degrees.

Theorem 4.1: *For every c.e. degree* $\mathbf{c}$*, the separating spectrum* $\{\mathbf{c}, \mathbf{0}'\}$ *is possible.*

Proof: Fix $C \in \mathbf{c}$. We build disjoint c.e. sets A and a B with $|\omega \setminus (A \cup B)| = \infty$ and meeting the following requirements, for all e and n:

R_e: If $\Phi_e^{W_e} = Z$, $A \subseteq Z$, and $|\overline{Z} \setminus B| = \infty$, then W_e computes K.
P_n: There is at most one i with $\langle n, i\rangle \in B$, and such i, if it exists, is bounded by $n^2 + 1$. Further, $n \in C \iff \exists i\,[\langle n, i\rangle \in B]$.

First, we argue that meeting these requirements suffices.

By the P_n, $B \equiv_T C$. For one direction, $n \in C \iff (\exists i \leq n^2+1)[\langle n, i\rangle \in B]$, which is bounded quantification. For the other, suppose we wish to determine whether $\langle n, j\rangle$ is in B. We first check whether $n \in C$; if not, $\langle n, j\rangle \notin B$. If so, we enumerate B until we see some $\langle n, i\rangle \in B$. Then $\langle n, j\rangle \in B \iff i = j$.

Now, suppose Z is a separator of c.e. degree. If $|\bar{Z} \setminus B| = \infty$, then by the appropriate R_e, Z is of degree $\mathbf{0}'$. If instead $\bar{Z} =^* B$, then Z is of degree $\mathbf{c}$. In particular, A itself is a separator of degree $\mathbf{0}'$, and $\overline{B}$ is a separator of degree $\mathbf{c}$.

We assume that for all i, $\langle n, i\rangle \geq n^3$.

Strategy for R_e: We construct a c.e. operator V_e, with the intention that $V_e^{W_e} = \overline{K}$ if $\Phi_e^{W_e}$ is as described. For each m, if $m \notin [K \cup V_e^{W_e}][s]$, we search for an x and a γ satisfying the following:

- For all $n \leq \max\{e, m\}$ and $i < n^2 + 1$, $x > \langle n, i\rangle$.
- $x \notin B_s$;
- For all $y \leq x$, $\Phi_e^{W_e \restriction \gamma}(y)[s]\downarrow$; and
- $\Phi_e^{W_e \restriction \gamma}(x)[s] = 0$.

If these exist, we fix the least such x and γ (by standard use assumptions, we can minimize these both simultaneously), and we define $m \in V_e^{W_e \restriction \gamma}[s]$. At every subsequent stage $t > s$, if $W_e \restriction_\gamma [t] = W_e \restriction_\gamma [s]$, then x is *blocked from B* at stage t.

If instead $m \in [K \cap V_e^{W_e}][s]$, we fix the x used in the definition of $m \in V_e^{W_e}[s]$. If $x \notin B_s$ (as we will later argue must be the case), we enumerate x into A.

Otherwise, do nothing for m.

Strategy for P_n: When n is enumerated into C, fix the least i such that $\langle n, i\rangle \notin A_s$ and $\langle n, i\rangle$ is not blocked from B at stage s by any of the R_e. We will later argue that $i < n^2 + 1$. Enumerate $\langle n, i\rangle$ into B.

Construction: Simply run all of the above strategies simultaneously. We take no care to ensure that different R-strategies are choosing different witnesses x.

Verification: By construction, we never enumerate an element into $A \cup B$, and so A and B are disjoint. Next, we show that our claim in the P_n-strategy holds.

Claim 4.2: *At any stage s, for every n,*

$$\#\{i < n^2 + 1 : \langle n, i\rangle \in A_s \vee \langle n, i\rangle \text{ is blocked from } B \text{ at stage } s\} \leq n^2.$$

Proof: The only way for $x = \langle n, i\rangle$ to be blocked or enumerated into A is for it to be selected by some R_e-strategy on behalf of some m. By our choice of such x in the R_e-strategy, it must be that $e, m < n$. Further, no pair (e, m) can contribute more than one x in this fashion: the R_e-strategy blocks at most one x at a time on behalf of each m, and if the strategy has enumerated an x on behalf of m, then $m \in C_s$ and so the strategy will not block or enumerate another element on behalf of m.

The claim follows. □

Thus, our P_n-strategy meets its requirement.

Claim 4.3: *For all k,*

$$|(A \cup B) \cap k^3| \leq O(k^2).$$

It follows that $|\omega \setminus (A \cup B)|$ is infinite.

Proof: By construction, since $\langle k, 0\rangle \geq k^3$, the only strategies which can enumerate elements into $A \cap k^3$ are R_e-strategies with $e < k$, and then only on behalf of some $m < k$. As in the previous claim, each pair (e, m) can contribute at most one such enumeration, and so $|A \cap k^3| \leq k^2$.

Also, the only strategies which can enumerate elements into $B \cap k^3$ are P_n-strategies with $n < k$, and each necessarily enumerates at most one element. So $|B \cap k^3| \leq k$.

The claim follows. □

Claim 4.4: *Each R_e-strategy meets its requirement.*

Proof: Fix m. Suppose first that $m \notin K$. Fix the least $x \notin (Z \cup B)$ such that $x > \langle n, i \rangle$, for all $n \leq \max\{e, m\}$ and $i < n^2 + 1$. Fix γ least such that $\Phi_e^{W_e \restriction \gamma} \supseteq Z \restriction_{x+1}$, and fix s_0 such that $W_e \restriction_\gamma = W_{e,s_0} \restriction_\gamma$. Then at any stage $s > s_0$ with $m \notin V_e^{W_e}[s]$, x and γ will be chosen for the new definition of $m \in V_e^{W_e \restriction \gamma}[s]$. By choice of s_0, this ensures $m \in V_e^{W_e}$.

If instead $m \in K$, then fix s_0 least with $m \in K_s$. By construction, we will never define $m \in V_e^{W_e}[s]$ for any $s \geq s_0$. So suppose $m \in V_e^{W_e}[s_0]$ because of our action at some stage $t < s_0$. Fix the witnessing x and γ. Then necessarily, $W_{e,r} \restriction_\gamma = W_{e,s_0} \restriction_\gamma$ for every $r \in [t, s_0]$, and so x was blocked from B at every such stage. Also, $x \notin B_t$. By construction, $x \notin B_{s_0}$. So x will be enumerated into A at stage s_0.

By assumption, $\Phi_e^{W_e \restriction \gamma}(x)[t] = \Phi_e^{W_e \restriction \gamma}(x)[s_0] = 0$. So if $\Phi_e^{W_e} = Z$ contains A, and in particular contains x, it must be that $W_{e,s_0} \restriction_\gamma \neq W_e \restriction_\gamma$. Since no future stage s will define $m \in V_e^{W_e}[s]$, it follows that $m \notin V_e^{W_e}$. □

This completes the proof. □

References

1. D. Cenzer and C. Jockusch, Π_1^0-classes - Structure and applications, in *Contemporary Mathematics* 257 (2000), 39–59.
2. B. Csima, R. Downey and K. M. Ng, *On the c.e. degrees realizable in Π_1^0 classes*, *Journal of Symbolic Logic*, in press.
3. R. Downey, *Abstract Dependence, Recursion Theory and the Lattice of Recursively Enumerable Filters* PhD Thesis, Monash University, 1982.
4. R. Downey, S. Lempp and J. Mileti, Ideals in computable rings, *Journal of Algebra* Vol. 314 (2007), 872–887.
5. R. Downey and N. Greenberg, *A Hierarchy of Turing Degrees*, Annals of Mathematics Studies, Vol. 206 Princeton University Press 2020.
6. R. Downey and D. Hirschfeldt, *Algorithmic Randomness and Complexity*, Springer-Verlag, 2010.
7. R. Downey, C. Jockusch and M. Stob, *Array nonrecursive sets and multiple permitting arguments*, in *Recursion Theory Week* (Ambos-Spies, Muller, Sacks, eds.) Lecture Notes in Mathematics 1432, Springer-Verlag, Heidelberg, 1990, 141–174.
8. R. Downey and A. Melnikov, A Note on Realization of Index Sets in Π_1^0 Classes, *Algebra i Logika.* Vol. 58 (2019), 659–663 (Russian).
9. H. Friedman, S. Simpson and R. Smith, Countable algebra and set existence axioms, *Annals of pure and applied logic*, vol. 25 (1983), pp. 141–181.
10. H. Friedman, S. Simpson and R. Smith, Addendum to Countable algebra and set existence axioms. *Annals of pure and applied logic*, vol. 28 (1985), pp. 319–320.

11. C. Jockusch and R. Soare, Π^0_1 classes and degrees of theories, *Trans. Amer. Math. Soc.*, 173 (1972), 33–56.
12. C. Jockusch, Jr. and R. Soare, A minimal pair of Π^0_1 classes, *J. Symbolic Logic*, Vol. 36 (1971), 66–78.
13. A. Lachlan, Lower bounds for pairs of recursively enumerable degrees, *Proc. London Math. Soc.*, Vol. 16 (1966), 537–569.
14. A. A. Muchnik, On the separability of recursively enumerable sets, *Dokl. Akad. Nauk SSSR (N.S.)*, Vol. 109 (1956), 29–32.
15. S. Simpson, *Subsystems of Second Order Arithmetic*, Second Edition, Perspectives in Logic, Association for Symbolic Logic, Cambridge University Press, 2009, XVI + 444 pages.
16. R. I. Soare, *Recursively Enumerable Sets and Degrees*, Springer-Verlag, 1987.

https://doi.org/10.1142/9789819806584_0003

On σ-complete Uniform Ultrafilters

Moti Gitik
Tel Aviv University, P.O. Box 39040, Tel Aviv 6997801, Israel
gitik@tauex.tau.ac.il

σ-complete uniform ultrafilters where extensively studied in early seventies. Many nice results were obtained, specially by J. Ketonen (see [10]).

Recently, G. Goldberg returned to the subject and proved a remarkable result that under The Ultrapower Axiom the first strongly compact cardinal is a supercompact.

The purpose of this chapter is to provide some concrete examples of σ-complete uniform ultrafilters.

1. Some Basic Definitions and Facts

Definition 1.1: Let $U \subseteq \mathcal{P}(\lambda)$ be an ultrafilter on λ.

(1) U is called *uniform* iff for every A of cardinality $< \lambda$, $\lambda \setminus A \in U$.
(2) U is called *κ-complete* iff the intersection of any less than κ members of U is in U.
(3) U is called *κ-complete exactly* iff U is κ-complete, but not κ^+-complete.

If κ is a strongly compact cardinal, then for every $\lambda \geq \kappa$ there is a uniform κ-complete ultrafilter over λ. By J. Ketonen [10] the opposite is true as well.

Here, we would like to examine the existence of σ-complete uniform ultrafilters over a cardinal λ which are exactly κ-complete for some $\kappa < \lambda$ under much weaker assumptions.

Note that if U is a σ-complete uniform ultrafilter over a cardinal λ and $j_U : V \to M_U \simeq \mathrm{Ult}(V, U)$ is the corresponding elementary embedding,

This chapter was partially supported by Israel Science Foundation Grants No. 58/14, 1216/18. We are grateful to the referee of this chapter for his long list of suggestions, remarks and corrections.

then

$$\sup(j_U{}''\lambda) \leq [id]_U < j_U(\lambda).$$

Also, if U is κ-complete for some $\kappa < \lambda$, then $\mathrm{cof}(\lambda) \geq \kappa$.

The following notion replaces the normality in the present context.

Definition 1.2: A uniform ultrafilter U over λ is called *weakly normal* iff for every $A \in U$ and every regressive $f : A \to \lambda$ there is $\alpha < \lambda$, such that the set $\{\nu \in A \mid f(\nu) < \alpha\}$ is in U.

Note that (for σ-complete ultrafilters) this is equivalent to the statement that $[id]_U$ is the least function which is not bounded by $[c_\gamma]_U$, for $\gamma < \lambda$, i.e. $[id]_U = \sup(j_U{}''\lambda)$.

Clearly, if U is exactly κ-complete ultrafilter over λ, then κ must be a measurable and the critical point of the corresponding elementary embedding $j_U : V \to M_U \simeq {}^\lambda V/U$.

Further by κ-completeness we will mean the exact κ-completeness.

Let us address the strength of the existence of such ultrafilters.

Proposition 1.3: *Assume that* $\neg 0^{¶}$.[1]

Suppose that there exists a κ*-complete uniform ultrafilter over* $\lambda > \kappa$. *Then the following hold:*

(1) κ *and* $\mathrm{cof}(\lambda)$ *are measurable cardinals in the core model,*
(2) if $\mathrm{cof}(\lambda) < \lambda$, *then either*

(a) λ *is a measurable cardinal in the core model,*
or
(b) λ *is singular limit of measurable cardinals in the core model.*

Proof: Let $\mathcal{U}$ be κ-complete uniform ultrafilter over $\lambda > \kappa$. Let $\mathcal{K}$ denotes the core model.

Consider $j_{\mathcal{U}} : V \to M_{\mathcal{U}} \simeq \mathrm{Ult}(V,\mathcal{U})$ the corresponding elementary embedding. By W. Mitchell [13], $j := j_{\mathcal{U}} \restriction \mathcal{K}$ is an iterated ultrapower of $\mathcal{K}$ by its measures or extenders.

Note that κ is a critical point of j, and so, it is a measurable cardinal in $\mathcal{K}$.

If $\mathrm{cof}(\lambda)$ is not a measurable in $\mathcal{K}$, then $\sup(j''cof(\lambda)) = j(\mathrm{cof}(\lambda))$, and so, $\sup(j''\lambda) = j(\lambda)$, which is impossible due to the uniformity of $\mathcal{U}$.

Suppose now that $\mathrm{cof}(\lambda) < \lambda$ and λ is not a measurable cardinal in $\mathcal{K}$.

[1]This means that there is no inner model with a sharp for a strong cardinal.

Assume that in $\mathcal{K}$, λ is not a limit of measurable cardinals. Let μ be a supremum of measurable cardinals of $\mathcal{K}$ below λ.

Consider $[id]_{\mathcal{U}}$. Then there are $n < \omega$ and $f : [\mu]^n \to On$ such that $j(f)(a) = [id]_{\mathcal{U}}$, for some $a \in [j(\mu)]^n = j(\text{dom}(f))$. So, $[id]_{\mathcal{U}} \in \text{rng}(j(f))$. Let $Z = \text{rng}(f)$, then $|Z| \leq \mu$. However, $[id]_{\mathcal{U}} \in j(Z) = \text{rng}(j(f))$ implies that $Z \in \mathcal{U}$ and this contradicts the uniformity of $\mathcal{U}$. □

2. Basic Constructions

Given two measurable cardinals $\kappa < \lambda$, it is easy to define a uniform ultrafilter U over λ which is exactly κ-complete.

Thus, fix some normal ultrafilters $\mathcal{V}$ on κ and $\mathcal{U}$ on λ. Let $j_{\mathcal{U}} : V \to M_{\mathcal{U}}$ be the ultrapower embedding by $\mathcal{U}$. Clearly, $\mathcal{V}$ remains a normal ultrafilter in $M_{\mathcal{U}}$. Let

$$j_{\mathcal{V}}^{M_{\mathcal{U}}} : M_{\mathcal{U}} \to N := M_{\mathcal{V}}^{M_{\mathcal{U}}}$$

be its ultrapower embedding there.

Set

$$i = j_{\mathcal{V}}^{M_{\mathcal{U}}} \circ j_{\mathcal{U}}.$$

Define a desired uniform κ-complete (exactly) ultrafilter $\mathcal{W}$ on λ as follows:

$$X \in \mathcal{W} \Leftrightarrow \lambda + \kappa \in i(X).$$

Such $\mathcal{W}$ is not weakly normal, since $i''\lambda = \lambda < [id]_W = \lambda + \kappa$.

It follows also that the weakly normal ultrafilter Rudin-Keisler below $\mathcal{W}$ is actually $\mathcal{U}$.

It is not hard to construct a model with a uniform weakly normal, κ-complete ultrafilter which is not normal.

Start with a GCH model with two measurable cardinals $\kappa < \lambda$.

Add a Cohen function $f_\alpha : \alpha \to \alpha$ for every inaccessible $\alpha, \kappa < \alpha \leq \lambda$ (with the Easton support iteration).

Extend the embedding i (of the previous construction) and change the value of $f_{i(\lambda)}(\lambda)$ to κ.

The resulting extension of $\mathcal{U}$ will be as desired, and, moreover it will be Rudin-Keisler equivalent to $\mathcal{W}$ defined as above using $\lambda + \kappa$.

3. Suslin Trees

It is possible (again from two measurables) to construct a model with a uniform κ-complete ultrafilter U over λ in which λ is not a measurable.

The basic idea goes back to K. Kunen's [12] construction of a model with a λ-saturated ideal over λ. Here we would like to add a λ-Suslin tree to λ and then to argue that in the ultrapower by a normal ultrafilter over κ λ-branches are added to its image. This will allow us to extend embeddings by measures over λ.

Theorem 3.1: *Assume GCH. Let $\kappa < \lambda$ be two measurable cardinals. Then there is a cofinality preserving generic extension which satisfies the following:*

(1) GCH,
(2) λ is not measurable,
(3) there is a uniform κ-complete ultrafilter over λ.
Moreover, every $\mathcal{U}$ which is a normal ultrafilter over λ in V extends to a uniform κ-complete ultrafilter over λ.

Proof: Fix some normal ultrafilters $\mathcal{V}$ on κ and $\mathcal{U}$ on λ. Let $j_{\mathcal{U}} : V \to M_{\mathcal{U}}$ be the ultrapower embedding by $\mathcal{U}$. Clearly, $\mathcal{V}$ remains a normal ultrafilter in $M_{\mathcal{U}}$. Let

$$i := j_{\mathcal{V}}^{M_{\mathcal{U}}} : M_{\mathcal{U}} \to N := M_{\mathcal{V}}^{M_{\mathcal{U}}}$$

be its ultrapower embedding there.

Our first attempt will be to add a λ-Suslin tree T such that its image under $j_{\mathcal{V}}$ has a branch in V.

Let $\eta, \kappa < \eta \leq \lambda$ be a 2-Mahlo cardinal. Define a forcing notion Q_η as follows:

$\langle T, \vec{f}\rangle \in Q_\eta$ iff

(1) $T \subseteq {}^{\eta>}2$ is a normal tree of a successor height $ht(T)$ below η,
(2) $\vec{f} = \langle f_\alpha \mid \alpha < ht(T)\rangle$, where
   (a) for every $\alpha < ht(T)$, $f_\alpha : \kappa \to Lev_\alpha(T)$,
   (b) for every $\alpha, \beta < ht(T)$, if $\alpha < \beta$, then the set $\{\nu < \kappa \mid f_\alpha(\nu) <_T f_\beta(\nu)\}$ is co-bounded in κ.
   The idea behind is that $[f_\alpha]_{\mathcal{V}}$ would eventually generate an η-branch in the ultrapower. □

The order on Q_η is defined in the natural fashion by taking end extensions.[2]

The following lemma is easy:

[2]We denote the fact that a condition p is stronger than a condition q by $p \geq q$.

Lemma 3.2: *The forcing Q_η is $< \eta$-strategically closed.*

Proof: Let $\tau < \eta$. Define a winning strategy for Player II (the one that plays even stages including limit ones) by induction.

Suppose that $\langle\langle T^\zeta, \vec{f^\zeta}\rangle \mid \zeta < \xi < \tau\rangle$, ξ even is a play according to this strategy. Define the ξth move of Player II. Let $T = \bigcup_{\zeta<\xi} T^\zeta$ and T^ξ will be the tree obtained from T by adding the last level. We just continue all maximal branches of T to this level.

Note that there are maximal branches, since trees in Q_η are normal, have successor heights and at each limit stage of the construction (below ξ) we do the same, i.e. all maximal branches of the union are continued.

Now, set $\vec{f}$ to be the union of all $\vec{f^\zeta}, \zeta < \xi$. Let ξ^* be the last level of T^ξ. Define f^{ξ^*} and then add it to $\vec{f}$.

Suppose first that ξ is a limit stage. Let $f^{\xi^*}(\nu), \nu < \kappa$, be the point at the level ξ^* which is a continuation of the branch $\{t \in T \mid \exists\zeta < \xi(\zeta$ is even and $(\vec{f^\zeta}(ht(T^\zeta))(\nu) \geq_T t)\}$. We assume by induction that all $\vec{f^\zeta}(ht(T^\zeta))(\nu)$, for even ζ are on the same branch.

In particular, for every two even ordinals $\zeta < \xi < \tau$, we will have $\vec{f^\zeta}(ht(T^\zeta))(\nu) \leq_T \vec{f^\xi}(ht(T^\xi))(\nu)$, for every $\nu < \kappa$.

Suppose now that ξ is a successor stage.

Set

$$X = \{\nu < \kappa \mid \vec{f}^{\xi-1}(ht(T^{\xi-1}))(\nu) \geq_T \vec{f}^{\xi-2}(ht(T^{\xi-2}))(\nu)\}.$$

For every $\nu \in X$, let $f^{\xi^*}(\nu)$ be a point at the level ξ^* which is above $\vec{f}^{\xi-1}(ht(T^{\xi-1}))(\nu)$.

For every $\nu \in \kappa \setminus X$, let $f^{\xi^*}(\nu)$ be a point at the level ξ^* which is above $\vec{f}^{\xi-2}(ht(T^{\xi-2}))(\nu)$.

Clearly such defined f^{ξ^*} is as desired. □

Let now $G \subseteq Q_\eta$ be generic and $T(G)$ be the η-tree added by G.

Lemma 3.3: *$T(G)$ is an η-Suslin tree.*

Proof: Work in V. Let $\underset{\sim}{A}$ be a name of a maximal antichain in $T(G)$. Using the strategy of the previous lemma define $\langle T, \vec{f}\rangle$ of a limit height $\delta < \eta$ of cofinality κ and an increasing continuous sequence $\langle\delta_\gamma \mid \gamma < \kappa\rangle$ with limit δ such that for every $\gamma \leq \gamma' < \kappa$,

(1) $\langle T \restriction \delta_\gamma + 1, \vec{f} \restriction \delta_\gamma + 1\rangle \in Q_\eta$,
(2) for every $t \in T \restriction \delta_\gamma + 1$ there is $t^* \in T \restriction \delta_{\gamma+1} + 1$, $t^* \geq_T t$ or $t \geq_T t^*$, $\langle T \restriction \delta_{\gamma+1} + 1, \vec{f} \restriction \delta_{\gamma+1} + 1\rangle \Vdash t^* \in \underset{\sim}{A}$.

(3) For every $\nu < \kappa$, $f_{\delta_\gamma}(\nu) \leq_T f_{\delta_{\gamma'}}(\nu)$.

Denote by A^* the set of all such t^*'s.

We would like now to freeze $\underset{\sim}{A}$. As usual, to do so we will continue to the level δ only branches that pass via an element of A^*. An additional thing to take care here is an extension of $\vec{f}$. Namely, we need to define a function f_δ from κ to the level δ of the tree which respects the item 2b of the definition of Q_η.

Let $f_{\delta_\nu} = \vec{f}(\delta_\nu)$, for every $\nu < \kappa$.

Consider first $f_{\delta_0}(0)$. Pick a δ-branch through T which passes via $f_{\delta_0}(0)$ and an element of A^*. Note that there is such a branch, since, following the strategy, we continue at limit stages of the construction all maximal branches. Continue it to the level δ and set $f_\delta(0)$ to be this continuation.

Proceed by induction on $\mu < \kappa$ and define $f_\delta(\mu)$ be above $f_{\delta_\mu}(\mu)$ and an element of A^*.

Note that $f_{\delta_\mu}(\mu) \geq_T f_{\delta_{\mu'}}(\mu)$, for every $\mu' \leq \mu$.

Then, for every $\mu < \kappa$, we will have that for every $\nu, \mu \leq \nu < \kappa$, $f_\delta(\nu) \geq_T f_{\delta_\mu}(\nu)$. Clearly, the last set is co-bounded, and so, we are done.

□

In particular, the natural forcing for adding a branch into $T(G)$ satisfies η-c.c. Consider the other forcing F associated with G.

Let f be in F iff there is $\langle T, \vec{f}\rangle \in G$ such that for some $\alpha < ht(T)$, $f = f_\alpha$. Set $f^1 \geq_F f^2$ iff for every $\nu < \kappa$, $f^1(\nu) \geq_{T(G)} f^2(\nu)$.

The intuition behind is that we would like to extend $G(T)$ to the level η. A generic for F allows to define η-branches and f_η, i.e. the function to the level η of such tree.

Let us argue that the forcing $\langle F, \leq_F \rangle$ satisfies η-c.c.

Lemma 3.4: *$\langle F, \leq_F \rangle$ satisfies η-c.c.*

Proof: Work in V. Let $\underset{\sim}{A}$ be a name of a maximal antichain in $\langle F, \leq_F\rangle$.

η is a 2-Mahlo cardinal,[3] so there an elementary submodel N of a large enough fragment of V such that $\underset{\sim}{A} \in N$ and $N \cap \eta = \eta'$, ${}^{\eta'>}N \subseteq N$, for some Mahlo cardinal $\eta' < \eta$.

We will construct by induction a pair $\langle T, \vec{f}\rangle$ with T of height η', using the strategy defined in Lemma 3.2. Player II will follow the strategy at

[3]It is possible to use weaker assumptions, but it does matter much since λ is a measurable, and so, there are many such cardinals below it.

even stages and at odd stages we will deal with $\underset{\sim}{A}$. Note that, according to this strategy, all maximal branches are continued at limit steps, so we need not worry what to do there. Let us concentrate on the construction at odd stages.

For every $s \in T$ we will define by induction an η'-branch b_s such that: for every $\eta'' < \eta'$, for every $\langle s_\rho \mid \rho < \kappa\rangle \in [T \restriction \eta'' + 1]^\kappa$ with the set $\{\rho < \kappa \mid b_{s_\rho}(\eta'') = f_{\eta''}(\rho)\}$ co-bounded, there is $\eta''', \eta'' \leq \eta''' < \eta'$ such that $\{(\rho, b_{s_\rho}(\eta''')) \mid \rho < \kappa\}$ is forced (by an initial segment of $\langle T, \vec{f}\rangle$) to be above an element of A.

Proceed as follows.

Start with a condition in $Q_\eta \cap N$. Extend it inside N to a condition $\langle T^0, \vec{f^0}\rangle$ which decides an element of A. Let g be this element and $\alpha_g < ht(T^0)$ its level.

Then the set $\{\nu < \kappa \mid g(\nu) <_{T^0} f^0_{ht(T^0)}(\nu)\}$ is co-bounded, since $g = f^0_{\alpha_g}$.

Branches $\langle b_{g(\nu)} \mid \nu < \kappa\rangle$ will be used to form f_α, for every even level $\alpha \leq \eta'$, i.e. at stages where Player II plays.

We continue by induction using the fact that $\eta'^\kappa = \eta' = 2^{<\eta'}$.

Thus, fix an enumeration $\pi : \eta' \to [2^{<\eta'}]^\kappa$ such that the pre-image of each element of $[2^{<\eta'}]^\kappa$ is stationary.

Suppose that for some $\eta'' < \eta'$, $\langle T \restriction \eta'' + 1, \vec{f} \restriction \eta'' + 1\rangle$ is defined.

Consider $\pi(\eta'')$. If $\pi(\eta'') \notin [T \restriction \eta'' + 1]^\kappa$, then nothing special is done. If $\pi(\eta'') \in [T \restriction \eta'' + 1]^\kappa$, then let $\pi(\eta'') = \langle s_\rho \mid \rho < \kappa\rangle$.

If the set $\{\rho < \kappa \mid b_{s_\rho}(\eta'') = f_{\eta''}(\rho)\}$ is not co-bounded, then again, do nothing special.

If the set $\{\rho < \kappa \mid b_{s_\rho}(\eta'') = f_{\eta''}(\rho)\}$ is co-bounded, but $\{(\rho, b_{s_\rho}(\eta'')) \mid \rho < \kappa\}$ is not forced by $\langle T \restriction \eta'' + 1, \vec{f} \restriction \eta'' + 1\rangle$ to be a condition, then do nothing special.

Suppose now that the set $\{\rho < \kappa \mid b_{s_\rho}(\eta'') = f_{\eta''}(\rho)\}$ is co-bounded and $\{(\rho, b_{s_\rho}(\eta'')) \mid \rho < \kappa\}$ is forced by $\langle T \restriction \eta'' + 1, \vec{f} \restriction \eta'' + 1\rangle$ to be a condition.

If $\{(\rho, b_{s_\rho}(\eta'')) \mid \rho < \kappa\}$ is forced by $\langle T \restriction \eta'' + 1, \vec{f} \restriction \eta'' + 1\rangle$ to be above an element of A, then do nothing special.

If $\{(\rho, b_{s_\rho}(\eta'')) \mid \rho < \kappa\}$ is not forced by $\langle T \restriction \eta'' + 1, \vec{f} \restriction \eta'' + 1\rangle$ to be above an element of A, then we extend $\langle T \restriction \eta''+1, \vec{f} \restriction \eta''+1\rangle$ to a condition $\langle T', \vec{f'}\rangle$ which decides an element of A above $\{(\rho, b_{s_\rho}(\eta'')) \mid \rho < \kappa\}$. Let $\eta''', \eta'' < \eta''' < ht(T')$ be the level of such element of A. Denote this element by $\langle a_\rho \mid \rho < \kappa\rangle$. Then for every $\rho < \kappa$, $b_{s_\rho}(\eta'') <_{T'} a_\rho$. Pick now $a'_\rho >_{T'} a_\rho$ at the level $ht(T') - 1$ such that $a'_\rho = f'_{ht(T')-1}(\rho)$, if $f'_{ht(T')-1}(\rho) >_{T'} a_\rho$, and arbitrary otherwise.[4]

[4]Note that for co-boundedly many ρ's, we will have $f'_{ht(T')-1}(\rho) >_{T'} a_\rho$.

Set $b_{s_\rho} \restriction ht(T') = \{x \in T' \mid x \leq a'_\rho\}$.

Define $b_s \restriction ht(T')$ to be an arbitrary extension of $b_s \restriction \eta'' + 1$, if $s \neq s_\rho$, for some $\rho < \kappa$.

Now we add to T the last level (the level η') by continuing branches $b_s, s \in T$ only. Set $f_{\eta'}(\nu)$ to be the continuation of $b_{g(\nu)}$ to the level η', for every $\nu < \kappa$.

Denote by $\langle T^*, \vec{f}\rangle$ such extension.

Let us show that the condition $\langle T^*, \vec{f}\rangle$ forces that $A \subseteq [T]^\kappa$. Suppose otherwise. Then there is $\langle S, \vec{g}\rangle \geq_{Q_\eta} \langle T^*, \vec{f}\rangle$ and $\xi, \eta' < \xi < ht(S)$ such that

$$\langle S, \vec{g}\rangle \Vdash g_\xi \in \underset{\sim}{A}.$$

For every $\rho < \kappa$, let a_ρ the predecessor of $g_\xi(\rho)$ at level η'. Then, for every $\rho < \kappa$, there is $s_\rho \in T$ such that $a_\rho = b_{s_\rho}$. Find $\eta'' < \eta'$ such that $\pi(\eta'') = \langle s_\rho \mid \rho < \kappa\rangle$ and $\langle s_\rho \mid \rho < \kappa\rangle \in [T \restriction \eta'' + 1]^\kappa$.

The set $\{\rho < \kappa \mid b_{s_\rho}(\eta'') = f_{\eta''}(\rho)\}$ co-bounded, since $b_{s_\rho}(\eta'')$ is the predecessor of $g_\xi(\rho)$ at the level η'' and the set $\{\rho < \kappa \mid g_\xi(\rho) >_S f_{\eta''}(\rho)\}$ co-bounded, by the definition of Q_η.

Then, by the construction, there is $\eta''', \eta'' \leq \eta''' < \eta'$ such that $\{(\rho, b_{s_\rho}(\eta''')) \mid \rho < \kappa\}$ is forced (by an initial segment of $\langle T, \vec{f}\rangle$) to be above an element of A. Which is impossible since A is an anti-chain.

Contradiction. □

We force in the interval (κ, λ) a η-Suslin tree $T(\eta)$ with the forcing Q_η, and then, we force with the forcing $\langle F_\eta, \leq_{F_\eta}\rangle$, defined above, branches trough $T(\eta)$, for every 2-Mahlo cardinal η.

More precisely, Easton support iteration is used between κ and λ. At all but 2-Mahlo cardinals the forcing is trivial and if $\eta, \kappa < \eta < \lambda$ is a 2-Mahlo cardinal, then we force with $Q_\eta * F_\eta$. Denote this iteration by $\mathbb{P}$.

Now, over λ itself, force only a λ-Suslin tree $T(\lambda)$ with Q_λ.

Note that $j_{\mathcal{V}}(\lambda) = \lambda$.

Let $G \subseteq \mathbb{P} * Q_\lambda$ be a generic.

We extend $j_{\mathcal{V}}$ in a natural fashion to $j^*_{\mathcal{V}} : V[G] \to M_{\mathcal{V}}[G^*]$. Namely, $j_{\mathcal{V}}{}''G$ will generate $G^* \subset j_{\mathcal{V}}(\mathbb{P} * Q_\lambda)$ generic over $M_{\mathcal{V}}$, since the forcing $\mathbb{P} * Q_\lambda$ is more than κ-strategically closed.

Then λ does not move and, by elementarity, $j^*_{\mathcal{V}}(T(\lambda))$ is a λ-Suslin tree in $M_{\mathcal{V}}[G^*]$.

Let $\langle f^G_\alpha \mid \alpha < \lambda\rangle$ be the part of $G \restriction Q_\lambda$ which consists of functions.

For every $\alpha < \beta < \lambda$, the set $\{\nu < \kappa \mid f^G_\alpha(\nu) \not<_{T(\lambda)} f^G_\beta(\nu)\}$ is bounded in κ by some $\delta(\alpha, \beta)$.

By elementarity, the set

$$\{\nu < j_{\mathcal{V}}(\kappa) \mid j^*_{\mathcal{V}}(f^G_\alpha)(\nu) <_{j^*_{\mathcal{V}}(T(\lambda))} j^*_{\mathcal{V}}(f^G_\beta)(\nu)\} \supseteq j_{\mathcal{V}}(\kappa) \setminus \delta(\alpha, \beta).$$

Note that there are unboundedly many $\alpha < \lambda$ such that $j_{\mathcal{V}}(\alpha) = \alpha$, for example, every inaccessible $\alpha, \kappa < \alpha < \lambda$ is a fixed point of $j_{\mathcal{V}}$.

Set

$$b_\nu := \{t \in j^*_{\mathcal{V}}(T(\lambda)) \mid \exists \alpha < \lambda (t <_{j^*_{\mathcal{V}}(T(\lambda))} j^*_{\mathcal{V}}(f^G_\alpha)(\nu))\},$$

for every $\nu, \kappa \leq \nu < j_{\mathcal{V}}(\kappa)$. Then the sequence $\langle b_\nu \mid \kappa \leq \nu < j_{\mathcal{V}}(\kappa)\rangle$ is in $V[G]$, and it is a sequence of λ-branches through $j^*_{\mathcal{V}}(T(\lambda))$.

Consider

$$H = \{f^{G^*}_\gamma \mid \gamma < \lambda, \text{ exists } \alpha, \gamma < \alpha < \lambda$$
$$\text{for every } \nu < j_{\mathcal{V}}(\kappa)(f^{G^*}_\gamma(\nu) <_{j^*_{\mathcal{V}}(T(\lambda))} j^*_{\mathcal{V}}(f^G_\alpha)(\nu))\}.$$

Then H is close to produce a $M_{\mathcal{V}}[G^*]$-generic set for the forcing $\langle F_\lambda, \leq_{F_\lambda}\rangle$ with the tree $j^*_{\mathcal{V}}(T(\lambda))$ in $M_{\mathcal{V}}[G^*]$. It follows from the fact that in $M_{\mathcal{V}}[G^*]$ the forcing satisfies λ-c.c. Thus, suppose that $A \subseteq F_\lambda$ is a maximal antichain. Then there is $\delta < \lambda$ such that each function in A acts at a level of $j^*_{\mathcal{V}}(T(\lambda))$ below δ. Take an inaccessible $\alpha, \delta < \alpha < \lambda$. Consider $j^*_{\mathcal{V}}(f^G_\alpha)$. It is compatible with an element of A, but since $j^*_{\mathcal{V}}(f^G_\alpha)$ acts at the level $\alpha > \delta$, it should be above this element. The only problem is that for $\beta, \alpha < \beta < \lambda$, $j^*_{\mathcal{V}}(f^G_\alpha)(\nu)$ need not be below $j^*_{\mathcal{V}}(f^G_\beta)(\nu)$ for every ν, but rather for $\nu \geq \kappa$. It is not hard to fix this and we do it below.

We would like to use $\mathcal{U}$ to extend further the elementary embedding and eventually to produce an extension of $\mathcal{U}$ which is as desired.

The problematic point is that $j^*_{\mathcal{V}}(T(\lambda))$ extended to the level λ by extending the branches $\langle b_\nu \mid \kappa \leq \nu < j_{\mathcal{V}}(\kappa)\rangle$ to the level λ is not normal. Namely, there will be many $t \in j^*_{\mathcal{V}}(T(\lambda))$ without a λ-branch through t.

In order to prevent this, let us revise the construction and use homogeneous trees instead of arbitrary ones.

Define a forcing notion Q^{hom}_η as follows:

$\langle T, \vec{f}\rangle \in Q^{hom}_\eta$ iff

(1) $T \subseteq {}^{\eta>}2$ is a normal tree of a successor height $ht(T)$ below η,
(2) T is homogeneous in the following sense:
if $s_0, s_1 \in T$ are on a same level, then $T_{s_0} = T_{s_1}$.[5]
(3) $\vec{f} = \langle f_\alpha \mid \alpha < ht(T)\rangle$, where

[5]If $s \in T$, then T_s denotes $\{t \restriction \eta \setminus |s| \mid t \in T, t \geq_T s\}$.

(a) for every $\alpha < ht(T)$, $f_\alpha : \kappa \to Lev_\alpha(T)$,
(b) for every $\alpha, \beta < ht(T)$, if $\alpha < \beta$, then the set $\{\nu < \kappa \mid f_\alpha(\nu) <_T f_\beta(\nu)\}$ is co-bounded in κ.

The order on Q_η^{hom} is defined in the natural fashion by taking end extensions.

The next lemma is as in the previous setting:

Lemma 3.5: *The forcing Q_η^{hom} is $< \eta$-strategically closed.*

Let now $G \subseteq Q_\eta^{hom}$ be generic and $T(G)$ be the η-tree added by G.

Lemma 3.6: *$T(G)$ is an η-Suslin tree.*

Proof: Work in V. Let $\underset{\sim}{A}$ be a name of a maximal antichain in $T(G)$.

Fix an increasing continuous elementary chain $\langle N_\gamma \mid \gamma < \kappa \rangle$ of elementary submodels of H_χ for χ large enough such that

(1) $Q_\eta^{hom}, \underset{\sim}{A} \in N_0$,
(2) for every $\gamma < \kappa$, $N_\gamma \cap \lambda = \delta_\gamma$ for some cardinal δ_γ,
(3) for every successor $\gamma < \kappa$, δ_γ is a regular cardinal and ${}^{\delta_\gamma >}N_\gamma \subseteq N_\gamma$.

For every successor $\gamma < \kappa$, we work inside N_γ and define by induction a homogeneous normal tree T^γ of height δ_γ, functions $\vec{f}^\gamma$ for T^γ and $\langle b_t^\gamma \mid t \in T^\gamma \rangle$ such that

(1) T^γ end-extends $T^{\gamma-1}$,
(2) $\vec{f}^\gamma$ end-extends $\vec{f}^{\gamma-1}$,
(3) for every $t \in T^\gamma$, b_t^γ is a δ_γ-branch through T^γ,
(4) if $t \in T^{\gamma-1}$, then $b_t^\gamma \restriction \delta_{\gamma-1} = b_t^{\gamma-1}$,
(5) $\langle T^\gamma, \vec{f}^\gamma \rangle$ decides $\underset{\sim}{A} \cap T^\gamma$,
(6) for every $t \in T^\gamma$, $\langle T^\gamma, \vec{f}^\gamma \rangle$ forces that an element of $\underset{\sim}{A}$ belongs to b_t^γ,
(7) for every $t, s \in T_\gamma$ which are on a same level $\alpha < \delta_\gamma$, $b_t^\gamma \setminus \alpha{+}1 = b_s^\gamma \setminus \alpha{+}1$.

Now, we extend *all* δ_γ-branches of T^γ to the level δ_γ, and not only the chosen b_t^γ's.

In addition, for every $\alpha \leq \delta_\gamma$, we will have a function $f_\alpha = f_\alpha^\gamma$ from κ to the αth level. They are defined together with trees as in Lemma 3.3.

Set $\delta_\kappa = \bigcup_{\gamma<\kappa} \delta_\gamma$, $T^\kappa = \bigcup_{\gamma<\kappa} T^\gamma$ and for every $t \in T^\kappa$, $b_t = \bigcup_{\gamma<\kappa} b_t^\gamma$.

Extend T^κ to the level δ_κ, but now continue only the branches b_t, $t \in T^\kappa$.

Let us define f_{δ_κ}, i.e. the function for the level δ_κ.

Consider first $f_{\delta_0}(0)$. Set $f_{\delta_\kappa}(0)$ be the continuation of $b_{f_{\delta_0}(0)}$ to the level δ_κ.

Given $\gamma < \kappa$, set $f_{\delta_\kappa}(\gamma)$ be the continuation of $b_{f_{\delta_\gamma}(\gamma)}$ to the level δ_κ.
Then for every $\gamma < \kappa$,

$$\{\nu < \kappa \mid f_{\delta_\kappa}(\nu) >_{T^\kappa} f_{\delta_\gamma}(\nu)\} \supseteq \kappa \setminus \gamma. \qquad \square$$

In particular, the natural forcing for adding a branch into $T(G)$ satisfies η-c.c. Consider the other forcing F associated with G.

Let f be in F iff there are $\xi_f < \kappa$ and $\langle T, \vec{f}\rangle \in G$ such that for some $\alpha < ht(T)$, $f = f_\alpha \setminus \xi_f$. Set $f^1 \geq_F f^2$ iff $\xi_{f_1} = \xi_{f_2}$ and for every $\nu, \xi_{f_1} \leq \nu < \kappa$, $f^1(\nu) \geq_{T(G)} f^2(\nu)$.

F can be viewed as a set of pairs $\langle f, \xi_f\rangle$.

The intuition behind this is that we would like to extend $G(T)$ to the level η. An η-branch (together with homogeneity) allows to extend the tree $T(G)$ and a generic for F allows to define f_η, i.e. the function to the level η of such tree.

The next lemma is a "homogeneous" analog of Lemma 3.4.

Lemma 3.7: *$\langle F, \leq_F\rangle$ satisfies η-c.c.*

Proof: Let $A \subseteq F$ be an anti-chain. Our aim is to show that $|A| < \eta$. We can split A into at most κ-many anti-chains according to the values of ξ_f's of its members. It is enough to show that each part of such splitting has cardinality less than η. So deal with them separately. Assume for simplicity that for every $f \in A$, $\xi_f = 0$.

Work in V. Let $\underset{\sim}{A}$ be a name of A.

Proceed as in Lemma 3.4.

η is a 2-Mahlo cardinal, so there an elementary submodel N of a large enough fragment of V such that $\underset{\sim}{A} \in N$ and $N \cap \eta = \eta'$, ${}^{\eta'>}N \subseteq N$, for some Mahlo cardinal $\eta' < \eta$.

We will construct by induction a pair $\langle T, \vec{f}\rangle$ with T of height η', using the strategy defined in Lemma 3.2. Player II will follow the strategy at even stages and at odd stages we will deal with $\underset{\sim}{A}$. Note that, according to this strategy, all maximal branches are continued at limit steps, so we need not worry what to do there. The main attention will be to the construction at odd stages.

We will associate a maximal branch b_s through s, for every $s \in T$. Only this branches will be continued to the final level η', as it was done in Lemma 3.4. This will insure the following:

for every $\eta'' < \eta'$, for every $\langle s_\rho \mid \rho < \kappa\rangle \in [T \restriction \eta'' + 1]^\kappa$ with the set $\{\rho < \kappa \mid b_{s_\rho}(\eta'') = f_{\eta''}(\rho)\}$ co-bounded, if there is an extension of $\langle T, \vec{f}\rangle$

which decides an element of A *above* $\{(\rho, b_{s_\rho}(\eta'')) \mid \rho < \kappa\}$, *then there is* $\eta''', \eta'' \leq \eta''' < \eta'$ *such that* $\{(\rho, b_{s_\rho}(\eta''')) \mid \rho < \kappa\}$ *is forced (by an initial segment of* $\langle T, \vec{f}\rangle$*) to be above an element of* A.

There is an additional requirement here — a homogeneity of the final tree. In order to insure this, the inductive construction of T will provide the following:

$(*)_T$ *for every* $s, t \in T$ *which are on a same level* $\alpha < \eta'$, *for every* b_x *with* $s \in b_x$ *there is* $y \in T$ *such that* $t \in b_y$ *and* $b_x \setminus \alpha = b_y \setminus \alpha$.

Proceed as follows.

Start with a condition $\langle T^0, \vec{f^0}\rangle \in Q_\eta \cap N$.

Let us define a sequence of maximal branches $\langle b_s^{T_0} \mid s \in T_0\rangle$ through T^0 which has the property $(*)_{T^0}$. Such $b_s^{T_0}$'s will be initial segments of the final branches b_s's through T. Let $\eta_0 + 1$ be the height of T^0. Proceed by induction on levels $\alpha \leq \eta_0$.

Consider the first level.

Suppose that $\langle 0\rangle$ is there. Pick some b at the last level of T^0 (i.e. a $0-1$ sequence of the length η_0) which extends $\langle 0\rangle$, or simply, its first element is 0. Set $b_{\langle 0\rangle}^{T_0} = b$. Also, set $b_s^{T_0} = b$ for every $s \in T^0$ which is on b, i.e. for every initial segment of b.

Now deal with the next element (if there is such) of the first level of T^0.

So consider $\langle 1\rangle$. By homogeneity of T^0, we have $T^0_{\langle 0\rangle} = T^0_{\langle 1\rangle}$. In particular, $b_1 := \langle 1\rangle^\frown b \setminus 1 \in T^0$. Set $b_{\langle 1\rangle}^{T_0} = b_1$ and let $b_s^{T_0} = b_1$, for every $s \in T^0$ which is on b_1, i.e. for every initial segment of b_1.

Now, suppose we got to a level α and there is s at this level with $b_s^{T_0}$ undefined. Set $b_s^{T_0} = s^\frown b_{\langle 0\rangle}^{T^0} \setminus \alpha$. Also, set $b_t^{T_0} = b_s^{T_0}$ for every $t \in T^0$ which is on $b_s^{T_0}$ and is above the level α.

Fix an enumeration $\pi : \eta' \to [2^{<\eta'}]^\kappa$ such that the pre-image of each element of $[2^{<\eta'}]^\kappa$ is stationary.

Suppose that for some $\eta'' < \eta'$, $\langle T \restriction \eta'' + 1, \vec{f} \restriction \eta'' + 1\rangle$ is defined.

Consider $\pi(\eta'')$. If $\pi(\eta'') \notin [T \restriction \eta'' + 1]^\kappa$, then nothing special is done. If $\pi(\eta'') \in [T \restriction \eta'' + 1]^\kappa$, then let $\pi(\eta'') = \langle s_\rho \mid \rho < \kappa\rangle$.

If the set $\{\rho < \kappa \mid b_{s_\rho}(\eta'') = f_{\eta''}(\rho)\}$ is not co-bounded, then again, do nothing special.

If the set $\{\rho < \kappa \mid b_{s_\rho}(\eta'') = f_{\eta''}(\rho)\}$ is co-bounded, but $\{(\rho, b_{s_\rho}(\eta'')) \mid \rho < \kappa\}$ is not forced by $\langle T \restriction \eta'' + 1, \vec{f} \restriction \eta'' + 1\rangle$ to be a condition, then do nothing special.

Suppose now that the set $\{\rho < \kappa \mid b_{s_\rho}(\eta'') = f_{\eta''}(\rho)\}$ is co-bounded and $\{(\rho, b_{s_\rho}(\eta'')) \mid \rho < \kappa\}$ is forced by $\langle T \restriction \eta'' + 1, \vec{f} \restriction \eta'' + 1\rangle$ to be a condition.

If $\{(\rho, b_{s_\rho}(\eta'')) \mid \rho < \kappa\}$ is forced by $\langle T \restriction \eta''+1, \vec{f} \restriction \eta''+1\rangle$ to be above an element of A, then do nothing special.

If $\{(\rho, b_{s_\rho}(\eta'')) \mid \rho < \kappa\}$ is not forced by $\langle T \restriction \eta''+1, \vec{f} \restriction \eta''+1\rangle$ to be above an element of A, and no extension $\langle T', \vec{f'}\rangle$ of $\langle T \restriction \eta''+1, \vec{f} \restriction \eta''+1\rangle$ decides an element of A above $\{(\rho, b_{s_\rho}(\eta'')) \mid \rho < \kappa\}$, then do nothing special. Note that now this may happens, since we deal with an antichain which is not maximal.

If $\{(\rho, b_{s_\rho}(\eta'')) \mid \rho < \kappa\}$ is not forced by $\langle T \restriction \eta''+1, \vec{f} \restriction \eta''+1\rangle$ to be above an element of A and there is an extension $\langle T', \vec{f'}\rangle$ of $\langle T \restriction \eta''+1, \vec{f} \restriction \eta''+1\rangle$ which decides an element of A above $\{(\rho, b_{s_\rho}(\eta'')) \mid \rho < \kappa\}$.

By extending if necessary, we can assume that $ht(T')-1$ is an inaccessible cardinal.

Let $\eta''', \eta'' < \eta''' < ht(T')$ be the level of such element of A. Denote this element by $\langle a_\rho \mid \rho < \kappa\rangle$. Then for every $\rho < \kappa$, $b_{s_\rho}(\eta'') <_{T'} a_\rho$. Pick now $a'_\rho >_{T'} a_\rho$ at the level $ht(T')-1$ such that $a'_\rho = f'_{ht(T')}(\rho)$, if $f'_{ht(T')-1}(\rho) >_{T'} a_\rho$, and arbitrary otherwise.[6]

Set $b^{T'}_{s_\rho} = b_{s_\rho} \restriction ht(T') = \{x \in T' \mid x \leq_{T'} a'_\rho\}$.

Set $b^{T'}_t = b_t \restriction ht(T') = \{x \in T' \mid x \leq_{T'} a'_\rho\}$, for every $t \in T', t \leq_{T'} a'_\rho$ which sits on a level $> \eta''$.

Apply the homogeneity of T' now.

For every $\rho, 0 < \rho < \kappa$ there is $a'_{\rho,0}$ at the last level of T' which is above $b_{s_0}(\eta'')$ and $a'_\rho \setminus \eta'' = a'_{\rho,0} \setminus \eta''$.

Proceed by induction on levels of T' above η'' and define $b^{T'}_t = b_t \restriction ht(T')$ for t's in $T'_{b_{s_0}(\eta'')}$ which are $\leq_{T'} a'_{\rho,0}, \rho < \kappa$.

Thus, suppose $t \in T'_{b_{s_0}(\eta'')}$ is like this and $b_t \restriction ht(T')$ is still undefined. Pick $\rho < \kappa$ to be the least such that $t \leq_{T'} a'_{\rho,0}$ and set $b^{T'}_t = b_t \restriction ht(T') = a'_{\rho,0}$.[7]

Repeat the same process above every $b_{s_\mu}(\eta'')$, $0 < \mu < \kappa$, i.e., with μ replacing 0.

There may be still s's in T' with $b^{T'}_s$ undefined. Proceed by induction on levels of T' starting from η''.

Thus, let, for example, $s \in Lev_{\eta''}(T) \setminus \{b_{s_\rho}(\eta'') \mid \rho < \kappa\}$.

Use the homogeneity of T'. For every $\rho < \kappa$ there is $a'_{s,\rho} >_{T'} s$ at the last level $ht(T')-1$ of T' such that $a'_{s,\rho} \setminus \eta'' = a'_\rho \setminus \eta''$. We set $b^{T'}_s = b_s \restriction ht(T') = a'_{s,0}$ and repeat the settings made above $b_{s_0}(\eta'')$ to those above s.

[6]Note that for co-boundedly many ρ's, we will have $f'_{ht(T')-1}(\rho) >_{T'} a_\rho$.

[7]Recall that nodes of T' are just sequences and we identify between them.

Let us argue that the choice of the branches $b_s^{T'}$ above gives the property $(*)_{T'}$.

Claim. $(*)_{T'}$ *holds.*

Proof: First, note that by the construction, for every $t \in T'$, there is $\rho < \kappa$ such that $b_t^{T'} \setminus ht(t) = a'_\rho \setminus ht(t)$.

The second point is that the branches which lead to a'_ρ's, $\rho < \kappa$, split completely before the final level $ht(T')-1$, i.e., there is $\tilde{\eta}, \eta'' < \tilde{\eta} < ht(T')-1$ such that $\langle a'_\rho(\tilde{\eta}) \mid \rho < \kappa\rangle$ are different points of the level $\tilde{\eta}$ of T'. This follows from an inaccessibility of $ht(T') - 1$.

So, by the construction, for every $t \in T', \rho < \kappa$, there will be $x \geq_{T'} t$ such that $b_x^{T'} \setminus ht(t) = a'_\rho \setminus ht(t)$.

Now $(*)_{T'}$ follows.

□ of the claim.

This completes the construction of T and $\langle b_s \mid s \in T\rangle$.

We continue only the branches $b_s, s \in T$ to the level η'.

The rest of the proof is as in Lemma 3.4. □

Now we are ready to complete the proof of Theorem 3.1.

Proceed as it was done after Lemma 3.4.

We force in the interval (κ, λ) a η-Suslin tree $T(\eta)$ with the forcing Q_η, and then, we force with the forcing $\langle F_\eta, \leq_{F_\eta}\rangle$, defined above, branches trough $T(\eta)$, for every 2-Mahlo cardinal η.

More precisely, Easton support iteration is used between κ and λ. At all but 2-Mahlo cardinals the forcing is trivial and if $\eta, \kappa < \eta < \lambda$ is a 2-Mahlo cardinal, then we force with $Q_\eta * F_\eta$. Denote this iteration by $\mathbb{P}$.

Now, over λ itself, force only a λ-Suslin tree $T(\lambda)$ with Q_λ.

Note that $j_{\mathcal{V}}(\lambda) = \lambda$.

Let $G \subseteq \mathbb{P} * Q_\lambda$ be a generic.

We extend $j_{\mathcal{V}}$ in a natural fashion to $j^*_{\mathcal{V}} : V[G] \to M_{\mathcal{V}}[G^*]$. Namely, $j_{\mathcal{V}}''G$ will generate $G^* \subset j_{\mathcal{V}}(\mathbb{P} * Q_\lambda)$ generic over $M_{\mathcal{V}}$, since the forcing $\mathbb{P} * Q_\lambda$ is more than κ-strategically closed.

Then λ does not move and, by elementarity, $j^*_{\mathcal{V}}(T(\lambda))$ is a λ-Suslin tree in $M_{\mathcal{V}}[G^*]$.

Let $\langle f^G_\alpha \mid \alpha < \lambda\rangle$ be the part of $G \restriction Q_\lambda$ which consists of functions.

For every $\alpha < \beta < \lambda$, the set $\{\nu < \kappa \mid f^G_\alpha(\nu) \not<_{T(\lambda)} f^G_\beta(\nu)\}$ is bounded in κ by some $\delta(\alpha, \beta)$.

By elementarity, the set

$$\{\nu < j_{\mathcal{V}}(\kappa) \mid j^*_{\mathcal{V}}(f^G_\alpha)(\nu) <_{j^*_{\mathcal{V}}(T(\lambda))} j^*_{\mathcal{V}}(f^G_\beta)(\nu)\} \supseteq j_{\mathcal{V}}(\kappa) \setminus \delta(\alpha,\beta).$$

Note that there are unboundedly many $\alpha < \lambda$ such that $j_{\mathcal{V}}(\alpha) = \alpha$, for example, every inaccessible $\alpha, \kappa < \alpha < \lambda$ is a fixed point of $j_{\mathcal{V}}$.

Set

$$b_\nu := \{t \in j^*_{\mathcal{V}}(T(\lambda)) \mid \exists \alpha < \lambda (t <_{j^*_{\mathcal{V}}(T(\lambda))} j^*_{\mathcal{V}}(f^G_\alpha)(\nu))\},$$

for every $\nu, \kappa \leq \nu < j_{\mathcal{V}}(\kappa)$. Then the sequence $\langle b_\nu \mid \kappa \leq \nu < j_{\mathcal{V}}(\kappa)\rangle$ is in $V[G]$, and it is a sequence of λ-branches through $j^*_{\mathcal{V}}(T(\lambda))$.

Consider

$$H = \{\langle f^{G^*}_\gamma, \kappa\rangle \mid \gamma < \lambda, \text{ exists } \alpha, \gamma < \alpha < \lambda, \text{ for every } \nu, \kappa \leq \nu < j_{\mathcal{V}}(\kappa)$$

$$(f^{G^*}_\gamma(\nu) <_{j^*_{\mathcal{V}}(T(\lambda))} j^*_{\mathcal{V}}(f^G_\alpha)(\nu))\}.$$

Let us argue now that such defined H is a $M_{\mathcal{V}}[G^*]$-generic set for the forcing $\langle F_\lambda, \leq_{F_\lambda}\rangle$ with the tree $j^*_{\mathcal{V}}(T(\lambda))$ in $M_{\mathcal{V}}[G^*]$, with $\xi_f = \kappa$.

First, we have that for every $\beta, \alpha < \beta < \lambda$, $j^*_{\mathcal{V}}(f^G_\alpha)(\nu)$ is below $j^*_{\mathcal{V}}(f^G_\beta)(\nu)$ for every ν, $\kappa \leq \nu < j_{\mathcal{V}}(\kappa)$.

Second, let us argue that H meets maximal antichains. It follows from the fact that in $M_{\mathcal{V}}[G^*]$ the forcing satisfies λ-c.c. Thus, suppose that $A \subseteq F_\lambda$ is a maximal anti-chain. Then there is $\delta < \lambda$ such that each function in A acts at a level of $j^*_{\mathcal{V}}(T(\lambda))$ below δ. Take an inaccessible $\alpha, \delta < \alpha < \lambda$.

Consider $\langle j^*_{\mathcal{V}}(f^G_\alpha), \kappa\rangle$. It is compatible with an element of A. Let $\langle g, \kappa\rangle$ be such an element. Then the level on which g acts should be below δ. However, $j^*_{\mathcal{V}}(f^G_\alpha)$ acts at the level $\alpha > \delta$.

Hence, for every $\nu, \kappa \leq \nu < j_{\mathcal{V}}(\kappa)$, $j^*_{\mathcal{V}}(f^G_\alpha) >_{j^*_{\mathcal{V}}(T(\lambda))} g(\nu)$.

So, $g \in H \cap A$.

Let $\mathcal{U}$ be a normal ultrafilter over λ in V. Consider its elementary embedding $j_{\mathcal{U}} : V \to M_U$. The critical point of $j_{\mathcal{U}}$ is λ, and so, $\mathcal{V}$ does not move.

Let $i : V \to M$ be the composition of $j_{\mathcal{U}}$ with $j^{M_{\mathcal{U}}}_{\mathcal{V}} : M_{\mathcal{U}} \to M \simeq \mathrm{Ult}(M_{\mathcal{U}}, \mathcal{V})$.

It is well-known that M is just the ultrapower by $\mathcal{U} \times \mathcal{V}$ (or equivalently by $\mathcal{V} \times \mathcal{U}$) and i is just the corresponding elementary embedding.

Consider the forcing $j_{\mathcal{U}}(\mathbb{P} * Q_\lambda)$ in $M_{\mathcal{U}}$. It splits into the part $\mathbb{P} * Q_\lambda * F_\lambda$ and the part $\mathbb{P}_{(\kappa, j_{\mathcal{U}}(\lambda))} * Q_{j_{\mathcal{U}}(\lambda)}$ above λ.

The part above λ is more than λ-strategically closed. So, we can construct a master condition sequence $\langle \underset{\sim}{q}_\zeta \mid \zeta < \lambda^+\rangle$.

Now use $j_{\mathcal{V}}^{M_{\mathcal{U}}}$ and move to M. Then $\langle j_{\mathcal{V}}^{M_{\mathcal{U}}}(\underset{\sim}{q}_{\varsigma}) \mid \zeta < \lambda^+\rangle$ will be a master condition sequence there. We refer to J. Cummings handbook article [5] for this type of arguments.

Finally, plugging in G^* and H, the embedding i extends to

$$i^* : V[G] \to M[G^*, H, R],$$

where R is generated by $\langle j_{\mathcal{V}}^{M_{\mathcal{U}}}(\underset{\sim}{q}_{\varsigma})_{G^* * H} \mid \zeta < \lambda^+\rangle$.

Define $\mathcal{U}^*$ in $V[G]$ as follows:

$$X \in \mathcal{U}^* \text{ iff } \lambda \in i^*(X).$$

Clearly, $\mathcal{U}^* \supseteq \mathcal{U}$, and so $\mathcal{U}^*$ is a uniform ultrafilter over λ.

It is not hard to see that the embedding i^* is actually the ultrapower embedding by $\mathcal{U}^*$, since the only generators of the original i are κ and λ, but we can obtain κ from H. So, only λ remains.

Clearly, $\sup(i''\lambda) = \lambda$. So, $i_{\mathcal{U}^*} = i^* \supseteq i$ implies that $\mathcal{U}^*$ is a weakly normal ultrafilter over λ.

Finally, λ is not a measurable cardinal in $V[G]$, since there exists a λ-Suslin tree. □

3.1. *Extensions with different degrees of completeness*

The construction above provides an extension of normal ultrafilter $\mathcal{U}$ over λ to a κ-complete uniform weakly normal ultrafilter.

Let us argue that from a bit stronger assumptions it is possible to obtain extensions with different degrees of completeness.

Theorem 3.8: *Assume GCH. Let $\kappa < \lambda$ be two measurable cardinals.*

Assume that κ is a limit of measurable cardinals. Then there is a cofinality preserving generic extension which satisfies the following:

(1) GCH,

(2) λ is not measurable,

(3) there is a uniform κ-complete ultrafilter over λ.
Moreover, every $\mathcal{U}$ which is a normal ultrafilter over λ in V extends to a uniform κ-complete ultrafilter over λ.

(4) For every $\delta < \kappa$ there is $\delta', \delta \leq \delta' < \kappa$ and a uniform δ'-complete ultrafilter over λ.
Moreover, every $\mathcal{U}$ which is a normal ultrafilter over λ in V extends to a uniform δ'-complete ultrafilter over λ.

Proof: We will combine the construction of the previous section with those of Theorem 3.1.

Fix some normal ultrafilters $\mathcal{V}$ on κ and $\mathcal{U}$ on λ. Fix in addition normal ultrafilters $\mathcal{V}_\delta$ over δ, or every $\delta < \kappa$

Iterate first the Cohen forcing and add function $f_\alpha : \alpha \to \alpha$ for every inaccessible $\alpha, \alpha \leq \kappa$.

Each of the ultrafilters above extends to a normal ultrafilter.

Let $\mathcal{V}'_\delta$ and $\mathcal{U}'$ be an extensions of $\mathcal{V}_\delta$ and $\mathcal{U}$, for every measurable $\delta < \kappa$.

Further we shall abuse the notation and refer to them as $\mathcal{V}_\delta$ and $\mathcal{U}$.

Over κ let us define extensions $\mathcal{V}(\delta)$ of $\mathcal{V}$, for every measurable $\delta \leq \kappa$. Pick $\mathcal{V}(\kappa)$ to be a normal ultrafilter which extends $\mathcal{V}$.

For every measurable $\delta < \kappa$, consider in V,

$$i_\delta := j_{\mathcal{V}} \circ j_{\mathcal{V}_\delta} : V \to M(\delta).$$

Extend the embedding i_δ to i^*_δ and change the value of $f_{i_\delta(\kappa)}(\kappa)$ to δ. Let $\mathcal{V}(\delta)$ be the corresponding extension of $\mathcal{V}$.

Then it is a uniform weakly normal δ-complete ultrafilter over κ.

We construct a Suslin tree as in Theorem 3.1. Now, any of $\mathcal{V}(\delta)$, $\delta \leq \kappa$, can be used to move κ, and so, to produce λ-branches in the tree.

This in turn will produce extensions of $\mathcal{U}$ with various degrees of completeness. □

J. Bagaria and M. Magidor [2] considered the following variation of strong compactness:

Every λ-complete filter can be extended to a κ-complete ultrafilter.

This clearly holds if λ is a strongly compact, but in general λ need not be even a measurable.

Thus J. Bagaria and M. Magidor [2] showed that it is possible to have such situation even with a singular λ.

Let us point out that our constructions can be used to provide additional examples.

Theorem 3.9: *Assume GCH. Let λ be a supercompact cardinal and $\kappa < \lambda$ be a measurable.*

Then there is a cofinality preserving generic extension which satisfies the following:

(1) GCH,
(2) λ is not measurable,
(3) every λ-complete filter can be extended to a κ-complete ultrafilter.

Proof: We use the construction of Theorem 3.1, only instead of using a normal ultrafilter $\mathcal{U}$ over λ and extending its embedding $j_{\mathcal{U}}$, supercompact ultrafilters over $\mathcal{P}_\lambda(\mu)$ for various μ's are used.

Suppose that in the extension F is a λ-complete filter on an ordinal and let μ be a regular large enough such that $|F| \leq \mu$. Let $F = \{Z_\alpha \mid \alpha < \mu\}$. Pick names $\underset{\sim}{F}$ and $\underset{\sim}{Z}_\alpha$'s for F and Z_α's. Assume that the weakest condition already forces all of this.

Work in V. Let $\mathcal{V}$ be a normal ultrafilter over κ, $\mathcal{U}$ a normal ultrafilter over $\mathcal{P}_\lambda(\mu)$ and $i = j^{M_\mathcal{V}}_{j_\mathcal{V}(\mathcal{U})}$.

Then $j_\mathcal{V}(\underset{\sim}{F})$ is forced to consist of $j_\mathcal{V}(\{\underset{\sim}{Z}_\alpha \mid \alpha < \mu\})$.

Set $j_\mathcal{V}(\{\underset{\sim}{Z}_\alpha \mid \alpha < \mu\}) = \{\underset{\sim}{Z}'_\alpha \mid \alpha < j_\mathcal{V}(\mu)\}$.

The ultrapower $M := M^{M_\mathcal{V}}_{j_\mathcal{V}(\mathcal{U})}$ of $M_\mathcal{V}$ by $j_\mathcal{V}(\mathcal{U})$ is closed, inside $M_\mathcal{V}$, under $j_\mathcal{V}(\mu)$-sequences of its elements. In particular, $\{i(\underset{\sim}{Z}'_\alpha) \mid \alpha < j_\mathcal{V}(\mu)\} \in M$.

We have $\sigma = i \circ j_\mathcal{V} : V \to M$.

It extends in the generic extension to $\sigma^* : V[G] \to M[G^*]$, as in Theorem 3.1.

By elementarity, $\sigma^*(F)$ is a $\sigma(\lambda) > j_\mathcal{V}(\mu)$ complete filter. Also, $Z''_\alpha := (i(\underset{\sim}{Z}'_\alpha))_{G^*} \in \sigma^*(F)$, for every $\alpha < j_\mathcal{V}(\mu)$.

Hence, there is some $\eta \in \bigcap_{\alpha < j_\mathcal{V}(\mu)} Z''_\alpha$.

Now we can define a desired extension F^* of F be setting

$$Z \in F^* \Leftrightarrow \eta \in \sigma^*(Z).$$

□

It is possible to modify the construction a bit in order to insure that λ is the least cardinal which satisfies (3), i.e. every λ-complete filter can be extended to a σ-complete ultrafilter.

We just follow Y. Kimchi and M. Magidor [11] and add non-reflecting stationary subsets of

$$\{\alpha \in (\gamma, \gamma^+) \mid \mathrm{cof}(\alpha) = \omega\},$$

for cofinally many $\gamma < \lambda$ preserving supercompactness of λ.

The next result is similar, only Theorem 3.8 is used instead of Theorem 3.1.

Theorem 3.10: *Assume GCH. Let λ be a supercompact cardinal and $\kappa < \lambda$ be a measurable.*

Assume that κ is a limit of measurable cardinals. Then there is a cofinality preserving generic extension which satisfies the following:

(1) GCH,

(2) λ is not measurable,
(3) every λ-complete filter can be extended to a κ-complete ultrafilter,
(4) for every $\delta < \kappa$ there is $\delta', \delta \leq \delta' < \kappa$ such that every λ-complete filter can be extended to a δ'-complete ultrafilter.

Again, using Y. Kimchi and M. Magidor [11] it is possible to insure that λ is the least such that every λ-complete filter can be extended to a σ-complete ultrafilter.

4. σ-complete Ultrafilters Over Singular Cardinals

Let κ be a measurable cardinal.

Note that if $\lambda > \kappa$ is a singular cardinal of cofinality κ, then it is easy to define a κ-complete ultrafilter over λ.

Namely, let $\langle \lambda_\alpha \mid \alpha < \kappa \rangle$ be a cofinal in λ sequence and let $\mathcal{V}$ be a κ-complete ultrafilter over κ.

Set $X \subseteq \lambda$ to be in $\mathcal{U}$ iff $\{\alpha < \kappa \mid \lambda_\alpha \in X\} \in \mathcal{V}$.

Clearly, such $\mathcal{U}$ is not uniform on λ, but rather concentrates on a set of cardinality κ and it is a copy of $\mathcal{V}$ to this set.

It is not hard to get a uniform ultrafilter.

Proposition 4.1: *Let κ be a measurable cardinal and $\lambda > \kappa$ a singular cardinal of cofinality κ. Suppose that λ is a limit of measurable cardinals. Then there exists a uniform κ-complete ultrafilter over λ.*

Proof: Fix a cofinal in λ sequence $\langle \lambda_\alpha \mid \alpha < \kappa \rangle$ whose members are measurable cardinals.

Let $\mathcal{V}$ be a κ-complete ultrafilter over κ and $\mathcal{V}_\alpha$ be a λ_α-complete ultrafilter over λ_α, for every $\alpha < \kappa$. Define $\mathcal{U}$ over λ as follows:

$$X \in \mathcal{U} \text{ iff } \{\alpha < \kappa \mid X \cap \lambda_\alpha \in \mathcal{V}_\alpha\} \in \mathcal{V}.$$

It is easy to see that $\mathcal{U}$ is as desired. □

The following is easy:

Proposition 4.2: *Suppose that U is a κ-complete weakly normal ultrafilter over a cardinal λ of cofinality κ.*

Then U concentrates on a set of cardinality κ. In particular, U is not uniform unless $\kappa = \lambda$.

Proof: By weak normality $[id]_U = \sup(j_U{}''\lambda)$.

Let $\langle \lambda_\alpha \mid \alpha < \kappa \rangle$ be an increasing continuous cofinal in λ sequence. Then in M_U, $\sup(\{j_U(\lambda_\alpha) \mid \alpha < \kappa\}) = \sup(j_U{}''\lambda) = [id]_U$.

So, $[id]_U \in j_U(\{\lambda_\alpha \mid \alpha < \kappa\})$, since the image of a closed set is a closed set. $\square$

Proposition 4.3: *Assume that κ is a measurable cardinal and $\lambda > \kappa$ with $o(\lambda) = \kappa$. Then there is a cardinal preserving forcing extension such that $\mathrm{cof}(\lambda) = \kappa$ and there is a κ-complete uniform ultrafilter over λ which extends a normal ultrafilter $\mathcal{U}$ over λ in V. In addition, $\mathcal{U}$ extends to a κ-complete ultrafilter over λ which concentrates on a set of cardinality κ.*

Proof: We will use the Magidor forcing and the idea similar to Bagaria-Magidor [2].

Assume GCH for simplicity. Let $\mathcal{V}$ be a normal ultrafilter over κ, $\vec{U} = \langle \mathcal{U}(\lambda, \alpha) \mid \alpha < \kappa \rangle$ be a coherent sequence of normal measures over λ. We refer, for example, to [6] for the relevant definitions.

We would like to change the cofinality of λ to κ using the Magidor forcing $\mathbb{M}_{\vec{U}}$.

Apply first $\mathcal{V}$. Let

$$i := j_{\mathcal{V}} : V \to M_{\mathcal{V}} = \mathrm{Ult}(V, \mathcal{V}).$$

Let $i(\vec{U}) = \langle \mathcal{U}'(\lambda, \alpha) \mid \alpha < i(\kappa) \rangle$. Then, $\mathcal{U}'(\lambda, \alpha) = \mathcal{U}(\lambda, \alpha) \cap M_{\mathcal{V}}$, for every $\alpha < \kappa$.

Now, in $M_{\mathcal{V}}$, each $\mathcal{U}'(\lambda, \alpha)$, $\alpha < i(\kappa)$, is a normal ultrafilter. In particular, $\mathcal{U}'(\lambda, \kappa)$ is such. So, if C is a closed unbounded subset of λ in V, then $i(C)$ is such in $M_{\mathcal{V}}$, and so, $i(C) \in \mathcal{U}'(\lambda, \kappa)$.

Still in $M_{\mathcal{V}}$, we apply $\mathcal{U}'(\lambda, \kappa)$:

$$j^{M_{\mathcal{V}}}_{\mathcal{U}'(\lambda,\kappa)} : M_{\mathcal{V}} \to M^{M_{\mathcal{V}}}_{\mathcal{U}'(\lambda,\kappa)}.$$

Denote $\mathcal{U}'(\lambda, \kappa)$ by $\mathcal{U}_1$, $j^{M_{\mathcal{V}}}_{\mathcal{U}'(\lambda,\kappa)}$ by j_1, $M^{M_{\mathcal{V}}}_{\mathcal{U}'(\lambda,\kappa)}$ by M_1 and $j_1(\lambda)$ by λ_1. Clearly, $crit(j_1) = \lambda$, and so, $\lambda < \lambda_1$. Also, $\lambda \in j_1(i(C)) \in j_1(\mathcal{U}_1)$.

Consider $j_2 := j^{M_1}_{j_1(\mathcal{U}'(\lambda,0))} : M_1 \to M_2 := M^{M_1}_{j_1(\mathcal{U}'(\lambda,0))}$.

Denote $j_2(\lambda_1)$ by λ_2.

Then, for every $X \subseteq \lambda$,

$$\begin{aligned} X \in \mathcal{U}(\lambda, 0) \Leftrightarrow i(X) \in \mathcal{U}'(\lambda, 0) &\Leftrightarrow j_1(i(X)) \in j_1(\mathcal{U}'(\lambda, 0)) \\ &\Leftrightarrow \lambda_1 \in j_2(j_1(i(X))). \end{aligned}$$

Denote $j_2 \circ j_1 \circ i$ by σ. Then $\sigma(\lambda) = \lambda_2$ and $o(\lambda_2) = i(\kappa)$ in M_2.

Force now with $\mathbb{M}_{\vec{U}}$. Let G be a generic.

Note that (identifying G with the Magidor sequence which it generates) it will be $\mathbb{M}_{\vec{\mathcal{U}}\restriction(\lambda,\kappa)}$-generic over $M_{\mathcal{V}}$.

Define U over λ in $V[G]$ as follows:

$X \in U$ iff, in M_2, there is $p \in \sigma(\mathbb{M}_{\vec{\mathcal{U}}\restriction(\lambda,\kappa)}) = \mathbb{M}_{\sigma(\vec{\mathcal{U}})\restriction(j_{\mathcal{U}}(\lambda),i(\kappa))}$ such that

$$p \restriction \lambda+1 \in G, p \setminus \lambda+1 \geq^* \langle \lambda_2, \lambda_2 \setminus \lambda_1 + 1\rangle \text{ and } p \Vdash (\lambda_1 \in \sigma(\underset{\sim}{X})).$$

The requirement $p \setminus \lambda+1 \geq^* \langle \lambda_2, \lambda_2 \setminus \lambda_1 + 1\rangle$ insures that the first element of the Magidor sequence in which is above λ will be above λ_1, as well. This will be used for a uniformity argument.

Then, U is a κ-complete ultrafilter which extends the normal ultrafilter $\mathcal{U}(\lambda, 0)$ over λ in V.

It is possible to argue that $j_U \restriction V$ is further iteration of σ starting with λ_2 and using the measures of the coherent sequence on it. The critical points of these embeddings will generate the Magidor sequence over the image of M_2.

Let us argue now that U is uniform.

Lemma 4.4: *U is a uniform ultrafilter over λ in $V[G]$.*

Proof: Suppose otherwise. Then there is a set $Z \subseteq \lambda$ of cardinality less than λ inside U. Let $|Z| = \delta < \lambda$.

It follows that in M_U, $|j_U(Z)| = j_U(\delta) < j_U(\lambda)$, but $j_U \restriction \lambda = i \restriction \lambda$, and so, in M_U, $|j_U(Z)| = j_U(\delta) = i(\delta) < \lambda$.

Now, λ_1 is a regular cardinal in M_U. Hence, $j_U(Z) \cap \lambda_1$ is bounded in λ_1. Pick some bound $\xi < \lambda_1$.

Then $\xi = j_U(h)(\kappa, \lambda)$, for some $h : \kappa \times \lambda \to \lambda, h \in V$. Without loss of generality, we can assume that $h : \lambda \to \lambda$, since otherwise ξ can be replaced by a bigger ordinal which is still below λ_1 but is represented by such a function.

Now, in $V[G]$, define a function $g : \lambda \to \lambda$ as follows:

$$g(\nu) = \min(Z \setminus h(\nu)).$$

Then, $j_U(g)(\lambda) = \lambda_1$.

Note that λ is the κth element of the Magidor sequence for $j_U(\lambda)$ in M_U.

Denote the Magidor sequence of G by $\langle \lambda_\nu \mid \nu < \kappa\rangle$.

We are interested in $\langle g(\lambda_\nu) \mid \nu < \kappa\rangle$.

Note that in M_U, the $\kappa+1$th element of the Magidor sequence for $j_U(\lambda)$ is already above λ_1. So, $g(\lambda_\nu) < \lambda_{\nu+1}$, for every $\nu < \kappa$.

Using standard arguments on Magidor forcing, it is possible to find $f : \lambda \to \lambda, f \in V$ such that $f(\lambda_\nu) \geq g(\lambda_\nu)$, for every $\nu < \kappa$.

Then, in M_U, $j_U(f)(\lambda) \geq j_U(g)(\lambda) = \lambda_1$.

But this is impossible, since $j_U \upharpoonright V$ starts with σ and $\sigma(f)(\lambda) = j_1(i(f)(\lambda)) < \lambda_1$. □

It is not hard to find in $V[G]$ an extension of $\mathcal{U}(\lambda, 0)$ to a κ-complete ultrafilter over λ which concentrates on a set of cardinality κ. Just note that the set $\{\lambda_{\nu+1} \mid \nu < \kappa\}$ is almost contained in every set in $\mathcal{U}(\lambda, 0)$. We can just copy $\mathcal{V}$ onto this set which will produce the desired extension. □

It is possible to obtain a similar result with $\lambda = \kappa^{+\kappa}$ in the extension.

Proposition 4.5: *Assume that κ is a measurable cardinal and $\lambda > \kappa$ is $\mathcal{P}^2(\lambda)$-hyper-measurable. Then in a forcing extension $\lambda = \kappa^{+\kappa}$ and there is a κ-complete uniform ultrafilter over λ which extends a measure over λ in V.*

Proof: We will use the Radin forcing with guiding generic for collapses (see for example J. Cummings [4]) and the idea similar to Bagaria-Magidor [2]. Note that the κ-complete ultrafilter over λ of Bagaria-Magidor [2] (with translation to the present context) is not uniform, but rather concentrates on a set of cardinality κ, namely on the ordinals of the Radin sequence. But we can use a different one which has an additional generator, i.e. $\langle \lambda, \mu \rangle$. The argument for showing uniformity, repeats basically those of Lemma 4.4. □

5. Blowing Up The Power of κ to λ

We would like to construct a model with $2^\kappa \geq \lambda$ and with a uniform κ-complete ultrafilter U over λ.

Note that it is easy to do so starting from a supercompact cardinal κ and blowing up its power preserving supercompactness. However, in such construction κ^+ will not be a continuity point of j_U, i.e. $j_U(\kappa^+) > \sup(j_U{}''\kappa^+)$.

We will start from a weaker assumptions and in our model κ^+ will be a continuity point of j_U.

Assume GCH. Let κ be a superstrong cardinal with a regular target, i.e. there is $j : V \to M$, $\operatorname{crit}(j) = \kappa$, $j(\kappa)$ is a regular cardinal in V and $M \supseteq V_{j(\kappa)}$. Assume that j is the embedding by an extender.

Let $\lambda, \kappa < \lambda < j(\kappa)$ be a measurable cardinal.

We can assume that $\lambda = j(h_\lambda)(\kappa)$ for some function $h_\lambda : \kappa \to \kappa$, see [6] for the argument.

We would like to add λ-many Cohen functions from κ to κ. Do it as follows: define the Easton support iteration $\langle P_\alpha, \underset{\sim}{Q}_\beta \mid \alpha \leq j(\kappa), \beta < j(\kappa) \rangle$

where

(1) $P_0 = \emptyset$;
(2) if β is not inaccessible or β is an inaccessible, but β is not closed under $j(h_\lambda)$, then let Q_β be a trivial forcing;
(3) if β is an inaccessible and β is closed under $j(h_\lambda)$, then let $Q_\beta = Cohen(\beta, j(h_\lambda)(\beta))$.

Back in V, let $\mathcal{U}$ be a normal ultrafilter over λ. Consider, in M, its image $j(\mathcal{U})$. Let $i : M \to N$ be its ultrapower embedding. Set $\sigma = i \circ j$.

Clearly,

$$X \in \mathcal{U} \Leftrightarrow j(X) \in j(\mathcal{U}) \Leftrightarrow j(\lambda) \in \sigma(X).$$

So,

$$\mathcal{U} = \{X \subseteq \lambda \mid j(\lambda) \in \sigma(X)\}.$$

We would like to extend the embedding σ. In order to do this, let us force Cohen subsets to $j(\kappa)$ over $V^{P_{j(\kappa)}}$. Force with $Cohen(j(\kappa), j(\lambda))$.

Note that $|j(\kappa)| = |j(\lambda)| = |j(\lambda^+)| = |\sigma(\lambda)|$, since j is the embedding by an extender. So, we can arrange $\sigma(\lambda)$-Cohen functions from $j(\kappa)$ to $j(\kappa)$.

Note that if $G \subseteq Cohen(j(\kappa), \sigma(\lambda))$-generic over $V^{P_{j(\kappa)}}$, G is not $Cohen(j(\kappa), \sigma(\lambda))$-generic over $M^{P_{j(\kappa)}}$.

Moreover, G is not a subset of $Cohen(j(\kappa), \sigma(\lambda))^{M^{P_{j(\kappa)}}}$. Namely, $j''\kappa^+ \notin M$, but, in V,

$$\langle j(\alpha), \langle 0, 0\rangle\rangle \mid \alpha < \kappa^+\rangle$$

is a legitimate condition in $Cohen(j(\kappa), \sigma(\lambda))$.

In order to overcome this, we proceed as follows. Let $\langle f_\alpha \mid \alpha < \sigma(\lambda)\rangle$ be a sequence that is $Cohen(j(\kappa), \sigma(\lambda))$-generic over $V^{P_{j(\kappa)}}$. Due to $j(\kappa)^+$-c.c. of the forcing, they will be $Cohen(j(\kappa), \sigma(\lambda))$-generic over $M^{P_{j(\kappa)}}$ as well.

Namely,

$$G' = \{p \in Cohen(j(\kappa), \sigma(\lambda))^{M^{P_{j(\kappa)}}} \mid \forall \alpha \in \mathrm{dom}(p)(p(\alpha) \subseteq f_\alpha)\}$$

will be $Cohen(j(\kappa), \sigma(\lambda))$-generic over $M^{P_{j(\kappa)}}$.

Thus, let in $M^{P_{j(\kappa)}}$, A be a maximal antichain in this forcing. Then $|A| \leq j(\kappa)$. So, there is $B \subseteq \sigma(\lambda)$ of size $j(\kappa)$ (in $M^{P_{j(\kappa)}}$) such that $A \subseteq Cohen(j(\kappa), B)^{M^{P_{j(\kappa)}}} = Cohen(j(\kappa), B)^{V^{P_{j(\kappa)}}}$.

But then $\langle f_\alpha \mid \alpha \in B\rangle$ will generate $G' \restriction B$ and $G' \restriction B$ will be $V^{P_{j(\kappa)}}$-generic for $Cohen(j(\kappa), B)$ and so will intersect A.

Finally, we need to change these functions a bit in order to include $\sigma''G(Cohen(\kappa,\lambda)) = j''G(Cohen(\kappa,\lambda))$, where $G(Cohen(\kappa,\lambda)) \subseteq Cohen(\kappa,\lambda)$ is generic. Note that for every set of ordinals A of cardinality less than $j(\kappa)$ in M, $|A \cap j''\lambda| \leq \kappa$, since the extender producing j is on κ. So, the usual argument of Woodin allows to make the desired change, see Cummings [5]. The rest of the argument is standard.

Let $\sigma^* : V[G(P_{j(\kappa)+1})] \to N[G^*]$ be the resulting extension of σ.

Define

$$U = \{X \subseteq \lambda \mid j(\lambda) \in \sigma^*(X)\}.$$

Then $U \supseteq \mathcal{U}$ be a weakly normal uniform κ-complete ultrafilter.

Note that $j_U(\kappa) < j(\kappa)$, since $|j_U(\kappa)| \leq \lambda^\kappa = \lambda < j(\kappa)$.

So only a part of the embedding σ^* is actually used.

Note that all regular cardinals in the interval (κ, λ) are continuity points of σ^* and j_U. Menachem Magidor (unpublished) gave an other very elegant construction of a uniform ultrafilter with this property. He started with a supercompact cardinal and used intersection of dense open sets for Cohen forcing to generate the desired ultrafilter.

If we use a (κ, λ^+)-extender E instead of $(\kappa, j(\kappa))$-extender, then the only missing element will be a generic for $Cohen(j_E(\kappa), j_E(\lambda))$ over M_E. H. Woodin used originally the forcing $Cohen(\kappa^+, \lambda^+)$ over V in order to produce the desired generic. Recently, Yoav Ben Shalom [3] showed that this additional Cohen forcing is unneeded, if we would like to blow up the power of κ below $\kappa^{+\kappa}$. It is unclear whether this is true in general, and, in particular in the case that we consider (i.e. blowing up the power of κ to a measurable λ).

However, if κ is λ^+-supercompact, then, using the standard methods, it is possible to blow up the power of κ to λ (or to λ^+). Moreover, if W is a normal measure over $\mathcal{P}_\kappa(\lambda^+)$, then the corresponding embedding $j_W : V \to M_W \simeq {}^{\mathcal{P}_\kappa(\lambda^+)}V/W$ extends in the generic extension to $j^*_W : V[G] \to M[G^*]$ (as well as W to W^* which gives the embedding). The only forcing performed at cardinals $\geq \kappa$ is $Cohen(\kappa, \lambda)$. Now, let us derive a (κ, λ^+)-extender E^* (actually it is equivalent to a measure over κ) from j^*_W:

$$X \in E^*_a \Leftrightarrow a \in j^*_W(X),$$

for every finite $a \subseteq \lambda^+$ and $X \subseteq [\kappa]^{|a|}$. Then E^* is a natural extension of a (κ, λ^+)-extender E (now not equivalent to a measure over κ) from j_W in the same fashion.

Now the ultrapower M_{E^*} by E^* need not be in general a generic extension of the corresponding ultrapower M_E by E. It depends on particular M_W-generics for $Cohen(j_W(\kappa), j_W(\lambda))$. However, it is possible always to find E^* which is (κ, τ)-extender for some τ of cardinality λ^+ which restriction E to V will be like this. For example pick an elementary submodel Y of large enough portion of M_W of cardinality $j_W(\lambda^+)$ such that $Y[G^*]$ is elementary in the extension $M_W[G^*]$. Then the transitive collapse of $Y \cap j_W(\kappa)$ will generate such E^* and E.

This means, in particular, that we have M_E-generic subsets of the missing Cohen part $Cohen(j_E(\kappa), j_E(\lambda))$.

We will explore this idea further in the next section.

6. On a Question of J. Hamkins

In [9], J. Hamkins introduced the following large cardinal:

Definition 6.1: A cardinal κ is called *strongly tall* iff for every $\delta > \kappa$ there is a uniform κ-complete ultrafilter U such that $j_U(\kappa) > \delta$.

Clearly, if κ is strongly compact then it is strongly tall. Hamkins [9] asked if the opposite is true as well.

In [1], a negative answer was claimed, however G. Goldberg found a gap in the argument.[8] Goldberg proved several very nice results in the opposite direction. Thus for example he showed that under GCH strong tallness implies strong compactness.

The next proposition shows that still — the answer to the question is negative.

Proposition 6.2: *Assume GCH and let $\kappa < \lambda$ be supercompact cardinals. Then there is a cofinality preserving extension in which the following hold:*

(1) κ is strongly tall, but not strongly compact,
(2) $2^\kappa = \lambda$,
(3) every λ-complete filter extends to a κ-complete ultrafilter.

Proof: The argument will be similar to the one of the previous section, only instead of a single normal ultrafilter over λ we will use normal ultrafilters over $\mathcal{P}_\lambda(\mu)$ for μ's above λ.

[8] The mistake is due solely to the second author.

We force the iteration similar the one of the previous section: the Easton support iteration

$$\langle P_\alpha, \underset{\sim}{Q}_\beta \mid \alpha \leq \kappa+1, \beta \leq \kappa\rangle,$$

where for every $\beta < \kappa$, $\underset{\sim}{Q}_\beta$ is trivial unless β is a strongly inaccessible in V^{P_β}, and if β is a strongly inaccessible in V^{P_β}, then we first force an ordinal $\bar{\beta} < \kappa$ with an atomic forcing and then force with the Cohen forcing $Cohen(\beta, \bar{\beta})$ to add $\bar{\beta}$-many Cohen subsets to β.

Finally, over κ, $\underset{\sim}{Q}_\kappa = Cohen(\kappa, \lambda)$.

Back in V, let $\mu \geq \lambda$ be a regular. Pick a normal ultrafilter $\mathcal{U}$ over $\mathcal{P}_\lambda(\mu)$.

Let $\mathcal{V}$ be a normal ultrafilter over $\mathcal{P}_\kappa(\mu^+)$. Proceed as in the previous section with $j = j_{\mathcal{V}}$ and $i = j^{M_{\mathcal{V}}}_{j(\mathcal{U})}$.

Now, as it was explained at the end of the previous section, by supercompactness of κ we will have an appropriate generic for $Cohen(j(\kappa), j(\mu^+))$.

This allows to define a uniform κ-complete ultrafilter $U(\mu)$ over μ, namely, it will extend the one in V generated by $\sup(j_{\mathcal{U}}{}''\mu)$, i.e. $\{X \subseteq \mu \mid \sup(j_{\mathcal{U}}{}''\mu) \in j_{\mathcal{U}}(X)\}$.

In order to destroy the strong compactness κ we just add in addition a non-reflecting stationary subset to κ^+ of some fixed cofinality below κ. Do this as a product and not an iteration, i.e. over V^{P_κ} we force with $Cohen(\kappa, \lambda) \times R$, where R denotes the forcing adding such a set.

Let us argue that in the final model every λ-complete filter extends to a κ-complete ultrafilter.

Suppose that F is such a filter on an ordinal and let μ be a regular large enough such that $|F| \leq \mu$. Let $F = \{Z_\alpha \mid \alpha < \mu\}$. Pick names $\underset{\sim}{F}$ and $\underset{\sim}{Z}_\alpha$'s for F and Z_α's. Assume that the weakest condition already forces all of this.

Work in V. Let $\mathcal{V}, \mathcal{U}, i$ be as above.

Then $j_{\mathcal{V}}(\underset{\sim}{F})$ is forced to consist of $j_{\mathcal{V}}(\{Z_\alpha \mid \alpha < \mu\})$.

Set $j_{\mathcal{V}}(\{Z_\alpha \mid \alpha < \mu\}) = \{\underset{\sim}{Z}'_\alpha \mid \alpha < j_{\mathcal{V}}(\mu)\}$.

The ultrapower $M := M^{M_{\mathcal{V}}}_{j_{\mathcal{V}}(\mathcal{U})}$ of $M_{\mathcal{V}}$ by $j_{\mathcal{V}}(\mathcal{U})$ is closed, inside $M_{\mathcal{V}}$, under $j_{\mathcal{V}}(\mu)$-sequences of its elements. In particular, $\{i(\underset{\sim}{Z}'_\alpha) \mid \alpha < j_{\mathcal{V}}(\mu)\} \in M$.

We have $\sigma = i \circ j_{\mathcal{V}} : V \to M$.

It extends in the generic extension to $\sigma^* : V[G] \to M[G^*]$, where G is a generic subset of $P_\kappa * (Cohen(\kappa, \lambda) \times R)$.

By elementarity, $\sigma^*(F)$ is $\sigma(\lambda) > j_{\mathcal{V}}(\mu)$ complete filter. Also, $Z''_\alpha := (i(\underset{\sim}{Z})'_\alpha)_{G^*} \in \sigma^*(F)$, for every $\alpha < j_{\mathcal{V}}(\mu)$.

Hence, there is some $\eta \in \bigcap_{\alpha<j_{\mathcal{V}}(\mu)} Z''_{\alpha}$.

Now we can define a desired extension F^* of F be setting

$$Z \in F^* \Leftrightarrow \eta \in \sigma^*(Z).$$

Claim 6.3: *κ is strongly tall in the generic extension.*

Proof: Let $\mu > \lambda$ be a cardinal of countable cofinality.

Let F be a λ-complete uniform ultrafilter over μ^+ in V with $j_F(\lambda) > \mu$, say $F = \{X \subseteq \mu^+ \mid \sup(j_{\mathcal{U}}{}''\mu^+) \in j_{\mathcal{U}}(X)\}$, where $\mathcal{U}$ is a normal ultrafilter over $\mathcal{P}_\lambda(\mu^+)$.

Then, F is still a λ-complete filter in $V[G]$, since the forcing used has λ-c.c. and so every $A \subseteq F, |A| < \lambda$ can be covered by some $B \in V, |B| < \lambda$.

Let $F^* \supseteq F$ be a κ-complete ultrafilter.

Note that $\kappa \leq \mathrm{crit}(j_{F^*}) \leq \lambda$ is a measurable cardinal and $2^\kappa = \lambda$, hence $\mathrm{crit}(j_{F^*}) = \kappa$.

In V, $j_F(\lambda) > \mu$. So there are $j_F(\lambda)$-many functions from μ^+ to λ increasing mod F. We have $F^* \supseteq F$, hence they will be such mod F^*. In particular, $j_{F^*}(\lambda) > \mu$. In $V[G]$, $2^\kappa = \lambda$, so by elementarity, in M_{F^*}, $2^{j_{F^*}(\kappa)} = j_{F^*}(\lambda) > \mu$.

Recall that we have GCH above λ in $V[G]$. Hence, $j_{F^*}(\kappa) > \mu$.

□ of the claim. □

7. An Example of a Not (κ, κ^+)-regular Uniform κ-complete Ultrafilter Over κ^+

In [10], J. Ketonen proved many interesting results concerning regularity of σ-complete ultrafilters. The existence of a not (κ, κ^+)-regular[9] uniform κ-complete ultrafilter over κ^+ remained open, at least to the best of our knowledge.

The example below shows that it is possible to have such ultrafilters.

Proposition 7.1: *Suppose that κ is a huge cardinal. Then there is an extension with a κ-complete uniform weakly normal ultrafilter U over κ^+ such that $\{\nu < \kappa^+ \mid \mathrm{cof}(\nu) = \kappa\} \in U$.*

In particular, by Ketonen [10], U is not (κ, κ^+)-regular.

[9] An ultrafilter U is called (κ, λ) regular iff there is subset of U of cardinality λ such that any κ-members of it have empty intersection.

Proof: Assume GCH and let $\mathcal{V}$ be a normal ultrafilter over $\mathcal{P}^\kappa(\kappa_1)$[10] witnessing hugeness of κ, with $j_{\mathcal{V}}(\kappa) = \kappa_1$.

Let us use the almost huge part of the embedding. Namely, for every $\lambda < \kappa_1$, set

$$W_\lambda = \{X \subseteq \mathcal{P}_\kappa(\lambda) \mid j_{\mathcal{V}}{}''\lambda \in j_{\mathcal{V}}(X)\}.$$

Let $j : V \to M$ be the corresponding direct limit embedding of the system $\langle W_\lambda \mid \lambda < \kappa_1\rangle$.

Note that the following hold:

(1) $j(\kappa) = \kappa_1$,
(2) $j \restriction \kappa_1 = j_{\mathcal{V}} \restriction \kappa_1$,
(3) ${}^{\kappa_1 >}M \subseteq M$,
(4) $j''\kappa_1 \notin M$,
(5) $j(\kappa_1) = \sup(j_{\mathcal{V}}{}''\kappa_1)$,
(6) $|\kappa_2| = \kappa_1$, where $\kappa_2 = j(\kappa_1)$,
(7) $j(\kappa_1^+) = \bigcup j''\kappa_1^+ = \kappa_1^+$.

By the elementarity of j, κ_1 is a huge cardinal with the target κ_2 in M, as witnessed by $j(\mathcal{V})$.

Form an ultrapower N of M using $j(\mathcal{V})$. Let $i : M \to N$ be the corresponding embedding.

Then $i(\kappa_1) = \kappa_2$, ${}^{\kappa_2}N \cap M \subseteq N$, $\delta_2 := \sup(i''\kappa_2) < \kappa_3 := i(\kappa_2)$. So, ${}^{\kappa_1 >}N \subseteq N$. In addition, note that in M, $|\kappa_3| = \kappa_2^+$, and so, in V, $\kappa_1^+ = j(\kappa_1^+) = (\kappa_2^+)^M < \kappa_3 < \kappa_1^{++}$.

Set $\sigma = i \circ j$. Then $\sigma : V \to N$, $\sigma(\kappa) = \kappa_2, \sigma(\kappa_1) = \kappa_3$.

$N \models \mathrm{cof}(\delta_2) = \kappa_2 = \sigma(\kappa)$.

Also, if $\xi < \delta_2$, then there is $\xi' < \kappa_2$ such that $\xi < i(\xi')$. But $\kappa_2 = \sup(j''\kappa_1)$, hence there is $\zeta < \kappa_1$ such that $\xi' < j(\zeta)$, and so,

$$\xi < i(\xi') < i(j(\zeta)) = \sigma(\zeta), \text{ which means that } \sigma''\kappa_1 \text{ is unbounded in } \delta_2.$$

Consider now

$$\mathcal{U} = \{Z \subseteq \kappa_1 \mid \delta_2 \in \sigma(Z)\}.$$

It is a uniform, weakly normal, κ-complete ultrafilter over κ_1 and $\{\nu < \kappa_1 \mid \mathrm{cof}(\nu) = \kappa\} \in \mathcal{U}$.

We would like now to force in order to turn κ_1 into κ^+.

Do this as follows.

[10] $\mathcal{P}^\kappa(\lambda) = \{X \subseteq \lambda \mid otp(X) = \kappa\}$.

Define by induction the Easton support iteration $\langle P_\alpha, Q_\beta \mid \alpha \leq \kappa, \beta < \kappa\rangle$. Suppose that $\alpha < \kappa$ and P_α is defined. Define Q_α. Let Q_α be trivial unless α is a strongly inaccessible in V^{P_α}.

Suppose that α is a strongly inaccessible in V^{P_α}. Then choose (force with an atomic forcing) an inaccessible cardinal $\bar{\alpha}, \alpha < \bar{\alpha} < \kappa$ and set $Q_\alpha = Col(\alpha, < \bar{\alpha})$.

In V^{P_κ}, we force finally with $Col(\kappa, < \kappa_1)$.

Let $G_\kappa * H$ be a generic subset of $P_\kappa * Col(\kappa, < \kappa_1)$.

Extend σ to $\sigma^* : V[G_\kappa * H] \to N^* = N[G_\kappa * H * G_{(\kappa_1,\kappa_2)} * H^*]$.

In order to get $G_{(\kappa_1,\kappa_2)} * H^*$, we will force with $(P_{(\kappa_1,\kappa_2)} * Col(\kappa_2, < \kappa_3))^N$ over $V[G_\kappa * H]$.

It is a $< \kappa_1$-closed forcing of cardinality κ_1^+, which is equivalent to $Col(\kappa_1, \kappa_1^+)$.

We have, in N^*, $\sigma^*(Col(\kappa_1, \kappa_1^+)) = Col(\kappa_2, \kappa_2^+)$.

So, using a generic for $Col(\kappa_1, \kappa_1^+)$, we can generate an N^*-generic for $Col(\kappa_2, \kappa_2^+)^{N^*}$.

Let R be $V[G_\kappa * H]$-generic subset of $Col(\kappa_1, \kappa_1^+)$ and R^* the corresponding to it N^*-generic for $Col(\kappa_2, \kappa_2^+)^{N^*}$.

Turn now to the master condition.

We need to insure that $\sigma'' G_\kappa * H * R \subseteq G_\kappa * H * G_{(\kappa_1,\kappa_2)} * H^* * R^*$. The problematic parts here are H^* and R^*. In general, it should not be the case that $\sigma'' H \subseteq H^*$ and $\sigma'' R \subseteq R^*$. However, it is possible, using the Woodin method, see [5], to change H^* and R^* a bit in order to satisfy the inclusions above.

The property used for this is that $|\sigma'' \kappa_1 \cap x| < \kappa_1$, for any $x \in N$ with $|x|^N < \kappa_2$.

Let us argue that this is true.

Suppose otherwise. Let x be a set of ordinals in N, $|x|^N < \kappa_2$ and $|\sigma'' \kappa_1 \cap x| = \kappa_1$.

Then there is $y \subseteq \kappa_1$ of cardinality κ_1 such that $\sigma'' y = \sigma'' \kappa_1 \cap x$. So, $\sigma'' y$ is covered by the set x.

We have $\sigma'' \kappa_1$ that is unbounded in δ_2 and that in M (and so in N) the cofinality of δ_2 is κ_2, as was shown above.

But then, also $\sigma'' y$ will be unbounded in δ_2, which implies that $x \cap \delta_2$ is unbounded in δ_2.

This is impossible, since $x \in N$ and $|x|^N < \kappa_2$.

After such changes, we will have an extension of σ:

$$\sigma^{**} : V[G_\kappa * H * R] \to N^* = N[G_\kappa * H * G_{(\kappa_1,\kappa_2)} * H^* * R^*].$$

Finally, in $V[G_\kappa * H * R]$, we define an extension of $\mathcal{U}$:

$$\mathcal{U}^* = \{Z \subseteq \kappa_1 \mid \delta_2 \in \sigma^{**}(Z)\}.$$

It is a uniform, weakly normal, κ-complete ultrafilter over κ_1 and $\{\nu < \kappa_1 \mid \mathrm{cof}(\nu) = \kappa\} \in \mathcal{U}^*$. However now $\kappa_1 = \kappa^+$. □

8. A Remark on The Prikry Forcing

Let $\mathcal{U}$ be a κ-complete uniform ultrafilter over a regular cardinal $\lambda > \kappa$.

In the classical paper [14], K. Prikry showed that $\mathcal{U}$ can be used in order to change the cofinality of λ to ω. Conditions are of the form $\langle t, T\rangle$, where $t \in [\lambda]^{<\omega}$ and $T \subseteq [\lambda]^{<\omega}$ is a tree with a trunk t with splittings in $\mathcal{U}$, i.e. for every $s \in T, s \geq_T t$, $Suc_T(s) \in \mathcal{U}$. Denote this forcing by $\mathcal{P}_{\mathcal{U}}$.

Let $G \subseteq \mathcal{P}_{\mathcal{U}}$ be generic.

Proposition 8.1: *The following hold in* $V[G]$*:*

(1) κ *and* λ *change their cofinality to* ω*,*
(2) if δ *is a regular cardinal in* V *and singular in* $V[G]$*, then* $\mathrm{cof}(\delta) = \omega$ *in* $V[G]$*,*
(3) a regular in V *cardinal changes its cofinality in* $V[G]$ *iff it is a discontinuity point of* $j_{\mathcal{U}}$*.*
(4) Suppose that $\delta, \kappa < \delta \leq \lambda$ *is a successor cardinal of* V *which changes its cofinality in* $V[G]$*. Then there is a limit cardinal* $\mu, \kappa \leq \mu < \delta$*, such that, in* $V[G]$*,*
   (a) μ *is a cardinal,*
   (b) $\mathrm{cof}(\mu) = \omega$*,*
   (c) all regular cardinals of V *in the interval* $[\mu, \delta]$ *change their cofinality to* ω*,*
   (d) $\mathrm{cof}^V(\mu) \geq \kappa$*.*

Proof: (1) is clear.

(2). Suppose that δ is a regular cardinal in V and singular in $V[G]$. Suppose for a moment that it has an uncountable cofinality $\mu < \delta$ there. Let $\langle \delta_\xi \mid \xi < \mu\rangle$ be a cofinal sequence. For every $\xi < \mu$, pick a condition $\langle t_\xi, A_\xi\rangle \in G$ which decides δ_ξ, i.e. forces that the ξth element of the cofinal sequence is δ_ξ. Then there a single t such that for μ-many ξ's, we have $t_\xi = t$. Denote by Y the set of all such ξ's. Clearly, $\mathrm{otp}(Y) = \mu, \bigcup_{\xi \in Y} \delta_\xi = \delta$.

But now in V, consider the set

$$Z = \{\tau \mid \exists \xi < \mu \exists A \in \mathcal{U}(\langle t, A\rangle \Vdash (\check{\tau} = \underset{\sim}{\delta}_\xi))\}.$$

Then $Z = Y$, which is impossible, since $Z \in V$ and $\delta > \mu$ is regular in V.

(3). Let us suppose now that a regular cardinal δ is a discontinuity point of $j_{\mathcal{U}}$, i.e. $\sup(j_{\mathcal{U}}{}''\delta) < j_{\mathcal{U}}(\delta)$. Set

$$\mathcal{U}_\delta = \{X \subseteq \delta \mid \sup(j_{\mathcal{U}}{}''\delta) \in j_{\mathcal{U}}(X)\}.$$

Then $\mathcal{U}_\delta$ is a uniform at least κ-complete weakly normal ultrafilter over δ. In addition it is Rudin-Keisler below $\mathcal{U}$, since any function which represents $\sup(j_{\mathcal{U}}{}''\delta)$ in $M_{\mathcal{U}}$ will project $\mathcal{U}$ onto $\mathcal{U}_\delta$. Let $\pi_\delta : \lambda \to \delta$ be such a function.

Now, if $\langle \lambda_n \mid n < \omega \rangle$ is a Prikry sequence for $\mathcal{U}$, then $\langle \pi_\delta(\lambda_n) \mid n < \omega \rangle$ will be a Prikry sequence for $\mathcal{U}_\delta$.

Suppose now that some regular in V cardinal δ changes its cofinality in $V[G]$. Let us argue that it must be a discontinuity point of $j_{\mathcal{U}}$.

Suppose otherwise, i.e. $\sup(j_{\mathcal{U}}{}''\delta) = j_{\mathcal{U}}(\delta)$.

Clearly, $\kappa < \delta < \lambda$. By the observation above, $\mathrm{cof}(\delta) = \omega$ in $V[G]$. Pick a cofinal sequence $\langle \delta_n \mid n < \omega \rangle$.

Use a standard argument to find a condition in G which non-direct extensions of a same length decide values for $\underset{\sim}{\delta}_n$'s. Just work in V and build inductively a tree with splittings in $\mathcal{U}$ which gradually decide $\underset{\sim}{\delta}_n$'s.

Namely, we construct a tree T such that for every $t \in T$, $Suc_T(t) \in \mathcal{U}$ and there is an increasing sequence $\langle m_n \mid n < \omega \rangle$ of natural numbers such that for every $n < \omega$ and $t \in T, |t| = m_n + 1$, $\langle t, T_t \rangle || \underset{\sim}{\delta}_n$, where $T_t = \{s \in T \mid s \geq_T t\}$.

Suppose for simplicity that $m_n = n$, for all $n < \omega$.

Define a function $f_0 : \lambda \to \delta$ which prescribes values of $\underset{\sim}{\delta}_0$.

Namely, we set $f_0(\nu) = \xi$, if $\langle \nu \rangle \in Lev_1(T)$ and $\langle \nu, T_{\langle \nu \rangle} \rangle \Vdash \check{\xi} = \underset{\sim}{\delta}_0$.

Now, by the elementarity, $j_{\mathcal{U}}(f_0)([id]_{\mathcal{U}}) < j_{\mathcal{U}}(\delta)$, but $\sup(j_{\mathcal{U}}{}''\delta) = j_{\mathcal{U}}(\delta)$, hence there will be $\alpha_0 < \delta$ such that $\mathrm{rng}(f_0) \subseteq \alpha_0 \pmod{\mathcal{U}}$.

Shrink T if necessary, and deal with the second level. Let $\nu_1 \in Lev_1(T)$. Consider $Suc_T(\nu_1)$. Define $f_{\nu_1 1} : Suc_T(\nu_1) \to \delta$ which prescribes values of $\underset{\sim}{\delta}_1$.

Namely, we set $f_{\nu_1 1}(\nu) = \xi$, if $\langle \nu \rangle \in Suc_T(\nu_1)$ and $\langle \nu_1, \nu, T_{\langle \nu_1, \nu \rangle} \rangle \Vdash \check{\xi} = \underset{\sim}{\delta}_1$.

As above, there will be $\alpha_{\nu_1 1} < \delta$ such that $\mathrm{rng}(f_{\nu_1 1}) \subseteq \alpha_{\nu_1 1} \pmod{\mathcal{U}}$.

Consider now the function $\nu_1 \mapsto \alpha_{\nu_1 1}$. By the same argument there is $\alpha_1 < \delta$ which contains its range $(\mathrm{mod}(\mathcal{U}))$.

But this means that α_1 will bound all the decision of $\underset{\sim}{\delta}_1$ made by any choice of $t \in T, |t| = 2$.

Proceed further in the similar fashion by induction on $n < \omega$ and define such $\alpha_n < \delta$ for $\underset{\sim}{\delta}_n$.

Finally set $\alpha := \bigcup_{n<\omega} \alpha_n$. It will be below δ and will bound all $\underset{\sim}{\delta}_n$'s, which is impossible. Contradiction. So, δ must be a discontinuity point of $j_{\mathcal{U}}$, and we are done.

Finally, let us deal with the last item (4).

So, suppose that $\delta, \kappa < \delta \leq \lambda$ is a successor cardinal of V which change a cofinality in $V[G]$.

Then the cofinality of δ must be ω in $V[G]$, and in particular it is not a cardinal. Let $\mu = |\delta|$ (in $V[G]$). Clearly, $\mu \geq \kappa$, since κ remains a cardinal.

Suppose for a moment that $\mathrm{cof}^{V[G]}(\mu) \neq \omega$.

By S. Shelah [15], (Lemma 4.9, p. 304), then $(\delta^+)^V$ is collapsed as well. Now, if $\delta = \lambda$, then we are done since $\mathcal{P}_{\mathcal{U}}$ satisfies λ^+-c.c.

Suppose that $\delta < \lambda$. Consider $\eta = (\mu^+)^V$. Then $\eta \leq \delta$ and η is collapsed, and so, changes its cofinality. Then, by item (2), $\mathrm{cof}(\eta) = \omega$ in $V[G]$ and by item (3), η is a discontinuity point of $j_{\mathcal{U}}$.

Consider the projection $\mathcal{U}_\eta$ of $\mathcal{U}$ to η:

$$\mathcal{U}_\eta = \{X \subseteq \eta \mid \sup(j_{\mathcal{U}}{}''\eta) \in j_{\mathcal{U}}(X)\}.$$

Then $\mathcal{U}_\eta$ is a uniform at least κ-complete weakly normal ultrafilter over η.

The Prikry forcing $\mathcal{P}_{\mathcal{U}_\eta}$ with $\mathcal{U}_\eta$ is a subforcing of $\mathcal{P}_{\mathcal{U}}$. Apply the argument to $\mathcal{P}_{\mathcal{U}_\eta}$. So both $\eta = (\mu^+)^V$ and μ should have cofinality ω in $V^{\mathcal{P}_{\mathcal{U}_\eta}}$, but the last model is contained in $V[G]$ and it is assumed that $\mathrm{cof}^{V[G]}(\mu) \neq \omega$.

Contradiction.

Finally, let us argue that $\mathrm{cof}^V(\mu) \geq \kappa$. Suppose otherwise. Then $\mathrm{cof}^V(\mu) = \omega$, since no cardinal below κ changes its cofinality.

Consider, as above, $\eta = (\mu^+)^V$. Then $\eta \leq \delta$ and η is collapsed, and so, changes its cofinality. Hence, η is a discontinuity point of $j_{\mathcal{U}}$. Then, by Ketonen [10], 1.7, a final segment of regular cardinals below μ consists of discontinuity points of $j_{\mathcal{U}}$, since $\eta = (\mu^+)^V$. So, all of them change cofinality, and hence μ is not a cardinal in $V[G]$.[11] Contradiction. $\square$

9. Moving Ordinals by σ-complete Uniform Ultrafilters

Answering a question of D. Fremlin, it was shown in [7] that:

if U_0, U_1 are κ-complete ultrafilters over a measurable cardinal κ, then $|j_{U_0}(\tau)| = |j_{U_1}(\tau)|$, for every ordinal τ.

[11]Basically, in $M_{\mathcal{U}}$, $\mathrm{cof}(\sup(j''_{\mathcal{U}}\eta)) < j_{\mathcal{U}}(\mu) = \bigcup_{n<\omega} j_{\mathcal{U}}(\mu_n)$, where $\langle \mu_n \mid n < \omega \rangle \in V$ is a cofinal in μ sequence. Hence for some $n^* < \omega$ we will have in $M_{\mathcal{U}}$, $\mathrm{cof}(\sup(j''_{\mathcal{U}}\eta)) < j_{\mathcal{U}}(\mu_{n^*})$. But then, every regular cardinal in the interval $[\mu_{n^*}, \mu)$ will be a discontinuity point of $j_{\mathcal{U}}$.

Let us note that this may break down for uniform ultrafilters with the same degree of completeness and over the same cardinal.

Suppose that $\mu < \lambda < \kappa$ are measurable cardinals. Pick a μ-complete ultrafilter $\mathcal{W}$ over μ, a λ-complete ultrafilter $\mathcal{V}$ over λ and a κ-complete ultrafilter $\mathcal{U}$ over κ.

Consider $U_0' = \mathcal{W} \times \mathcal{V} \times \mathcal{U}$ and $U_1' = \mathcal{W} \times \mathcal{U}$.

Then U_0' is a uniform μ-complete ultrafilter over $\mu \times \lambda \times \kappa$ and U_1' is a uniform μ-complete ultrafilter over $\mu \times \kappa$.

Let U_0 be an equivalent to U_0' ultrafilter over κ and U_1 be an equivalent to U_1' ultrafilter over κ.

Then $j_{U_1}(\lambda) = \lambda$, but $j_{U_1}(\lambda) > \lambda^+$.

References

1. A. Apter and M. Gitik, On tall cardinals and some related generalizations, Israel J. of Math, 1–31, 2014.
2. J. Bagaria and M. Magidor, Group radicals and strongly compact cardinals. Transactions of the American Mathematical Society, vol. 366 (2014), pp. 1857–1877.
3. Y. Ben Shalom, On the Woodin construction of failure of GCH at a measurable cardinal, M. Sc. thesis, Tel Aviv, 2017, arXiv:1706.08143.
4. J. Cummings, A model in which GCH holds at successors but fails at limits. Transactions of the American Mathematical Society 329 (1992) 1–39.
5. J. Cummings, Iterated forcing and elementary embeddings, in M. Foreman and A. Kanamori eds., Handbook of Set Theory, 775–884, 2010.
6. M. Gitik, Prikry type forcings, in Handbook of Set Theory, Foreman, Kanamori, eds., Springer 2010, 1351–1448.
7. M. Gitik and S. Shelah, Applications of pcf for mild large cardinals to elementary embeddings. Annals of Pure and Applied Logic Volume 164, Issue 9, September 2013, pp. 855–865.
8. G. Goldberg, Compactness and Comparison, arxiv:1801.09153.
9. J. Hamkins, Tall cardinals, Math. Logic Quarterly 55, 2009, 3025–3029.
10. J. Ketonen, Strong compactness and other cardinal sins, Annals of Math. Logic 5, (1972) 47–76.
11. Y. Kimchi and M. Magidor, The independence between the concepts of compactness and supercompactness. Circulated manuscript (1980).
12. K. Kunen, Saturated ideals, J. of Symbolic Logic 43(1), 1978, 65–76. Vol. 43, No. 1 (Mar. 1978), pp. 65–76.
13. W. Mitchell, Beginning Inner Model Theory, in M. Foreman and A. Kanamori eds., Handbook of Set Theory, 1449–1497, 2010.
14. K. Prikry, Changing measurable into accessible cardinals. Dissertationes Mathematicae 68:55, 1970.
15. S. Shelah, Cardinal Arithmetic, v.29 Oxford Logic Guides, Oxford University Press, 1994.

https://doi.org/10.1142/9789819806584_0004

More on Bases of Uncountable Free Abelian Groups

Noam Greenberg
School of Mathematics and Statistics, Victoria University of Wellington, Wellington, New Zealand
http://homepages.mcs.vuw.ac.nz/~greenberg/
greenberg@msor.vuw.ac.nz

Linus Richter
Department of Mathematics, National University of Singapore Singapore 119076
https://blog.nus.edu.sg/linusrichter/
richter@nus.edu.sg

Saharon Shelah
The Hebrew University of Jerusalem
Einstein Institute of Mathematics
Edmond J. Safra Campus, Givat Ram
Jerusalem 91904, Israel
Department of Mathematics
Hill Center-Busch Campus
Rutgers, The State University of New Jersey
110 Frelinghuysen Road
Piscataway, NJ 08854-8019 USA
shelah@math.huji.ac.il

Dan Turetsky
School of Mathematics and Statistics, Victoria University of Wellington, Wellington, New Zealand
https://homepages.ecs.vuw.ac.nz/~dan/
dan.turetsky@msor.vuw.ac.nz

Greenberg was partially supported by a Marsden grant from the Royal Society of New Zealand, and by a Lady Davis Visiting Professorship. Shelah was partially supported by the European Research Council grant 338821. Paper 1188 on Shelah's list. The results of this chapter are also part of Richter's PhD thesis [11] written at Victoria University of Wellington.

We extend results found by Greenberg, Turetsky and Westrick in [7] and investigate effective properties of bases of uncountable free abelian groups. Assuming $V = L$, we show that if κ is a regular uncountable cardinal and X is a $\Delta^1_1(L_\kappa)$ subset of κ, then there is a κ-computable free abelian group whose bases cannot be effectively computed by X. Unlike in [7], we give a direct construction.

1. Introduction

The recursive construction of a basis of a vector space is taught to first-year students of linear algebra: at each step, add some vector which does not lie in the span of the previous ones chosen. This "algorithm" works equally well in finite, countable and uncountable-dimensional vector spaces. In the uncountable case, this can be made precise using the framework introduced by Greenberg and Knight for uncountable computable structure theory, using admissible computability on uncountable cardinals [6]. In fact, just like the countable case, a single Turing jump (relative to the diagram of the vector space) suffices. In particular, for every cardinal κ, every κ-computable vector space (over a κ-computable field) has a basis which is definable in L_κ.

What happens when we consider free abelian groups, objects which appear just as simple as vector spaces? In the countable case, based on a strong notion of independence introduced by Pontryagin [10], a similar algorithm can be performed; Downey and Melnikov [1] used this notion to show that a single Turing jump suffices to build a basis of a countable free abelian group. This algorithm, however, fails badly in the uncountable case, because difficulties can arise in limit steps: it is possible to generate an ω-sequence $b_0, b_1, \ldots$ of elements of a free abelian group such that every finite initial segment of the sequence is extendible to a basis of the group, but the whole sequence cannot. That is, the construction, which appears perfectly fine at every finite step, "explodes" at the limit step.

In [7], the authors showed that this difficulty is fundamental: in general, there is no way to construct bases of free abelian groups in a step-by-step recursive, or definable, construction. Formally, what they showed (under the standard assumption that $V = L$) is that for most uncountable regular cardinals κ, there are κ-computable free abelian groups which have no bases definable over L_κ; in fact, no single κ-Turing degree which is $\Delta^1_1(L_\kappa)$ can serve as an oracle which computes bases for all κ-computable free abelian groups. This result has two shortcomings:

(1) It does not apply to all regular uncountable cardinals κ;
(2) It does not rely on a direct construction.

These two are related. The argument in [7] relies on the complexity of *identifying* which groups are free. The authors show that if κ is not weakly compact, then the collection of free abelian groups is $\mathbf{\Sigma}^1_1(L_\kappa)$-complete, and in many cases (such as successor cardinals, or the least inaccessible), is (lightface) $\Sigma^1_1(L_\kappa)$-complete. In these cases, this completeness result implies that searching for bases cannot be restricted to the sets computable from a fixed $\Delta^1_1(L_\kappa)$ oracle. It is not clear from the construction, though, how to directly build, given a $\Delta^1_1(L_\kappa)$ oracle X, a κ-computable free abelian group G with no X-computable basis. And this approach cannot work for weakly compact cardinals κ; the very compactness property of these cardinals implies that it is in fact not very difficult to identify which groups are free; in fact, a single jump suffices. This leaves the original question, of the complexity of bases, open for these cardinals.

In this chapter, we fully answer the original question:

Theorem 1.1: ($V = L$). *Let κ be regular and uncountable. For any $X \in 2^\kappa$ which is $\Delta^1_1(L_\kappa)$ there is a κ-computable free abelian group, none of whose bases is X-computable.*

The proof of the theorem is done by a direct construction.

2. Preliminaries, and the Plan of the Proof

The proof of Theorem 1.1 is the combination of two distinct parts: we will first reduce the problem to one in "computable set theory", one which abstracts away the algebraic part; and then give a construction solving the reduced problem. Throughout this chapter, we assume $V = L$. We refer the reader to [6] for the basic definitions of κ-computability. A general reference for torsion-free abelian groups is [5].

We start by recalling some basic concepts that were used in [7].

All groups we will discuss are abelian and torsion-free. A group G is *free abelian* if it has a basis: a subset B which is linearly independent ($\sum m_i b_i = 0$ implies each $m_i = 0$, where $m_i \in \mathbb{Z}$ and $b_i \in B$) and spans G (every element of G is of the form $\sum m_i b_i$ for some $m_i \in \mathbb{Z}$ and $b_i \in B$). We will omit the adjective "abelian" and just call these groups free.

Definition 2.1: Let G be a group. A subgroup H of G *detaches* in G if there is a G-subgroup K such that $G = H \oplus K$. We write $H \mid G$.

Any subgroup of a free abelian group is free. A subgroup H of a free abelian group G detaches in G if and only if some (equivalently any) basis of H can be extended to a basis of G, if and only if the quotient G/H is free. If H detaches in G then H detaches in any subgroup K of G containing H.

Definition 2.2: A sequence $\langle G_\alpha \rangle_{\alpha<\gamma}$ of groups of some ordinal length γ is *increasing* if $\alpha < \beta$ implies $G_\alpha \subseteq G_\beta$; it is *continuous* if for all limit $\alpha < \gamma$, $G_\alpha = \bigcup_{\beta<\alpha} G_\beta$. A *filtration* of a group G is a sequence $\overline{G} = \langle G_\alpha \rangle$ such that $\overline{G}\,\hat{}\,\langle G \rangle$ is increasing and continuous, and $|G_\alpha| \leqslant |\alpha|$ for all α.

If γ is regular and G is a group of universe γ then all filtrations of G agree on a club; in fact, for club many α, $G_\alpha = G \cap \alpha$.

Definition 2.3: Let $\overline{G} = \langle G_\alpha \rangle_{\alpha<\gamma}$ be increasing and continuous. The *detachment set* of $\overline{G}$ is

$$\mathrm{Div}(\overline{G}) = \{\alpha < \gamma \,:\, \forall \beta \in (\alpha, \gamma)\ (G_\alpha \mid G_\beta)\}\,.$$

If γ is regular and $\overline{G}, \overline{G}'$ are two filtrations of a group of universe γ, then $\mathrm{Div}(\overline{G})$ and $\mathrm{Div}(\overline{G}')$ agree on a club.

The following can essentially be found in [2]; see [3, IV.1.7].

Proposition 2.4: *Suppose γ is a limit ordinal and let $\overline{G} = \langle G_\alpha \rangle_{\alpha<\gamma}$ be a filtration of a group G_γ. Suppose G_α is free for all $\alpha < \gamma$.*

(1) If $\mathrm{Div}(\overline{G})$ contains a γ-club, then G_γ is free.
(2) If γ is regular and G_γ is free, then $\mathrm{Div}(\overline{G})$ contains a γ-club.

Remark 2.5: Let γ be a limit ordinal; let $\overline{G} = \langle G_\alpha \rangle_{\alpha<\gamma}$ be a filtration of a group G_γ. Suppose that $\mathrm{Div}(\overline{G})$ contains a club of γ. Then

$$\mathrm{Div}(\overline{G}) = \{\alpha < \gamma \,:\, G_\alpha \mid G_\gamma\}\,.$$

The proof of Proposition 2.4 is effective. This gives the following:

Proposition 2.6: *Let κ be regular and uncountable; $\overline{G} = \langle G_\alpha \rangle_{\alpha<\kappa}$ be a filtration of a free group G (with universe κ). The bases of G and the club subsets of $\mathrm{Div}(\overline{G})$ are equicomputable modulo $\overline{G}$. That is:*

- *If C is a club subset of $\mathrm{Div}(\overline{G})$ then there is a basis B of G such that $B \leqslant_\kappa C \oplus \overline{G}$;*
- *If B is a basis of G then there is a club subset C of $\mathrm{Div}(\overline{G})$ such that $C \leqslant_\kappa B \oplus \bar{G}$.*

The reductions in both directions are uniform. Briefly, given a club subset $C = \{\alpha_i : i < \kappa\}$ of $\mathrm{Div}(\overline{G})$, we can present G as the direct sum of κ-finite groups H_i with $G_{\alpha_i} = \bigoplus_{j<i} H_j$, and combine bases of the groups H_i to a basis of G. In the other direction, given a basis B of G, we let C be the collection of α such that $B \cap \alpha$ generates G_α.

From now, we fix a regular uncountable cardinal κ. Recall the lexicographic ordering on 2^κ: $S < T$ if for the greatest $\delta < \kappa$ such that $S \upharpoonright \delta = T \upharpoonright \delta$ we have $\delta \in T \setminus S$. If $\langle S_i \rangle_{i \leqslant i^*}$ is a lexicographically nondecreasing sequence (for some ordinal i^*), $i < j \leqslant i^*$ and $S_i \upharpoonright \delta = S_j \upharpoonright \delta$ for some $\delta < \kappa$, then $S_r \upharpoonright \delta$ is constant for all $r \in [i, j]$.

The lexicographic ordering is complete. A lexicographically nondecreasing sequence $\langle S_i \rangle_{i \leqslant i^*}$ is *continuous* if for all limit $j \leqslant i^*$, S_j is the least upper bound of $\langle S_i \rangle_{i<j}$. This means that if $j \leqslant i^*$ is a limit, $\delta < \kappa$ and $S_i \upharpoonright \delta$ is constant for some final segment of $i < j$, then $S_i \upharpoonright \delta + 1$ is also constant for some final segment of $i < j$.

A *nondecreasing approximation* of a set $S \subseteq \kappa$ is a uniformly κ-computable sequence $\langle S_i \rangle_{i<\kappa}$ which is lexicographically nondecreasing such that $S = \sup_{i<\kappa} S_i$. Sets which have nondecreasing approximations are called *κ-left-c.e.* If S is κ-left-c.e. then it has a nondecreasing approximation which is also continuous.

We say that a set $S \subseteq \kappa$ is *nowhere stationary* if for all $\alpha \leqslant \kappa$ of uncountable cofinality, $S \cap \alpha$ is nonstationary in α.

To avoid repetition, we make the following definition.

Definition 2.7: We say that a nowhere stationary, κ-left-c.e. set $S \subseteq \kappa$ is *nicely thin* if there is a continuous nondecreasing approximation $\langle S_i \rangle_{i<\kappa}$ of S such that for all $i < \kappa$, S_i is nowhere stationary, and every $\alpha \in S_i$ is a limit ordinal of countable cofinality.

As promised, the following proposition distills the algebraic aspects of our construction.

Proposition 2.8: *If $S \subset \kappa$ is nicely thin, then there is a κ-computable free abelian group G and a $\Delta^0_2(L_\kappa)$ filtration $\overline{G}$ of G such that* $\mathrm{Div}(\overline{G}) = \kappa \setminus S$.

The second part of the proof is:

Proposition 2.9: *Let X be $\Delta^1_1(L_\kappa)$. There is a nicely thin set $S \subseteq \kappa$ which intersects every X-computable club set.*

Proof: [Proof of Theorem 1.1, given Propositions 2.8 and 2.9] Let X be $\Delta^1_1(L_\kappa)$; we may assume that $X \geqslant_\kappa \emptyset'$. Let $S \subseteq \kappa$ be a nicely thin set given by Proposition 2.9. Let G and $\overline{G}$ be given by Proposition 2.8 from this set S. Since $\overline{G}$ is $\Delta^0_2(L_\kappa)$ and $X \geqslant_\kappa \emptyset'$, we have $X \geqslant_\kappa \overline{G}$. Suppose that B is a basis of G. By Proposition 2.6, there is a club set $C \subseteq \mathrm{Div}(\overline{G})$ such that $C \leqslant_\kappa B \oplus \overline{G}$. So C is disjoint from $S = \kappa \setminus \mathrm{Div}(\overline{G})$, whence $C \nleqslant_\kappa X$. Hence $B \nleqslant_\kappa X$ as well. □

The next two sections are devoted to the proofs of Propositions 2.8 and 2.9, respectively.

3. Building Groups from Left-c.e. Sets

In this section, we prove Proposition 2.8. The main tool we use is the "twisting" of an increasing ω-sequence of free groups, which is Proposition 2.16 of [7].

Proposition 3.1: *There is a κ-computable process which given a κ-finite increasing ω-sequence $\langle H_n \rangle$ of free groups such that $H_n \mid H_{n+1}$ for all n, produces a κ-finite free group $G \supseteq H_\omega = \bigcup_n H_n$ (with $|G| = |H_\omega|$) such that every H_n detaches in G but H_ω does not detach in G.*

This construction relies on a copy G of $\mathbb{Z}^{(\omega+1)}$ such that $G/\mathbb{Z}^{(\omega)}$ is not free; Pontryagin's criterion implies that each finitely generated subgroup of $\mathbb{Z}^{(\omega)}$ detaches in G. In Proposition 3.1, note that each H_n detaches in H_ω.

First, we show how to prove Proposition 2.8 in the simpler case when S is κ-computable (rather than merely κ-left-c.e.). We can then produce both G and $\overline{G}$ to be κ-computable. The construction follows the proof of [7, Theorem 3.1]. We define the sequence $\langle G_\alpha \rangle_{\alpha<\kappa}$ by effective κ-recursion (Σ_1 recursion over L_κ). During the construction we ensure that each G_α is free, and for all $\beta < \alpha$, if $\beta \notin S$ then $G_\beta \mid G_\alpha$. The construction has three cases.

1. α is a limit: We must define $G_\alpha = \bigcup_{\beta<\alpha} G_\beta$. Since $S \cap \alpha$ is nonstationary in α and $\mathrm{Div}(\langle G_\beta \rangle_{\beta<\alpha}) \supseteq \alpha \setminus S$, G_α is free (Proposition 2.4). Further, if $\beta \in \alpha \setminus S$ then as $\beta \in \mathrm{Div}(\langle G_\beta \rangle_{\beta<\alpha})$, we have $G_\beta \mid G_\alpha$ (Remark 2.5).

2. Defining $G_{\alpha+1}$ when $\alpha \notin S$: We let $G_{\alpha+1} = G_\alpha \oplus \mathbb{Z}$. Since G_α is free, so is $G_{\alpha+1}$. We certainly have $G_\alpha \mid G_{\alpha+1}$, and for all $\beta \in \alpha \setminus S$, by induction, $G_\beta \mid G_\alpha$, and so $G_\beta \mid G_{\alpha+1}$ as well.

3. Defining $G_{\alpha+1}$ when $\alpha \in S$: Since $\alpha \in S$, $\mathrm{cf}(\alpha) = \omega$, so we can effectively find a cofinal ω-sequence $\langle \beta_n \rangle$ in α; by replacing each β_n by its successor, we may assume that each $\beta_n \notin S$. Thus, by induction, for all n, $G_{\beta_n} \mid G_{\beta_{n+1}}$, and $G_\alpha = \bigcup_n G_{\beta_n}$. We then apply Proposition 3.1 to the sequence $\langle G_{\beta_n}\rangle$ to obtain a free group $G_{\alpha+1}$ extending G_α such that $G_\alpha \nmid G_{\alpha+1}$ yet for all n, $G_{\beta_n} \mid G_{\alpha+1}$. If $\beta \leqslant \alpha$ and $\beta \notin S$, then $\beta < \alpha$; for sufficiently large n we have $\beta_n > \beta$. By induction, $G_\beta \mid G_{\beta_n}$, and by construction, $G_{\beta_n} \mid G_{\alpha+1}$, so $G_\beta \mid G_{\alpha+1}$ as required.

We define $G = G_\kappa = \bigcup_{\alpha<\kappa} G_\alpha$. It is κ-computable since it is the union of a κ-computable sequence of κ-finite groups. Case 1 above holds for κ as well: $\operatorname{Div}(\langle G_\alpha\rangle) \supseteq \kappa \setminus S$; since S is nonstationary in κ, G is free. On the other hand, for all $\alpha \in S$, by construction, $G_\alpha \nmid G_{\alpha+1}$, so $\beta \notin \operatorname{Div}(\langle G_\alpha\rangle)$; overall, we see that $\operatorname{Div}(\langle G_\alpha\rangle) = \kappa \setminus S$ as required.

We now consider the general case, when S is κ-left-c.e. The construction is an elaboration on the proof of [7, Theorem 3.2]. The idea is that if at stage $s < \kappa$ we see a change in S at some small $\alpha < s$, then we keep the group that we have constructed so far, but "squash" the filtration so that the old G_s becomes a subgroup of the new G_α. The fact that the smallest such change is into S allows us to introduce a twist (otherwise we would need to remove a twist, which is impossible).

Let $S \subset \kappa$ be nicely thin, and let $\langle S_t\rangle_{t<\kappa}$ be an approximation satisfying Definition 2.7. For brevity, we let $S_\kappa = S$.

We now construct, by recursion on $t \leqslant \kappa$, filtrations $\overline{G^t} = \langle G^t_\alpha\rangle_{\alpha\leqslant\gamma(t)}$ (for ordinals $\gamma(t) \leqslant t$), satisfying:

(a) $G^s_{\gamma(s)} \subseteq G^t_{\gamma(t)}$ for $s \leqslant t \leqslant \kappa$, and if t is a limit then $G^t_{\gamma(t)} = \bigcup_{s<t} G^s_{\gamma(s)}$. That is, the sequence $\langle G^s_{\gamma(s)}\rangle_{s\leqslant\kappa}$ is increasing and continuous.

(b) $G^t_{\gamma(t)}$ is free and $\operatorname{Div}(\overline{G^t}) = \gamma(t) \setminus S_t$ for all $t \leqslant \kappa$. Indeed, for all $\beta < \gamma(t)$, if $\beta \notin S_t$ then $G^t_\beta \mid G^t_{\gamma(t)}$, while if $\beta \in S_t$ then $G^t_\beta \nmid G^t_{\beta+1}$.

(c) For all $s \leqslant t \leqslant \kappa$ and $\alpha \leqslant \gamma(s)$, if $S_s \restriction \alpha = S_t \restriction \alpha$ then $\alpha \leqslant \gamma(t)$ and $G^s_\alpha = G^t_\alpha$.

(d) For all $s < t \leqslant \kappa$ and $\alpha \leqslant \gamma(t)$, if $S_s \restriction \alpha \neq S_t \restriction \alpha$ then $G^s_{\gamma(s)} \subseteq G^t_\alpha$.

We first consider the successor case. Suppose that $\overline{G^s}$ has been defined for all $s \leqslant t$; we show how to define $\overline{G^{t+1}}$. First, we let

$$\delta = \max\{\beta \leqslant \gamma(t) \;:\; S_t \restriction \beta = S_{t+1} \restriction \beta\}.$$

For $\beta \leqslant \delta$ we let $G^{t+1}_\beta = G^t_\beta$. We let $\gamma(t+1) = \delta + 1$, and so we need to define $G^{t+1}_{\delta+1}$. There are two sub-cases. If $\delta \notin S_{t+1}$, then $\delta \notin S_t$, and so the maximality of δ shows that $\delta = \gamma(t)$. In this sub-case, we simply let $G^{t+1}_{\delta+1} = G^t_\delta \oplus \mathbb{Z}$.

Suppose that $\delta \in S_{t+1}$. Either $\delta = \gamma(t)$, or $\delta \notin S_t$ (by the maximality of δ). In either case, $G^t_\delta \mid G^t_{\gamma(t)}$. Write $G^t_{\gamma(t)} = G^t_\delta \oplus K$ (where K may be trivial). Since $\mathrm{cf}(\delta) = \omega$ (as $\delta \in S_{t+1}$), we choose a sequence $\langle \beta_n \rangle$ cofinal in δ and disjoint from S_{t+1}. By (b), $G^t_{\beta_n} \mid G^t_{\beta_{n+1}}$ for all $n < \omega$, so we appeal to Proposition 3.1 to get a free group $G \supset G^t_\delta$ in which every G_{β_n} detaches, but $G^t_\delta \nmid G$. We let $G^{t+1}_{\delta+1} = G \oplus K$.

By design, $G^t_{\gamma(t)} \subseteq G^{t+1}_{\delta+1} = G^{t+1}_{\gamma(t+1)}$, verifying (a) for stage $t+1$. We verify (b). For $\beta < \delta$, if $\beta \in S_{t+1}$ then $\beta \in S_t$ and then $G^t_\beta \nmid G^t_{\beta+1}$, which shows that $\beta \notin \mathrm{Div}(\overline{G^{t+1}})$; if $\beta \notin S_{t+1}$ then $\beta \notin S_t$. If $\delta \notin S_{t+1}$ then $G^t_\beta \mid G^t_\delta \mid G^{t+1}_{\delta+1}$, showing that $\beta \in \mathrm{Div}(\overline{G^{t+1}})$. If $\delta \in S_{t+1}$ then for some n, $\beta < \beta_n$, and $G^t_\beta \mid G^t_{\beta_n} \mid G^{t+1}_{\delta+1}$, showing that $\beta \in \mathrm{Div}(\overline{G^{t+1}})$. (c) between t and $t+1$ follows from the definition of δ. In general, for $s < t$, if $\alpha \leqslant \gamma(s)$ and $S_s \restriction \alpha = S_{t+1} \restriction \alpha$, then $S_s \restriction \alpha = S_t \restriction \alpha$ as well; by induction, $\alpha \leqslant \gamma(t)$, and so $\alpha \leqslant \delta$ and $G^s_\alpha = G^t_\alpha = G^{t+1}_\alpha$. For (d), let $s < t+1$, $\alpha \leqslant \gamma(t+1)$, and suppose that $S_s \restriction \alpha \neq S_{t+1} \restriction \alpha$; we need to show that $G^s_{\gamma(s)} \subseteq G^{t+1}_\alpha$. In case $\alpha = \gamma(t+1)$, this follows from (a). Otherwise, $\alpha \leqslant \delta$, and so $\alpha \leqslant \gamma(t)$, and $G^{t+1}_\alpha = G^t_\alpha$. By the definition of δ, it must be that $S_s \restriction \alpha \neq S_t \restriction \alpha$. Then by induction, $G^s_{\gamma(s)} \subseteq G^t_\alpha$.

Now suppose that $t \leqslant \kappa$ is a limit ordinal, and suppose that the filtrations $\overline{G^s}$ have been defined for all $s < t$, and satisfy the conditions described. We now describe how to define $\overline{G^t}$. Let

$$\Delta = \{\beta : \exists s < t \ [\beta \leqslant \gamma(s) \ \& \ (S_s \restriction \beta = S_t \restriction \beta)]\}.$$

If $s < t$ witnesses that $\beta \in \Delta$, then s witnesses that every $\epsilon < \beta$ is in Δ as well, and so Δ is actually an ordinal (a set of ordinals closed downwards). We let $\gamma(t) = \Delta$.

We argue that Δ is in fact a limit ordinal. This relies on the approximation $\langle S_s \rangle$ being continuous. Suppose, for a contradiction, that $\delta = \max \Delta = \Delta - 1$ exists. Let $s < t$ witness that $\delta \in \Delta$. Then $S_r \restriction \delta$ is constant for $r \in [s,t]$. By definition of Δ we have $\delta \leqslant \gamma(s)$. Since (c) holds by induction, for all $r \in [s,t)$ we have $\delta \leqslant \gamma(r)$, that is, every $r \geqslant s$ also witnesses that $\delta \in \Delta$. Since the approximation is continuous, for sufficiently late r, we have $S_r(\delta) = S_t(\delta)$, that is, $S_r \restriction \delta+1$ is constant on a final segment of $r < t$. So we may assume that $S_s \restriction \delta + 1 = S_t \restriction \delta + 1$. Now by our instructions for the successor case, we have $\gamma(s+1) \geqslant \delta+1$, so stage $s+1$ witnesses that $\delta + 1 \in \Delta$, a contradiction.

Now for all $\beta < \gamma(t) = \Delta$, we define $G^t_\beta = G^s_\beta$ for any $s < t$ witnessing that $\beta \in \Delta$; again, as $S_r \upharpoonright \beta = S_t \upharpoonright \beta$ for all $r \in [s,t)$, by (c) G^r_β is constant for all such r, so G^t_β is well-defined, and $\langle G^t_\beta \rangle_{\beta<\gamma(t)}$ is increasing and continuous; we let $G^t_{\gamma(t)} = \bigcup_{\beta<\gamma(t)} G^t_\beta$, and this defines $\overline{G^t}$, which is indeed a filtration. We verify that the conditions above hold for t.

(c) is by construction. For (d), let $s < t$ and $\alpha \leqslant \gamma(t)$, and suppose that $S_s \upharpoonright \alpha \neq S_t \upharpoonright \alpha$. As above, we now assume that $\alpha < \gamma(t)$, as the case $\alpha = \gamma(t)$ will follow from (a), which we will soon verify. Since $\alpha \in \Delta$, find some $r < t$ such that $S_r \upharpoonright \alpha = S_t \upharpoonright \alpha$ and $\alpha \leqslant \gamma(r)$. Then $s < r < t$. By induction, $G^s_{\gamma(s)} \subseteq G^r_\alpha$, and $G^r_\alpha = G^t_\alpha$.

For (a), let $s < t$. Since $\gamma(t) \notin \Delta$, and $\gamma(t)$ is a limit, there is some $\beta < \gamma(t)$ such that $S_s \upharpoonright \beta \neq S_t \upharpoonright \beta$. Again, let $r \in (s,t)$, such that $S_r \upharpoonright \beta = S_t \upharpoonright \beta$ and $\beta \leqslant \gamma(r)$. By induction,

$$G^s_{\gamma(s)} \subseteq G^r_\beta = G^t_\beta \subseteq G^t_{\gamma(t)}$$

as required.

Finally, we verify (b). Let $\beta < \gamma(t)$. Suppose that $\beta \in S_t$. Take $s < t$ witnessing that $\beta + 1 \in \Delta$. Then $\beta \in S_s$ and by induction, $G^s_\beta \nmid G^s_{\beta+1}$; and $G^t_\beta = G^s_\beta$ and $G^t_{\beta+1} = G^s_{\beta+1}$. Suppose that $\beta \notin S_t$, and let $\delta \in (\beta, \gamma(t))$. Let $s < t$ witness that $\delta \in \Delta$. Then $\delta \notin S_s$, and so $G^s_\beta \mid G^s_\delta$; so $G^t_\beta \mid G^t_\delta$, showing that $\beta \in \mathrm{Div}(\langle G^t_\alpha \rangle_{\alpha<\gamma(t)})$. Now by assumption, the set S_t is nonstationary in $\gamma(t)$, and so $G^t_{\gamma(t)}$ is free, and $\mathrm{Div}(\overline{G^t}) = \gamma(t) \setminus S_t$.

This completes the construction. There is one thing left to show: that $\gamma(\kappa) = \kappa$. This fundamentally follows from the regularity of κ, which implies that the approximation $\langle S_t \rangle$ of $S = S_\kappa$ is *tame*: for all $\beta < \kappa$ there is some $t < \kappa$ such that $S_t \upharpoonright \beta = S_\kappa \upharpoonright \beta$. [This is proved by induction on β; if this is known for all $\alpha < \beta$, then the function taking such α to the least t for which $S_t \upharpoonright \alpha = S_\kappa \upharpoonright \alpha$ must be bounded below κ.] So now we show by induction on $\beta < \kappa$ that $\beta \leqslant \gamma(s)$ for some s for which $S_s \upharpoonright \beta = S_\kappa \upharpoonright \beta$ (which by (c) for s and κ implies that $\beta \leqslant \gamma(\kappa)$). If this is known for β, then by taking a sufficiently late s, we may assume that $S_s \upharpoonright \beta + 1 = S_\kappa \upharpoonright \beta + 1$ as well; then by construction, $\beta + 1 \leqslant \gamma(s+1)$. Now suppose that β is a limit ordinal. For all $\alpha < \beta$, let s_α be the least s for which $\alpha \leqslant \gamma(s)$ and $S_s \upharpoonright \alpha = S_\kappa \upharpoonright \alpha$. Then the sequence $\langle s_\alpha \rangle_{\alpha<\beta}$ is nondecreasing; let $s^* = \sup_{\alpha<\beta} s_\alpha$. Then $S_{s^*} \upharpoonright \beta = S_\kappa \upharpoonright \beta$, and by (c), for all $\alpha < \beta$, $\alpha \leqslant \gamma(s^*)$, so $\beta \leqslant \gamma(s^*)$.

The restriction of the construction to $t < \kappa$ is κ-computable. In particular, the sequence $\langle G^s_{\gamma(s)} \rangle$ is κ-computable, and so its union, which is

G^κ_κ, is κ-computable; by (b) at $t = \kappa$, it is free. The filtration $\langle G^\kappa_\alpha \rangle_{\alpha<\kappa}$ is $\Delta^0_2(L_\kappa)$ and by (b), $\mathrm{Div}(\overline{G^\kappa}) = \kappa \setminus S_\kappa$ as required. This completes the proof of Proposition 2.8.[1]

4. Constructing Fat Thin Sets

In this section, we prove Proposition 2.9. As in the previous section, we first consider the construction under some simplifying assumptions. First, we review the basic tools of fine-structure theory that were used in [7] and will use again below. They are taken from Jensen's original paper [8].

Definition 4.1: The class E consists of all the singular ordinals α such that for some $\beta > \alpha$:

- $L_\beta \models \mathrm{ZF}^-$;
- α is the greatest cardinal of L_β;
- there exists $p \in L_\beta$ such that L_β is the least fully elementary substructure $M \prec L_\beta$ with $p \in M$ and $M \cap \alpha$ transitive.

Each $\alpha \in E$ has countable cofinality; in fact, if $\alpha \in E$, witnessed by β, then there is a cofinal ω-sequence in α definable over $L_{\beta+1}$. Thus, once we see that an ordinal α is singular, we can effectively tell if $\alpha \in E$ or not. Thus, for any regular κ, $E \cap \kappa$ is κ-c.e.; if κ is a successor cardinal, then $E \cap \kappa$ is κ-computable.

The following lemma is used to produce elements of E:

Lemma 4.2: *Let κ be regular and uncountable; let $q \in L_{\kappa^+}$. Let M be the least elementary substructure of L_{κ^+} such that $q \in M$ and $M \cap \kappa$ is transitive. Let $\pi\colon M \to L_\beta$ be the Mostowski collapse; let $\alpha = \pi(\kappa) = M \cap \kappa$. Then $\alpha \in E$, witnessed by β.*

The deep fact about E that is used throughout is [8, Theorem 5.1]:

Theorem 4.3: (Jensen). *The class E does not reflect at any singular ordinal. That is, if α is singular then $E \cap \alpha$ is nonstationary in α.*

On the other hand, E is stationary in every regular cardinal.

Toward the full proof of Proposition 2.9, we give the proof when κ is a successor cardinal. In that case, $E \cap \kappa$ is κ-computable, and so we can make

[1] Note that there is no reason to believe that $\mathrm{Div}(\langle G^s_{\gamma(s)} \rangle) = \kappa \setminus S$; they agree on a club, but that club may fail to be κ-computable, it is merely $\Delta^0_2(L_\kappa)$.

the desired set S κ-computable as well. To simplify even further, rather than meeting X-computable clubs for some possibly quite complicated X, we consider a simpler collection of sets, namely the first-order definable ones. We thus prove:

Proposition 4.4: *Let κ be a successor cardinal. There is a κ-computable set S which is nowhere stationary but intersects every club of κ which is first-order definable over L_κ.*

This implies that there is a κ-computable free group with no first-order definable basis.

Proof: The set S is constructed by recursion. At stage $\delta < \kappa$, we will have already defined $S \restriction \delta$. If $\delta \notin E$ then $\delta \notin S$. Suppose that $\delta \in E$. Then we let $\delta \in S$ if and only if there is some $C \subseteq \delta$, closed and unbounded in δ, which is first-order definable over L_δ, which is disjoint from $S \cap \delta$.

Because E is κ-computable, the construction is κ-computable, and so S is κ-computable.

Let us show that S intersects every club of κ, first-order definable over L_κ. Let C be such a club. Let M be the smallest elementary substructure of L_{κ^+} which contains the parameters used for the definition of C, and such that $M \cap \kappa$ is transitive. Let $\pi\colon M \to L_\beta$ be the Mostowski collapse, and let $\alpha = \pi(\kappa) = M \cap \kappa$. By Lemma 4.2, $\alpha \in E$. Now $\pi(C) = C \cap \alpha$ is a club of α which is first-order definable over $L_\alpha = \pi(L_\kappa)$. Also note that $\alpha \in C$ (as C is closed). If $\pi(C) \cap (S \cap \alpha) \neq \emptyset$ then some $\gamma < \alpha$ is an element of $S \cap C$, in which case we are done. Otherwise, by construction, we put α into S, so in this case $\alpha \in C \cap S$.

It remains to show that S is nowhere stationary. By construction, $S \subseteq E$, and E does not reflect at any singular ordinal (Theorem 4.3), so it suffices to show that $S \cap \lambda$ is nonstationary in λ, for every regular cardinal $\lambda \leqslant \kappa$.

By constructing an increasing, continuous sequence of elementary submodels of L_{κ^+}, we obtain a closed set $D \subseteq \kappa$ such that $D \cap \lambda$ is unbounded in λ for every regular $\lambda \leqslant \kappa$, and such that for all $\alpha \in D$ there is a model $M_\alpha \prec L_{\kappa^+}$ with $\alpha = M_\alpha \cap \kappa$. It suffices to show that $D \cap S = \emptyset$. Let $\alpha \in D$; let $\pi\colon M_\alpha \to L_\beta$ be the Mostowski collapse. Let $C \subseteq \alpha$ be a club of α, first-order definable over L_α. Then $C \in L_\beta$, and $\pi^{-1}(C)$ is a club of κ, first-order definable over L_κ. We have just proved that $S \cap \pi^{-1}(C) \neq \emptyset$. Note that $S \in M_\alpha$ (as it is definable over L_κ). By elementarity of M_α, there is some $\gamma < \alpha$ in $\pi^{-1}(C) \cap S$. Then $\gamma \in C \cap (S \cap \alpha)$. Thus, $S \cap \alpha$ meets all clubs of α which are first-order definable over L_α, and so even if $\alpha \in E$, the construction would instruct us to keep α out of S. $\square$

Now for the full proof of Proposition 2.9, we need to overcome two obstacles:

- If κ is not a successor cardinal, we need to deal with the fact that $E \cap \kappa$ is not κ-computable, but merely κ-c.e.
- We need to diagonalise against all X-computable clubs for some X which is $\Delta^1_1(L_\kappa)$, not just first-order definable ones.

For the first difficulty, we fix a κ-computable enumeration $\langle E_t \rangle_{t<\kappa}$ of E, and repeat the construction above at each stage $t \leqslant \kappa$, giving us a set S_t for each such t; we will show that this is an approximation as required. For the second, we use a technique of approximating (or reflecting) Δ^1_1 sets that was used by Johnston [9, Theorem 4.43] and S. Friedman and his co-authors [4, Lemma 2.5].

Fix a $\Delta^1_1(L_\kappa)$ set X. Thus, there are two first-order formulas φ and ψ, with parameters in L_κ, such that $(\exists Y)\psi(-,Y)$ and $(\exists Y)\varphi(-,Y)$ define X and its complement respectively, where the variable Y ranges over subsets of κ, and for each $Y \subseteq \kappa$, the formula $\varphi(-,Y)$ is evaluated in the structure $(L_\kappa; \in, Y)$ (with Y interpreting a unary predicate).

Fix ordinals $\alpha < \beta$ such that $L_\beta \models \mathrm{ZF}^-$ and α is regular in L_β and is the largest cardinal of L_β. Also assume that the parameters used for the formulas φ and ψ are elements of L_α. We let

$$A^\beta_\alpha = \{\gamma < \alpha : (\exists y \in L_\beta)\ y \subseteq \alpha\ \&\ L_\alpha \models \varphi(\gamma, y)\}$$

and

$$B^\beta_\alpha = \{\gamma < \alpha : (\exists y \in L_\beta)\ y \subseteq \alpha\ \&\ L_\alpha \models \psi(\gamma, y)\}.$$

If $\alpha \in E$ then we write A_α and B_α for A^β_α and B^β_α for the unique β which witnesses that $\alpha \in E$. We say that $\alpha \in E$ is *good* if $B_\alpha = \alpha \setminus A_\alpha$. In this case, A_α is our guess for X at level α. The guess is correct on a club. This follows from:

Lemma 4.5: *Suppose that M is an elementary substructure of L_{κ^+} (containing the parameters for φ and ψ), and $M \cap \kappa \in \kappa$. Let $\alpha = M \cap \kappa$, and let L_β be the Mostowski collapse of M. Then $B^\beta_\alpha = \alpha \setminus A^\beta_\alpha$, and $A^\beta_\alpha = X \cap \alpha$.*

We now provide the proof of Proposition 2.9. Let κ be regular; fix X, φ and ψ as above. Let $\langle E_t \rangle$ be a κ-computable enumeration of $E \cap \kappa$: for all

$t \leqslant \kappa$, E_t is the collection of $\alpha < t$ which are witnessed to be in E by some $\beta < t$. The facts about this enumeration that we use are:

(1) If $s \leqslant t \leqslant \kappa$ then $E_s \subseteq E_t$; if $t \leqslant \kappa$ is limit then $E_t = \bigcup_{s<t} E_s$; $E_\kappa = E \cap \kappa$.
(2) For any cardinal $\lambda < \kappa$, for all $t \geqslant \lambda$, $E_t \cap \lambda = E \cap \lambda$.

Now for each $t \leqslant \kappa$ we define a set $S_t \subseteq E_t$ by recursion. For $\alpha < t$, if $S_t \restriction \alpha$ has already been defined, then we set $\alpha \in S_t$ if and only if $\alpha \in E_t$, α is good, and there is a club of α, disjoint from $S_t \cap \alpha$, which is Δ^0_1 definable in the structure $(L_\alpha; \in, A_\alpha)$.

We first observe that S_κ intersects every X-computable club. Let C be an X-computable club. As above, let M be a minimal elementary submodel of L_{κ^+} containing both the parameters for the definitions of X and its complement, and for the reduction of C to X, such that $M \cap \kappa \in \kappa$; let $\alpha = M \cap \kappa$. Let $\pi \colon M \to L_\beta$ be the Mostowski collapse. By Lemma 4.2, $\alpha \in E$, witnessed by β. The reduction of C to X, and the fact that $A_\alpha = X \cap \alpha$ (Lemma 4.5) shows that $C \cap \alpha$ is Δ^0_1-definable in $(L_\alpha; \in, A_\alpha)$. Thus either $C \cap S_\kappa \cap \alpha$ is nonempty, or the construction puts α into S_κ, and so $\alpha \in S_\kappa \cap C$.

Next, we show that S_κ is nowhere stationary. Since $S_\kappa \subseteq E$, it again suffices to consider regular cardinals $\lambda \leqslant \kappa$, and we use the same club D as above: $D \cap \lambda$ is a club of λ for all regular $\lambda \leqslant \kappa$, and for all $\alpha \in D$ there is some $M_\alpha \prec L_{\kappa^+}$ with $\alpha = M_\alpha \cap \kappa$. We show that $D \cap S_\kappa = \emptyset$. Let $\alpha \in D$. To verify that $\alpha \notin S$, we may assume that $\alpha \in E$ and is good. Let β witness that $\alpha \in E$, and let $\pi \colon M_\alpha \to L_\gamma$ be the Mostowski collapse. Since L_γ thinks that α is regular (even with respect to sequences definable over L_γ), we have $\gamma \leqslant \beta$. By Lemma 4.5, $B^\gamma_\alpha = \alpha \setminus A^\gamma_\alpha$ (and $A^\gamma_\alpha = X \cap \alpha$). On the other hand, because both A_α and B_α are defined by existential formulas, they are upwards absolute, which implies that $A^\gamma_\alpha \subseteq A^\beta_\alpha = A_\alpha$ and $B^\gamma_\alpha \subseteq B^\beta_\alpha = B_\alpha$. Since α is good, we must have $A_\alpha = A^\gamma_\alpha = X \cap \alpha$.

Let C be a club of α which is Δ^0_1-definable over $(L_\alpha; \in, A_\alpha)$. Using the same definition over L_κ with X replacing A_α results in a club $\tilde{C}$ of κ which is X-computable and such that $\pi(\tilde{C}) = \tilde{C} \cap \alpha = C$. Now $S_\kappa \in M_\alpha$ and $\pi(S_\kappa) = S_\kappa \cap \alpha$; as $S_\kappa \cap \tilde{C} \neq \emptyset$ and $M_\alpha \prec L_{\kappa^+}$, as above we have $C \cap (S_\kappa \cap \alpha) \neq \emptyset$. So our instructions ensure that $\alpha \notin S$.

It follows that for all $t < \kappa$, S_t is nowhere stationary as well. Since $S_t \subseteq E_t \subseteq E$, it again suffices to check that $S_t \cap \lambda$ is nonstationary in λ, when $\lambda \leqslant \kappa$ is regular. Now if $\lambda > t$ then $S_t \subseteq t$ is bounded in λ, and so nonstationary. If $\lambda \leqslant t$ then as $E_t \cap \lambda = E \cap \lambda$, we can see that $S_t \cap \lambda = S_\kappa \cap \lambda$ (by induction on $\alpha < \lambda$ we see that $S_t \cap \alpha = S_\kappa \cap \alpha$). And we have already observed that $S_\kappa \cap \lambda$ is nonstationary in λ.

It remains to show that $\langle S_t \rangle_{t \leqslant \kappa}$ is lexicographically nondecreasing and is continuous. Let $s < t \leqslant \kappa$. Let $\delta < \kappa$ be such that $S_s \upharpoonright \delta = S_t \upharpoonright \delta$. If $\delta \in S_s$ then $\delta \in E_s$ and so $\delta \in E_t$; the same calculation that put δ into S_s holds for S_t, so $\delta \in S_t$. Hence $S_s \leqslant S_t$ lexicographically.

Let $t \leqslant \kappa$ be a limit ordinal. Let $\delta < \kappa$ and $s < t$ such that $S_s \upharpoonright \delta = S_t \upharpoonright \delta$. Suppose that $\delta \in S_t$. So $\delta \in E_t$. Since $E_t = \bigcup_{r<t} E_r$, for sufficiently late r, we have $\delta \in E_r$; if $\delta \in E_r$ and $S_r \upharpoonright \delta = S_t \upharpoonright \delta$ then the same calculation that put δ into S_t also puts δ into S_r. This shows that $S_t = \sup_{r<t} S_r$, and completes the proof of Proposition 2.9, and so of Theorem 1.1.

5. Further Work

Theorem 1.1 is optimal: R. Johnston has noticed (see [9]) that every κ-computable free group has a basis which is $\Delta^1_1(L_\kappa)$; namely, the $<_L$-least basis is $\Delta^1_1(L_\kappa)$.

However, one can study the general question of the "degree spectra" of bases of κ-computable free groups: what collections of κ-degrees can be realised as the collection of degrees computing bases for some κ-computable free group? In [7], the authors show that one cannot code much into bases; depending on κ, the most that can be coded into all bases of some group is either $\emptyset'$ or $\emptyset''$. One can ask, though, for example, in view of Proposition 4.4, whether there is a κ-computable free group G such that a κ-degree computes a basis for G if and only if it is not first-order definable over L_κ (not computable from $\emptyset^{(n)}$ for any n), and similarly, whether the non-arithmetic degrees (those not below $\emptyset^{(\alpha)}$ for any $\alpha < \kappa$) can be similarly realised.

References

1. Rodney Downey and Alexander G. Melnikov. Effectively categorical abelian groups. *J. Algebra*, 373:223–248, 2013.
2. Paul C. Eklof. Methods of logic in abelian group theory, pp. 251–269. Lecture Notes in Math., Vol. 616, 1977.
3. Paul C. Eklof and Alan H. Mekler. *Almost free modules*, volume 65 of *North-Holland Mathematical Library*. North-Holland Publishing Co., Amsterdam, revised edition, 2002. Settheoretic methods.
4. Ekaterina Fokina, Sy-David Friedman, Julia Knight and Russell Miller. Classes of structures with universe a subset of ω_1. *J. Logic Comput.*, 23(6):1249–1265, 2013.

5. László Fuchs. *Infinite abelian groups. Vol. I.* Pure and Applied Mathematics, Vol. 36. Academic Press, New York-London, 1970.
6. Noam Greenberg and Julia F. Knight. Computable structure theory on ω_1 using admissibility. In *Effective mathematics of the uncountable*, volume 41 of *Lect. Notes Log.*, pp. 50–80. Assoc. Symbol. Logic, La Jolla, CA, 2013.
7. Noam Greenberg, Dan Turetsky and Linda Brown Westrick. Finding bases of uncountable free abelian groups is usually difficult. *Trans. Amer. Math. Soc.*, 370(6):4483–4508, 2018.
8. Ronald Björn Jensen. The fine structure of the constructible hierarchy. *Ann. Math. Logic*, 4:229–308; erratum, ibid. 4 (1972), 443, 1972. With a section by Jack Silver.
9. Reese Johnston. Computability in uncountable binary trees. *J. Symb. Log.*, 84(3):1049–1098, 2019.
10. L. Pontrjagin. The theory of topological commutative groups. *Ann. of Math. (2)*, 35(2):361–388, 1934.
11. L. Richter. On the Definability and Complexity of Sets and Structures. PhD thesis, Victoria University of Wellington, 2024.

https://doi.org/10.1142/9789819806584_0005

Turing Degrees of Hyperjumps

Hayden R. Jananthan
Lincoln Laboratory Supercomputing Center
Massachusetts Institute of Technology
Lexington, MA 02421, USA
https://jananthan.mit.edu
hayden.r.jananthan@gmail.edu

Stephen G. Simpson
Department of Mathematics
Vanderbilt University
Nashville, TN 37203, USA
https://sgslogic.net
sgslogic@gmail.com

The Posner-Robinson Theorem states that for any reals Z and A such that $Z \oplus 0' \leq_{\mathrm{T}} A$ and $0 <_{\mathrm{T}} Z$, there exists B such that $A \equiv_{\mathrm{T}} B' \equiv_{\mathrm{T}} B \oplus Z \equiv_{\mathrm{T}} B \oplus 0'$. Consequently, any nonzero Turing degree $\deg_{\mathrm{T}}(Z)$ is a Turing jump relative to some B. Here we prove the hyperarithmetical analog, based on an unpublished proof of Slaman, namely that for any reals Z and A such that $Z \oplus \mathcal{O} \leq_{\mathrm{T}} A$ and $0 <_{\mathrm{HYP}} Z$, there exists B such that $A \equiv_{\mathrm{T}} \mathcal{O}^B \equiv_{\mathrm{T}} B \oplus Z \equiv_{\mathrm{T}} B \oplus \mathcal{O}$. As an analogous consequence, any nonhyperarithmetical Turing degree $\deg_{\mathrm{T}}(Z)$ is a hyperjump relative to some B.

Contents

1. Introduction

Our starting point is the *Friedberg Jump Theorem*:

Theorem 1.1: (Friedberg Jump Theorem). *[10, Theorem 13.3.IX, p. 265] Suppose A is a real such that* $0' \leq_{\mathrm{T}} A$*. Then there exists B such that*

$$A \equiv_{\mathrm{T}} B' \equiv_{\mathrm{T}} B \oplus 0'.$$

There are several refinements of the Friedberg Jump Theorem. One such extension shows that B can be taken to be an element of any special Π^0_1 class $P \subseteq \{0,1\}^{\mathbb{N}}$. Here *special* means that P is nonempty and has no recursvie elements.

Theorem 1.2: *[6, following Theorem 3.1, p. 37] Suppose* $P \subseteq \{0,1\}^{\mathbb{N}}$ *is a special* Π^0_1 *class and A is a real such that* $0' \leq_{\mathrm{T}} A$*. Then there exists* $B \in P$ *such that*

$$A \equiv_{\mathrm{T}} B' \equiv_{\mathrm{T}} B \oplus 0'.$$

Another refinement is the *Posner-Robinson Theorem*:

Theorem 1.3: (Posner-Robinson Theorem). *[8, Theorem 1, p. 715] [5, Theorem 3.1, p. 1228] Suppose Z and A are reals such that* $Z \oplus 0' \leq_{\mathrm{T}} A$ *and* $0 <_{\mathrm{T}} Z$*. Then there exists B such that*

$$A \equiv_{\mathrm{T}} B' \equiv_{\mathrm{T}} B \oplus Z \equiv_{\mathrm{T}} B \oplus 0'.$$

In this chapter, we prove hyperarithmetical analogs of Theorem 1.2 and Theorem 1.3. The hyperarithmetical analog of Theorem 1.1 is due to Macintyre [7, Theorem 3, p. 9]. In these hyperarithmetical analogs, the Turing jump operator $X \mapsto X'$ is replaced by the hyperjump operator $X \mapsto \mathcal{O}^X$ and Π^0_1 classes are replaced by Σ^1_1 classes. A feature of [7, Theorem 3, p. 9] and of our results is that they involve Turing degrees rather than hyperdegrees, so for instance $\mathcal{O}^B$ is not only hyperarithmetically equivalent to A, but in fact Turing equivalent to A.

Here is an outline of this chapter:

In §2, we prove the following basis theorem for uncountable Σ^1_1 classes $K \subseteq \{0,1\}^{\mathbb{N}}$.

Theorem 1.4: *Suppose $K \subseteq \{0,1\}^{\mathbb{N}}$ is an uncountable Σ^1_1 class and Z and A are reals such that $Z \oplus \mathcal{O} \leq_{\mathrm{T}} A$ and $0 <_{\mathrm{HYP}} Z$. Then there exists $B \in K$ such that*

$$A \equiv_{\mathrm{T}} \mathcal{O}^B \equiv_{\mathrm{T}} B \oplus \mathcal{O}$$

and $Z \not\leq_{\mathrm{HYP}} B$.

In §3, we prove the following analog of Theorem 1.3, which is essentially due to Slaman [13].

Theorem 1.5: *Suppose Z and A are reals such that $Z \oplus \mathcal{O} \leq_{\mathrm{T}} A$ and $0 <_{\mathrm{HYP}} Z$. Then there exists B such that*

$$A \equiv_{\mathrm{T}} \mathcal{O}^B \equiv_{\mathrm{T}} B \oplus Z \equiv_{\mathrm{T}} B \oplus \mathcal{O}.$$

The remainder of this section fixes notation and terminology.

$g\colon \subseteq A \to B$ denotes a partial function with domain $\operatorname{dom} g \subseteq A$ and codomain B. For $a \in A$, if $a \in \operatorname{dom} g$ then we say '$g(a)$ converges' or '$g(a)$ is defined' and write $g(a)\downarrow$. Otherwise, we say '$g(a)$ diverges' or '$g(a)$ is undefined' and write $g(a)\uparrow$. If f and g are two partial functions $\subseteq A \to B$ and $a \in A$, then $f(a) \simeq g(a)$ means $(f(a)\downarrow \wedge g(a)\downarrow \wedge f(a) = g(a)) \vee (f(a)\uparrow \wedge g(a)\uparrow)$. We write $f(a)\downarrow = b$ to mean that $f(a)\downarrow$ and $f\colon a \mapsto b$.

$\mathbb{N}^{\mathbb{N}}$ and $\{0,1\}^{\mathbb{N}}$ denote the Baire and Cantor spaces, respectively, whose elements we sometimes call *reals*. We identify $\{0,1\}^{\mathbb{N}}$ and the powerset $\mathcal{P}(\mathbb{N})$ in the usual manner.

If S is a set, then S^* is the set of strings of elements from S. If $s_0, \ldots, s_{n-1} \in S$, then $\sigma = \langle s_0, \ldots, s_{n-1} \rangle \in S^*$ denotes the string of *length* $|\sigma| := n$ defined by $\sigma(k) = s_k$. If $\langle s_0, \ldots, s_{n-1} \rangle$, $\langle t_0, \ldots, t_{m-1} \rangle \in S^*$, then their *concatenation* is $\langle s_0, \ldots, s_{n-1} \rangle^\frown \langle t_0, \ldots, t_{m-1} \rangle := \langle s_0, \ldots, s_{n-1}, t_0, \ldots, t_{m-1} \rangle$. If $\sigma, \tau \in S^*$, then σ is an *initial segment* of τ (equivalently, τ is an extension of σ) written $\sigma \subseteq \tau$, if $\tau\restriction|\sigma| = \sigma$. If $f\colon \mathbb{N} \to S$ then $\sigma \in S^*$ is an *initial segment* of f (equivalently, f is an *extension* of σ), written $\sigma \subset f$, if $f\restriction|\sigma| = \sigma$. $\sigma, \tau \in S^*$ are *incompatible* if neither is an initial segment of the other. If $\leq$ is a partial order on S, then the *lexicographical ordering* $\leq_{\mathrm{lex}}$ on S^* is defined by setting $\sigma \leq_{\mathrm{lex}} \tau$ if $\sigma \subseteq \tau$ or, where k is the least index at which $\sigma(k) \neq \tau(k)$, then $\sigma(k) < \tau(k)$.

$\varphi_e^{(k)}$ denotes the e-th partial recursive function $\subseteq\mathbb{N}^k \to \mathbb{N}$; e is called an *index* of $\varphi_e^{(k)}$. Likewise, if $f \in \mathbb{N}^{\mathbb{N}}$ then $\varphi_e^{(k),f}$ denotes the e-th partial function $\varphi_e^{(k),f}\colon \subseteq\mathbb{N}^k \to \mathbb{N}$ which is partial recursive in f; e is again called an *index* of $\varphi_e^{(k),f}$, while f is called an *oracle* of $\varphi_e^{(k),f}$.

$\leq_{\mathrm{T}}$ denotes Turing reducibility while $\equiv_{\mathrm{T}}$ denotes Turing equivalence. $\leq_{\mathrm{HYP}}$ denotes hyperarithmetical reducibility while $\equiv_{\mathrm{HYP}}$ denotes hyperarithmetical equivalence. For $X \in \{0,1\}^{\mathbb{N}}$, X' denotes the Turing jump of X and $\mathcal{O}^X$ denotes the hyperjump of X. $\mathcal{O}$ denotes Kleene's $\mathcal{O}$. For $f, g \in \mathbb{N}^{\mathbb{N}}$, their *join* $f \oplus g \in \mathbb{N}^{\mathbb{N}}$ is defined by $(f \oplus g)(2n) = f(n)$ and $(f \oplus g)(2n+1) = g(n)$.

P_e denotes the e-th Π^0_1 set $\{f \in \mathbb{N}^{\mathbb{N}} \mid \varphi_e^{(1),f}(0)\downarrow\} \subseteq \mathbb{N}^{\mathbb{N}}$. P_e^* denotes the e-th Σ^1_1 class $\{X \in \{0,1\}^{\mathbb{N}} \mid \exists f\ (f \oplus X \in P_e)\}$.

2. A Basis Theorem for Σ^1_1 Classes

The following theorem includes the Gandy Basis Theorem [11, Theorem III.1.4, p. 54], the Kreisel Basis Theorem for Σ^1_1 Classes [11, Theorem III.7.2, p. 75], and Macintyre's Hyperjump Inversion Theorem [7, Theorem 3, p. 9].

Theorem 2.1: *Suppose $K \subseteq \{0,1\}^{\mathbb{N}}$ is an uncountable Σ^1_1 class and Z and A are reals such that $Z \oplus \mathcal{O} \leq_{\mathrm{T}} A$ and $0 <_{\mathrm{HYP}} Z$. Then there exists $B \in K$ such that*

$$A \equiv_{\mathrm{T}} \mathcal{O}^B \equiv_{\mathrm{T}} B \oplus \mathcal{O}$$

and $Z \not\leq_{\mathrm{HYP}} B$.

To prove Theorem 2.1, we use Gandy-Harrington forcing (first introduced by Harrington in an unpublished manuscript [2]; see, e.g., [11, Theorem IV.6.3, p. 108]), forming a descending sequence of uncountable Σ^1_1 classes

$$K = K_0 \supseteq K_1 \supseteq \cdots \supseteq K_n \supseteq \cdots$$

where an element of the intersection $\bigcap_{n=0}^{\infty} K_n$ has the desired property. Unlike in the case of Π^0_1 subsets of $\{0,1\}^{\mathbb{N}}$, compactness cannot be used to easily show that the intersection $\bigcap_{n=0}^{\infty} K_n$ is nonempty. Instead, some care must be taken to show that this is the case.

Proposition 2.2:

(a) *Given a Σ^1_1 predicate $K \subseteq \{0,1\}^{\mathbb{N}} \times \mathbb{N}^k$, there is a primitive recursive function $f\colon \mathbb{N}^k \to \mathbb{N}$ such that*

$$P^*_{f(x_1,\ldots,x_k)}(X) \equiv K(X, x_1, \ldots, x_k).$$

(b) *Suppose $X \in \{0,1\}^{\mathbb{N}}$. Then $\{e \in \mathbb{N} \mid X \notin P_e^*\} \equiv_{\mathrm{T}} \mathcal{O}^X$.*

(c) *$\{e \in \mathbb{N} \mid P_e^* = \emptyset\} \equiv_{\mathrm{T}} \mathcal{O}$.*

Proof: Straight-forward. □

Corollary 2.3: *There exist primitive recursive functions* v, u, *and* U *such that for all* $n, m \in \mathbb{N}$ *and* $\sigma, \tau \in N^*$ *and* $I \in \mathcal{P}_{\text{fin}}(\mathbb{N})$,

$$\begin{aligned} P^*_{v(n,m)} &= P^*_n \cap P^*_m, \\ P^*_{u(e,\sigma,\tau)} &= P^*_e[\sigma, \tau] = \{X \in \{0,1\}^{\mathbb{N}} \mid \sigma \subset X \wedge \exists g\ (X \oplus g \in P_e \wedge \tau \subset g)\}, \\ P^*_{U(I,\sigma,\langle \tau_0,\ldots,\tau_{n-1}\rangle)} &= \bigcap_{k \in I \wedge k < n} P^*_k[\sigma, \tau_k]. \end{aligned}$$

Proposition 2.4: *The following partial functions are* $\mathcal{O}$*-recursive:*

(a) The partial function $\rho(\sigma, e) \simeq \langle \sigma_0, \sigma_1 \rangle$ *where* σ_0, σ_1 *are minimal incompatible extensions of* σ *which have extensions in* P^*_e *and* σ_0 *is lexicographically less than* σ_1, *whenever* σ *has at least two extensions in* P^*_e, *otherwise diverging.*

(b) The partial function $\text{ext}(\langle e_1, \ldots, e_N \rangle, \sigma, \langle \tau_1, \ldots, \tau_N \rangle) \simeq (\tilde{\sigma}, \langle \tilde{\tau_1}, \ldots, \tilde{\tau_N} \rangle)$ *where* $(\tilde{\sigma}, \langle \tilde{\tau}, \ldots, \tilde{\tau_N} \rangle)$ *is the lexicographically least pair such that*

1. $\sigma \subsetneq \tilde{\sigma}$ *and* $\tau_k \subsetneq \tilde{\tau_k}$ *for* $1 \leq k \leq N$ *and*
2. $\bigcap_{k=1}^N P^*_{e_k}[\tilde{\sigma}, \tilde{\tau_k}] \neq \emptyset$

whenever $\bigcap_{k=1}^N P^*_{e_k}[\sigma, \tau_k] \neq \emptyset$, *otherwise diverging.*

Proof:

(a) Using $\mathcal{O}$, search for the first string ν such that $P^*_e[\sigma^\frown \nu^\frown \langle i \rangle, \langle\rangle] \neq \emptyset$ for $i = 0, 1$. Once such ν has been found, $\rho(\sigma, e)\downarrow = \langle \sigma^\frown \nu^\frown \langle 0 \rangle, \sigma^\frown \nu^\frown \langle 1 \rangle \rangle$.

(b) Using $\mathcal{O}$, search for the first of $i = 0, 1$ for which $\bigcap_{k=1}^N P^*_{e_k}[\sigma^\frown \langle i \rangle, \tau_k] \neq \emptyset$, then search for the lexicographically least $\langle j_1, \ldots, j_N \rangle \in \{0,1\}^N$ such that $\bigcap_{k=1}^N P^*_{e_k}[\sigma^\frown \langle i \rangle, \tau_k{}^\frown \langle j_k \rangle] \neq \emptyset$. If no such i or $j_1, \ldots, j_N$ are found, then diverge. Otherwise, $\text{ext}(\langle e_1, \ldots, e_N \rangle, \sigma, \langle \tau_1, \ldots, \tau_N \rangle)\downarrow = (\sigma^\frown \langle i \rangle, \langle \tau_1{}^\frown \langle j_1 \rangle, \ldots, \tau_N{}^\frown \langle j_N \rangle \rangle)$. □

Let ρ_0, ρ_1 be defined by

$$\rho(\sigma, e) \simeq \langle \rho_0(\sigma, e), \rho_1(\sigma, e) \rangle.$$

We use the ordinal notation description of $\mathcal{O}$ (and, more generally, $\mathcal{O}^Y$ for $Y \in \{0,1\}^{\mathbb{N}}$) described in [11] and use the following well-known lemma to describe hyperarithmetical reducibility in terms of H-sets.

Notation. For $X \in \{0,1\}^{\mathbb{N}}$ and $n \in \mathbb{N}$, define

$$(X)_n := \{x \in \mathbb{N} \mid 2^n \cdot 3^x \in X\}.$$

Lemma 2.5: *Suppose X and Y are reals in $\{0,1\}^{\mathbb{N}}$. Then $X \leq_{\mathrm{HYP}} Y$ if and only if there exists $b \in \mathcal{O}^Y$ and $n \in \mathbb{N}$ such that $X = (H_b^Y)_n$.*

Proof: Suppose $X \leq_{\mathrm{HYP}} Y$, so that there is $b \in \mathcal{O}^Y$ such that $X \leq_{\mathrm{T}} H_b^Y$. Let e be the index of such a Turing reduction, i.e., let e be such that $X = \varphi_e^{(1),H_b^Y}$. By definition [11], $2^b \in \mathcal{O}^Y$ and

$$H_{2^b}^Y := \{2^n 3^x \mid \varphi_n^{(1),H_b^Y}(x)\downarrow\}.$$

Let f be an index such that

$$\varphi_f^{(1),H_b^Y}(x)\downarrow \iff \varphi_e^{(1),H_b^Y}(x)\downarrow = 1.$$

Then

$$\begin{aligned}(H_{2^b}^Y)_f &= \{x \in \mathbb{N} \mid \varphi_f^{(1),H_b^Y}(x)\downarrow\} \\ &= \{x \in \mathbb{N} \mid \varphi_e^{(1),H_b^Y}(x)\downarrow = 1\} \\ &= X.\end{aligned}$$

Conversely, suppose there is $b \in \mathcal{O}^Y$ and $n \in \mathbb{N}$ such that $X = (H_b^Y)_n$. Let e be an index such that

$$\varphi_e^{(1),Z}(x) = \begin{cases} 1 & \text{if } x \in (Z)_n \\ 0 & \text{if } x \notin (Z)_n \end{cases}$$

for any $Z \in \{0,1\}^{\mathbb{N}}$. Then $\varphi_e^{(1),H_b^Y} = X$, showing that $X \leq_{\mathrm{T}} H_b^Y$. □

Proof: [Proof of Theorem 2.1] By the Gandy Basis Theorem [11, Theorem III.1.4, p. 54], assume without loss of generality that $\omega_1^Y = \omega_1^{\mathrm{CK}}$ for all $Y \in K$.

In order to control the hyperjump $\mathcal{O}^B$, we choose B to be an element of an intersection of Σ_1^1 subsets

$$K = K_0 \supseteq K_1 \supseteq \cdots \supseteq K_n \supseteq \cdots .$$

In order for B to be an element of $K_n = P_{j(n)}^*$ for each n, there must be $g_n \in \mathbb{N}^{\mathbb{N}}$ such that $B \oplus g_n \in P_{j(n)}$, where $j(n)$ is some index of K_n. Such

g_n depend on B. Thus, we additionally define sequences of strings

$$\begin{array}{ccccc} \sigma_0 \subseteq \sigma_1 \subseteq \cdots \subseteq \sigma_n \subseteq \cdots \\ \tau_{0,0} \subseteq \tau_{1,0} \subseteq \cdots \subseteq \tau_{n,0} \subseteq \cdots \\ \tau_{0,1} \subseteq \tau_{1,1} \subseteq \cdots \subseteq \tau_{n,0} \subseteq \cdots \\ \tau_{0,2} \subseteq \tau_{1,2} \subseteq \cdots \subseteq \tau_{n,0} \subseteq \cdots \\ \vdots \quad \vdots \quad \ddots \quad \vdots \quad \ddots \end{array}$$

so that $B = \bigcup_{n\in\omega} \sigma_n$ and $g_k = \bigcup_{n\in\omega} \tau_{n,k}$. We also define a sequence of finite subsets of $\mathbb{N}$

$$I_0 \subseteq I_1 \subseteq \cdots \subseteq I_n \subseteq \cdots$$

encoded as finite sequences $\{e_1, \ldots, e_N\} \mapsto \langle e_1, \ldots, e_N\rangle$ which keep track of the indices e of Σ^1_1 classes we have committed to intersecting, so that $K_n = \bigcap_{k\in I_n} P^*_k[\sigma_n, \tau_{n,k}]$. A function $j\colon \mathbb{N} \to \mathbb{N}$ keeps track of the index of K_n, i.e.,

$$K_n = P^*_{j(n)}.$$

In the course of the proof, we assume that j encodes all of the information from previous steps (i.e., a course-of-value computation) though we avoid making this precise to ease the burden of notation.

To ease in the notation and exposition, we set the following temporary definitions. An *intersection system* consists of the following data:

(i) a finite subset $I \subseteq \mathbb{N}$,
(ii) a string σ, and
(iii) a sequence of strings $\langle \tau_k \mid k \in I\rangle$

subject to the constraint that $\bigcap_{k\in I} P^*_k[\sigma, \tau_k]$ is nonempty. If $k \notin I$, then we assign the value $\langle\rangle$ to τ_k.

By *adding P^*_e to the intersection system $I, \sigma, \langle \tau_k \mid k \in I\rangle$*, we mean the following procedure, where $K = \bigcap_{k\in I} P^*_k[\sigma, \tau_k]$:

Case 1: $K \cap P^*_e = \emptyset$. Let $\tilde{I} = I$, $\tilde{K} = K$, $\tilde{\sigma} = \sigma$, and $\tilde{\tau_k} = \tau_k$ for each k.
Case 2: $K \cap P^*_e \neq \emptyset$. Let $\tilde{I} = I \cup \{e\}$, and let $\tilde{\sigma}$ and, simultaneously for all $k \in \tilde{I}$, $\tilde{\tau_k}$ be the lexicographically least proper extensions of σ and τ_k, respectively, such that $\bigcap_{k\in \tilde{I}} P^*_k[\tilde{\sigma}, \tilde{\tau_k}] \neq \emptyset$.

The resulting intersection system is $\tilde{I}, \tilde{\sigma}, \langle \tilde{\tau_k} \mid k \in \tilde{I}\rangle$. Note that from $I, \sigma, \langle \tau_k \mid k \in I\rangle$ and e, the new intersection system $\tilde{I}, \tilde{\sigma}, \langle \tilde{\tau_k} \mid k \in \tilde{I}\rangle$

can be determined in a uniform way recursively in $\mathcal{O}$: representing I as $\langle e_1, \ldots, e_N \rangle$ and writing $e_{N+1} = e$, then

$$\tilde{I} = \begin{cases} \langle e_1, \ldots, e_N, e_{N+1} \rangle & \text{if } K \cap P_e^* \neq \emptyset, \\ I & \text{otherwise,} \end{cases}$$

$$(\tilde{\sigma}, \langle \tilde{\tau_k} \mid k \in \tilde{I} \rangle) = \begin{cases} ext(\tilde{I}, \sigma, \langle \tau_{e_1}, \ldots, \tau_{e_N}, \langle\rangle \rangle) & \text{if } K \cap P_e^* \neq \emptyset, \\ (\sigma, \langle \tau_k \mid k \in I \rangle) & \text{otherwise.} \end{cases}$$

In particular, the index $U(\tilde{I}, \tilde{\sigma}, \langle \tilde{\tau_k} \mid k < \max I \rangle)$ of $\tilde{K}$ can be determined uniformly from the intersection system $I, \sigma, \langle \tau_k \mid k \in I \rangle$ using $\mathcal{O}$ as an oracle.

Now we proceed with the construction. As K is Σ_1^1, there is e_0 such that $K = P_{e_0}^*$.

Stage $n = 0$: Define

$$K_0 := K, \qquad \sigma_0 := \langle\rangle, \qquad \tau_{0,k} := \langle\rangle, \qquad j(0) := e_0, \qquad I_0 := \{e_0\}.$$

Note that $P_{j(0)}^* = K_0 = \bigcap_{k \in I_0} P_k^*[\sigma_0, \tau_{0,k}]$.

Stage $n = 3e + 1$: Let $I_n, \sigma_n, \langle \tau_{n,k} \mid k \in I_n \rangle$ be the result of adding P_e^* to the intersection system $I_{n-1}, \sigma_{n-1}, \langle \tau_{n-1,k} \mid k \in I_{n-1} \rangle$, and let $K_n := \bigcap_{k \in I_n} P_k^*[\sigma_n, \tau_{n,k}]$ and $j(n)$ be an index for K_n.

Stage $n = 3e + 2$: At this stage we encode $A(e)$ into B.

By construction,

$$P_{j(n-1)}^* = K_{n-1} = \bigcap_{k \in I_{n-1}} P_k^*[\sigma_{n-1}, \tau_{n-1,k}] \neq \emptyset.$$

As K_{n-1} is uncountable, there are infinitely many pairwise-incompatible extensions of σ_{n-1} which extend to elements of K_{n-1}. Thus, let

$$\sigma_n := \rho_{A(e)}(\sigma_{n-1}, j(n-1)).$$

Define

$$\begin{aligned} K_n &:= \bigcap_{k \in I_{n-1}} P_k^*[\sigma_n, \tau_{n-1,k}] = P_{U(\sigma_n, I_{n-1}, \langle \tau_{n-1,0}, \ldots, \tau_{n-1,n-1} \rangle)}, \\ \tau_{n,k} &:= \tau_{n-1,k}, \qquad \text{(for all } k) \\ I_n &:= I_{n-1}, \\ j(n) &:= U(\sigma_n, I_{n-1}, \langle \tau_{n-1,0}, \ldots, \tau_{n-1,n-1} \rangle). \end{aligned}$$

Stage $n = 3^{b+1} \cdot 5^e \cdot 7^f$: Suppose $b \in \mathcal{O}$. Let $m \in \mathbb{N}$ be the least natural number for which there are $Y_1, Y_2 \in K_{n-1}$ such that $\varphi_f^{(1),H_b^{Y_1}}(2^e \cdot 3^m)$ and $\varphi_f^{(1),H_b^{Y_2}}(2^e \cdot 3^m)$ are both defined and unequal. For $i \in \{0,1\}$, let

$$K_{n-1}^i = \{Y \in K_{n-1} \mid \varphi_f^{(1),H_b^{Y_1}}(2^e \cdot 3^m)\downarrow = i\}.$$

Because $K_{n-1}^0 \cap K_{n-1}^1 = \emptyset$, there is a least $k \in \mathbb{N}$ and $i \in \{0,1\}$ such that $\{Y \in K_{n-1}^0 \mid Y(k) = i\}$ and $\{Y \in K_{n-1}^1 \mid Y(k) \neq i\}$ are nonempty. Let $i_0 = i$ and $i_1 = 1 - i$.

Let $I_n, \sigma_n, \langle \tau_{n,k} \mid k \in I_n \rangle$ be the result of adding the (uniformly in b, e, f, m, k,and i, given $Z(m)$) Σ_1^1 class $\{Y \in \{0,1\}^{\mathbb{N}} \mid \varphi_f^{(1),H_b^Y}(2^e \cdot 3^m)\downarrow \neq Z(m) \wedge Y(k) \neq i_{Z(m)}\}$ to the intersection system $I_{n-1}, \sigma_{n-1}, \langle \tau_{n-1,k} \mid k \in I_{n-1} \rangle$, and let $K_n := \bigcap_{k \in I_n} P_k^*[\sigma_n, \tau_{n,k}]$ and $j(n)$ be an index for K_n.

If $b \notin \mathcal{O}$ or no such m exists, do nothing, i.e., let

$$K_n := K_{n-1}, \qquad \sigma_n := \sigma_{n-1}, \qquad \tau_{n,k} := \tau_{n-1,k},$$
$$j(n) := j(n-1), \qquad I_n := I_{n-1}.$$

All Other Stages n: Do nothing, i.e., let

$$K_n := K_{n-1}, \qquad \sigma_n := \sigma_{n-1}, \qquad \tau_{n,k} := \tau_{n-1,k},$$
$$j(n) := j(n-1), \qquad I_n := I_{n-1}.$$

This completes the construction.

Define

$$B := \bigcup_{n \in \mathbb{N}} \sigma_n \quad \text{and} \quad g_k := \bigcup_{n \in \mathbb{N}} \tau_{n,k}.$$

We start by claiming $B \in \bigcap_{n \in \mathbb{N}} K_n$: by construction, for $k \in \bigcap_{n \in \mathbb{N}} I_n$, we have $B \oplus g_k \in P_k$, showing $B \in P_k^*$. Additionally, by construction $B \in P_k^*[\sigma_n, \tau_{n,k}]$ for every n and every $k \in \bigcap_{n \in \mathbb{N}} I_n$, so $B \in \bigcap_{k \in I_n} P_k^*[\sigma_n, \tau_{n,k}] = K_n$. Thus, $B \in \bigcap_{n \in \mathbb{N}} K_n$. In particular, $B \in K_0 = K$, so $\omega_1^B = \omega_1^{\mathrm{CK}}$.

If $Z \leq_{\mathrm{HYP}} B$, then Lemma 2.5 shows there are $c \in \mathcal{O}^B$ and $e \in \mathbb{N}$ such that $Z = (H_b^B)_e$. Because $\omega_1^B = \omega_1^{\mathrm{CK}}$, there exists $b \in \mathcal{O}$ such that $|b| = |c|$ and hence $H_b^B \equiv_{\mathrm{T}} H_c^B$ by Spector's Uniqueness Theorem [11, Corollary II.4.6, p. 40]. Let f be an index such that $\varphi_f^{(1),H_b^B} = H_c^B$, so that $Z = (\varphi_f^{(1),H_b^B})_e$. By construction, at Stage $n = 3^{b+1} \cdot 5^e \cdot 7^f$ it must have been the case that no m and $Y_1, Y_2 \in K_{n-1}$ existed with $\varphi_f^{(1),H_b^{Y_1}}(2^e \cdot 3^m)$ and $\varphi_f^{(1),H_b^{Y_2}}(2^e \cdot 3^m)$ both defined and unequal. In particular, $\varphi_f^{(1),H_b^B}$ is a Σ^1_1 singleton, and so hyperarithmetical. But then $H^B_c \equiv_{\mathrm{T}} H^B_b$ is hyperarithmetical, hence $Z = (H^B_c)_e$ is hyperarithmetical, a contradiction. Thus, $Z \not\leq_{\mathrm{HYP}} B$.

We now make the following observations: assuming $j(n-1)$ is known (and utilizing the implicit course-of-values procedure to yield $I_{n-1}, \sigma_{n-1}, \langle \tau_{n-1,k} \rangle_{k\in\mathbb{N}}$), then...

...in Stage $n = 3e+1$, the determination of $I_n, \sigma_n, \langle \tau_{n,k} \rangle_{k\in\mathbb{N}}$ (and hence also $j(n)$) is recursive in $\mathcal{O}$ by Proposition 2.4.

...in Stage $n = 3e+2$, the determination of $I_n, \sigma_n, \langle \tau_{n,k} \rangle_{k\in\mathbb{N}}$ (and hence also $j(n)$) is recursive in A (by construction) or $B \oplus \mathcal{O}$ (by determining the unique i for which $\rho_i(\sigma_{n-1}, j(n-1)) \subset B$) by Proposition 2.4.

...in Stage $n = 3^{b+1} \cdot 5^e \cdot 7^f$, the determination of $I_n, \sigma_n, \langle \tau_{n,k} \rangle_{k\in\mathbb{N}}$ (and hence also $j(n)$) is recursive in $B \oplus \mathcal{O}$ (the determination of whether $b \in \mathcal{O}$ and whether there exists an m and $Y_1, Y_2 \in K_{n-1}$ for which $\varphi_f^{(1),H_b^{Y_1}}(2^e \cdot 3^m)$ and $\varphi_f^{(1),H_b^{Y_2}}(2^e \cdot 3^m)$ are both defined and unequal may be performed recursively in $\mathcal{O}$ since it corresponds to checking whether a particular Σ^1_1 class is nonempty, and once the least such m is found, we may determine the least k and $i \in \{0,1\}$ for which $\{Y \in K^0_{n-1} \mid Y(k) = i\}$ and $\{Y \in K^1_{n-1} \mid Y(k) = 1-i\}$ are nonempty; finally, checking whether $B(k) = i$ or $B(k) = 1-i$ determines whether we intersected $\{Y \in \{0,1\}^{\mathbb{N}} \mid \varphi_f^{(1),H_b^Y}(2^e \cdot 3^m) \downarrow = 0 \wedge Y(k) = i\}$ or $\{Y \in \{0,1\}^{\mathbb{N}} \mid \varphi_f^{(1),H_b^Y}(2^e \cdot 3^m) \downarrow = 1 \wedge Y(k) = 1-i\}$, respectively) or A (as before, the determination of whether $b \in \mathcal{O}$ and of the existence of such an m may be done recursively in $\mathcal{O} \leq_{\mathrm{T}} A$, and $Z \leq_{\mathrm{T}} A$).

...in all other Stages n, the determination of $I_n, \sigma_n, \langle \tau_{n,k} \rangle_{k\in\mathbb{N}}$ (and hence also $j(n)$) is recursive.

In particular, $j \leq_{\mathrm{T}} A$ and $j \leq_{\mathrm{T}} B \oplus \mathcal{O}$.

We make the following final observations:

- $A \leq_{\mathrm{T}} j \oplus \mathcal{O}$ as $A(e) = i$ if and only if $j(n) = U(\rho_i(\sigma_{n-1}, j(n-1)), I_{n-1}, \langle \tau_{n-1,0}, \ldots, \tau_{n-1,n-1} \rangle)$, where $n = 3e+2$.
- $\mathcal{O}^B \leq_{\mathrm{T}} j \oplus \mathcal{O}$ as $B \in P^*_e$ if and only if $v(j(n-1), e) \notin \{i \mid P^*_i = \emptyset\} \equiv_{\mathrm{T}} \mathcal{O}$. The determination $v(j(n-1), e) \notin \{i \mid P^*_i = \emptyset\} \equiv_{\mathrm{T}} \mathcal{O}$ can be made recursively in $j \oplus \mathcal{O}$.

Thus, we find that

$$A \leq_{\mathrm{T}} j \oplus \mathcal{O} \leq_{\mathrm{T}} B \oplus \mathcal{O} \leq_{\mathrm{T}} \mathcal{O}^B \leq_{\mathrm{T}} j \oplus \mathcal{O} \leq_{\mathrm{T}} A$$

so we have Turing equivalence throughout. □

The following corollary is originally due to Macintyre [7, Theorem 3, p. 9].

Corollary 2.6: *Suppose A is a real such that $\mathcal{O} \leq_{\mathrm{T}} A$. Then there exists B such that*

$$A \equiv_{\mathrm{T}} \mathcal{O}^B \equiv_{\mathrm{T}} B \oplus \mathcal{O}.$$

The following corollary is "folklore", being unpublished but known to researchers and stated in [1, Exercise 2.5.6, p. 40] without proof or references. Other than [1, Exercise 2.5.6, p. 40] we have not seen any statement of Corollary 2.7 in the literature.

Corollary 2.7: *Suppose K is a nonempty Σ^1_1 class. Then there exists $B \in K$ such that $\mathcal{O} \equiv_{\mathrm{T}} \mathcal{O}^B \equiv_{\mathrm{T}} B \oplus \mathcal{O}$.*

Proof: If K is uncountable, then we apply Theorem 2.1 with $Z = A = \mathcal{O}$.

If K is countable, then its elements are hyperarithmetical [11, Theorem III.6.2, p. 72] and so any $B \in K$ satisfies $\mathcal{O} \equiv_{\mathrm{T}} \mathcal{O}^B \equiv_{\mathrm{T}} B \oplus \mathcal{O}$. □

We can generalize Theorem 2.1, replacing the real Z by a sequence of reals, as follows.

Theorem 2.8: *Suppose K is an uncountable Σ^1_1 class and Z and A are reals such that $Z \oplus \mathcal{O} \leq_{\mathrm{T}} A$ and $0 <_{\mathrm{HYP}} (Z)_k$ for each $k \in \mathbb{N}$. Then there exists $B \in K$ such that*

$$A \equiv_{\mathrm{T}} \mathcal{O}^B \equiv_{\mathrm{T}} B \oplus \mathcal{O}$$

and $(Z)_k \not\leq_{\mathrm{HYP}} B$ for all k.

Proof: The proof of Theorem 2.1 may be adapted by replacing Stage $n = 3^{b+1} \cdot 5^e \cdot 7^f$ with $n = 3^{b+1} \cdot 5^e \cdot 7^f \cdot 11^k$ and replacing therein Z with $(Z)_k$. □

3. Posner-Robinson for Turing Degrees of Hyperjumps

Theorem 3.1: (Posner-Robinson for Turing Degrees of Hyperjumps). *Suppose Z and A are reals such that $Z \oplus \mathcal{O} \leq_{\mathrm{T}} A$ and $0 <_{\mathrm{HYP}} Z$. Then there exists B such that*

$$A \equiv_{\mathrm{T}} \mathcal{O}^B \equiv_{\mathrm{T}} B \oplus Z \equiv_{\mathrm{T}} B \oplus \mathcal{O}.$$

Theorem 3.1 is essentially due to Slaman [13]. The rest of this section is devoted to a proof of Theorem 3.1. The key to the proof is a forcing notion known as Kumabe-Slaman forcing, which was originally introduced in [12].

3.1. *Kumabe-Slaman forcing*

In order to prove Theorem 3.1, we will use Turing functionals and an associated notion of forcing to construct the desired B.

Definition 3.2: (Turing Functionals). [9,12] A **Turing functional** Φ is a set of triples $(x, y, \sigma) \in \mathbb{N} \times \{0,1\} \times \{0,1\}^*$ (called **computations in** Φ) such that if $(x, y_1, \sigma_1), (x, y_2, \sigma_2) \in \Phi$ and σ_1 and σ_2 are compatible, then $y_1 = y_2$ and $\sigma_1 = \sigma_2$.

A Turing functional Φ is **use-monotone** if:

(i) For all (x_1, y_1, σ_1) and (x_2, y_2, σ_2) are elements of Φ and $\sigma_1 \subset \sigma_2$, then $x_1 < x_2$.
(ii) For all x_1 and $(x_2, y_2, \sigma_2) \in \Phi$ where $x_2 > x_1$, then there are y_1 and σ_1 such that $\sigma_1 \subseteq \sigma_2$ and $(x_1, y_1, \sigma_1) \in \Phi$.

Remark 3.3: Despite the terminology, a Turing functional Φ is not assumed to be recursive or even recursively enumerable.

Definition 3.4: (Computations along a Real). [9,12] Suppose Φ is a Turing functional and $X \in \{0,1\}^{\mathbb{N}}$. Then $(x, y, \sigma) \in \Phi$ is a **computation along** X if $\sigma \subset X$, in which case we write $\Phi(X)(x) = y$. If for every $x \in \mathbb{N}$ there exists $y \in \{0,1\}$ and $\sigma \subset X$ such that $(x, y, \sigma) \in \Phi$, then $\Phi(X)$ defines an element of $\{0,1\}^{\mathbb{N}}$ (otherwise it is a partial function).

Lemma 3.5: *Suppose Φ is a Turing functional, $X \in \{0,1\}^{\mathbb{N}}$, and $\Phi(X) \in \{0,1\}^{\mathbb{N}}$. Then*

$$\Phi(X) \leq_{\mathrm{T}} \Phi \oplus X.$$

Proof: Obvious from the definition of $\Phi(X)$. □

Definition 3.6: (Kumabe-Slaman Forcing). [9,12] Define the poset $(\mathbb{P}, \leq)$ as follows:

(i) Elements of $\mathbb{P}$ are pairs $(\Phi, \mathbf{X})$ where Φ is a finite use-monotone Turing functional and $\mathbf{X}$ is a finite subset of $\{0,1\}^{\mathbb{N}}$.
(ii) If $p = (\Phi_p, \mathbf{X}_p)$ and $q = (\Phi_q, \mathbf{X}_q)$ are in $\mathbb{P}$, then $p \leq q$ if
  (a) $\Phi_p \subseteq \Phi_q$ and for all $(x_q, y_q, \sigma_q) \in \Phi_q \setminus \Phi_p$ and all $(x_p, y_p, \sigma_p) \in \Phi_p$, the length of σ_q is greater than the length of σ_p.
  (b) $\mathbf{X}_p \subseteq \mathbf{X}_q$.
  (c) For every x, y, and $X \in \mathbf{X}$, if $\Phi_q(X)(x) = y$, then $\Phi_p(X)(x) = y$.

In other words, a stronger condition than p can add longer computations to Φ_p, provided they don't apply to any element of $\mathbf{X}_p$.

In the remainder of §3, we will be discussing Kumabe-Slaman forcing over countable ω-models of ZFC.[1] Unlike in the forcing constructions in axiomatic set theory, it will be important here that the countable ground model M is *not* well-founded. We now introduce some conventions for discussing such models.

Let M be a countable non-well-founded ω-model of ZFC. Let $\theta(x_1,\ldots,x_n)$ be a sentence in the language of ZFC with parameters $x_1,\ldots,x_n$ from M. We write $\theta^M(x_1,\ldots,x_m)$ or $M \models \theta(x_1,\ldots,x_n)$ to mean that $\theta(x_1,\ldots,x_n)$ holds in M. In particular, $x_1 \in^M x_2$ means that $M \models x_1 \in x_2$, etc. We tacitly identify the natural number system of M with the standard natural number system, the reals of M with standard reals, etc. In particular, let $\mathbb{P}^M$ be the set of pairs (Φ, X) such that $M \models$ "(Φ, X) is a Kumabe-Slaman forcing condition". In this case, Φ is identified with a finite Turing functional, X is identified with a finite set of reals belonging to M, etc., so (Φ, X) actually *is* a Kumabe-Slaman forcing condition.

The key property of Kumabe-Slaman Forcing is the following:

Lemma 3.7: *[9, based on Lemma 3.10, p. 23] Suppose M is an ω-model of* ZFC*, $D \in M$ is dense in $\mathbb{P}^M$, and $X_1,\ldots,X_n \in \{0,1\}^{\mathbb{N}}$. Then for any $p \in \mathbb{P}^M$, there exists $q \geq p$ such that $q \in D$ and Φ_q does not add any new computations along any X_k.*

Proof: Suppose $p = (\Phi_p, \mathbf{X}_p) \in \mathbb{P}^M$. Say that an n-tuple of strings $\vec{\tau}$ is *essential* for (p, D) if $q > p$ and $q \in D$ implies the existence of $(x, y, \sigma) \in \Phi_q \setminus \Phi_p$ such that σ is compatible with some component of $\vec{\tau}$, i.e., any extension of p in D adds a computation along a string compatible with a component of $\vec{\tau}$. Being essential for (p, D) is definable in M.

Define

$$T_n(p, D) := \{\vec{\tau} \in (\{0,1\}^*)^n \mid \vec{\tau} \text{ is essential for } (p, D) \text{ and } |\tau_1| = \cdots = |\tau_n|\}.$$

Being essential for (p, D) is closed under taking (component-wise) initial segments, so $T_n(p, D)$ is a finitely branching tree in M.

[1]Here ZFC denotes Zermelo-Fraenkel Set Theory with the Axiom of Choice. However, for the purposes of this chapter, our ω-models need not satisfy ZFC but only a small subsystem of ZFC or actually of second-order arithmetic.

Suppose for the sake of a contradiction that for every $q > p$, either $q \notin D$ or else q adds a new computation along some X_k. We claim that $\langle X_1 \restriction m, \ldots, X_n \restriction m \rangle$ is essential for (p, D) for all $m \in \mathbb{N}$. Given $q > p$ with $q \in D$, by hypothesis there is some computation $(x, y, \sigma) \in \Phi_q \setminus \Phi_p$ along some X_k. This means that $\sigma \subset X_k$ (outside of M), so σ is compatible with $X_k \restriction m$.

This shows that $T_n(p, D)$ is infinite. As M is a model of ZFC, it follows that $T_n(p, D)$ has a path through it. The requirement that the components of any element of $T_n(p, D)$ are of the same length implies that such a path is of the form $(Y_1, \ldots, Y_n)$ for $Y_1, \ldots, Y_n \in M \cap \{0, 1\}^{\mathbb{N}}$.

Consider $p_1 = (\Phi_p, \mathbf{X}_p \cup \{Y_1, \ldots, Y_n\})$. Suppose $q \geq p_1$ and $q \in D$. Each n-tuple $\langle Y_1 \restriction m, \ldots, Y_n \restriction m \rangle$ is essential for (p, D) for each m, so there exists $(x_m, y_m, \sigma_m) \in \Phi_q \setminus \Phi_p$ such that σ_m is compatible with $Y_k \restriction m$ for some k. As Φ_q is finite, letting m be sufficiently large shows that there is $(x, y, \sigma) \in \Phi_q \setminus \Phi_p$ for which σ is an initial segment of Y_k for some k. However, this is not possible since $q \leq p_1$ implies $Y_k \in \mathbf{X}_q$. This provides the needed contradiction. □

Suppose G is an M-generic filter for $\mathbb{P}^M$. Then for every X

$$X \subseteq^M \mathbb{N} \iff \text{there is } p \in G \text{ with } X \in \mathbf{X}_p$$

since for any $X \subseteq^M \mathbb{N}$, the set $\{p \in \mathbb{P}^M \mid (\emptyset, \{X\}) \leq p\}^M$ is a dense open subset of $\mathbb{P}^M$ in M. Thus, the essential parts of an M-generic filter G are the Turing functionals Φ_p for $p \in G$.

Definition 3.8: A Turing functional Φ is M**-generic for** $\mathbb{P}^M$ if and only if there exists an M-generic filter G such that

$$(x, y, \sigma) \in \Phi \iff \text{there exists } p \in G \text{ such that } (x, y, \sigma) \in^M \Phi_p.$$

Φ may be identified with an element $(\dot{\Phi})_G$ in $M[G]$, where

$$M \models \dot{\Phi} = \{(p, \dot{c}) \mid p \in \mathbb{P}^M \wedge c \in \Phi_p\}$$

and $\dot{c}$ is a *canonical name* for $c \in M$, defined by transfinite recursion in M to be the unique element in M for which

$$M \models \dot{c} = \mathbb{P}^M \times \{\dot{b} \mid b \in c\}.$$

By defining an M-generic Turing functional Φ for $\mathbb{P}^M$ by means of approximations, Lemmas 3.7, and 3.9 allow us to meet dense sets without affecting $\Phi(Z)$, which can then be arranged independently.

3.2. *Proof of Posner-Robinson for hyperjumps*

Now we proceed with the proof of Theorem 3.1:

Lemma 3.9: *Suppose Z and A are reals such that $Z \oplus \mathcal{O} \leq_{\mathrm{T}} A$ and $0 <_{\mathrm{HYP}} Z$. Then there exists a (code for a) countable ω-model M of* ZFC *such that $\mathcal{O}^M \equiv_{\mathrm{T}} A$ and $Z \notin M$.*

Proof: The set of codes for countable ω-models of ZFC is Σ^1_1, so the existence of a code of such an M follows from Theorem 2.1. □

Proof: [Proof of Theorem 3.1] The main idea of the proof is due to Slaman [13].

We shall construct an M-generic Turing functional Φ with $B = \Phi$ the desired real. Assume without loss of generality that no initial segment of Z is an initial segment of $\mathcal{O}$. By arranging for $\Phi(Z) \in \{0,1\}^{\mathbb{N}}$ and $\Phi(Z) = \mathcal{O}^{\Phi}$ and $\Phi(\mathcal{O}) = A$, this will complete the proof.

By Lemma 3.9, there exists a countable ω-model M of ZFC such that $\mathcal{O}, Z \notin M$ and $\mathcal{O}^M \equiv_{\mathrm{T}} A$. Without loss of generality, $M = \langle \omega, E \rangle$.

Let $D_0, D_1, D_2, \ldots$ be an enumeration, recursive in A, of the dense open subsets of $\mathbb{P}^M$ in M (M is countable and $\mathcal{O}^M \equiv_{\mathrm{T}} A$, so this is possible). To construct our M-generic Φ, we approximate it by the finite use-monotone Turing functional parts of Kumabe-Slaman conditions

$$p_0 \leq p_1 \leq \cdots \leq p_n \leq \cdots .$$

During our construction, we alternate between meeting dense sets, arranging for $\Phi(\mathcal{O}) = A$, and arranging for $\Phi(Z) \equiv_{\mathrm{T}} \mathcal{O}^{\Phi}$.

Stage $n = 0$: Define $p_0 := (\emptyset, \emptyset)$.

Stage $n = 2^m$: Suppose p_{n-1} has been constructed. By Lemma 3.7, there exists $q \in D_n$ extending p_{n-1} which does not add any new computations along Z or $\mathcal{O}$. Let p_n be the least such condition.

Stage $n = 2^m \cdot 3$: We extend p_{n-1} to p_n by adding $(m, A(m), \sigma)$ where $\sigma \subset \mathcal{O}$ is a sufficiently long initial segment of $\mathcal{O}$ (i.e., the shortest initial segment of $\mathcal{O}$ which is longer than any existing strings in elements of $\Phi_{p_{n-1}}$).

Stage $n = 2^m \cdot 5$: Suppose p_{n-1} has been constructed. By construction, there is no y and $\sigma \subset Z$ such that $(m, y, \sigma) \in \Phi_{p_{n-1}}$. Now proceed as follows:

Substage 1: Consider the set D (in M) containing all $q \in \mathbb{P}^M$ such that one of the following conditions hold:

(i) $q \Vdash (m$ encodes a Φ-recursive linear order on $\omega \wedge m \in \mathcal{O}^{\Phi} \wedge \exists \alpha\ (\alpha \in \mathrm{Ord}^M \wedge |m| = \alpha))$,

(ii) $q \Vdash (m$ encodes a Φ-recursive linear order on $\omega \wedge m \notin \mathcal{O}^{\Phi})$, or

(iii) $q \Vdash \neg(m$ encodes a Φ-recursive linear order on $\omega)$.

D is dense. By Lemma 3.7, there exists $q \in D$ extending p_{n-1} which does not add any new computations along Z or $\mathcal{O}$. Let q be minimal with that property.

Substage 2: Extend q to p_n by adding (m, y, σ), where $\sigma \subset Z$ is a sufficiently long initial segment of Z (i.e., the shortest initial segment of Z which is longer than any existing strings in elements of Φ_q) and y depends on the following cases:

Case 1: If $q \Vdash (m$ encodes a Φ-recursive linear order on $\omega \wedge m \in \mathcal{O}^{\Phi} \wedge \exists \alpha\ (\alpha \in \mathrm{Ord}^M \wedge |m| = \alpha))$, then we break into two subcases:

Case 1a: If α is in the standard part of Ord^M, then α is actually an ordinal and m *does* encode a Φ-recursive linear order on ω. Thus, set $y := 1$.

Case 1b: If α is not in the standard part of Ord^M, then α is not actually well-ordered (it is only well-ordered when viewed in M) so m *does not* encode a Φ-recursive linear order on ω. Thus, set $y := 0$.

Case 2: If $q \Vdash (m$ encodes a Φ-recursive linear order on $\omega \wedge m \notin \mathcal{O}^{\Phi})$, then m cannot encode a Φ-recursive well-ordering of ω. Thus, set $y := 0$.

Case 3: If $q \Vdash \neg(m$ encodes a Φ-recursive linear order on $\omega)$, then set $y := 0$.

All Other Stages n: Let $p_n = p_{n-1}$.

Define Φ to be the unique set such that

$$(x, y, \sigma) \in \Phi \iff \text{there exists } n \in \mathbb{N} \text{ such that } (x, y, \sigma) \in^M \Phi_{p_n}.$$

Thanks to Stages $n = 2^m$ and the observation that $G := \{q \in \mathbb{P}^M \mid \exists n\ (q \leq p_n)\}$ is consequently M-generic, Φ is an M-generic Turing functional. Thanks to Stages $n = 2^m \cdot 3$, $\Phi(\mathcal{O}) = A$. Thanks to Stages $n = 2^m \cdot 5$, $\Phi(Z) = \mathcal{O}^{\Phi}$.

We also note that in the above construction of Φ, (assuming p_{n-1} is given)...

...Stage $n = 2^m$ is recursive in $\mathcal{O}^M \oplus Z \oplus \mathcal{O} \leq_{\mathrm{T}} A$,

... Stage $n = 2^m \cdot 3$ is recursive in $\mathcal{O} \leq_{\mathrm{T}} A$,
... Stage $n = 2^m \cdot 5$ (Substage 1) is recursive in $\mathcal{O} \oplus \mathcal{O}^M \oplus Z \leq_{\mathrm{T}} A$,
... Stage $n = 2^m \cdot 5$ (Substage 2) is recursive in $Z \leq_{\mathrm{T}} A$, and
... Stage n (for all other n) is recursive.

Thus,

$$\Phi \leq_{\mathrm{T}} M \oplus (\mathcal{O}^M \oplus Z) \oplus A \leq_{\mathrm{T}} A.$$

Applying Lemma 3.5, we find

$$A = \Phi(\mathcal{O}) \leq_{\mathrm{T}} \mathcal{O} \oplus \Phi \leq_{\mathrm{T}} \mathcal{O}^{\Phi} \equiv_{\mathrm{T}} \Phi(Z) \leq_{\mathrm{T}} Z \oplus \Phi \leq_{\mathrm{T}} Z \oplus A \equiv_{\mathrm{T}} A$$

so we have Turing equivalence throughout. $B = \Phi$ is hence the desired real. □

Theorem 3.1 can be generalized, replacing the real Z by a sequence of reals.

Theorem 3.10: *Suppose Z and A are reals such that $Z \oplus \mathcal{O} \leq_{\mathrm{T}} A$ and $0 <_{\mathrm{HYP}} (Z)_k$ for every $k \in \mathbb{N}$. Then there exists B such that for every $k \in \mathbb{N}$*

$$A \equiv_{\mathrm{T}} \mathcal{O}^B \equiv_{\mathrm{T}} B \oplus (Z)_k \equiv_{\mathrm{T}} B \oplus \mathcal{O}.$$

Proof: The proof of Theorem 3.1 may be adapted by making the following adjustments. First, we replace the use of Lemma 3.9 with the following lemma:

Lemma 3.11: *Suppose Z and A are reals such that $Z \oplus \mathcal{O} \leq_{\mathrm{T}} A$ and $0 <_{\mathrm{HYP}} (Z)_k$ for every $k \in \mathbb{N}$. Then there exists a (code for a) countable ω-model M of* ZFC *such that $\mathcal{O}^M \equiv_{\mathrm{T}} A$ and $(Z)_k \notin M$ for every $k \in \mathbb{N}$.*

Proof: Replace the usage of Theorem 2.1 in the proof of Lemma 3.9 with Theorem 2.8. □

This yields a (code for a) countable ω-model M of ZFC such that $\mathcal{O}, (Z)_0, (Z)_1, \ldots \notin M$ and $\mathcal{O}^M \equiv_{\mathrm{T}} A$. We assume without loss of generality that $\mathcal{O} \neq (Z)_k$ for each k.

The adjustments to the construction are the following:

- In Stages $n = 2^m$ and $n = 2^m \cdot 3$, we avoid adding new computations along $(Z)_0, \ldots, (Z)_n$ and $\mathcal{O}$.

- Replace Stage $n = 2^m \cdot 5$ with Stages $n = 2^m \cdot 5^{k+1}$, and at the beginning of Stage $n = 2^m \cdot 5^{k+1}$, first check if there exists y and $\sigma \subset (Z)_k$ such that $(m, y, \sigma) \in \Phi_{p_{n-1}}$. If such a y and σ are found, do nothing and proceed to the next stage. Otherwise, proceed as in Stage $n = 2^m \cdot 5$ of the proof of Theorem 3.1, with the same adjustment of avoiding adding new computations along $(Z)_0, \ldots, (Z)_n$ and $\mathcal{O}$ as above.

Note that it is no longer necessarily the case that $\Phi((Z)_k) = \mathcal{O}^\Phi$ for every $k \in \mathbb{N}$, as early stages may have added computations to Φ which make $\Phi((Z)_k)$ disagree with $\mathcal{O}^\Phi$. However, after Stage k, no other stages add new computations along $(Z)_k$ except for those purposely added (i.e., in Stages $n = 2^m \cdot 5^{k+1}$). It follows that $\Phi((Z)_k)$ and $\mathcal{O}^\Phi$ differ only on a finite set of indices, so $\Phi((Z)_k) \equiv_{\mathrm{T}} \mathcal{O}^\Phi$.

In the resulting construction of Φ, (assuming p_{n-1} is given)

...Stage $n = 2^m$ is recursive in $\mathcal{O}^M \oplus \bigoplus_{i=0}^n (Z)_i \oplus \mathcal{O} \leq_{\mathrm{T}} A$,
...Stage $n = 2^m \cdot 3$ is recursive in $\mathcal{O} \leq_{\mathrm{T}} A$,
...Stage $n = 2^m \cdot 5^{k+1}$ (Substage 1) is recursive in $\mathcal{O} \oplus \mathcal{O}^M \oplus \bigoplus_{i=0}^n (Z)_i \leq_{\mathrm{T}} A$,
...Stage $n = 2^m \cdot 5^{k+1}$ (Substage 2) is recursive in $(Z)_k \leq_{\mathrm{T}} A$, and
...Stage n (for all other n) is recursive.

Thus,

$$\Phi \leq_{\mathrm{T}} M \oplus (\mathcal{O}^M \oplus Z) \oplus A \equiv_{\mathrm{T}} A.$$

The proof concludes as in the proof of Theorem 3.1. □

4. Open Problems

In light of Theorems 2.1, and 3.1, it is natural to ask whether they can be combined into one theorem. In other words, for which uncountable Σ_1^1 classes $K \subseteq \{0,1\}^{\mathbb{N}}$ do the following properties hold?

Property 4.1: Suppose Z and A are reals such that $Z \oplus \mathcal{O} \leq_{\mathrm{T}} A$ and $0 <_{\mathrm{HYP}} (Z)_k$ for every $k \in \mathbb{N}$. Then there exists $B \in K$ such that

$$A \equiv_{\mathrm{T}} \mathcal{O}^B \equiv_{\mathrm{T}} B \oplus Z \equiv_{\mathrm{T}} B \oplus \mathcal{O}.$$

Property 4.2: Suppose Z and A are reals such that $Z \oplus \mathcal{O} \leq_{\mathrm{T}} A$ and $0 <_{\mathrm{HYP}} (Z)_k$ for every $k \in \mathbb{N}$. Then there exists $B \in K$ such that for every k

$$A \equiv_{\mathrm{T}} \mathcal{O}^B \equiv_{\mathrm{T}} B \oplus (Z)_k \equiv_{\mathrm{T}} B \oplus \mathcal{O}.$$

The following theorem answers some special cases of this problem.

Theorem 4.3: *Let $L_{\mathrm{T}} = \{X \mid \mathcal{O}^X \equiv_{\mathrm{T}} X \oplus \mathcal{O}\}$. Suppose K is an uncountable Σ^1_1 class which is Turing degree upward closed in L_{T}, i.e., whenever $X, Y \in L_{\mathrm{T}}$, $X \in K$, and $X \leq_{\mathrm{T}} Y$, then there is $Y_0 \in K$ such that $Y \equiv_{\mathrm{T}} Y_0$. Then K has Properties 4.1, and 4.2.*

Proof: This theorem is analogous to [3, Lemma 3.3]. By Theorem 2.1, let C be such that

$$A \equiv_{\mathrm{T}} \mathcal{O}^C \equiv_{\mathrm{T}} C \oplus Z \equiv_{\mathrm{T}} C \oplus \mathcal{O}. \tag{$*$}$$

Theorem 2.1, relativized to C, yields $B_0 \in K$ such that

$$\mathcal{O}^C \equiv_{\mathrm{T}} \mathcal{O}^{B_0 \oplus C} \equiv_{\mathrm{T}} B_0 \oplus \mathcal{O}^C. \tag{$\dagger$}$$

Combining (†) and ($*$) shows that $\mathcal{O}^{B_0 \oplus C} \equiv_{\mathrm{T}} B_0 \oplus C \oplus \mathcal{O}$. As $B_0 \leq_{\mathrm{T}} B_0 \oplus C$, there is $B \in K$ such that $B \equiv_{\mathrm{T}} B_0 \oplus C$ by hypothesis. In particular,

$$\mathcal{O}^C \equiv_{\mathrm{T}} \mathcal{O}^B \equiv_{\mathrm{T}} B \oplus \mathcal{O}.$$

Moreover, in combination with ($*$),

$$A \equiv_{\mathrm{T}} \mathcal{O}^B \equiv_{\mathrm{T}} B \oplus \mathcal{O} \equiv_{\mathrm{T}} B \oplus Z.$$

This shows that K has Property 4.1.

To show that K has Property 4.2, repeat the above argument using Theorem 2.8 instead of Theorem 2.1. □

Remark 4.4: The proof of Theorem 4.3 is easily adapted to prove the same result with $L_{\mathrm{T}} = \{X \mid \mathcal{O}^X \equiv_{\mathrm{T}} X \oplus \mathcal{O}\}$ replaced by $L_{\mathrm{HYP}} = \{X \mid \mathcal{O}^X \equiv_{\mathrm{HYP}} X \oplus \mathcal{O}\}$.

The hyperarithmetical analog of the Pseudojump Inversion Theorem [4, Theorem 2.1, p. 601] also remains open. Namely, suppose V_e^X is an effective enumeration of the $\Pi^{1,X}_1$ predicates, uniformly in X. Define the **e-th pseudo-hyperjump** by

$$\mathrm{HJ}_e(X) := X \oplus V_e^X.$$

Does the following result hold?

Conjecture 4.5: *Suppose $e \in \mathbb{N}$ and A is a real such that $\mathcal{O} \leq_{\mathrm{T}} A$. Then there exists B such that*

$$A \equiv_{\mathrm{T}} \mathrm{HJ}_e(B) \equiv_{\mathrm{T}} B \oplus \mathcal{O}. \tag{4.1}$$

Even if Conjecture 4.5 holds, this leaves open the question of characterizing the Σ^1_1 classes $K \subseteq \{0,1\}^{\mathbb{N}}$ with the following properties:

Property 4.6: Suppose $e \in \mathbb{N}$ and A is a real such that $\mathcal{O} \leq_{\mathrm{T}} A$. Then there exists $B \in K$ such that Eq. (4.1) holds.

Property 4.7: Suppose $e \in \mathbb{N}$ and Z and A are reals such that $Z \oplus \mathcal{O} \leq_{\mathrm{T}} A$ and $0 <_{\mathrm{HYP}} (Z)_k$ for each $k \in \mathbb{N}$. Then there exists $B \in K$ such that Eq. (4.1) holds and $(Z)_k \not\leq_{\mathrm{HYP}} B$ for every $k \in \mathbb{N}$.

Property 4.8: Suppose $e \in \mathbb{N}$ and Z and A are reals such that $Z \oplus \mathcal{O} \leq_{\mathrm{T}} A$ and $0 <_{\mathrm{HYP}} Z$. Then there exists $B \in K$ such that

$$A \equiv_{\mathrm{T}} \mathrm{HJ}_e(B) \equiv_{\mathrm{T}} B \oplus Z \equiv_{\mathrm{T}} B \oplus \mathcal{O}.$$

Acknowledgements

We thank Patrick Lutz for pointing out an error in the original manuscript, which we have corrected.

References

1. Chi Tat Chong and Liang Yu, *Recursion Theory: Computational Aspects of Definability*, Walter de Gruyter GmbH & Co KG, 2015, xiii + 306 pages.
2. Leo Harrington, *A powerless proof of a theorem of Silver*, unpublished manuscript, 8 pages, 1976.
3. Hayden R. Jananthan and Stephen G. Simpson, *Pseudojump inversion in special r.b. Π^0_1 classes*, 2020, 19 pages, submitted for publication in *Lecture Notes Series, Institute for Mathematical Sciences, National University of Singapore*, 2020.
4. Carl G. Jockusch and Richard A. Shore, *Pseudojump operators. I: The r.e. case*, Transactions of the American Mathematical Society **275** (1983), no. 2, 599–609.
5. ———, *Pseudo-jump operators. II: Transfinite iterations, hierarchies and minimal covers*, The Journal of Symbolic Logic **49** (1984), no. 4, 1205–1236.
6. Carl G. Jockusch and Robert I. Soare, *Π^0_1 classes and degrees of theories*, Transactions of the American Mathematical Society **173** (1972), 35–56.
7. John M. Macintyre, *Transfinite extensions of Friedberg's completeness criterion 1*, The Journal of Symbolic Logic **42** (1977), no. 1, 1–10.
8. David B. Posner and Robert W. Robinson, *Degrees joining to* $0'$, The Journal of Symbolic Logic **46** (1981), no. 4, 714–722.
9. Jan Reimann and Theodore A. Slaman, *Effective randomness for continuous measures*, 2018, preprints, arXiv:1808.10102v1, 2018, 57 pages, and arXiv:1808.10102v2, 2020, 62 pages.

10. Hartley Rogers, Jr, *Theory of Recursive Functions and Effective Computability*, McGraw-Hill Series in Higher Mathematics, McGraw-Hill, 1967, xix + 482 pages.
11. Gerald E. Sacks, *Higher Recursion Theory*, Perspectives in Mathematical Logic, Springer-Verlag, Berlin, 1990, xv + 344 pages. MR 1080970.
12. Richard A. Shore and Theodore A. Slaman, *Defining the Turing jump*, Mathematical Research Letters **6** (1999), no. 5/6, 711–722.
13. Theodore A. Slaman, private communication, November 2018.

https://doi.org/10.1142/9789819806584_0006

Pseudojump Inversion in Special Π_1^0 Classes of Sets

Hayden R. Jananthan
Lincoln Laboratory Supercomputing Center
Massachusetts Institute of Technology
Lexington, MA 02421, USA
https://jananthan.mit.edu
hayden.r.jananthan@gmail.edu

Stephen G. Simpson
Department of Mathematics
Vanderbilt University
Nashville, TN 37203, USA
https://sgslogic.net
sgslogic@gmail.com

The Friedberg Jump Theorem says that for every real A there is a real B such that $A \oplus 0' \equiv_T B' \equiv_T B \oplus 0'$. A well-known refinement of this says that, given a special Π_1^0 class Q, we can choose B to be an element of Q. In this chapter, we consider analogous refinements of two other well-known theorems: the Posner-Robinson Theorem — for all reals A and Z such that $0 <_T Z$, there is a real B such that $A \oplus Z \oplus 0' \equiv_T B' \equiv_T B \oplus Z \equiv_T B \oplus 0'$ — and the Pseudojump Inversion Theorem — for all $e \in \mathbb{N}$ and all reals A, there is a real B such that $A \oplus 0' \equiv_T B \oplus J_e(B) \equiv_T B \oplus 0'$. We show that in these theorems, B can be found in some special Π_1^0 classes but not in others.

Contents

1. Introduction

Our starting point is the *Friedberg Jump Theorem*:

Theorem 1.1: (Friedberg Jump Theorem). *[8, Theorem 13.3.IX, p. 265] For every real A, there exists B such that*

$$A \oplus 0' \equiv_{\mathrm{T}} B' \equiv_{\mathrm{T}} B \oplus 0'.$$

There are several refinements of the Friedberg Jump Theorem. One such refinement is the *Posner-Robinson Theorem*:

Theorem 1.2: (Posner-Robinson Theorem). *[7, Theorem 1, p. 715] [2, Theorem 2.1, p. 601] Suppose Z and A are reals such that $0 <_{\mathrm{T}} Z \leq_{\mathrm{T}} A \oplus 0'$. Then there exists B such that*

$$A \oplus 0' \equiv_{\mathrm{T}} B' \equiv_{\mathrm{T}} B \oplus Z \equiv_{\mathrm{T}} B \oplus 0'.$$

Another refinement shows that in the Friedberg Jump Theorem the Turing jump operator can be replaced with any pseudojump operator:

Theorem 1.3: (Pseudojump Inversion Theorem). *[2, Theorem 2.1, p. 601] For every real A and $e \in \mathbb{N}$, there exists B such that*

$$A \oplus 0' \equiv_{\mathrm{T}} \mathrm{J}_e(B) \equiv_{\mathrm{T}} B \oplus 0'.$$

By a *special Π^0_1 class*, we mean a nonempty Π^0_1 subset of $\{0,1\}^{\mathbb{N}}$ with no recursive elements. Our results will also hold for nonempty, recursively bounded Π^0_1 subsets of $\mathbb{N}^{\mathbb{N}}$ containing no recursive members, as they are recursively homeomorphic to special Π^0_1 subsets of $\{0,1\}^{\mathbb{N}}$.

It is known [3, following Theorem 2.1, p. 37] that the in Friedberg Jump Theorem the real B may be taken to be a member of an arbitrary special Π^0_1 class:

Theorem 1.4: *[3, following Theorem 2.1, p. 37] If Q is a special Π^0_1 class, then for every real A there exists $B \in Q$ such that*

$$A \oplus 0' \equiv_{\mathrm{T}} B' \equiv_{\mathrm{T}} B \oplus 0'.$$

This motivates the question as to which special Π^0_1 classes Q have the following properties:

Property 1.5: (Posner-Robinson Property). Suppose Z and A are reals such that $0 <_\mathrm{T} Z$. Then there exists $B \in Q$ such that

$$A \oplus Z \oplus 0' \equiv_\mathrm{T} B' \equiv_\mathrm{T} B \oplus Z \equiv_\mathrm{T} B \oplus 0'.$$

Property 1.6: (Pseudojump Inversion Property). For every real A and $e \in \mathbb{N}$, there exists $B \in Q$ such that

$$A \oplus 0' \equiv_\mathrm{T} \mathrm{J}_e(B) \equiv_\mathrm{T} B \oplus 0'.$$

We present partial results in the direction of answering these questions. Namely, we show that CPA, and more generally any Medvedev complete Π^0_1 class, satisfies both the Posner-Robinson Property and the Pseudojump Inversion Property. Next, we show that a Π^0_1 class consisting of Martin-Löf random reals satisfies the Pseudojump Inversion Property but fails the Posner-Robinson Property. Finally, we construct a special Π^0_1 class which fails both the Posner-Robinson Property and the Pseudojump Inversion Property. We do not know whether there exists a special Π^0_1 class which has the Posner-Robinson Property but not the Pseudojump Inversion Property.

Here is an outline of this chapter:

In §2, we present the classical proofs of Theorems 1.2, and 1.3.

In §3, we prove (Theorem 3.3) that CPA has both the Posner-Robinson Property and the Pseudojump Inversion Property.

In §4, we prove some basic results about how satisfaction of the Posner-Robinson Property and the Pseudojump Inversion Property interacts with strong notions of 'equivalence' of special Π^0_1 classes. We then apply these results to CPA in §4.1 to show that any Muchnik complete Π^0_1 class has both the Posner-Robinson Property and the Pseudojump Inversion Property. (On the other hand, every nonzero Muchnik or Medvedev degree contains *a* special Π^0_1 class having both the Posner-Robinson Property and the Pseudojump Inversion Property.)

In §5, we show that special Π^0_1 classes consisting only of Martin-Löf random reals have the Pseudojump Inversion Property (Theorem 5.3) but not the Posner-Robinson Property (Proposition 5.2, due to [4]).

In §6, we construct (using a finite-injury priority argument involving tree maps) a special Π^0_1 class which has neither the Posner-Robinson Property nor the Pseudojump Inversion Property (Corollary 6.5).

The remainder of this section fixes notation and terminology.

$g\colon \subseteq A \to B$ denotes a partial function with domain $\operatorname{dom} g \subseteq A$ and codomain B. For $a \in A$, if $a \in \operatorname{dom} g$ then we say '$g(a)$ converges' or '$g(a)$ is defined' and write $g(a) \downarrow$. Otherwise, we say '$g(a)$ diverges' or '$g(a)$ is undefined' and write $g(a) \uparrow$. If f and g are two partial functions $\subseteq A \to B$ and $a \in A$, then $f(a) \simeq g(a)$ means $(f(a) \downarrow \wedge g(a) \downarrow \wedge f(a) = g(a)) \vee (f(a) \uparrow \wedge g(a) \uparrow)$. We write $f(a) \downarrow= b$ to mean that $f(a) \downarrow$ and $f : a \mapsto b$.

$\mathbb{N}^{\mathbb{N}}$ and $\{0,1\}^{\mathbb{N}}$ denote the Baire and Cantor spaces, respectively, whose elements we sometimes call *reals*. We identify $\{0,1\}^{\mathbb{N}}$ and $\mathcal{P}(\mathbb{N})$ in the usual manner.

If S is a set, then S^* is the set of strings of elements from S. If $s_0, \ldots, s_{n-1} \in S$, then $\sigma := \langle s_0, \ldots, s_{n-1}\rangle \in S^*$ denotes the string of *length* $|\sigma| := n$ defined by $\sigma(k) = s_k$. If $\langle s_0, \ldots, s_{n-1}\rangle$, $\langle t_0, \ldots, t_{m-1}\rangle \in S^*$, then their *concatenation* is $\langle s_0, \ldots, s_{n-1}\rangle^\frown\langle t_0, \ldots, t_{m-1}\rangle := \langle s_0, \ldots, s_{n-1}, t_0, \ldots, t_{m-1}\rangle$. If $\sigma, \tau \in S^*$, then σ is an *initial segment* of τ (equivalently, τ is an extension of σ) written $\sigma \subseteq \tau$, if $\tau \restriction |\sigma| = \sigma$. If $f : \mathbb{N} \to S$ then $\sigma \in S^*$ is an *initial segment* of f (equivalently, f is an *extension* of σ), written $\sigma \subset f$, if $f \restriction |\sigma| = \sigma$. $\sigma, \tau \in S^*$ are *incompatible* if neither is an initial segment of the other. If $\leq$ is a partial order on S, then the *lexicographical ordering* $\leq_{\text{lex}}$ on S^* is defined by setting $\sigma \leq_{\text{lex}} \tau$ if $\sigma \subseteq \tau$ or, where k is the least index at which $\sigma(k) \neq \tau(k)$, then $\sigma(k) \leq \tau(k)$.

$\varphi_e^{(k)}$ denotes the e-th partial recursive function $\subseteq \mathbb{N}^k \to \mathbb{N}$; e is called a *Gödel number* or *index* of $\varphi_e^{(k)}$. Likewise, if $f \in \mathbb{N}^{\mathbb{N}}$ then $\varphi_e^{(k),f}$ denotes the e-th partial function $\varphi_e^{(k),f} :\subseteq \mathbb{N}^k \to \mathbb{N}$ which is partial recursive in f; e is again called a *Gödel number* or *index* of $\varphi_e^{(k),f}$, while f is called a *oracle* of $\varphi_e^{(k),f}$.

$\leq_{\mathrm{T}}$ denotes Turing reducibility while $\equiv_{\mathrm{T}}$ denotes Turing equivalence. $\leq_{\mathrm{tt}}$ denotes truth table reducibility. For $X \in \{0,1\}^{\mathbb{N}}$, X' denotes the Turing jump of X. For $f, g \in \mathbb{N}^{\mathbb{N}}$, their *join* $f \oplus g \in \mathbb{N}^{\mathbb{N}}$ is defined by $(f \oplus g)(2n) = f(n)$ and $(f \oplus g)(2n+1) = g(n)$. $\mathrm{J}_e(X) := X \oplus W_e^X$ denotes the e-th pseudojump of X, where W_e^X is the e-th $\Sigma_1^{0,X}$ subset of $\mathbb{N}$.

P_e denotes the e-th Π_1^0 class $\{X \in \{0,1\}^{\mathbb{N}} \mid \varphi_e^{(1),X}(0) \uparrow\}$.

2. Posner-Robinson and Pseudojump Inversion

In this section, we give the classical proofs of the Posner-Robinson Theorem and Pseudojump Inversion Theorem that we build on later.

Theorem 2.1: (Posner-Robinson Theorem). *Suppose Z and A are reals* *such that* $0 <_{\mathrm{T}} Z$. *Then there exists* B *such that*

$$A \oplus Z \oplus 0' \equiv_{\mathrm{T}} B' \equiv_{\mathrm{T}} B \oplus Z \equiv_{\mathrm{T}} B \oplus 0'.$$

Proof: By replacing A with $A \oplus Z \oplus 0'$, assume without loss of generality that $0' \leq_{\mathrm{T}} A$ and $Z \leq_{\mathrm{T}} A$.

As $0 <_{\mathrm{T}} Z$, it follows that Z is not Δ^0_1, and hence either not Π^0_1 or not Σ^0_1. As $Z \equiv_{\mathrm{T}} Z^{\mathrm{c}}$, we may assume without loss of generality that Z is not Π^0_1.

Lemma 2.2: *If* $Z \subseteq \mathbb{N}$ *is not* Π^0_1, *then for any* Σ^0_1 $W \subseteq \mathbb{N}$ *there are infinitely many* i *for which* $i \in Z$ *if and only if* $i \in W$.

Proof: Suppose not, so that there are finitely many i for which $i \in Z \leftrightarrow i \in W$, say $i_1, \ldots, i_k$. Then

$$i \in Z \equiv \bigvee_{i_j \in Z} (i = i_j) \vee \left(\bigwedge_{j=1}^{k} (i \neq i_j) \wedge i \notin W \right)$$

shows that Z is Π^0_1, a contradiction. □

We define a sequence of strings in $\{0,1\}^*$

$$\sigma_0 \subseteq \sigma_1 \subseteq \cdots \subseteq \sigma_n \subseteq \cdots.$$

The function $B := \bigcup_{n=0}^{\infty} \sigma_n$ will be the desired real.

Stage 0: Let $\sigma_0 := \langle\rangle$.
Stage $n = 2m+1$: Define

$$W_m := \{i \in \mathbb{N} \mid \exists \sigma (\sigma \supseteq \sigma_{n-1}{}^\frown \langle 0 \rangle^{i\frown} \langle 1 \rangle \wedge \varphi^{(1),\sigma}_{m,|\sigma|}(0) \downarrow)\}$$

which is Σ^0_1. By Lemma 2.2, there are infinitely many i such that $i \in Z \leftrightarrow i \in W_m$. Recursively in A, we choose the first such i for which $i \in Z \leftrightarrow i \in W_m$ holds. We consider two cases:

Case 1: $i \in Z$. By definition, there is $\sigma \supseteq \sigma_{n-1}{}^\frown \langle 0 \rangle^{i\frown} \langle 1 \rangle$ such that $\varphi^{(1),\sigma}_{m,|\sigma|}(0) \downarrow$, so let σ_n be the least such σ.
Case 2: $i \notin Z$. Let $\sigma_n = \sigma_{n-1}{}^\frown \langle 0 \rangle^{i\frown} \langle 1 \rangle$.

Stage $n = 2m+2$: Let $\sigma_n = \sigma_{n-1}{}^\frown \langle A(m) \rangle$.

This completes the construction.

We now make the following observations: assuming σ_{n-1} is known, then

...in Stage $n = 2m + 1$, the determination of i is recursive in either A (as $0' \oplus Z \leq_{\mathrm{T}} A$) or B (there is exactly one i such that $\sigma_{n-1}{}^\frown\langle 0\rangle^{i\frown}\langle 1\rangle \subset B$). The resulting case analysis is recursive in either Z (to test if $i \in Z$) or $0'$ (to determine the existence of a $\sigma \supseteq \sigma_{n-1}{}^\frown\langle 0\rangle^{i\frown}\langle 1\rangle$ for which $\varphi^{(1),\sigma}_{m,|\sigma|}(0)\downarrow$). Thus, σ_n can either be determined recursively in A, $B \oplus Z$, or $B \oplus 0'$.

...in Stage $n = 2m + 2$, the determination of σ_n can be determined recursively in either A or B (and hence also by $B \oplus Z$ and $B \oplus 0'$).

These observations and the uniformity of the processes described therein show that $\langle \sigma_n \rangle_{n\in\mathbb{N}} \leq_{\mathrm{T}} A, B \oplus Z, B \oplus 0'$.

Some additional observations:

- As $B = \bigcup_{n=0}^{\infty} \sigma_n$ and $Z \leq_{\mathrm{T}} A$ by hypothesis, we find $B \oplus Z \leq_{\mathrm{T}} A$.
- $A \leq_{\mathrm{T}} \langle \sigma_n \rangle_{n\in\mathbb{N}}$ as $A(m) = \sigma_{2n+2}(|\sigma_{2n+2}| - 1)$.
- $B' \leq_{\mathrm{T}} \langle \sigma_n \rangle_{n\in\mathbb{N}}$ as the finite-use principle and our construction of B implies $m \in B'$ if and only if $\varphi^{(1),\sigma_{2m+1}}_{m,|\sigma_{2m+1}|}(0)\downarrow$.

Bringing these observations together shows

$$A \leq_{\mathrm{T}} \langle \sigma_n \rangle_{n\in\mathbb{N}} \leq_{\mathrm{T}} B \oplus Z \leq_{\mathrm{T}} A \quad \text{and} \quad B' \leq_{\mathrm{T}} \langle \sigma_n \rangle_{n\in\mathbb{N}} \leq_{\mathrm{T}} B \oplus 0' \leq_{\mathrm{T}} B'$$

yielding

$$A \equiv_{\mathrm{T}} B' \equiv_{\mathrm{T}} B \oplus Z \equiv_{\mathrm{T}} B \oplus 0'.$$

□

Definition 2.3: (Pseudojump Operator). Given $X \in \{0,1\}^{\mathbb{N}}$, let $W_0^X, W_1^X, W_2^X, \ldots$ be an effective enumeration of the $\Sigma_1^{0,X}$ subsets of $\mathbb{N}$ uniform in X. (E.g., $W_e^X := \{n \in \mathbb{N} \mid \varphi_e^{(1),X}(n)\downarrow\}$.)

The e**-th pseudojump operator** J_e is defined by

$$\mathrm{J}_e(X) := X \oplus W_e^X.$$

Theorem 2.4: (Pseudojump Inversion). *For every real A and $e \in \mathbb{N}$, there exists B such that*

$$A \oplus 0' \equiv_{\mathrm{T}} \mathrm{J}_e(B) \equiv_{\mathrm{T}} B \oplus 0'.$$

Proof: By replacing A with $A \oplus 0'$, assume without loss of generality that $0' \leq_{\mathrm{T}} A$.

We shall construct a sequence of strings in $\{0,1\}^*$

$$\sigma_0 \subseteq \sigma_1 \subseteq \cdots \subseteq \sigma_n \subseteq \cdots.$$

The function $B := \bigcup_{n=0}^{\infty} \sigma_n$ will be the desired real.

Let $W_e^\sigma := \{n \in \mathbb{N} \mid \varphi^{(1),\sigma}_{e,|\sigma|}(n)\downarrow\}$.

Stage 0: Let $\sigma_0 = \langle\rangle$.
Stage $n = 2m+1$: Let σ_n be the least string $\sigma \supseteq \sigma_{n-1}$ for which $n \in W_e^\sigma$, if such a σ exists. Otherwise, let $\sigma_n = \sigma_{n-1}$.
Stage $n = 2m+2$: Let $\sigma_n = \sigma_{n-1}{}^\frown\langle A(n)\rangle$.

This completes the construction.

We now make the following observations: assuming σ_{n-1} is known, then...

...in Stage $n = 2m+1$, the existence of such a σ can be determined recursively from either $\mathrm{J}_e(B)$ or $0'$ (and hence also $A \geq_{\mathrm{T}} 0'$ and $B \oplus 0'$). Once existence is established, σ_n can be found recursively.
...in Stage $n = 2m+2$, σ_n can be found recursively in either A or B (and hence also $\mathrm{J}_e(B) \geq_{\mathrm{T}} B$ and $B \oplus 0'$).

These observations and the uniformity of the processes described therein show that $\langle\sigma_n\rangle_{n\in\mathbb{N}} \leq_{\mathrm{T}} A, \mathrm{J}_e(B), B \oplus 0'$.

Some additional observations:

- $A \leq_{\mathrm{T}} \langle\sigma_n\rangle_{n\in\mathbb{N}}$ as $A(m) = \sigma_{2m+1}(|\sigma_{2m+1}| - 1)$.
- $\mathrm{J}_e(B) \leq_{\mathrm{T}} \langle\sigma_n\rangle_{n\in\mathbb{N}}$ as the finite-use principle implies $m \in W_e^B$ if and only if $\varphi^{(1),\sigma_{2m+1}}_{e,|\sigma_{2m+1}|}(m)\downarrow$.
- As $B = \bigcup_{n=0}^\infty \sigma_n$ and $0' \leq_{\mathrm{T}} A$ by hypothesis, we find $B \oplus 0' \leq_{\mathrm{T}} A$.

Bringing these observations together shows

$$A \leq_{\mathrm{T}} \langle\sigma_n\rangle_{n\in\mathbb{N}} \leq_{\mathrm{T}} B \oplus 0' \leq_{\mathrm{T}} A \quad \text{and} \quad \mathrm{J}_e(B) \leq_{\mathrm{T}} \langle\sigma_n\rangle_{n\in\mathbb{N}} \leq_{\mathrm{T}} \mathrm{J}_e(B)$$

yielding

$$A \equiv_{\mathrm{T}} \mathrm{J}_e(B) \equiv_{\mathrm{T}} B \oplus 0'.$$

□

3. Completions of Peano Arithmetic

We now show that CPA has the Posner-Robinson Property. Recall that

$$\mathrm{CPA} := \{X \in \{0,1\}^{\mathbb{N}} \mid X \text{ encodes a complete, consistent extension of } \mathsf{PA}\}.$$

Definition 3.1: (PA-degree). $X \in \{0,1\}^{\mathbb{N}}$ is of **PA-degree** if X is Turing equivalent to a member of CPA.

Lemma 3.2: *[1, Theorem 2.21.3, p. 85] [9, Corollary 6.6, p. 12] Suppose $X, Y \in \{0,1\}^{\mathbb{N}}$ and $X \leq_{\mathrm{T}} Y$. If X is of* PA*-degree, then Y is of* PA*-degree.*

Theorem 3.3: CPA *has the Posner-Robinson Property.*

Proof: Let A and Z be as in the statement of the Posner-Robinson Property, and, by replacing A with $A \oplus Z \oplus 0'$, assume without loss of generality that $0' \leq_{\mathrm{T}} A$ and $Z \leq_{\mathrm{T}} A$.

By Theorem 1.2, there exists a C such that

$$A \equiv_{\mathrm{T}} C' \equiv_{\mathrm{T}} C \oplus Z \equiv_{\mathrm{T}} C \oplus 0'.$$

By Low Basis Theorem [3, Theorem 2.1, p. 36] relativized to C, there exists $B_0 \in \mathrm{CPA}$ such that

$$C' \equiv_{\mathrm{T}} (B_0 \oplus C)' \equiv_{\mathrm{T}} (B_0 \oplus C) \oplus 0'. \qquad (*)$$

By Lemma 3.2, $B_0 \leq_{\mathrm{T}} B_0 \oplus C$ implies $\deg_{\mathrm{T}}(B_0 \oplus C)$ is a PA-degree. Thus, there is $B \in \mathrm{CPA}$ such that $B \equiv_{\mathrm{T}} B_0 \oplus C$. Then Eq. $(*)$ becomes

$$C' \equiv_{\mathrm{T}} B' \equiv_{\mathrm{T}} B \oplus 0'.$$

Next, $C' \equiv_{\mathrm{T}} C \oplus Z$ implies

$$C' \equiv_{\mathrm{T}} C \oplus Z \leq_{\mathrm{T}} B \oplus Z \leq_{\mathrm{T}} C' \oplus Z \equiv_{\mathrm{T}} C'$$

and hence

$$A \equiv_{\mathrm{T}} C' \equiv_{\mathrm{T}} B' \equiv_{\mathrm{T}} B \oplus 0' \equiv_{\mathrm{T}} B \oplus Z. \qquad \square$$

To show that CPA satisfies the Pseudojump Inversion Property, we show that CPA exhibits a *splitting property.*

Notation. Suppose $\mathcal{L}$ is a language and T is a consistent $\mathcal{L}$-theory. We define

$$\mathrm{CMP}_{\mathsf{T}} := \{X \in \{0,1\}^{\mathrm{Sent}(\mathcal{L})} \mid X \text{ is a complete, consistent extension of } \mathsf{T}\}$$

where $\mathrm{Sent}(\mathcal{L})$ is the set of sentences in the language $\mathcal{L}$.

Lemma 3.4: *Let $\mathcal{L}$ be a language, T is a consistent $\mathcal{L}$-theory, P is a nonempty closed subset of $\{0,1\}^{\mathrm{Sent}(\mathcal{L})}$, and $P \subseteq \mathrm{CMP}_{\mathsf{T}}$. Then*

$$P = \mathrm{CMP}_{\tilde{\mathsf{T}}}$$

where

$$\tilde{\mathsf{T}} = \{\theta \in \mathrm{Sent}(\mathcal{L}) \mid \varphi \in X \textit{ for all } X \in P\} = \bigcap P.$$

Proof: By construction, $\tilde{\mathsf{T}} \subseteq \mathsf{T}$, so every element of P is a complete, consistent extension of T and hence also of $\tilde{\mathsf{T}}$. Thus, $P \subseteq \mathrm{CMP}_{\tilde{\mathsf{T}}}$.

Conversely, suppose $Y \in \mathrm{CMP}_{\tilde{\mathsf{T}}}$. The basic open sets of $2^{\mathrm{Sent}(\mathcal{L})}$ are of the form $N_\Psi = \{X \in \{0,1\}^{\mathrm{Sent}(\mathcal{L})} \mid \Psi \subseteq X\}$ where Ψ is a finite subset of Sent($\mathcal{L}$). If $Y \in N_\Psi$, then $\tilde{\Phi} \cup \Psi$ is consistent, being contained in the consistent theory Y. In particular, $\neg \bigwedge \Psi \notin \tilde{\Phi}$, so there must exist $X \in P$ such that $\bigwedge \Psi \in X$. In other words, $N_\Psi \cap P \neq \emptyset$, showing that $Y \in \mathrm{cl}(P) = P$. Thus, $\mathrm{CMP}_{\tilde{\mathsf{T}}} \subseteq P$. □

We choose a suitable Gödel numbering $\# : \mathrm{Sent}(\mathcal{L}) \to \mathbb{N}$, where $\mathcal{L}$ is the language of arithmetic, and hence identify $\{0,1\}^{\mathrm{Sent}(\mathcal{L})}$ with $\{0,1\}^{\mathbb{N}}$.

Lemma 3.5: *[11, Lemma 9.9, p. 19] Suppose* $S \subseteq \{0,1\}^{\mathbb{N}} \times (\mathbb{N}^{\mathbb{N}})^k \times \mathbb{N}^\ell$.

(a) If S *is* Π^0_1*, then* $\exists X\, S(X, -)$ *is* Π^0_1.
(b) If S *is* Σ^0_1*, then* $\forall X\, S(X, -)$ *is* Σ^0_1.

Lemma 3.6:

(a) Given a Π^0_1 *predicate* $Q \subseteq \{0,1\}^{\mathbb{N}} \times \mathbb{N}^k$*, there is a primitive recursive function* $f : \mathbb{N}^k \to \mathbb{N}$ *such that*

$$P_{f(x_1,\ldots,x_k)}(X) \equiv Q(X, x_1, \ldots, x_k).$$

(b) $\{e \in \mathbb{N} \mid P_e = \emptyset\} \equiv_{\mathrm{T}} 0'$.

Proof: Straight-forward. □

Corollary 3.7: *There exist partial recursive functions* $v, w : \mathbb{N}^2 \to \mathbb{N}$ *such that*

$$P_{v(n,m)} = P_n \cap P_m$$
$$P_{w(e,n)} = \{X \in \{0,1\}^{\mathbb{N}} \mid n \notin \mathrm{J}_e(X)\}.$$

Proposition 3.8: *There exists a primitive recursive function* $s : \mathbb{N} \times \{0,1\} \to \mathbb{N}$ *such that whenever* P_e *is a nonempty subset of* CPA *then* $P_{s(e,0)}$ *and* $P_{s(e,1)}$ *are disjoint non-empty subsets of* P_e.

Proof: Suppose $e \in \mathbb{N}$, and define

$$\mathsf{T} := \{\alpha \mid \forall X\, (X \in P_e \to \#\alpha \in X)\}.$$

T is recursively enumerable by Lemma 3.5, and the proof therein yields an index i for a recursive function enumerating the Gödel numbers of elements of T.

The Gödel-Rosser Theorem [5, Proposition 3.38, p. 207] shows that there is a primitive recursive function p such that for all i as above, $p(i) = \#(\psi)$ for some sentence ψ. If $\mathsf{PA} \subseteq \mathsf{T}$, then $\mathsf{T} \nvdash \psi$ and $\mathsf{T} \nvdash \neg\psi$.

Thus, using Lemma 3.6, define $s : \mathbb{N} \times \{0,1\} \to \mathbb{N}$ to be a primitive recursive function satisfying

$$P_{s(e,0)} = P_e \cap \mathrm{CMP}_{\mathsf{T}\cup\{\neg\psi\}},$$
$$P_{s(e,1)} = P_e \cap \mathrm{CMP}_{\mathsf{T}\cup\{\psi\}}.$$

In particular, if $P_e \subseteq \mathrm{CPA}$, then $\mathsf{PA} \subseteq \mathsf{T}$ and hence $P_{s(e,0)}$ and $P_{s(e,1)}$ will be disjoint and nonempty. □

Using this splitting property, we can imitate the proof of Theorem 1.3 within CPA.

Theorem 3.9: CPA *has the Pseudojump Inversion Property.*

Proof: Let A and e be as in the statement of the Pseudojump Inversion Property, and, by replacing A with $A \oplus 0'$, assume without loss of generality that $0' \leq_{\mathrm{T}} A$. Write $Q = \mathrm{CPA}$.

We shall define a sequence

$$Q = Q_0 \supseteq Q_1 \supseteq Q_2 \supseteq \cdots \supseteq Q_n \supseteq \cdots$$

of nonempty Π^0_1 subsets of Q. Compactness will imply that the intersection $\bigcap_{n=0}^{\infty} Q_n$ is nonempty, and an element of that intersection will be our desired $B \in Q$. We shall also define a function $j : \mathbb{N} \to \mathbb{N}$ for which

$$Q_n = P_{j(n)}$$

for each n.

Let e_0 be an index for Q, i.e. $Q = P_{e_0}$. Let s be as in Proposition 3.8.

Stage 0: Let $Q_0 = Q$ and $j(0) = e_0$.
Stage $n = 2m + 1$: Define

$$Q_n := \begin{cases} Q_{n-1} \cap P_{w(e,m)} & \text{if this intersection is nonempty,} \\ Q_{n-1} & \text{otherwise,} \end{cases}$$

$$j(n) := \begin{cases} v(j(n-1), w(e,m)) & \text{if } Q_{n-1} \cap P_{w(e,m)} \neq \emptyset, \\ j(n-1) & \text{otherwise.} \end{cases}$$

Stage $n = 2m + 2$: Define

$$Q_n := P_{s(j(n-1),A(m))}$$
$$j(n) := s(j(n-1), A(m)).$$

By the compactness of $\{0,1\}^{\mathbb{N}}$, the intersection $\bigcap_{n=0}^{\infty} Q_n$ is nonempty, so let B be an element of that intersection.

We now make the following observations: assuming $j(n-1)$ is known, then

...in Stage $n = 2m+1$, $j(n)$ can be found recursively in $0'$ (to check the nonemptiness of $P_{j(n-1)} \cap P_{w(e,m)}$, as in Lemma 3.6(b)) or $\mathrm{J}_e(B)$ (to check if $n \notin J_e(B)$ and hence whether $B \in Q_{n-1} \cap P_{w(e,m)}$, which is equivalent to the nonemptiness of $P_{j(n-1)} \cap P_{w(e,m)}$).

...in Stage $n = 2m+2$, $j(n)$ can be found recursively in B (finding the unique k such that $B \notin P_{s(j(n-1),k)}$) or A.

These observations and the uniformity of the processes described therein show that $j \leq_{\mathrm{T}} A, B \oplus 0', \mathrm{J}_e(B)$.

Some additional observations:

We now make the following observations:

- $A \leq_{\mathrm{T}} j$, as $A(m) = i$ if and only if $j(2m+2) = s(j(2m+1), i)$.
- $J_e(B) \leq_{\mathrm{T}} j$ as $m \in J_e(X)$ if and only if $j(2m+1) = j(2m)$.
- $B \oplus 0' \leq_{\mathrm{T}} j$ as $0' \leq_{\mathrm{T}} A \leq_{\mathrm{T}} j$ and $B \leq_{\mathrm{T}} J_e(B) \leq_{\mathrm{T}} j$.

Bringing these observations together shows

$$j \equiv_{\mathrm{T}} A \equiv_{\mathrm{T}} J_e(B) \equiv_{\mathrm{T}} B \oplus 0'. \qquad \square$$

Remark 3.10: Using the splitting property of CPA, we could analogously imitate the proof of Theorem 1.2 within CPA.

One difference is that instead of using a splitting function $s : \mathbb{N} \times \{0,1\} \to \mathbb{N}$, we would use a splitting function $\tilde{s} : \mathbb{N} \times \mathbb{N} \to \mathbb{N}$ with the property that for every nonempty Π^0_1 subclass $P_e \subseteq \mathrm{CPA}$ we obtain nonempty and pairwise disjoint subclasses $P_{e,0}, P_{e,1}, \ldots \subseteq P_e$. We could then consider the Σ^0_1 set $W_m = \{i \in \mathbb{N} \mid \forall X (X \in P_{\tilde{s}(j(n-1),i)} \to \varphi_m^{(1),X}(0)\downarrow)\}$.

4. Turing Degree Isomorphisms

A major difference between the Posner-Robinson Property and the Pseudojump Inversion Property is that the former is invariant under Turing equivalence, while the latter is not.

Definition 4.1: (Turing Degree Isomorphic Π^0_1 Classes). Suppose P and Q are Π^0_1 classes. P and Q are **Turing degree isomorphic** if

$$\{\deg_{\mathrm{T}}(X) \mid X \in P\} = \{\deg_{\mathrm{T}}(X) \mid X \in Q\}$$

or equivalently that every member of P is Turing equivalent to a member of Q, and every member of Q is Turing equivalent to a member of P.

Proposition 4.2: *If P and Q are Turing degree isomorphic Π^0_1 classes, then P has the Posner-Robinson Property if and only if Q does.*

Proof: Suppose P has the Posner-Robinson Property, and let A and Z be as its statement. Then there exists $B \in P$ such that

$$A \oplus 0' \equiv_{\mathrm{T}} B' \equiv_{\mathrm{T}} B \oplus Z \equiv_{\mathrm{T}} B \oplus 0'.$$

Because P and Q are Turing degree isomorphic, there exists $C \in Q$ such that $\deg_{\mathrm{T}}(B) = \deg_{\mathrm{T}}(C)$. Then $B' \equiv_{\mathrm{T}} C'$, $B \oplus Z \equiv_{\mathrm{T}} C \oplus Z$, and $B \oplus 0' \equiv_{\mathrm{T}} C \oplus 0'$, so

$$A \oplus 0' \equiv_{\mathrm{T}} C' \equiv_{\mathrm{T}} C \oplus Z \equiv_{\mathrm{T}} C \oplus 0'.$$

As A and Z were arbitrary reals satisfying $0 <_{\mathrm{T}} Z \leq_{\mathrm{T}} A \oplus 0'$, it follows that Q has the Posner-Robinson Property.

The symmetry of the definition of Turing degree isomorphism implies the converse. □

In contrast, pseudojump operators are not necessarily invariant under Turing equivalence, i.e., $X \equiv_{\mathrm{T}} Y$ does not imply $\mathrm{J}_e(X) \equiv_{\mathrm{T}} \mathrm{J}_e(Y)$. Thus, it is not clear whether Proposition 4.2 holds for the Pseudojump Inversion Property. In other words, the following question is open.

Question 4.3: Suppose P and Q are Turing degree isomorphic special Π^0_1 classes. If P has the Pseudojump Inversion Property, is it necessary that Q has the Pseudojump Inversion Property?

We will show that the answer is in the affirmative in the case when P is Muchnik complete or when P has positive measure.

Under a stronger notion of *equivalence* of Π^0_1 classes, satisfaction of the Pseudojump Inversion Property is preserved.

Definition 4.4: (Recursively Homeomorphism). Suppose P and Q are Π^0_1 classes. A **recursive homeomorphism** is a recursive functional $\Phi : P \to Q$ which is bijective and for which $\Phi^{-1} : Q \to P$ is recursive. If there is a recursive homeomorphism $P \to Q$, then P and Q are **recursively homeomorphic**.

Proposition 4.5: *If P and Q are recursively homeomorphic Π^0_1 classes, then P has the Pseudojump Inversion Property if and only if Q does.*

Proof: Suppose $\Phi : P \to Q$ is a recursive homeomorphism. For $e \in \mathbb{N}$, the partial functional $\langle X, n\rangle \mapsto \varphi_e^{\Phi(X)}(n)$ is partial recursive, so by the s^m_n Theorem, there exists $k \in \mathbb{N}$ such that

$$\varphi_k^X(n) \simeq \varphi_e^{\Phi(X)}(n)$$

for all $n \in \mathbb{N}$ and all $X \in$ *cantor*. In particular, for $X \in P$,

$$W_k^X = \{n \in \mathbb{N} \mid \varphi_k^X(n)\downarrow\} = \{n \in \mathbb{N} \mid \varphi_e^{\Phi(X)}(n)\downarrow\} = W_e^{\Phi(X)}.$$

Suppose $e \in \mathbb{N}$ and $A \in \{0,1\}^{\mathbb{N}}$. If P has the Pseudojump Inversion Property, then there exists $C \in Q$ such that

$$A \oplus 0' \equiv_{\mathrm{T}} \mathrm{J}_k(C) = C \oplus W_k^C \equiv_{\mathrm{T}} C \oplus 0'.$$

Let $B = \Phi(C)$. Because Φ is a recursive homeomorphism, $C \equiv_{\mathrm{T}} \Phi(C) = B$. Thus,

$$\mathrm{J}_e(B) = B \oplus W_e^B = B \oplus W_e^{\Phi(C)} = B \oplus W_k^C \equiv_{\mathrm{T}} C \oplus W_k^C = J_k(C)$$

and hence

$$A \oplus 0' \equiv_{\mathrm{T}} J_e(B) = B \oplus W_k^B \equiv_{\mathrm{T}} B \oplus 0'.$$ □

One connection between Turing degree isomorphisms and recursive homeomorphisms is the following result.

Proposition 4.6: *Suppose P and Q are Turing degree isomorphic Π^0_1 classes. Then P and Q contain recursively homeomorphic Π^0_1 classes.*

To prove Proposition 4.6, we must introduce the notion of *hyperimmune free reals.* For $f, g \in \mathbb{N}^{\mathbb{N}}$, f is *dominated* by g if $f(n) < g(n)$ for all sufficiently large n. $X \in \{0,1\}^{\mathbb{N}}$ is *hyperimmune-free* if every function computable from X is dominated by some computable function. The *Hyperimmune-free Basis Theorem* [3, Theorem 2.4, p. 37] [11, Theorem 12.4, p. 24] states that any nonempty Π^0_1 class P contains a hyperimmune-free real.

Lemma 4.7: *Suppose Y is a hyperimmune-free real. If $X \leq_{\mathrm{T}} Y$, then $X \leq_{\mathrm{tt}} Y$.*

Proof: Because $X \leq_{\mathrm{T}} Y$, there exists e such that $\varphi_e^{(1),Y} = X$. Define $g \in \mathbb{N}^{\mathbb{N}}$

$$g(n) := \text{least } s \text{ such that } \varphi_{e,s}^{(1),Y}(n)\downarrow$$

which is Turing reducible to Y. Because Y is hyperimmune-free, there exists $f \in \mathbb{N}^{\mathbb{N}}$ recursive such that f dominates g. Then the functional $\Psi : \mathbb{N}^{\mathbb{N}} \to \mathbb{N}^{\mathbb{N}}$ defined by

$$\Psi(Z)(n) := \varphi^{(1),Z}_{e,f(n)}(n)$$

is recursive. By construction, $\Psi(Y) = X$, so $X \leq_{\mathrm{tt}} Y$. □

Proof: [Proof of Proposition 4.6.] By the Hyperimmune-free Basis Theorem, there exists $X_0 \in P$ which is hyperimmune-free. By hypothesis, there is $Y_0 \in Q$ such that $X_0 \equiv_{\mathrm{T}} Y_0$. Consequently, Y_0 is hyperimmune-free (if $Z \leq_{\mathrm{T}} Y_0$, then $Z \leq_{\mathrm{T}} X_0$ so Z is dominated by a recursive function). Thus, Lemma 4.7 implies there are total recursive functionals Φ and Ψ such that $\Phi(X_0) = Y_0$ and $\Psi(Y_0) = X_0$. Define

$$\tilde{P} := \{X \in P \mid \Psi(\Phi(X)) = X\}$$
$$\tilde{Q} := \{Y \in Q \mid \exists X(X \in P \wedge Y = \Phi(X))\}$$

which are Π^0_1 (for $\tilde{Q}$, we use Lemma 3.5) and nonempty, as $X_0 \in \tilde{P}$ and $Y_0 \in \tilde{Q}$. By construction, $\tilde{P}$ and $\tilde{Q}$ are recursively homeomorphic as witnessed by $\Phi|_{\tilde{P}} : \tilde{P} \to \tilde{Q}$ and $\Psi|_{\tilde{Q}} : \tilde{Q} \to \tilde{P}$. □

A final general result which follows immediately from the statement of the Posner-Robinson Property and the Pseudojump Inversion Property is the following.

Lemma 4.8: *Suppose P is a Π^0_1 class and $\tilde{P}$ is a Π^0_1 subclass of P. If $\tilde{P}$ has either the Posner-Robinson Property or the Pseudojump Inversion Property, then P has the Posner-Robinson Property or Pseudojump Inversion Property as well, respectively.*

Proof: Immediate, as a witness $B \in \tilde{P}$ of the satisfaction of the Posner-Robinson Property or the Pseudojump Inversion Property will be an element of P. □

4.1. *Muchnik complete Π^0_1 classes*

As an application, we examine the case of Muchnik and Medvedev complete Π^0_1 classes, which are the Π^0_1 classes Turing degree isomorphic and recursively homoeomorphic to CPA, respectively.

We recall the relevant definitions and some useful properties.

Definition 4.9: (Muchnik and Medvedev Reducibility). Suppose P and Q are subsets of $\mathbb{N}^{\mathbb{N}}$. Then...

...P is **Muchnik reducible** to Q, written $P \leq_{\mathrm{w}} Q$, if for every $X \in Q$ there is $Y \in P$ such that $Y \leq_{\mathrm{T}} X$.
...P and Q are **Muchnik equivalent**, written $P \equiv_{\mathrm{w}} Q$, if $P \leq_{\mathrm{w}} Q$ and $Q \leq_{\mathrm{w}} P$.
...P is **Medvedev reducible** to Q, written $P \leq_{\mathrm{s}} Q$, if there exists a recursive functional $\Phi : Q \to P$.
...P and Q are **Medvedev equivalent**, written $P \equiv_{\mathrm{s}} Q$, if $P \leq_{\mathrm{s}} Q$ and $Q \leq_{\mathrm{s}} P$.

Definition 4.10: (Muchnik and Medvedev Complete Π^0_1 Classes). Suppose Q is a nonempty Π^0_1 class. Q is...

...**Muchnik complete** if $P \leq_{\mathrm{w}} Q$ for every non-empty Π^0_1 class P.
...**Medvedev complete** if $P \leq_{\mathrm{s}} Q$ for every non-empty Π^0_1 class P.

Proposition 4.11: *[9, Theorem 6.3, p. 11 & Corollary 6.10, p. 13] Suppose P and Q are Π^0_1 classes.*

(a) If P is Muchnik complete, then Q is Muchnik complete if and only if P and Q are Turing degree isomorphic.
(b) If P is Medvedev complete, then Q is Medevedev complete if and only if P and Q are recursively homeomorphic.

Remark 4.12: If T is any consistent and recursively enumerable theory extending Robinson's Q, then the Π^0_1 class

$$\mathrm{CMP}_{\mathsf{T}} := \{X \in \{0,1\}^{\mathbb{N}} \mid X \text{ encodes a complete, consistent extension of } \mathsf{T}\}$$

is Muchnik complete. By Proposition 4.11, it is then Turing degree isomorphic to CPA. Thus, PA-degrees could equally well be called ZF-*degrees*, Q-*degrees*, etc.

Corollary 4.13: *Suppose Q is a Muchnik complete Π^0_1 class. Then Q has the Posner-Robinson Property.*

Proof: By Proposition 4.11, Q is Turing degree isomorphic to CPA, which has the Posner-Robinson Property by Theorem 3.3. Thus, Proposition 4.2 implies Q has the Posner-Robinson Property. □

Corollary 4.14: *Suppose Q is a Medvedev complete Π^0_1 class. Then Q has the Pseudojump Inversion Property.*

Proof: By Proposition 4.11, Q is recursively homeomorphic to CPA, which has the Pseudojump Inversion Property by Theorem 3.9. Thus, Proposition 4.5 implies Q has the Pseudojump Inversion Property. $\square$

In other words, Medvedev complete Π^0_1 classes have both the Posner-Robinson Property and the Pseudojump Inversion Property. What about an arbitrary Muchnik complete Π^0_1 class?

Lemma 4.15: *[9, Lemma 6.9, p. 12] Suppose P and Q are Π^0_1 classes and $P \leq_{\mathrm{w}} Q$. Then there is a non-empty Π^0_1 subclass $\tilde{Q} \subseteq Q$ such that $P \leq_{\mathrm{s}} \tilde{Q}$.*

Proof: By the Hyperimmune-Free Basis Theorem, there exists $X \in Q$ hyperimmune-free. Since $P \leq_{\mathrm{w}} Q$, there exists $Y \in P$ such that $Y \leq_{\mathrm{T}} X$. By Lemma 4.7, $Y \leq_{\mathrm{tt}} X$, so there is a total recursive functional $\Phi : \{0,1\}^{\mathbb{N}} \to \{0,1\}^{\mathbb{N}}$ such that $\Phi(X) = Y$. Define

$$\tilde{Q} := \{Z \in Q \mid \Phi(Z) \in P\}$$

which is Π^0_1 and nonempty, containing X. $\Phi|_{\tilde{Q}} : \tilde{Q} \to P$ shows that $P \leq_{\mathrm{s}} \tilde{Q}$. $\square$

Proposition 4.16: *Suppose P is a Muchnik complete Π^0_1 class. Then P contains a Medvedev complete Π^0_1 class.*

Proof: Because P is Muchnik complete, CPA $\leq_{\mathrm{w}} P$. Lemma 4.15 implies there is $\tilde{P} \subseteq P$ which is non-empty and Π^0_1 such that CPA $\leq_{\mathrm{s}} \tilde{P}$. Because CPA is Medvedev complete, $\tilde{P}$ is as well. $\square$

Corollary 4.17: *Suppose P is a Muchnik complete Π^0_1 class. Then P has the Pseudojump Inversion Property.*

Proof: By Proposition 4.16, there exists $\tilde{P} \subseteq P$ which is a Medvedev complete Π^0_1 class. By Corollary 4.14 shows that $\tilde{P}$ has the Pseudojump Inversion Property, and so by Lemma 4.8 P also has the Pseudojump Inversion Property. $\square$

Remark 4.18: If P is any special Π^0_1 class, then $P \cup \mathrm{CPA}$ has both the Posner-Robinson Property and the Pseudojump Inversion Property while $\deg_{\mathrm{w}}(P) = \deg_{\mathrm{w}}(P \cup \mathrm{CPA})$ and $\deg_{\mathrm{s}}(P) = \deg_{\mathrm{s}}(P \cup \mathrm{CPA})$. Thus, the content of Corollaries 4.13, and 4.17 is that *every* Π^0_1 class P of Muchnik degree $\deg_{\mathrm{w}}(\mathrm{CPA})$ or Medvedev degree $\deg_{\mathrm{s}}(\mathrm{CPA})$ has the Posner-Robinson Property and the Pseudojump Inversion Property.

5. Π^0_1 Classes of Martin-Löf Reals

It has been observed by Lutz [4] that Π^0_1 classes with positive measure do not necessarily have the Posner-Robinson Property. We make use of the notion of Martin-Löf randomness and LR-reducibility. See, e.g., [1].

Lemma 5.1: *Suppose Z is low-for-random, i.e., $Z \leq_{\mathrm{LR}} 0$, and B is Martin-Löf random. Then $B \oplus Z \leq_{\mathrm{LR}} B$.*

Proof: If Y is 1-random relative to B, then $B \oplus Y$ is 1-random by van Lambalgen's Theorem [1, Corollary 6.9.3, p. 258]. Because Z is K-trivial, $B \oplus Y$ is 1-random relative to Z. van Lambalgen's Theorem, relative to Z, implies Y is 1-random relative to $B \oplus Z$. Thus, $B \oplus Z \leq_{\mathrm{LR}} B$. □

Proposition 5.2: *[4] If P is a non-empty Π^0_1 class whose elements are Martin-Löf random reals, then P does not have the Posner-Robinson Property.*

Proof: Let Z be low-for-random. By Lemma 5.1, Z is low-for-random relative to B for any $B \in P$. Relativizing the proof of [1, Theorem 11.4.1, p. 518] shows that Z is low relative to B, i.e., $(B \oplus Z)' \equiv_{\mathrm{T}} B'$. Thus, we cannot have $B \oplus Z \equiv_{\mathrm{T}} B'$. □

On the other hand, Π^0_1 classes with positive measure *do* have the Pseudojump Inversion Property.

Theorem 5.3: *[6, Theorem 6.3.9, p. 254] Suppose $e \in \mathbb{N}$, $0' \leq_{\mathrm{T}} A$, and Q is a Π^0_1 class of positive measure. Then there exists $B \in Q$ such that*

$$A \equiv_{\mathrm{T}} \mathrm{J}_e(B) \equiv_{\mathrm{T}} B \oplus 0'.$$

An alternate proof can be found in [10, Theorem 5.1, p. 9].

It can be shown that having positive measure is sufficient for Turing degree isomorphisms to preserve satisfaction of the Pseudojump Inversion Property.

Theorem 5.4: *Suppose P and Q are Turing degree isomorphic Π^0_1 classes such that P has positive measure. Then Q has the Pseudojump Inversion Property.*

Proof: By Proposition 4.6, there exists nonempty Π^0_1 subclasses $\tilde{P} \subseteq P$ and $\tilde{Q} \subseteq Q$ such that $\tilde{P}$ and $\tilde{Q}$ are recursively homeomorphic. Moreover, we may choose $\tilde{P}$ so that it has positive measure: P contains a Π^0_1 subclass P_0 of Martin-Löf random reals and hence a hyperimmune-free Martin-Löf random real. Letting that real be our choice of the hyperimmune-free X_0 in the proof of Proposition 4.6, we can ensure that $\tilde{P}$ contains a Martin-Löf random real, and hence has positive measure.

By Theorem 5.3, $\tilde{P}$ has the Pseudojump Inversion Property. Proposition 4.5 implies that $\tilde{Q}$ has the Pseudojump Inversion Property, and hence Q does as well by Lemma 4.8. □

6. Π^0_1 Classes Constructed by Priority Arguments

In this section, we construct a Π^0_1 class which has neither the Posner-Robinson Property nor the Pseudojump Inversion Property.

Notation. If $P \subseteq \{0,1\}^{\mathbb{N}}$ and $n \in \mathbb{N}$, then we write $P^n := \{X_1 \oplus \cdots \oplus X_n \mid X_1, \ldots, X_n \in P\}$.

Theorem 6.1: *There exists a special Π^0_1 class P with the following properties.*

(i) For all $n, e \in \mathbb{N}$ and $X \in P^n$, there exists $\sigma \subset X$ such that either $P_e \cap N_\sigma \cap P^n = \emptyset$ or $P^c_e \cap N_\sigma \cap P^n = \emptyset$.
(ii) Moreover, such a σ can be computed uniformly from e and n using $X \oplus 0'$ as an oracle.

Notation. Suppose $\sigma_1, \ldots, \sigma_n \in \{0,1\}^*$. Let $\ell = \min_{1 \le k \le n} |\sigma_k|$. Define $\sigma_1 \oplus \cdots \oplus \sigma_n$ to be the string of length $n\ell$ such that

$$(\sigma_1 \oplus \cdots \oplus \sigma_n)(i) := \sigma_r(q) \qquad \text{where } i = qn + r < n\ell.$$

If $\sigma \in \{0,1\}^*$, $n \ge 1$, and $k \le n$, then $(\sigma)_{k,n}$ is the string of length $\lfloor |\sigma|/n \rfloor$ such that

$$(\sigma)_{k,n}(i) := \sigma(in + k).$$

In particular, if n divides $|\sigma|$, then $\sigma = (\sigma)_{0,n} \oplus \cdots \oplus (\sigma)_{n-1,n}$.

Lemma 6.2: *Suppose P is a Π^0_1 class such that if $X_1, \ldots, X_n$ are pairwise-distinct elements of P and $e \in \mathbb{N}$, then there exists $\sigma \subseteq X_1 \oplus \cdots \oplus X_n$ such that either $P_e \cap N_\sigma \cap P = \emptyset$ or $P^c_e \cap N_\sigma \cap P = \emptyset$. Then if $X \in P$ is recursive, it is an isolated point of P.*

Proof: Let $S = \{0,1\}^{\mathbb{N}} \setminus \{X\}$. As X is recursive and hence is a Π^0_1 singleton, S is Σ^0_1. Thus, there exists $\sigma \subset X$ such that either $N_\sigma \cap P \subseteq S = \{0,1\}^{\mathbb{N}} \setminus$

$\{X\}$ or $N_\sigma \cap P \subseteq S^c = \{X\}$. But $X \in N_\sigma \cap P$, so the latter case occurs. In particular, $N_\sigma \cap P = \{X\}$, so X is an isolated point of P. □

Proposition 6.3: *The special Π^0_1 class P described in Theorem 6.1 has the following properties:*

(a) For all $X_0, X_1, \ldots, X_n \in P$, if X_0 is distinct from $X_1, \ldots, X_n$, then $X_0 \not\leq_T X_1 \oplus \cdots \oplus X_n$.
(b) Every hyperimmune-free real in P is of minimal Turing degree in P.
(c) P is thin, i.e., if Q is a Π^0_1 subclass of P, then $Q = P \cap (N_{\sigma_1} \cup \cdots \cup N_{\sigma_n})$ for some $n \in \mathbb{N}$ and $\sigma_1, \ldots, \sigma_n \in \{0,1\}^*$.
(d) For all $X_1, \ldots, X_n \in P$, $X_1 \oplus \cdots \oplus X_n$ is generalized-low.

Proof:

(a) Suppose for the sake of a contradiction that there exists $X_0, X_1, \ldots, X_n \in P$ such that $X_0 \leq_T X_1 \oplus \cdots \oplus X_n$. Assume without loss of generality that $X_0, X_1, \ldots, X_n$ are pairwise distinct.
Let $e \in \mathbb{N}$ be such that $X_0 = \varphi_e^{(1), X_1 \oplus \cdots \oplus X_n}$. Define

$$S := \{Y_0 \oplus \cdots \oplus Y_n \in \{0,1\}^{\mathbb{N}} \mid \exists i\, (\varphi_e^{(1), Y_1 \oplus \cdots \oplus Y_n}(i) \downarrow \neq Y_1(i))\}$$

which is a Σ^0_1 class. Because P is 1-generic in itself, there exists $\sigma \subset X_0 \oplus \cdots \oplus X_n$ such that either $P^{n+1} \cap N_\sigma \subseteq S$ or $P^{n+1} \cap N_\sigma \subseteq S^c$. By hypothesis, $X_0 \oplus \cdots \oplus X_n \in S^c$, so $P^{n+1} \cap N_\sigma \subseteq S^c$.
Because X_0 is not recursive, it isn't an isolated point in P. In particular, there is $Y \in P$ such that $(\sigma)_{0,n+1} \subset Y$ and $X_0 \neq Y$. But then $Y \oplus X_1 \oplus \cdots \oplus X_n \in N_\sigma \cap P^{n+1} \subseteq S^c$, implying $\varphi_e^{(1), X_1 \oplus \cdots \oplus X_n} = Y$. This contradicts the hypothesis that $\varphi_e^{(1), X_1 \oplus \cdots \oplus X_n} = X_0$ and $X_0 \neq Y$.

(b) Suppose $Y \in P$ is hyperimmune-free, $X \in P$, and that $X <_T Y$. We claim that X is recursive.
By Lemma 4.7, there exists a recursive functional $\Psi : \mathbb{N}^{\mathbb{N}} \to \mathbb{N}^{\mathbb{N}}$ such that $\Psi(Y) = X$. Define $S \subseteq \{0,1\}^{\mathbb{N}}$ by

$$S := \{Z \oplus W \in \{0,1\}^{\mathbb{N}} \mid \exists n\, (\Psi(W)(n) \neq Z(n))\}$$

which is Σ^0_1. Because P is 1-generic in itself, there exists $\sigma \subset X \oplus Y$ such that either $N_\sigma \cap P^2 \subseteq S$ or $N_\sigma \cap P^2 \subseteq S^c$. As $\Psi(Y) = X$, the latter case occurs, i.e. $Z \oplus W \in N_\sigma \cap P^2$ implies $\Psi(W) = Z$. There are two cases:

Case 1: X is isolated in P. If X is isolated in P, then X is recursive.

Case 2: X is not isolated in P. If X is not isolated in P, then there exists Z such that $(\sigma)_{0,2} \subset Z$ and $Z \neq X$. But $Z \oplus Y \in N_\sigma \cap P^2 \subseteq S^c$ so that $\Psi(Y) = Z$. This is a contradiction, as $\Psi(Y) = X$ and $X \neq Z$.

(c) Suppose $Q \subseteq P$ is Π^0_1. Considering P with the subspace topology inherited from $\{0,1\}^{\mathbb{N}}$, Q is a closed subset of P. As P is compact (being a closed subset of the compact space $\{0,1\}^{\mathbb{N}}$), Q is compact.
Given $X \in Q$, there exists $\sigma \subset X$ such that $P \cap N_\sigma \subseteq \{0,1\}^{\mathbb{N}} \setminus Q$ or $P \cap N_\sigma \subseteq Q$. The former case cannot occur since $X \in Q$, so $X \in P \cap N_\sigma \subseteq Q$. Letting $\mathcal{U}$ be the collection of sets $P \cap N_\sigma$ which are contained within Q, this shows $\mathcal{U}$ is an open cover of Q. As Q is compact, there exist $\sigma_1, \ldots, \sigma_n$ such that

$$Q \subseteq (P \cap N_{\sigma_1}) \cup \cdots \cup (P \cap N_{\sigma_n}).$$

The reverse inclusion holds by the hypothesis that elements of $\mathcal{U}$ are contained in Q. Hence,

$$Q = P \cap (N_{\sigma_1} \cup \cdots \cup N_{\sigma_n}).$$

(d) Assume without loss of generality that $X_1, \ldots, X_n$ are pairwise-distinct. Let $f : \mathbb{N}^2 \to \{0,1\}^*$ be a function such that $f(e,n) \subset X_1 \oplus \cdots \oplus X_n$ and either $N_{f(e,n)} \cap P^n \subseteq P_e$ or $N_{f(e,n)} \cap P^n \subseteq P_e^c$ and for which $f \leq_T X_1 \oplus \cdots \oplus X_n \oplus 0'$, whenever $X_1, \ldots, X_n$ are pairwise-distinct elements of P.
To determine whether $\varphi_e^{(1), X_1 \oplus \cdots \oplus X_n}(0) \downarrow$, it suffices to check whether $N_{f(e)} \cap P^n \cap P_e = \emptyset$, which can be determined using $0'$ as an oracle. As $f \leq_T X_1 \oplus \cdots \oplus X_n \oplus 0'$, it follows that

$$X_1 \oplus \cdots \oplus X_n \oplus 0' \leq_T (X_1 \oplus \cdots \oplus X_n)' \leq_T f \oplus 0' \leq_T X_1 \oplus \cdots \oplus X_n \oplus 0'$$

so $X_1 \oplus \cdots \oplus X_n$ is generalized low. □

Remark 6.4: Only property (i) of the Π^0_1 class of Theorem 6.1 is needed for the proofs of Proposition 6.3(a)–(c), with property (ii) being needed only for the proof of Proposition 6.3(d).

Corollary 6.5: *There exists a special Π^0_1 class having neither the Posner-Robinson Property nor the Pseudojump Inversion Property.*

Proof: Let P be the Π^0_1 class of Theorem 6.1. By Proposition 6.3(d), $X_1 \oplus \cdots \oplus X_n \oplus 0' \equiv_T (X_1 \oplus \cdots \oplus X_n)'$ for every $X_1, \ldots, X_n \in P$.

To see that P does not have the Posner-Robinson Property, note that by the Low Basis Theorem [3, Theorem 2.1, p. 36] there exists $Z \in P$ such that $Z' \equiv_{\mathrm{T}} 0'$. Thus, for any $B \in P$, $B \oplus Z$ is generalized low. Now, if $0' \equiv_{\mathrm{T}} B' \equiv_{\mathrm{T}} B \oplus Z$, then

$$B \oplus Z \equiv_{\mathrm{T}} B \oplus Z \oplus 0' \equiv_{\mathrm{T}} (B \oplus Z)'$$

a contradiction.

To see that P does not have the Pseudojump Inversion Property, let e be such that for all $X \in \{0,1\}^{\mathbb{N}}$, $X <_{\mathrm{T}} \mathrm{J}_e(X)$ and $\mathrm{J}_e(X)' \equiv_{\mathrm{T}} X'$, and suppose for the sake of a contradiction that $B \in P$ and $0' \equiv_{\mathrm{T}} \mathrm{J}_e(B)$. Then $B <_{\mathrm{T}} 0'$, so $0' \equiv_{\mathrm{T}} \mathrm{J}_e(B) \equiv_{\mathrm{T}} B \oplus 0'$. Because B is generalized-low, $\mathrm{J}_e(B) \equiv_{\mathrm{T}} B \oplus 0' \equiv_{\mathrm{T}} B'$. But $\mathrm{J}_e(B)' \equiv_{\mathrm{T}} B'$, giving a contradiction. □

We've shown that there are special Π^0_1 classes which has both the Posner-Robinson Property and the Pseudojump Inversion Property, which has the Pseudojump Inversion Property but not the Posner-Robinson Property, and which has neither the Posner-Robinson Property nor the Pseudojump Inversion Property. This suggests the following open question.

Question 6.6: Does there exist a Π^0_1 class P which has the Posner-Robinson Property but not the Pseudojump Inversion Property?

The remainder of the section is devoted to proving Theorem 6.1. The proof involves a finite-injury priority method using tree maps.

6.1. *Tree maps*

Definition 6.7: (Tree Map). A **tree map** is a map $T : \{0,1\}^* \to \{0,1\}^*$ such that for every $\sigma \in \{0,1\}^*$ and $i \in \{0,1\}$

$$T(\sigma^\frown\langle i\rangle) \supseteq T(\sigma)^\frown\langle i\rangle.$$

The **tree associated with** T is the tree

$$\mathrm{Tree}(T) := \{\tau \in \{0,1\}^* \mid \tau \subseteq T(\sigma) \text{ for some } \sigma \in \{0,1\}^*\}.$$

For a tree map T, let $[T]$ be the set of paths through $\mathrm{Tree}(T)$.

Proposition 6.8: *Suppose T is a tree map and* $\mathrm{Tree}(T)$ *is the tree associated with T.*

(a) $|T(\sigma)| \geq |\sigma|$ *for every* $\sigma \in \{0,1\}^*$.
(b) If $\sigma \subseteq \tau$, *then* $T(\sigma) \subseteq T(\tau)$.

(c) Tree(T) *is recursive in* T.
(d) $[T]$ *is perfect.*

Proof:

(a) $|T(\sigma^\frown\langle i\rangle)| \geq |T(\sigma)^\frown\langle i\rangle| = |T(\sigma)| + 1$, so use induction on the length of σ.
(b) Induction on the length of ρ shows that $T(\sigma)^\frown\rho \subseteq T(\sigma^\frown\rho)$. Thus, if $\sigma \subseteq \tau$, there is ρ such that $\sigma^\frown\rho = \tau$, from which we find

$$T(\sigma) \subseteq T(\sigma)^\frown\rho \subseteq T(\sigma^\frown\rho) = T(\tau).$$

(c) To determine if $\tau \subseteq T(\sigma)$ for some $\sigma \in \{0,1\}^*$, it suffices to check whether $\tau \subseteq T(\sigma)$ for some $\sigma \in \{0,1\}^*$ for which $|\sigma| \leq |\tau|$.
(d) Suppose $X \in [T]$ is isolated, so there is $\sigma \in \{0,1\}^*$ such that $T(\sigma) \subset X$ and X is the unique path in Tree(T) extending $T(\sigma)$. Consider

$$Y_0 := \bigcup_{n\in\omega} T(\sigma^\frown\langle 0\rangle^n) \quad \text{and} \quad Y_1 := \bigcup_{n\in\omega} T(\sigma^\frown\langle 1\rangle^n).$$

By (b) above, Y_0 and Y_1 are reals, and by construction are paths through Tree(T). That $Y_0 \supset T(\sigma)^\frown\langle 0\rangle$ and $Y_1 \supset T(\sigma)^\frown\langle 1\rangle$ implies $Y_0 \neq Y_1$. Thus, at least one of Y_0 and Y_1 are unequal to X, contradicting the hypothesis that X was an isolated element of $[T]$. □

Our finite-injury argument involves constructing a sequence $(T_s)_{s\in\mathbb{N}}$ of tree maps which satisfy the conditions

(i) $\operatorname{im} T_{s+1} \subseteq \operatorname{im} T_s$ and
(ii) $\lim_{s\to\infty} T_s(\sigma)$ exists for each $\sigma \in \{0,1\}^*$.

Condition (i) implies Tree(T_{s+1}) $\subseteq$ Tree(T_s), and our desired Π^0_1 class will be $\bigcap_{s\in\mathbb{N}}[T_s]$. Condition (ii) shows that there are only finitely many injuries in the construction.

Proposition 6.9: *Suppose* $(T_s)_{s\in\mathbb{N}}$ *is a sequence of tree maps satisfying*

(i) $\operatorname{im} T_{s+1} \subseteq \operatorname{im} T_s$ *and*
(ii) $\lim_{s\to\infty} T_s(\sigma)$ *exists for each* $\sigma \in \{0,1\}^*$.

Define $T_\infty : \{0,1\}^* \to \{0,1\}^*$ *by* $T_\infty(\sigma) := \lim_{s\to\infty} T_s(\sigma)$.

(a) T_∞ *is a tree map.*
(b) If the tree maps T_s *are uniformly recursive in* X*, then* T_∞ *is recursive in* X'.
(c) $\bigcap_{s\in\mathbb{N}}[T_s] = [T_\infty]$.

Proof:

(a) Suppose $\sigma \in \{0,1\}^*$ and $i \in \{0,1\}$. Then for sufficiently large s, $T_s(\sigma^\frown\langle i\rangle) = T_\infty(\sigma^\frown\langle i\rangle)$ and $T_s(\sigma) = T_\infty(\sigma)$, so

$$T_\infty(\sigma^\frown\langle i\rangle) = T_s(\sigma^\frown\langle i\rangle) \supseteq T_s(\sigma)^\frown\langle i\rangle = T_\infty(\sigma)^\frown\langle i\rangle.$$

(b) Define

$$f(\sigma) = \text{least } s \text{ such that } T_s(\sigma) = T_t(\sigma) \text{ for all } t \geq s$$

which is total (by hypothesis) and recursive in X'. Then

$$T_\infty(\sigma) = T_{f(\sigma)}(\sigma)$$

is recursive in $X \oplus X' \equiv_{\mathrm{T}} X'$.

(c) Suppose X is a path through $\mathrm{Tree}(T_s)$ for every s. Consider the collection of strings in $\{0,1\}^*$ of length n. There are only finitely many such strings, so there exists an s such that $T_s(\sigma) = T_\infty(\sigma)$ whenever $|\sigma| = n$. As X is a path through $\mathrm{Tree}(T_s)$, there are arbitrarily long ρ such that $T_s(\rho) \subset X$. But then there is such a ρ and a σ of length n such that $\sigma \subseteq \rho$ and $T_s(\rho) \subset X$. Proposition 6.8(b) shows that

$$T_\infty(\sigma) = T_s(\sigma) \subseteq T_s(\rho) \subset X.$$

As n is arbitrary, Proposition 6.8(a) shows that arbitrarily large initial segments of X are in $\mathrm{Tree}(T_\infty)$, and hence $X \in [T_\infty]$.
Conversely, suppose X is a path through $\mathrm{Tree}(T_\infty)$. $X \upharpoonright n \in \mathrm{Tree}(T_\infty)$, so there is σ such that $X \upharpoonright n \subseteq T_\infty(\sigma)$. There exists s_0 such that $T_\infty(\sigma) = T_t(\sigma)$ for every $t \geq s_0$. This shows that $X \upharpoonright n \in \mathrm{Tree}(T_t)$ for every $t \geq s_0$. As $\mathrm{im}\, T_{s+1} \subseteq \mathrm{im}\, T_s$, it follows that $\mathrm{Tree}(T_{s+1}) \subseteq \mathrm{Tree}(T_s)$. Thus, $X \upharpoonright n \in \mathrm{Tree}(T_s)$ for every s. As n was arbitrary, it follows that $X \in [T_s]$ for every s. □

6.2. *Proof of Theorem 6.1*

Given a Π^0_1 class P, $X \in \{0,1\}^{\mathbb{N}}$, and $W \subseteq \{0,1\}^{\mathbb{N}}$, temporarily say that X is *good for W in P* if there exists $\sigma \subset X$ such that $N_\sigma \cap P^n \subseteq W$ or $N_\sigma \cap P^n \subseteq W^{\mathrm{c}}$. Likewise, if $S \subseteq \{0,1\}^*$, then X is *good for S in P* if there exists $\sigma \subseteq X$ such that either $\sigma \in S$ or there exists no $\tau \supseteq \sigma$ such that both $\tau \in S$ and $\exists Y \in P^n\,(\tau \subset Y)$ hold.

Proposition 6.10: *Suppose P is a Π^0_1 class, $X \in \{0,1\}^{\mathbb{N}}$, $S \subseteq \{0,1\}^*$, and $W = \{Y \in \{0,1\}^{\mathbb{N}} \mid \exists\sigma\,(\sigma \subset Y \wedge \sigma \in S)\}$. Then the following are equivalent:*

(i) X *is good for* S *in* P.
(ii) X *is good for* $\hat{S} = \{\tau \in \{0,1\}^* \mid \exists \sigma\, (\sigma \in S \wedge \sigma \subseteq \tau)\}$ *in* P.
(iii) X *is good for* W *in* P.

Proof:

(i) $\implies$ *(ii)*: Take $\sigma \subset X$ such that either $\sigma \in S$ or there does not exist $\tau \supseteq \sigma$ such that both $\tau \in S$ and $\exists Y \in P^n\, (\tau \subset Y)$.
If the former condition $\sigma \in S$ holds, then $\sigma \in \hat{S}$ as $S \subseteq \hat{S}$.
If the latter condition holds, then suppose there is $\tau \supseteq \sigma$ with $\tau \in \hat{S}$ and $Y \in P^n \cap N_\tau$. By definition, there is $\tau_0 \subseteq \tau$ such that $\tau_0 \in S$, so in particular τ_0 is compatible with σ. If $\sigma \subseteq \tau_0$, then this contradicts our hypothesis, so $\tau_0 \subseteq \sigma$ and hence $\tau_0 \subset X$ and $\tau_0 \in S \subseteq \hat{S}$.
(ii) $\implies$ *(iii)*: Note that

$$\begin{aligned} W &= \{Y \in \{0,1\}^{\mathbb{N}} \mid \exists \sigma\, (\sigma \subset Y \wedge \sigma \in S)\} \\ &= \{Y \in \{0,1\}^{\mathbb{N}} \mid \exists \sigma\, (\sigma \subset Y \wedge \sigma \in \hat{S})\}. \end{aligned}$$

Take $\sigma \subset X$ such that either $\sigma \in \hat{S}$ or there does not exist $\tau \supseteq \sigma$ such that both $\tau \in \hat{S}$ and $\exists Y \in P^n(\tau \subset Y)$ hold.
If the former condition $\sigma \in \hat{S}$ holds, then any $Y \in N_\sigma \cap P^n$ is an element of W, i.e., $N_\sigma \cap P^n \subseteq W$.
If the latter condition holds, then $Y \in N_\sigma \cap P^n \cap W$ implies that there is $\tau \subset Y$ such that $\tau \in \hat{S}$. That $Y \in N_\sigma$ implies τ and σ are compatible. But then the fact that $\hat{S}$ is upward closed and $\sigma \notin \hat{S} \wedge \tau \in \hat{S}$ implies $\sigma \subseteq \tau$, contradicting our hypothesis.
(iii) $\implies$ *(i)*: Take $\sigma \subset X$ such that either $N_\sigma \cap P^n \subseteq W$ or $N_\sigma \cap P^n \subseteq W^c$.
If the former condition $N_\sigma \cap P^n \subseteq W$ holds, then in particular $X \in W$. Thus, there is $\sigma \subset X$ such that $\sigma_0 \in S$.
If the latter condition $N_\sigma \cap P^n \cap W = \emptyset$ holds, then if there is $\tau \supseteq \sigma$ and $Y \in P^n$ such that $\tau \subset Y$ and $\tau \in S$, then $Y \in N_\tau \cap P^n \cap W$, a contradiction. □

We need some additional notation:

Notation. For $e \in \mathbb{N}$, define

$$S_e := \{\sigma \in \{0,1\}^* \mid \varphi_e^{(1),\sigma}(0) \downarrow\}.$$

Additionally, for $s \in \mathbb{N}$, define

$$S_e\!\downarrow_s := \{\sigma \in \{0,1\}^* \mid \varphi_{e,s}^{(1),\sigma}(0) \downarrow\}.$$

Remark 6.11:

- $S \subseteq \{0,1\}^*$ is Σ^0_1 and upwards closed, i.e., $\sigma \in S$ and $\sigma \subseteq \tau$ implies $\tau \in S$, if and only if there is e such that $S = S_e$.
- For each $e, s \in \mathbb{N}$, $S_e{\downarrow_s}$ is recursive for each s and $S_e = \bigcup_{s\in\mathbb{N}} S_e{\downarrow_s}$.

Our proof generalizes that of [3, Theorem 4.7, p. 48].

Proof: [Proof of Theorem 6.1.] We assume without loss of generality that for each e, there are infinitely-many e' such that $S_e = S_{e'}$.

We define a recursive sequence $(T_s)_{s\in\mathbb{N}}$ of recursive tree maps such that

(i) $\operatorname{im} T_{s+1} \subseteq \operatorname{im} T_s$ for all $s \in \mathbb{N}$ and
(ii) $\lim_{s\to\infty} T_s(\sigma)$ exists for each $\sigma \in \{0,1\}^*$.

We let $T_\infty(\sigma) := \lim_{s\to\infty} T_s(\sigma)$. Condition (i) above implies $\operatorname{Tree}(T_{s+1}) \subseteq \operatorname{Tree}(T_s)$. Our desired P will be $[T_\infty] = \bigcap_{s\in\mathbb{N}} [T_s]$.

Our goal will be to arrange for $T_\infty(\sigma_1) \oplus \cdots \oplus T_\infty(\sigma_n) \in S_e$ whenever possible. An *injury at stage s* is a quadruple e, n, σ, σ' such that

(1) $e \leq s$ and n divides both $|\sigma|$ and $|\sigma'|$ and $n < s$,
(2) $|\sigma| = n \cdot 2^n \cdot 3^e$ and $|\sigma'|^* = n \cdot 2^s \cdot 3^e$,
(3) $\sigma \subseteq \sigma'$,
(4) for each $0 \leq k < \ell < n$, $(\sigma)_{k,n} \neq (\sigma)_{\ell,n}$, and
(5) $T_s((\sigma)_{0,n}) \oplus \cdots \oplus T_s((\sigma)_{n-1,n}) \notin S_e{\downarrow_s}$ but $T_s((\sigma')_{0,n}) \oplus \cdots \oplus T_s((\sigma')_{n-1,n}) \in S_e{\downarrow_s}$.

With this goal in mind, we define the tree maps T_s as follows:

Stage 0: Let $T_0 := \operatorname{id}_{\{0,1\}^*}$.
Stage $s+1$: If there is no injury at stage s, let $T_{s+1} = T_s$.

Otherwise, let e, n, σ, σ' be the minimal injury, ordered lexicographically. Define T_{s+1} as follows:

$$T_{s+1}(\nu) := \begin{cases} T_s((\sigma')_{k,n}{}^\frown\nu') & \text{if } \nu = (\sigma)_{k,n}{}^\frown\nu' \text{ for some } k, \\ T_s(\nu) & \text{otherwise.} \end{cases}$$

If $T_{s+1}(\nu) \neq T_s(\nu)$, then we say that e, n, σ, σ' *injures ν at Stage s*, or simply *injures ν*.

This completes the construction.

We make the following observations:

Claim 1: The functions T_s are uniformly recursive.

Proof: Induction on s. T_0 is clearly recursive, so suppose T_s is recursive.
Because of the bounds on e, n, $|\sigma|$, and $|\sigma'|$, checking whether there are injuries at Stage s is recursive.
Once the minimal injury at Stage s is known, the definition of T_{s+1} is clearly recursive. □

Claim 2: T_s is a tree map for each s.

Proof: Induction on s. Suppose $\nu \in \{0,1\}^*$ and $i \in \{0,1\}$. If $T_{s+1} = T_s$, then T_{s+1} is a tree map by our inductive hypothesis. Otherwise, take e, n, σ, σ' to be the minimal injury at Stage s. Then

$$\begin{aligned}
T_{s+1}(\nu^\frown\langle i\rangle) &= \begin{cases} T_s((\sigma')_{k,n}{}^\frown\nu'^\frown\langle i\rangle) & \text{if } \nu = (\sigma)_{k,n}{}^\frown\nu' \text{ for some } k,\\ T_s(\nu^\frown\langle i\rangle) & \text{if } (\sigma)_{k,n} \not\subseteq \nu \text{ for all } k,\\ T_s((\sigma')_{k,n}) & \text{if } \nu^\frown\langle i\rangle = (\sigma)_{k,n} \text{ for some } k,\end{cases}\\
&\supseteq \begin{cases} T_s((\sigma')_{k,n}{}^\frown\nu')^\frown\langle i\rangle & \text{if } \nu = (\sigma)_{k,n}{}^\frown\nu' \text{ for some } k,\\ T_s(\nu)^\frown\langle i\rangle & \text{if } (\sigma)_{k,n} \not\subseteq \nu \text{ for all } k,\\ T_s(\nu^\frown\langle i\rangle) & \text{if } \nu^\frown\langle i\rangle = (\sigma)_{k,n} \text{ for some } k,\end{cases}\\
&\supseteq \begin{cases} T_s((\sigma')_{k,n}{}^\frown\nu')^\frown\langle i\rangle & \text{if } \nu = (\sigma)_{k,n}{}^\frown\nu' \text{ for some } k,\\ T_s(\nu)^\frown\langle i\rangle & \text{if } (\sigma)_{k,n} \not\subseteq \nu \text{ for all } k,\\ T_s(\nu)^\frown\langle i\rangle & \text{if } \nu^\frown\langle i\rangle = (\sigma)_{k,n} \text{ for some } k,\end{cases}\\
&= T_{s+1}(\nu)^\frown\langle i\rangle.
\end{aligned}$$
□

Claim 3: $\operatorname{im} T_{s+1} \subseteq \operatorname{im} T_s$.

Proof: By construction. □

Claim 4: $\lim_{s\to\infty} T_s(\nu)$ exists for each $\nu \in \{0,1\}^*$. In other words, the sequence $(T_s(\nu))_{s\in\mathbb{N}}$ is eventually constant.

Proof: Induction on the length of ν.
Suppose first that $\nu = \langle\rangle$. If $T_{s+1}(\langle\rangle) \neq T_s(\langle\rangle)$, then there is a minimal injury at Stage s, e, n, σ, σ', such that $(\sigma)_{k,n} \subseteq \langle\rangle$ for some k. But the minimal possible length of $(\sigma)_{k,n}$ is 1. Thus, $T_s(\langle\rangle) = T_{s+1}(\langle\rangle)$ for every s.

Now suppose for our induction hypothesis that $\lim_{s\to\infty} T_s(\sigma)$ exists for every σ of length strictly less than $|\nu|$.

Let ν be the first string of length $|\nu|$ for which $T_{s+1}(\nu) \neq T_s(\nu)$ and $T_s(\sigma) = T_\infty(\sigma)$ for all σ with $|\sigma| < |\nu|$. Let e, n, σ, σ' be the minimal injury at Stage s. As ν is injured by e, n, σ, σ' at Stage s, we must have $(\sigma)_{k,n} \subseteq \nu$ for some k. By the hypothesis that $T_{s+1}(\rho) = T_s(\rho)$ for every ρ of length strictly less than $|\nu|$, it must be that $(\sigma)_{k,n} = \nu$ for some k. Consequently, $T_{s+1}(\nu) = T_s((\sigma')_{k,n})$.

Now suppose a string ρ of length $|\nu|$ is injured at Stage $t > s$ by some i, m, τ, τ'. Thus, there is some ℓ for which $(\tau)_{\ell,n} \subseteq \rho$. As before, we have in fact that $(\tau)_{\ell,n} = \rho$, so in particular $|\tau| = n \cdot |\rho| = n \cdot |\nu| = |\sigma|$, implying $e = i$ and $n = m$. Suppose for the sake of a contradiction that $\sigma = \tau$. Then the hypothesis that $i = e, m = n, \tau = \sigma, \tau'$ is an injury at Stage t implies $T_t((\sigma)_{0,n}) \oplus \cdots \oplus T_t((\sigma)_{n-1,n}) \notin S_e{\downarrow}_t$ but $T_t((\tau')_{0,n}) \oplus \cdots \oplus T_t((\tau')_{n-1,n}) \in S_e{\downarrow}_t$. Because $T_s(\rho) = T_\infty(\rho)$ for all ρ with $|\rho| < |\nu|$, if $|\rho| = |\nu|$ then $T_t(\rho)$ is monotonically increasing as a function of t for $t \geq s$. The fact that $T_{s+1}((\sigma)_{0,n}) \oplus \cdots \oplus T_{s+1}((\sigma)_{n-1,n}) \in S_e{\downarrow}_s$ implies $T_t((\sigma)_{0,n}) \oplus \cdots \oplus T_t((\sigma)_{n-1,n}) \in S_e{\downarrow}_s \subseteq S_e{\downarrow}_t$. Contradiction.

Continuing in this way shows that any injuries i, m, τ, τ' of a string ρ of length $|\nu|$ must satisfy $i = e$ and $m = n$, and that moreover all such τ must be pairwise distinct. It follows that the strings of length $|\nu|$ can only be injured finitely-many times over the course of the construction. This completes the proof by induction. □

Let $P = [T_\infty] = \bigcap_{s\in\mathbb{N}} [T_s]$, which is Π^0_1 as the intersection of uniformly Π^0_1 classes (using the hypothesis that each T_s is recursive and hence T_s is recursive by Proposition 6.8(c)) and non-empty by compactness.

Suppose $X_1, \ldots, X_n \in P$ are pairwise distinct and $e \in \mathbb{N}$. We wish to show that there exists an m such that either $\sigma := (X_1 \upharpoonright m) \oplus \cdots \oplus (X_n \upharpoonright m) \in S_e$ or else there is no $\tau \supseteq \sigma$ such that $\tau \in S_e$ and $(\tau)_{0,n}, \ldots, (\tau)_{n-1,n} \in \mathrm{im}(T_\infty)$.

To this effect, choose σ such that $|\sigma| = n \cdot 2^n \cdot 3^e$ and such that $T_\infty((\sigma)_{0,n}) \oplus \cdots \oplus T_\infty((\sigma)_{n-1,n}) \subset X_1 \oplus \cdots \oplus X_n$. By replacing with a larger e as necessary, we may assume without loss of generality that $(\sigma)_{0,n}, \ldots, (\sigma)_{n-1,n}$ are distinct. Choose s large enough so that there are no injuries at Stage s of the form e, n, σ, σ'.

Suppose $T_s((\sigma)_{0,n}) \oplus \cdots \oplus T_s((\sigma)_{n-1,n}) \notin S_e$ and suppose for the sake of a contradiction that there exists $\sigma' \supseteq \sigma$ such that $T_\infty((\sigma')_{0,n}) \oplus \cdots \oplus T_\infty((\sigma')_{n-1,n}) \in S_e$. By potentially making s larger, we may assume without loss of generality that $|\sigma'| = n \cdot 2^s \cdot 3^e$ and that $T_s((\sigma')_{0,n}) \oplus \cdots \oplus T_s((\sigma')_{n-1,n}) \in S_e{\downarrow}_s$. But this means that e, n, σ, σ' is an injury at Stage s, a contradiction. Thus, such a σ' cannot exist. Note that the above determination of σ is uniformly recursive in $X_1 \oplus \cdots \oplus X_n \oplus 0'$ as a function of e and n.

Finally, Proposition 6.8(d) implies P has no isolated points. By Lemma 6.2, this shows P is special.

This completes the proof. □

References

1. Rodney G. Downey and Denis R. Hirschfeldt, *Algorithmic Randomness and Complexity*, Theory and Applications of Computability, Springer-Verlag, New York, 2010, xxviii + 855 pages.
2. Carl G. Jockusch and Richard A. Shore, *Pseudojump operators. I: The r.e. case*, Transactions of the American Mathematical Society **275** (1983), no. 2, 599–609.
3. Carl G. Jockusch and Robert I. Soare, Π^0_1 *classes and degrees of theories*, Transactions of the American Mathematical Society **173** (1972), 35–56.
4. Patrick Lutz, private communications, August 9 2019.
5. Elliott Mendelson, *Introduction to Mathematical Logic*, 5th ed., Chapman & Hall/CRC, 2009, xxiv + 469 pages.
6. Andre Nies, *Computability and randomness*, Oxford University Press, Inc., New York, NY, USA, 2009, xv + 434 pages.
7. David B. Posner and Robert W. Robinson, *Degrees joining to* $0\prime$, The Journal of Symbolic Logic **46** (1981), no. 4, 714–722.
8. Hartley Rogers, Jr, *Theory of Recursive Functions and Effective Computability*, McGraw-Hill Series in Higher Mathematics, McGraw-Hill, 1967, xix + 482 pages.
9. Stephen G. Simpson, *Mass problems and randomness*, Bulletin of Symbolic Logic **11** (2005), no. 1, 1–27.
10. ———, *Mass problems and almost everywhere domination*, Mathematical Logic Quarterly **53** (2007), no. 4-5, 483–492.
11. ———, *Degrees of Unsolvability. Lectures Notes for "Mass Problems", Spring 2009 at the Pennsylvania State University*, 2009, URL: `https://sgslogic.net/t20/notes/dou.pdf`. Last visited on 2019/04/19.

https://doi.org/10.1142/9789819806584_0007

A Tractable Case of the Turing Automorphism Problem: Bi-uniformly E_0-invariant Cantor Homeomorphisms

Bjørn Kjos-Hanssen

Department of Mathematics, University of Hawaii at Manoa
2565 McCarthy Mall, Honolulu HI 96822, USA
bjoernkh@hawaii.edu

A function $F : 2^\omega \to 2^\omega$ is an E_0-isomorphism if for all $x, y \in 2^\omega$, we have $xE_0y \leftrightarrow f(x)E_0f(y)$, where $xE_0y \leftrightarrow (\exists a)(\forall n \geq b)x(n) = y(n)$. If such witnesses a for xE_0y and for $f(x)E_0f(y)$ depend on each other but not on x, y, then F is called bi-uniform. It is shown that a homeomorphism of Cantor space which is a bi-uniform E_0-isomorphism can induce only the trivial automorphism of the truth-table degrees.

Dedicated to the celebration of the work of
Theodore A. Slaman and W. Hugh Woodin

Contents

1. Introduction

Let $\mathscr{D}_\mathrm{T}$ denote the set of Turing degrees and let $\leq$ denote its ordering. This chapter gives a partial answer to the following famous question.

This chapter was partially supported by grants from the Simons Foundation (#315188 and #704836 to Bjørn Kjos-Hanssen) and by the Institute for Mathematical Sciences, National University of Singapore.

Question 1.1: Does there exist a nontrivial automorphism of $\mathscr{D}_{\mathrm{T}}$?

Definition 1.2: A bijection $\pi : \mathscr{D}_{\mathrm{T}} \to \mathscr{D}_{\mathrm{T}}$ is an *automorphism* of $\mathscr{D}_{\mathrm{T}}$ if for all $\mathbf{x}, \mathbf{y} \in \mathscr{D}_{\mathrm{T}}$, $\mathbf{x} \le \mathbf{y}$ iff $\pi(\mathbf{x}) \le \pi(\mathbf{y})$. If moreover there exists an $\mathbf{x}$ with $\pi(\mathbf{x}) \ne \mathbf{x}$ then π is *nontrivial.*

Question 1.1 has a long history. Already in 1977, Jockusch and Solovay [3] showed that each jump-preserving automorphism of the Turing degrees is the identity above $\mathbf{0}^{(4)}$. Nerode and Shore [5] showed that each automorphism (not necessarily jump-preserving) is equal to the identity on some cone $\{\mathbf{a} : \mathbf{a} \ge \mathbf{b}\}$. Slaman and Woodin [6, 7] showed that each automorphism is equal to the identity on the cone above $\mathbf{0}''$ and that $\mathrm{Aut}(\mathcal{D})$ is countable.

There is an obstacle to reducing the base of the cone to $\mathbf{0}'$ and ultimately $\mathbf{0}$: Turing reducibility is Σ^0_3, but not Π^0_2 or Σ^0_2 in the sense of descriptive set theory.

In the other direction, S. Barry Cooper [1] claimed to construct a nontrivial automorphism, induced by a discontinuous function on ω^ω, itself induced by a function on $\omega^{<\omega}$. That claim was not independently verified. In [4] we attacked the problem by ruling out a certain simple but natural possibility: automorphisms induced by permutations of finite objects. We showed that no permutation of ω represents a nontrivial automorphism of the Turing degrees. That proof was too complicated, in a way, and did not extend from D_T down to D_m. Here we give a more direct proof using the shift map $n \mapsto n+1$. Our proof here will generalize to a certain class of homeomorphisms, distinct from the class of such homeomorphisms that the result in [4] generalizes to.

2. Excluding Permutations by Recursion

Lemma 2.1: *Suppose $\theta : \omega \to \omega$ is a bijection such that $\theta^{-1} \circ S \circ \theta$ is computable, where S is the successor function given by $S(n) = n+1$. Then θ is computable.*

Proof: Let $k(n) = (\theta^{-1} \circ S \circ \theta)(n)$. Then for any m, $k(\theta^{-1}(m)) = \theta^{-1}(S(m))$ and so we compute θ^{-1} by recursion:

$$\theta^{-1}(m+1) = k(\theta^{-1}(m)). \qquad \square$$

We write $\sigma \preceq \tau$ if σ is a prefix of τ.

Lemma 2.2: *Suppose $\sigma \in 2^{<\omega}$, $g : \omega \to \omega$, and Φ is a Turing functional, satisfying*

$$(\forall n)(\forall \tau \succeq \sigma)(\exists \rho \succeq \tau)(\Phi^\rho(n) \downarrow \text{ and } \rho \circ g(n) = \Phi^\rho(n)). \tag{2.1}$$

For any $\rho \succeq \sigma$ and n, if $\Phi^\rho(n) \downarrow$, then $g(n) < |\rho|$.

Proof: If instead $g(n) \geq |\rho|$ then $\rho(g(n))$ is undefined. So let $\tau \succeq \rho$, $\tau(g(n)) = 1 - \Phi^\rho(n)$. This τ violates (2.1). □

Lemma 2.3: *If $g : \omega \to \omega$ is injective and Φ is a Turing functional such that*

$$\{B : B \circ g = \Phi^B\}$$

is nonmeager, then g is computable.

Proof: By assumption, it is not the case that

$$(\forall \sigma)(\exists n)(\exists \tau \succeq \sigma)(\forall \rho \succeq \tau)(\Phi^\rho(n) \downarrow \to \rho \circ g(n) \neq \Phi^\rho(n)).$$

So we have

$$(\exists \sigma)(\forall n)(\forall \tau \succeq \sigma)(\exists \rho \succeq \tau)(\Phi^\rho(n) \downarrow \text{ and } \rho \circ g(n) = \Phi^\rho(n)).$$

Pick such a σ: then Φ cannot make a mistake above σ, and we can always extend to get the right answer.

As finite data we assume we know the values of n and $g(n)$ for which $g(n) < |\sigma|$.

We compute the value $g(n)$ as follows.

Check the finite database of $\{(k, g(k)) : g(k) < |\sigma|\}$, and output $g(n)$ if found. Otherwise we know $g(n) \geq |\sigma|$.

By dovetailing computations, find a $\rho_0 \succeq \sigma$ such that $\Phi^{\rho_0}(n) \downarrow$. By Lemma 2.2, we have that $g(n) < |\rho_0|$. Thus, $g(n) \in I$ where I is the closed interval $[|\sigma|, |\rho_0| - 1]$. Let $a \in I, b \in I, a < b$. It suffices to show how to eliminate either a or b as a candidate for being equal to $g(n)$.

Let $\tau \succ \sigma$ be such that $\tau(a) \neq \tau(b)$ and let $\rho \succeq \tau$ be such that $\Phi^\rho(n) \downarrow$. Then mark as eliminated whichever $c \in \{a, b\}$ makes $\rho(c) \neq \Phi^\rho(n)$. Thus we one-by-one eliminate all $a \in I$ until only one candidate remains. □

Definition 2.4: We say that a permutation $\theta : \omega \to \omega$ *induces* the automorphism π of $\mathcal{D}_r$ if, letting $\Pi = \Pi_\theta$ be defined by $\Pi(A) = A \circ \theta$ for all A, we have $\pi([X]_r) = [\Pi(X)]_r$ for all X.

Lemma 2.5: *If θ is a permutation of ω and induces an automorphism π of $\mathcal{D}_r$ then θ^{-1} induces π^{-1}.*

Proof: We must show that $\pi^{-1}([X]_r) = [\Pi_{\theta^{-1}}(X)]_r$. We have $\Pi_\theta^{-1} = \Pi_{\theta^{-1}}$, since

$$\Pi_{\theta^{-1}}(\Pi_\theta(X)) = (X \circ \theta) \circ \theta^{-1} = X = (X \circ \theta^{-1}) \circ \theta = \Pi_\theta(\Pi_{\theta^{-1}}(X)).$$

We have the well-definedness condition $X \equiv_r Y \Rightarrow \Pi(X) \equiv_r \Pi(Y)$. Moreover, since π is injective, it must be that $X \equiv_r Y \Leftrightarrow \Pi(X) \equiv_r \Pi(Y)$. Since $\Pi_\theta^{-1} = \Pi_{\theta^{-1}}$, we also have

$$\Pi^{-1}(X) \equiv_r \Pi^{-1}(Y) \Leftrightarrow X \equiv_r Y \leftrightarrow \Pi_{\theta^{-1}}(X) \equiv_r \Pi_{\theta^{-1}}(Y),$$

so that θ^{-1} also induces the automorphism π^{-1}:

$$[\Pi^{-1}(Y)]_r = [X]_r \Leftrightarrow \Pi^{-1}(Y) \equiv_r X \Leftrightarrow Y \equiv_r \Pi(X)$$
$$\Leftrightarrow [Y]_r = [\Pi(X)]_r \Leftrightarrow [Y]_r = \pi([X]_r) \Leftrightarrow \pi^{-1}([Y]_r) = [X]_r. \qquad \square$$

Theorem 2.6: *No permutation of the integers can induce a nontrivial automorphism of $\mathcal{D}_r$, for any reducibility $\leq_r$ between $\leq_1$ and $\leq_T$.*

Proof: Suppose $\theta : \omega \to \omega$ is a permutation (bijection) and that θ induces an automorphism of $\mathcal{D}_r$. Thus, letting $\Pi(A) = A \circ \theta$ and $\pi([A]_r) = [\Pi(A)]_r$, π is well-defined and is an automorphism of $\mathcal{D}_r$.

Recall that S from Definition 2.1 is simply the successor function given by $S(n) = n+1$. For any B, let $A = B \circ \theta^{-1} \circ S$ (so $x \in A$ iff $S(x) \in B \circ \theta^{-1}$). We have $A \leq_1 B \circ \theta^{-1}$ and so by assumption $A \circ \theta \leq_T B \circ \theta^{-1} \circ \theta = B$. This gives $(B \circ \theta^{-1} \circ S) \circ \theta = \Phi^B$ for some Turing functional Φ. Let $g = \theta^{-1} \circ S \circ \theta$. Since for each B there exists such a Φ, there must be some Φ such that the G_δ set

$$\{B : B \circ g = \Phi^B\}$$

is nonmeager. By Lemma 2.3, g is computable. By Lemma 2.1, θ is computable.

But this means that for any A, $A \circ \theta \leq_1 A$, so that the map Π is everywhere-decreasing: $\Pi(X) \leq_1 X$, and in particular

$$\pi([X]_r) \leq [X]_r \tag{2.2}$$

where $\leq$ is the ordering of $\mathcal{D}_r$. By Lemma 2.5, π^{-1} is also induced by a permutation, namely θ^{-1}. Applying (2.2) to θ^{-1} we get

$$\pi^{-1}([X]_r) \leq [X]_r$$

for all X, and so $\pi([X])_r = [X]_r$ for all X. $\square$

3. Excluding Bi-uniformly E_0-invariant Homeomorphisms

Making computability-theoretic uniformity assumptions is one way to rule out certain possible Turing automorphisms. We will instead focus on uniformity of a simpler, combinatorial kind.

Let $[a, \infty) = \{n \in \omega : a \leq n\}$. As usual, we write $X =^* Y$ if $\{n \in \omega : X(n) \neq Y(n)\}$ is finite. The equivalence relation $=^*$ is also known as E_0. We refine this to

$$X =^*_n Y \iff X \upharpoonright [n, \infty) = Y \upharpoonright [n, \infty).$$

As a relation, $=^*$ is the union $\bigcup_n =^*_n$. A map F is E_0-invariant, or an E_0-endomorphism, if for all X, Y, if $X =^* Y$ then $F(X) =^* F(Y)$. In terms of the refinements $=^*_n$, this means that for all X, Y, and a, if $X =^*_a Y$ then there is an b such that $F(X) =^*_b F(Y)$. If this b only depends on a we have a certain uniformity:

Definition 3.1: A function $F : 2^\omega \to 2^\omega$ is a *uniformly E_0-invariant* if for each a there is a b such that for all X, Y, if $X =^*_a Y$ then $F(X) =^*_b F(Y)$. If F is invertible and both F and F^{-1} are uniformly E_0-invariant then F is said to be a *bi-uniformly E_0-invariant.*

Some continuous maps $F : 2^\omega \to 2^\omega$ induce maps $\tilde{F} : 2^\omega/{=^*} \to 2^\omega/{=^*}$ but do not have the uniform property:

Example 3.2: Let $F(A) = A \setminus \min(A)$, if $A \neq \emptyset$, and $F(A) = A$ otherwise. We shall show that for $a = 1$ there is no b as required in Definition 3.1. Let b be given. Let Y be any set with $0 < \min(Y) < \infty$ and $\min(Y) \geq b$, and let $X = Y \cup \{0\}$. Thus $X =^*_1 Y$. We compute that $F(X) = Y$ and $F(Y) = Y \setminus \min(Y)$. Suppose $F(X) =^*_b F(Y)$. Then $\min(Y) < b$, a contradiction. Thus F is not uniformly E_0-invariant.

On the other hand, F is E_0-invariant. Indeed, let $d_H : 2^\omega \times 2^\omega \to \omega + 1$ denote the Hamming distance function, whereby $d_H(X, Y) = |\{n : X(n) \neq Y(n)\}|$. We have $d_H(F(X), X) \leq 1$, so that if $X =^*_a Y$ then

$$d_H(F(X), F(Y)) \leq d_H(F(X), X) + d_H(X, Y) + d_H(Y, F(Y)) \leq a + 2$$

and hence $F(X) =^* F(Y)$.

Lemma 3.3: *If $F(A) = A \circ f$ for a permutation $f : \omega \to \omega$ then F is uniformly E_0-invariant.*

Proof: Let $a \in \omega$. Let $b = \max\{f^{-1}(m) : m < a\}$. Suppose $X =^*_a Y$, i.e., $X(m) = Y(m)$ for all $m \geq a$. Let $n > b$ and let m be such that $n = f^{-1}(m)$.

By definition of b, we have $m \geq a$, so that

$$X \circ f(n) = X(f(f^{-1}(m))) = X(m) = Y(m) = Y(f(f^{-1}(m))) = Y \circ f(n),$$

giving $X \circ f =^*_{b+1} Y \circ f$. □

Define the $\star$ operator by $f^\star(A)(n) = A(f(n))$. Again, recall the successor function S from Definition 2.1.

Lemma 3.4: *Suppose $\Theta : 2^\omega \to 2^\omega$ is a homeomorphism such that the function $\Theta^{-1} \circ S^\star \circ \Theta$ is computable. Then Θ is computable.*

Proof: Let $\pi_n : \omega \to \omega$ be the constant n function. For $\pi_n^\star : 2^\omega \to 2^\omega$, note that $\pi_n^\star(A)(u) = (A \circ \pi_n)(u) = A(n)$, so $\pi_n^\star(A) \in \{0^\omega, 1^\omega\}$. Note that

$$\begin{aligned}(\pi_n^\star \circ S^\star)(A)(u) &= \pi_n^\star(S^\star(A))(u) = ((S^\star(A)) \circ \pi_n)(u) = (S^\star(A))(n) \\ &= A(S(n)) = A(n+1) = (A \circ \pi_{n+1})(u) = (\pi_{n+1}^\star(A))(u)\end{aligned}$$

so $\pi_n^\star \circ S^\star = \pi_{n+1}^\star$. Also, $S \circ \pi_n = \pi_{n+1}$. Let $\Phi = \Theta^{-1} \circ S^\star \circ \Theta$. By assumption, Φ is computable. Then

$$\begin{aligned}\Theta \circ \Phi &= S^\star \circ \Theta, \\ \text{hence} \quad (\pi_n^\star \circ \Theta) \circ \Phi &= \pi_{n+1}^\star \circ \Theta.\end{aligned}$$

Since homeomorphisms have finite use, $\pi_0^\star \circ \Theta$ is just a finite amount of information. Thus, we can recursively compute $\pi_{n+1}^\star \circ \Theta$ this way. For a concrete case, the reader may wish to inspect Example 3.8. □

Definition 3.5: For a real X and a string σ of length n,

$$(\sigma \searrow X)(n) = \begin{cases} \sigma(n) & \text{if } n < |\sigma|, \\ X(n) & \text{otherwise.} \end{cases}$$

Lemma 3.6: *Suppose $F : 2^\omega \to 2^\omega$ is a uniformly E_0-invariant continuous function. Suppose $F(X) = \Phi^X$ is forced above $\sigma \in 2^{<\omega}$, where Φ is a truth-table functional. Then F is computable.*

Proof: Indeed, let σ force $F(X) = \Phi^X$. Let $a = |\sigma|$. Let b be as in Definition 3.1. Let the truth tables for F for $n < b$ be given as a finite database. For $n \geq b$,

$$F^X(n) = F^{\sigma \searrow X}(n) = \Phi^{\sigma \searrow X}(n). \qquad \square$$

Theorem 3.7: *Let π be an automorphism of $\mathcal{D}_r$, for any reducibility $\leq_r$ between $\leq_1$ and $\leq_{tt}$. Suppose that π is induced by a homeomorphism Θ of*

2^ω that is a bi-uniform E_0-isomorphism.[1] *Then Θ is computable and π is trivial.*

Proof: The overall proof strategy mirrors that for Theorem 2.6. Suppose that letting $\pi([A]_r) = [\Theta^A]_r$ makes π well-defined and makes π an automorphism of $\mathcal{D}_r$.

Again, let S be the successor function, $S(n) = n + 1$. For any B, let $x \in A$ iff $S(x) \in \Theta^{-1}(B)$. We have $A \leq_1 \Theta^{-1}(B)$ and so by assumption $\Theta(A) \leq_r \Theta(\Theta^{-1}(B)) = B$. This gives $\Theta^A = \Phi^B$ for some truth-table functional Φ. Let $\Gamma = \Theta \circ S^\star \circ \Theta^{-1}$, which is also uniformly E_0-invariant. Since for each B there exists such a Φ, there must be some Φ such that the G_δ set

$$\{B : \Gamma(B) = \Phi^B\}$$

is nonmeager. By Lemma 3.6, Γ is computable. By Lemma 3.4, Θ is computable.

But this means that for any A, $\Theta(A) \leq_{tt} A$, and in particular

$$\pi([X]_r) \leq [X]_r \tag{3.1}$$

where $\leq$ is the ordering of $\mathcal{D}_r$. Now by assumption, π^{-1} is also induced by a homeomorphism, namely Θ^{-1} and so applying (3.1) to Θ^{-1} we get

$$\pi^{-1}([X]_r) \leq [X]_r$$

for all X, and so $\pi([X])_r = [X]_r$ for all X. □

Example 3.8: [The inductive procedure in Lemma 3.4.] Suppose Φ is the truth table reduction given by $\Phi^A(n) = A(2n) \cdot A(2n+1)$ for all A and n. Suppose we know the first truth-table for Θ, in that we know that $\Theta^A(0) = A(2) \to A(3)$ for all A. Then

$$\begin{aligned}\Theta^A(1) = \Theta^{\Phi^A}(0) &= \Phi^A(2) \to \Phi^A(3)\\ &= A(4)A(5) \to A(6)A(7).\end{aligned}$$

Next,

$$\begin{aligned}\Theta^A(2) &= \Theta^{\Phi^A}(1)\\ &= \Phi^A(4)\Phi^A(5) \to \Phi^A(6)\Phi^A(7)\\ &= A(8)A(9)A(10)A(11) \to A(12)A(13)A(14)A(15).\end{aligned}$$

[1] The non-redundancy of the notion of bi-uniformly E_0-invariant Cantor homeomorphism has been pointed out by Salo [2].

Remark 3.9: Woodin mentioned on June 6, 2019 that he and Slaman may have shown the following result in unpublished work from the 1990s: Each automorphism of $\mathcal{D}_a$, the degrees of arithmetical reducibility, is represented by a continuous function (outright). This gives some extra interest in a possible future $\mathcal{D}_a$ version of our results.

References

1. S. Barry Cooper. The Turing universe is not rigid. University of Leeds Pure Mathematics Preprint Series 1997, no. 16 (revised February 1998).
2. Ville Salo (`https://mathoverflow.net/users/123634/ville-salo`). Homeomorphisms and "mod finite". MathOverflow. URL: `https://mathoverflow.net/q/364844` (version: 2020-07-04).
3. Carl G. Jockusch, Jr. and Robert M. Solovay. Fixed points of jump preserving automorphisms of degrees. *Israel J. Math.*, 26(1):91–94, 1977.
4. Bjørn Kjos-Hanssen. Permutations of the integers induce only the trivial automorphism of the Turing degrees. *Bull. Symb. Log.*, 24(2):165–174, 2018.
5. Anil Nerode and Richard A. Shore. Reducibility orderings: theories, definability and automorphisms. *Ann. Math. Logic*, 18(1):61–89, 1980.
6. Theodore A. Slaman. Global properties of the Turing degrees and the Turing jump. In *Computational prospects of infinity. Part I. Tutorials*, volume 14 of *Lect. Notes Ser. Inst. Math. Sci. Natl. Univ. Singap.*, pp. 83–101. World Sci. Publ., Hackensack, NJ, 2008.
7. Theodore A. Slaman and Hugh Woodin. Definability in degree structures. Online draft, July 2005. URL: `https://math.berkeley.edu/~slaman/talks/sw.pdf`.

https://doi.org/10.1142/9789819806584_0008

An Introduction to AD^+

Paul B. Larson
Department of Mathematics, Miami University
Oxford, Ohio 45056, USA
larsonpb@miamioh.edu

This survey is based on notes taken by Luke Serafin, on lectures given by the author on June 4th, 6th, 11th and 13th of 2019, at the Workshop on Set Theory, Recursion Theory and their Interaction, held at the Institute for Mathematical Sciences at the University of Singapore. It covers an introductory portion of the classical theory of Woodin's AD^+. There is significant overlap between the material presented here and in the paper [9]. The lectures were based on parts of the author's book [13].

1. Determinacy Axioms and Forms of Choice

Woodin's axiom AD^+ is a strengthening of the classical Axiom of Determinacy (AD), which is inconsistent with the Axiom of Choice (AC). We define AD below, but refer the reader to [4,6,12] for a more thorough introduction, including an account of the history of this axiom. Before giving the definition of AD^+ we will quickly review AD and its strengthening $\mathsf{AD}_{\mathbb{R}}$, and forms of the Axiom of Choice consistent with AD, such as DC and $\mathsf{Uniformization}$. We will also discuss structural consequences of determinacy, such as the Wadge hierarchy and Moschovakis Coding Lemma, and ways of defining subsets of the real line using sets of ordinals, such as Suslin representations and ∞-Borel codes. With these definitions in hand we define AD^+ in Section 3.3. In the same section, we list some of the major open questions surrounding AD^+, and discuss some of the original motivation for its study. For now, we note that part of the motivation for studying AD^+ is that

The research of the author is supported in part by NSF grant DMS-1764320.

it allows the lifting of some of the consequences of AD in $L(\mathbb{R})$ to larger models of determinacy.

Collectively these notes present preliminary material toward some of Woodin's major theorems of the 1990's. Since then significant progress in the study of AD^+ has been made by applying the stationary tower and inner model theory; we do not discuss these results here. Some of the theorems we do discuss are listed in Section 3.3. For the most part we sketch this material at a high level, giving short proofs when appropriate. Complete proofs of most of the theorems discussed here appear in the author's book.

Our base theory throughout this chapter is ZF. Additional axioms will be stated as needed.

1.1. AD *and* $\mathsf{AD}_{\mathbb{R}}$

Let X be an arbitrary set and let A be a subset of X^ω. In the game $\mathcal{G}_X(A)$, two players, I and II, alternate playing elements of X. Player I plays a first element $x(0)$, player II a second element $x(1)$, and so on, with a play at each natural number "time". The entire run of the game produces an $x \in X^\omega$. Player I wins if and only if $x \in A$ (so player II wins if $x \in X^\omega \setminus A$).

I	$x(0)$		$x(2)$		$x(4)$		$\ldots$	
II		$x(1)$		$x(3)$		$\ldots$		

The game $\mathcal{G}_X(A)$; I wins if and only if x is in A.

A *strategy* is a function $\sigma : X^{<\omega} \to X$, which we think of as determining a move for a player given the history of the game up to the point where the strategy is applied. Given a strategy σ and an $x \in X^\omega$, $\sigma * x$ is the result a run of $\mathcal{G}_X(A)$ where player I plays according to σ and x lists player II's moves. That is, $(\sigma * x)(0) = \sigma(\langle\rangle)$ and, for all $k \in \omega$, $(\sigma * x)(2k+1) = x(k)$ and $(\sigma * x)(2k+2) = \sigma((\sigma * x) \restriction 2k+1)$. The result of a run where x lists player I's moves and player II plays according to σ is defined analogously, and denoted $x*\sigma$. A strategy σ is *winning* for player I if and only if $\sigma*x \in A$ for all $x \in X^\omega$, and winning for player II if $x * \sigma \notin A$, for all such x. The game $\mathcal{G}_X(A)$ is said to be *determined* if and only if one of the players has a winning strategy.

We list below three forms of determinacy defined in terms of the framework just given. The first two are consistent with ZF, relative to the consistency of certain large cardinal axioms (see Theorems 1.1 and 1.2). The second implies the first, since ω can be injected into ω^ω. The third is refuted by ZF, as we discuss at the beginning of Section 3.2.

- AD: $\mathcal{G}_\omega(A)$ is determined for all $A \subseteq \omega^\omega$
- AD$_\mathbb{R}$: $\mathcal{G}_{\omega^\omega}(A)$ is determined for all $A \subseteq (\omega^\omega)^\omega$
- AD$_{\omega_1}$: $\mathcal{G}_{\omega_1}(A)$ is determined for all $A \subseteq (\omega_1)^\omega$.

We refer the reader to [4, 6] for discussion of the following standard consequences of AD. The first two of the these will be used in these notes, at least implicitly, but not the third.

- All sets of reals have the perfect set property.
- All sets of reals have the property of Baire.
- All sets of reals are Lebesgue measurable.

The assertion that all sets of reals have the perfect set property implies that there are no injections from ω_1 into ω^ω. We will write this latter statement as $\aleph_1 \not\leq 2^{\aleph_0}$. We will also make use of (generalizations of) Solovay's theorem that, under AD, ω_1 is a measurable cardinal (see [4, 6]).

Large cardinals imply the existence of inner models of AD and AD$_\mathbb{R}$, as shown by the two following theorems (due to Martin, Steel and Woodin; see [11, 15, 21]).

Theorem 1.1: *If there are infinitely many Woodin cardinals with a measurable cardinal above them all, then* $L(\mathbb{R}) \models$ AD.

Theorem 1.2: *If there exists a cardinal κ which is a limit of Woodin cardinals and of strong-to-κ cardinals and there is a measurable cardinal $\lambda > \kappa$, then there is a $\Gamma \subseteq \mathcal{P}(\omega^\omega)$ such that* $L(\Gamma, \mathbb{R}) \models$ AD$_\mathbb{R}$.

Since strategies in integer games can be coded by real numbers, AD implies that AD holds in every inner model containing the reals, for instance, in the inner model $L(\mathbb{R})$. Similarly, since strategies in real games can be coded by sets of real numbers, AD$_\mathbb{R}$ implies that AD$_\mathbb{R}$ holds in every inner model containing $\mathcal{P}(\mathbb{R})$. However, a model of the form $L(A, \mathbb{R})$, where A is a subset of $L(\mathbb{R})$, can never satisfy AD$_\mathbb{R}$, since such a model can never satisfy both Uniformization and $\aleph_1 \not\leq 2^{\aleph_0}$, which are consequences of AD$_\mathbb{R}$.

1.2. *Uniformization*

Given a set $A \subseteq \omega^\omega \times \omega^\omega$, and an $x \in \omega^\omega$, we let A_x denote the *section* of A at x, i.e., $\{y \in \omega^\omega : (x, y) \in A\}$. Uniformization is the assertion that every subset of the plane contains a function with the same domain:

$$\forall A \subseteq \omega^\omega \times \omega^\omega \, \exists B \subseteq A \, \forall x \in \omega^\omega \, (A_x \neq \emptyset \Rightarrow |B_x| = 1).$$

Uniformization follows from the Axiom of Choice (AC). It also follows from $\mathsf{AD}_\mathbb{R}$ via a game with one round, where player I plays $x \in \omega^\omega$, player II plays $y \in \omega^\omega$, and II wins if and only if $A_x = \emptyset$ or $y \in A_x$. Player I cannot have a winning strategy in this game, and a winning strategy for player II gives a uniformizing function for A.

A diagonal argument (given in the proof of Theorem 1.3) shows that the statement Uniformization fails in models of the form $L(A, \mathbb{R})$ for any set $A \subseteq L(\mathbb{R})$, assuming $\aleph_1 \not\leq 2^{\aleph_0}$. It follows that models of this form cannot satisfy $\mathsf{AD}_\mathbb{R}$. For any set or class X, we let OD_X denote the class of sets which are definable from an ordinal and a finite sequence from X (i.e., which are *ordinal definable* from a finite sequence from X). We write OD for OD_X when $X = \emptyset$. A standard definability-order argument shows that if OD_X contains a wellordering of X then it contains a wellordering of each element of $\mathrm{OD}X$.

Theorem 1.3: (Solovay). *If $\aleph_1 \not\leq 2^{\aleph_0}$ and A is a set contained in $L(\mathbb{R})$, then $L(A, \mathbb{R}) \models \neg$Uniformization.*

Proof: For each set a in $L(A, \mathbb{R})$ there is an $x \in \omega^\omega$ such that a is in $\mathrm{OD}_{\{A,x\}}$. Since $\aleph_1 \not\leq 2^{\aleph_0}$, each set of the form $\omega^\omega \cap \mathrm{OD}_{\{A,x\}}$ is countable. Let B be the set of $(x, y) \in \omega^\omega \times \omega^\omega$ such that y is not in $\mathrm{OD}_{\{A,x\}}$. Any uniformizing function f for B would be in $\mathrm{OD}_{\{A,x\}}$ for some $x \in \omega^\omega$, which would mean that $f(x)$ would also be in $\mathrm{OD}_{\{A,x\}}$. □

Uniformization is then one difference between AD and $\mathsf{AD}_\mathbb{R}$, as it is implied by the latter and not the former. There is a sense in which it may be the only difference, as results of Becker, Martin and Woodin show that AD + DC + Uniformization implies $\mathsf{AD}_\mathbb{R}$ (see Theorem 3.12). We now briefly review DC, which is a weak form of the Axiom of Choice.

1.3. $\mathsf{CC}_\mathbb{R}$, DC *and* $\mathsf{DC}_\mathbb{R}$

Countable Choice for Reals ($\mathsf{CC}_\mathbb{R}$) is the restriction of the Axiom of Choice to countable subsets of $\mathcal{P}(\mathbb{R})$. Mycielski [16] proved that AD implies $\mathsf{CC}_\mathbb{R}$. For each set X, DC_X is the statement that each nonempty tree $T \subseteq X^{<\omega}$ without terminal nodes has an infinite branch. The Axiom of Dependent Choice, DC, is the statement that DC_X holds for every X.

Uniformization implies $\mathsf{DC}_{\mathbb{R}}$, since for any such tree $T \subseteq (\omega^\omega)^{<\omega}$ it implies the existence of a function choosing a successor for each node. It follows that $\mathsf{AD}_{\mathbb{R}}$ implies $\mathsf{DC}_{\mathbb{R}}$. Whether or not AD implies $\mathsf{DC}_{\mathbb{R}}$ is an open question. At the end of Section 11, we will review Solovay's theorem showing that $\mathsf{AD}_{\mathbb{R}}$ does not imply that DC holds in $L(\mathcal{P}(\mathbb{R}))$.

Since an infinite path through a tree on ω^ω can be coded by a single real, if $\mathsf{DC}_{\mathbb{R}}$ holds then it holds in any transitive model M containing $\mathbb{R}$. In particular, it holds in $L(\mathbb{R})$ if the Axiom of Choice holds in the full universe V.

While our current goal is the formulation of AD^+, we digress briefly to review some important structural properties of $\mathcal{P}(\omega^\omega)$ under AD.

2. The Wadge Hierarchy

Given $A, B \subseteq \omega^\omega$, we write $A \leq_{\mathrm{W}} B$ (and say that A is *Wadge below* B) to mean that there is a continuous function $f : \omega^\omega \to \omega^\omega$ such that $A = f^{-1}[B]$. We write $A \equiv_{\mathrm{W}} B$ for the conjunction $A \leq_{\mathrm{W}} B \wedge B \leq_{\mathrm{W}} A$. The equivalence class $[A]_{\mathrm{W}}$ of A with respect to $\equiv_{\mathrm{W}}$ is the *Wadge degree* of A. The Wadge order gives an almost linear hierarchy on the set of Wadge degrees, in the following sense

Theorem 2.1: (Wadge). AD *implies that, for all* $A, B \subseteq \omega^\omega$, $A \leq_W B$ *or* $B \leq_W \omega^\omega \setminus A$.

Proof: Consider the game where players I and II play nonempty finite sequences of integers in each turn, with I building $x \in \omega^\omega$, and II building y. Say that I wins if and only if the statement $x \in B \Leftrightarrow y \in A$ holds. A winning strategy for I induces a witness to the statement $A \leq_{\mathrm{W}} B$. A winning strategy for II witnesses $B \leq_{\mathrm{W}} \omega^\omega \setminus A$. □

Determinacy for the class of games in the proof just given is called Wadge Determinacy. It is an open question whether Wadge Determinacy implies AD.

Martin proved that the Wadge order is wellfounded, assuming Wadge Determinacy, $\mathsf{DC}_{\mathbb{R}}$ and the assumption that every set of reals has the Baire property. It is not known if $\mathsf{DC}_{\mathbb{R}}$ is needed for this result, or if AD alone implies that the Wadge hierarchy is wellfounded. The wellfoundedness of the Wadge order induces the notion of Wadge rank, where the Wadge rank of $A \subseteq \omega^\omega$ is the ordinal rank of $[A]_{\mathrm{W}}$ in the Wadge order.

A Wadge class $[A]_W$ is said to be *selfdual* if $[A]_W = [\omega^\omega \setminus A]_W$. We mention a few important results from the highly-developed theory of Wadge classes (see, for instance [22]).

A standard diagonalization along the lines of Cantor's theorem shows that there is no largest Wadge degree. The proof below uses the following notation, which appears also in the next section. Using a fixed recursive bijection between ω and $\omega^{<\omega} \times \omega^{<\omega}$, we can think of an element c of ω^ω as listing a sequence of basic open sets in $\omega^\omega \times \omega^\omega$, and let fc be the complement of the union of these open sets. Letting CF be the (projective) set of c for which f_c is a continuous function from ω^ω to ω^ω, we have that $\{f_c : c \in CF\}$ is the set of all such functions. In particular, there is a surjection from ω^ω to the set of continuous functions from ω^ω to ω^ω.

Theorem 2.2: *There is no largest Wadge degree.*

Proof: Given $A \subseteq \omega^\omega$, let B be the set of $c \in CF$ for which $f_c(c) \notin A$. Then B is not Wadge below A. □

For each pair of complementary non-selfdual degrees there is a minimal selfdual degree above them.

Theorem 2.3: *For a non-selfdual class $[A]_W$, there is a minimal Wadge degree above A and $\omega^\omega \setminus A$.*

Proof: Let B be the set of $x \in \omega^\omega$ of the form $0^\frown x$ for $x \in A$ or $1^\frown x$ for $x \notin A$. Then $A \leq_W B$ and $(\omega^\omega \setminus A) \leq_W B$ and, for any $C \subseteq \omega^\omega$, if $A \leq_W C$ and $(\omega^\omega \setminus A) \leq_W C$ then $B \leq_W C$. □

The corresponding fact for selfdual degrees follows from the wellfoundedness of the Wadge hierarchy.

Theorem 2.4: *If* $\mathsf{AD} + \mathsf{DC}_\mathbb{R}$ *holds then each selfdual Wadge degree has a non-selfdual successor in the Wadge order.*

Theorem 2.5 (whose proof includes contributions from Steel and Van Wesep) says when a Wadge degree of limit rank is selfdual.

Theorem 2.5: *Assuming Wadge Determinacy and that all subsets of ω^ω have the property of Baire, limit classes under the Wadge order are selfdual precisely when the Wadge order below them has countable cofinality.*

2.1. *The ordinal Θ*

The ordinal Θ is defined to be the least ordinal which is not a surjective image of ω^ω. Clearly, Θ is a cardinal; the Moschovakis Coding Lemma (see Section 4) implies that it is a limit cardinal if AD holds. Assuming that $A \subseteq \omega^\omega$ has Wadge rank α, there is a surjection from ω^ω to $\alpha+1$, assigning each $c \in CF$ the Wadge rank of $f_c^{-1}[A]$. Thus, assuming $\mathrm{AD} + \mathrm{DC}_{\mathbb{R}}$, each $A \subseteq \omega^\omega$ has Wadge rank less than Θ. Similarly, one can use a surjection from ω^ω to an ordinal α to build a set of reals of Wadge rank at least α. In fact, ZF proves that for every ordinal $\alpha < \Theta$ there is a $<_W$-increasing sequence of length α, where $<_W$ is the Wadge order. This gives the following theorem, where we let $\mathrm{WR}(A)$ denote the Wadge rank of a set $A \subseteq \omega^\omega$.

Theorem 2.6: (Solovay). *Assuming* $\mathsf{AD} + \mathsf{DC}_{\mathbb{R}}$, $\Theta = \{\mathrm{WR}(A) : A \subseteq \mathcal{P}(\mathbb{R})\}$.

We return now to our goal of defining AD$^+$.

3. The AD$^+$ Axioms

In Sections 3.1 and 3.2, we define ∞-Borel sets and $<\Theta$-Determinacy. With these definitions in hand we finally define AD$^+$ in Section 3.3. Section 3.4 discusses strong partition cardinals, which can be used to show that $<\Theta$-Determinacy follows from AD plus the assumption that every subset of ω^ω is Suslin.

3.1. *Suslin sets and ∞-Borel sets*

Given a set X and a tree $T \subseteq X^{<\omega}$, $[T]$ denotes the set of infinite paths through T. Given sets X and Y and a tree $T \subseteq (X \times Y)^{<\omega}$, the *projection* $\mathrm{p}[T]$ is defined to be the set of $x \in X^\omega$ for which there exists an $f \in Y^\omega$ such that $(x, f) \in [T]$.

Definition 3.1: *Given an ordinal γ and a set X, a set $A \subseteq X^\omega$ is γ-*Suslin *if $A = \mathrm{p}[T]$ for some tree $T \subseteq (X \times \gamma)^{<\omega}$. The set A is* Suslin *if it is γ-Suslin for some ordinal γ, and* co-Suslin *if its complement in X^ω is Suslin.*

Continuous preimages of γ-Suslin sets are γ-Suslin, so the Suslin sets form an initial segment of the Wadge hierarchy.

If $A \subseteq \omega^\omega \times \omega^\omega$ (identified with the corresponding subset of $(\omega \times \omega)^\omega$) is Suslin then it can be uniformized. To see this, suppose that $A = \mathrm{p}[T]$, where $T \subseteq (\omega \times \omega \times \gamma)^{<\omega}$. For each $x \in A$ one can recursively find the lexicographically least $y \in (\omega \times \gamma)^{<\omega}$ such that $(x, y) \in [T]$. Models of the form $L(A, \mathbb{R})$ for $A \subseteq L(\mathbb{R})$ will not satisfy the assertion that all sets of reals are Suslin as they don't satisfy Uniformization.

The following theorems, whose proofs are beyond the scope of these notes, relate the Suslin property and Uniformization with $\mathsf{AD}_{\mathbb{R}}$. The first is a combination of results of Becker and Woodin. The second is due to Martin and Woodin independently.

Theorem 3.2: AD + DC + Uniformization *implies all subsets of* ω^ω *are Suslin.*

Theorem 3.3: *If* AD *holds and all subsets of* ω^ω *are Suslin, then* $\mathsf{AD}_{\mathbb{R}}$ *holds.*

We say that a tree T is wellfounded if it does not have an infinite path, i.e., if $[T]$ is the emptyset. Assuming DC (or the wellorderability of T), wellfoundedness of a tree T is witnessed by a ranking function ρ from T to the ordinals, where $\rho(s) > \rho(t)$ whenever t extends s. If T is wellfounded, then a ranking function witnessing this can be constructed recursively in the model $L[T]$. It follows that, given an ordinal γ, a tree T on $\omega \times \gamma$, and an $x \in \omega^\omega$, $x \in \mathrm{p}[T]$ if and only if $L[T, x] \models x \in p[T]$. This property is generalized in the following definition. In Section 6.1, we discuss an equivalent definition which is more closely related to the usual notion of Borel set.

Definition 3.4: *A set* $A \subseteq \omega^\omega$ *is* ∞-Borel *if and only if there exist a formula* ϕ *and a set* $S \subseteq \mathrm{Ord}$ *such that* $A = \{x \in \omega^\omega : L[S, x] \models \phi(S, x)\}$. *We call the pair* (S, ϕ) *an* ∞*-Borel* code *for* A.

Again, Suslin sets are ∞-Borel. As with the Suslin sets, the ∞-Borel sets form an initial segment of the Wadge hierarchy, which can be seen as follows. If (S, ϕ) is an ∞-Borel code for A, and c is in CF, then there is an ∞-Borel code for $f_c^{-1}[A]$ of the form (S', ϕ'), where S' codes the pair (S, c) and $\phi'(S', x)$ is the formula $\phi(S, f_c(x))$ (phrased in terms of the coding of f_c by c).

If $A \subseteq \omega^\omega$ is ∞-Borel, then this is witnessed by a pair (S, ϕ) with S a bounded subset of Θ. This can be shown by mapping ω^ω onto an elementary submodel (containing ω^ω) of a suitable model of the form $L_\alpha(T, \mathbb{R})$, where (T, ϕ) is a given ∞-Borel code for A. Since every element of $L_\alpha(T, \mathbb{R})$ is definable in $L_\alpha(T, \mathbb{R})$ from T, an ordinal and an element of ω^ω, such a submodel can be built assuming only ZF.

If A is ∞-Borel and $\aleph_1 \not\leq 2^{\aleph_0}$, then A has the property of Baire and is Lebesgue measurable. To see this for the property of Baire, let $\mathbb{P}$ be Cohen forcing and let $\dot{c}$ be the canonical $\mathbb{P}$-name for the generic Cohen real. Suppose that (S, ϕ) is an ∞-Borel code for A. Since $\aleph_1 \not\leq 2^{\aleph_0}$ and $L[S] \models \mathsf{AC}$, $\omega^\omega \cap L[S]$ is countable, so the set of reals which are Cohen-generic over $L[S]$ is comeager. For each such real x, let g_x be the filter for which $\dot{c}_{g_x} = x$. Let A' be the set of x which are Cohen-generic over $L[S]$ for which there is a condition $p \in g_x$ forcing that $L[S, \dot{c}] \models \phi(S, \dot{c})$. Then A' is Borel, and the symmetric difference $A \triangle A'$ is contained in the set of reals which are not Cohen-generic over $L[S]$. For Lebesgue measurability the argument is the same, using random forcing instead of Cohen forcing. One can use the same argument with Mathias forcing to show that A (under the same assumptions) satisfies the Ramsey property. Unlike the cases of the Baire property and Lebesgue measurability, it is still unknown whether AD alone implies that every subset of 2^ω has the Ramsey property.

One part of AD^+ is the statement that all $A \subseteq \omega^\omega$ are ∞-Borel. We discuss proofs of the following theorems (due to Woodin) in Sections 5 and 6.2.

Theorem 3.5: $\mathsf{AD} + V = L(\mathbb{R})$ *implies that all subsets of* ω^ω *are* ∞*-Borel.*

Theorem 3.6: *If* $\mathsf{AD} + \mathsf{DC}_\mathbb{R}$ *holds and* A *is* ∞*-Borel then it is* ∞*-Borel in* $L(A, \mathbb{R})$.

We list three theorems of Woodin from the 1990's on ∞-Borel sets, whose proofs are too involved to include in these notes. Proofs of the first two appear in the author's book. Since $\mathsf{DC}_\mathbb{R}$ implies that DC holds in models of the form $L(S, \mathbb{R})$ whenever S is a set of ordinals, Theorem 3.9 implies Theorem 3.8.

Theorem 3.7: AD + Uniformization *implies that all subsets of* ω^ω *are* ∞*-Borel.*

Theorem 3.8: *Assuming* $\mathsf{AD} + \mathsf{DC}_\mathbb{R}$, *for* $A \subseteq \omega^\omega$, A *is* ∞*-Borel if and only if there is an* $S \subseteq \mathrm{Ord}$ *with* $A \in L(S, \mathbb{R})$.

Theorem 3.9: *Assuming* AD + DC, *for every ordinal* $\lambda < \Theta$, *all subsets of* ω^ω *in* $L(\mathcal{P}(\lambda))$ *are* ∞*-Borel.*

There is still one part of AD^+ left to be defined, concerning the determinacy of games on the ordinals.

3.2. γ-determinacy

Let γ be an ordinal, and $f\colon \gamma^\omega \to \omega^\omega$ be a continuous function, where γ^ω is given the product topology induced by the discrete topology on γ. Given $A \subseteq \omega^\omega$, we let $\mathcal{G}(f, A)$ be the game where players I and II alternately choose the values of some $x \in \gamma^\omega$, with player I winning if and only if $f(x) \in A$. We let γ-$\mathsf{Determinacy}$ be the statement that each such game $\mathcal{G}(f, A)$ is determined. For all $\gamma \geq 2$, γ-$\mathsf{Determinacy}$ implies AD.

I	$x(0)$		$x(2)$		$x(4)$		...
II		$x(1)$		$x(3)$		...	

The game $\mathcal{G}(f, A)$; I wins if and only if $f(x)$ is in A.

Although we call this statement γ-$\mathsf{Determinacy}$, we do not consider arbitrary subsets of γ^ω as payoff sets. The reason for this is that ZF implies the existence of a nondetermined game on ω_1^ω (ie., that AD_{ω_1} is false, as mentioned above). To see this, consider the game where player I plays a countable ordinal α in round 1 and makes no subsequent plays, and player II in her turns plays natural numbers x_i $(i \in \omega)$ collectively forming $x \in \omega^\omega$, with player I wins if x codes a wellordering of ω in ordertype α. Since player I cannot have a winning strategy, determinacy of this game implies the existence of an injection from ω_1 to ω^ω, which implies the existence of a set of reals without the perfect set property, and therefore contradicts AD, as noted in Section 1.1. One interesting aspect of this argument is that it does not produce a single game on ω_1 which is provably undetermined (in particular, player II has a winning strategy in the game just described if AC holds). Building on work of Neeman [17], Woodin has shown that in fact determinacy can hold for all definable games on ω_1.

A typical game of the type considered in the statement of γ-$\mathsf{Determinacy}$ is when A is a subset of ω^ω, τ is a $\mathrm{Col}(\omega, \gamma)$-name for an element of ω^ω, and players I and II play a descending sequence of (codes for) conditions in $\mathrm{Col}(\omega, \gamma)$ collectively realizing τ, with player I winning if and only if the given realization of τ is in the set A.

We let $<\Theta$-$\mathsf{Determinacy}$ be the statement that γ-$\mathsf{Determinacy}$ holds for every ordinal $\gamma < \Theta$.

3.3. AD$^+$

Having defined all the parts of AD$^+$ we finally state its definition.

Definition 3.10: AD$^+$ *is the conjunction of the following three statements.*

- $\mathsf{DC}_{\mathbb{R}}$
- *Every subset of* ω^ω *is* ∞*-Borel.*
- $<\Theta$-Determinacy.

Some authors (including Woodin) do not include $\mathsf{DC}_{\mathbb{R}}$ in the statement of AD$^+$, but consider AD$^+$ (defined instead as the conjunction of the other two parts) only in the context of $\mathsf{DC}_{\mathbb{R}}$. The axiom $\mathsf{DC}_{\mathbb{R}}$ is different from the other two parts of AD$^+$ in that it is witnessed by individual real numbers, as opposed to bounded subsets of Θ.

Here is a brief listing of implications, known non-implications, and open questions regarding the statements discussed so far.

- $\mathsf{AD}_{\mathbb{R}}$ implies AD + Uniformization (Sections 1.1 and 1.2), which implies that all subsets of ω^ω are ∞-Borel (Theorem 3.7).
- $\mathsf{AD}_{\mathbb{R}}$ + DC implies that all subsets of ω^ω are Suslin (Theorem 3.2).
- AD plus "all subsets of ω^ω are Suslin" implies AD$^+$ (Theorem 3.19 for $<\Theta$-Determinacy).
- If all subsets of ω^ω are Suslin, then Uniformization holds (Section 1.2).
- Uniformization implies $\mathsf{DC}_{\mathbb{R}}$ (Section 1.2).
- $<\Theta$-Determinacy implies AD, so AD$^+$ does, too (Section 3.2).
- AD$^+$ does not imply Uniformization (Theorem 1.3 and Section 5).
- It is open whether AD implies any or all of the parts of AD$^+$.
- It is open whether $\mathsf{AD}_{\mathbb{R}}$ implies $<\Theta$-Determinacy (it does imply the other two parts of AD$^+$).
- It is open whether $<\Theta$-Determinacy implies the other two parts of AD$^+$ (the other two parts are consequences of AC, so they don't imply $<\Theta$-Determinacy).

The study of AD$^+$ was inspired by the following question. Suppose that $M \subseteq N$ are models of ZF + AD with the same reals, and that every set of reals in M is Suslin in N. What consequences does this entail for M? As a partial answer, we have that in this situation $M \models \mathsf{AD}^+$. To see this for the case of $<\Theta$-Determinacy, we use the Moschovakis Coding Lemma and the fact that AD implies the existence of cofinally many strong partition cardinals below Θ (see Sections 4 and 3.4). Similar arguments give the following theorem. As seen above, the corresponding theorem fails for $\mathsf{AD}_{\mathbb{R}}$.

Theorem 3.11: (Woodin). *If* AD^+ *holds, then it holds in every inner model containing* ω^ω.

More generally, for all three parts of AD^+, if $A \subseteq \omega^\omega$ and $L(A, \mathbb{R}) \subseteq M$ where A satisfies the given part of AD^+ in M, then it satisfies this part in $L(A, \mathbb{R})$ also. For $\mathsf{DC}_{\mathbb{R}}$, this follows from the fact, observed above, that $\mathsf{DC}_{\mathbb{R}}$ is witnessed by individual real numbers. For $<\Theta$-Determinacy, this is discussed at the end of Section 4. The corresponding fact for ∞-Borel sets is covered in Section 6.2.

Combining these remarks with Theorems 3.2 and 3.3, we have the following.

Theorem 3.12: *Each of the following statements implies the ones below it, and the first two are equivalent. Assuming* DC *all four statements are equivalent.*

(1) AD + *"all subsets of* $\mathcal{P}(\omega^\omega)$ *are Suslin"*
(2) AD^+ + *"all subsets of* $\mathcal{P}(\omega^\omega)$ *are Suslin"*
(3) $\mathsf{AD}_{\mathbb{R}}$
(4) AD + $\mathsf{Uniformization}$.

While the question of whether AD implies AD^+ is open, it is known that the consistency strength of $\mathsf{AD} + \neg\mathsf{AD}^+$ is greater than that of AD. To illustrate this, let Γ_+ be the set of $A \subseteq \omega^\omega$ for which $L(A, \mathbb{R}) \models \mathsf{AD}^+$. By Theorem 3.11, Γ_+ is an initial segment of the Wadge hierarchy. Since the implication $\mathsf{AD} \Rightarrow \mathsf{AD}^+$ is known to hold in $L(\mathbb{R})$ (see Section 5), AD implies that Γ_+ is nonempty. The following theorem (whose proof we will not discuss in these notes) shows that a model of $\mathsf{AD} + \neg\mathsf{AD}^+$ would contain a model of $\mathsf{AD}_{\mathbb{R}}$.

Theorem 3.13: (Woodin). *Assuming* $\mathsf{AD} + \mathsf{DC}_{\mathbb{R}}$, $L(\Gamma_+, \mathbb{R}) \models \mathsf{AD}^+$. *If* $\Gamma_+ \neq \mathcal{P}(\omega^\omega)$ *then* $L(\Gamma_+, \mathbb{R}) \models \mathsf{DC} + \mathsf{AD}_{\mathbb{R}}$.

Woodin's Derived Model Theorem (Theorem 3.14) is the standard method for producing models of AD^+. The model $L(\Gamma, \mathbb{R}^*)$ from the statement of the theorem is called the *derived model.* For any set or class X, the inner model HOD_X is the class of all sets a such that the transitive closure of $\{a\}$ is contained in OD_X. When X is ordinal definable from a finite sequence from X, HOD_X is a model of ZF. If a wellordering of X is ordinal definable from a finite sequence from X, HOD_X is a model of ZFC. A proof of an early version of the theorem below can be found in [21].

Theorem 3.14: (Woodin). *Let δ be a limit of Woodin cardinals and let G be a V-generic filter for* $\mathrm{Col}(\omega, < \delta)$. *Let*

$$\mathbb{R}^* = \bigcup\{\mathbb{R}^{V[G\restriction\lambda]} : \lambda < \delta\}$$

and

$$\Gamma = \{A \subseteq \mathbb{R}^* : A \in \mathrm{HOD}_{V,\mathbb{R}^*},\ L(A, \mathbb{R}) \models \mathsf{AD}^+\}.$$

Then $L(\Gamma, \mathbb{R}^*) \models \mathsf{AD}^+$.

This is the standard way to produce models of AD^+, and in a sense it is the only way, since the following theorem, a reversal of the Derived Model Theorem, says that every model of AD^+ arises as a derived model.

Theorem 3.15: (Woodin). *Assuming* $\mathsf{AD} + \mathsf{DC}_{\mathbb{R}}$, *$L(\Gamma_+, \mathbb{R})$ is the derived model of some inner model which exists in some forcing extension of V.*

Finally, we mention Theorem 3.16 below, which is a generalization of the Solovay Basis Theorem (Theorem 5.3). Among other things, the forward direction of the theorem enables one to carry out the basic analysis of Woodin's $\mathbb{P}_{\max}$ forcing assuming AD^+ (see [23]).

Theorem 3.16: (Woodin). *Assuming* AD, AD^+ *is equivalent to the statement that every true $\utilde{\Sigma}^2_1$ statement has a witness which is Suslin and co-Suslin.*

In the theorem above, a sentence is $\utilde{\Sigma}^2_1$ if it has the form $\exists A \subseteq \omega^\omega \phi(A, a)$, where a is an element of ω^ω and all quantifiers in ϕ range over ω^ω (i.e., ϕ is *projective*). More generally, given a set $B \subseteq \omega^\omega$, a set X of some Polish space Y is $\utilde{\Sigma}^2_1(B)$ if there exist a projective formula ϕ and an $a \in \omega^\omega$ such that

$$X = \{y \in Y : \exists A \subseteq \omega^\omega \phi(A, B, x, a)\}.$$

When $B = \emptyset$ we write $\utilde{\Sigma}^2_1$. The set X is $\utilde{\Delta}^2_1(B)$ if X and $Y \setminus X$ are both $\utilde{\Sigma}^2_1(B)$. The ordinal $\utilde{\delta}^2_1(B)$ is defined to be the supremum of the order-types of the $\utilde{\Delta}^2_1(B)$ prewellorderings, where a *prewellordering* is a reflexive, transitive, and wellfounded binary relation.

3.4. *Strong partition cardinals*

Given ordinals α and β, we let $[\alpha]^\beta$ be the collection of subsets of α of ordertype β. The existence of an infinite κ satisfying the following definition contradicts the Axiom of Choice. It appears to be an open question whether one gets an equivalent definition by restricting to the case $\gamma = 2$.

Definition 3.17: *A cardinal κ is a* strong partition cardinal *if for every $\gamma < \kappa$ and every $f : [\kappa]^\kappa \to \gamma$ there is an $A \in [\kappa]^\kappa$ such that $f \restriction [A]^\kappa$ is constant.*

The converse of the following theorem holds in $L(\mathbb{R})$, but not in general [3]. The proof of the theorem involves a deep analysis of pointclasses under AD.

Theorem 3.18: (Kechris-Kleinberg-Moschovakis-Woodin [8]). *The Axiom of Determinacy implies that the strong partition cardinals are cofinal in Θ, and in particular that, for each $A \subseteq \omega^\omega$, $\underset{\sim}{\delta}^2_1(A)^{L(A,\mathbb{R})}$ is a strong partition cardinal.*

The existence of strong partition cardinals, in conjunction with Suslin representations, can be used to prove $<\Theta$-$\mathsf{Determinacy}$. In particular, assuming $\mathsf{AD} + \mathsf{DC}_\mathbb{R}$, $<\Theta$-determinacy holds for games $\mathcal{G}(f, A)$ where A is Suslin and co-Suslin.

Theorem 3.19: (Moschovakis, Woodin). *Suppose that $\mathsf{AD}+\mathsf{DC}_\mathbb{R}$ holds. Let $\gamma < \kappa < \lambda$ be ordinals, with λ a strong partition cardinal. Let $A \subseteq \omega^\omega$ be such that A and $\omega^\omega \setminus A$ are both κ-Suslin and let $f : \gamma^\omega \to \omega^\omega$ be continuous. Then $\mathcal{G}(f, A)$ is determined.*

The proof of Theorem 3.19 uses the Moschovakis Coding Lemma, which is discussed in Section 4. With the Coding Lemma in hand we will outline in Section 5 a proof that the implication $\mathsf{AD} \Rightarrow \mathsf{AD}^+$ holds in $L(\mathbb{R})$.

4. The Moschovakis Coding Lemma

A prewellordering of a set X is the order induced by a surjection from X to some ordinal. We call this ordinal the *length* of the prewellording. For each $x \in X$, the image of x under this surjection is called the *rank* of x in the prewellordering. We write $\mathrm{rank}_\leq(x)$ for the rank of x in $\leq$ when $\leq$ is a prewellordering of a set with x as a member. We shall be interested in prewellorderings of subsets of ω^ω. Note that Θ is the supremum of the lengths of prewellorderings of ω^ω.

The following simplified version of the Moschovakis Coding Lemma is sufficient for our purposes. A more precise version replaces "projective" with a Σ^1_1 condition in parameters derived from the prewellordering $\leq$.

Theorem 4.1: (Moschovakis). *Assume that* AD *holds and let* $\leq$ *be a prewellordering of* ω^ω *of length* γ. *Then there is a projective formula* ϕ *with* $\leq$ *as its only parameter such that for every* $A \subseteq \gamma$ *there is* $x \in \omega^\omega$ *such that*

$$A = \{\mathrm{rank}_{\leq}(y) : y \in \omega^\omega \wedge \phi(x, y)\}.$$

One important corollary of the Moschovakis Coding Lemma is that, assuming AD, whenever $A \subseteq \omega^\omega$ and γ is an ordinal below $\Theta^{L(A,\mathbb{R})}$, every subset of γ is in $L(A, \mathbb{R})$. Moreover, $\mathcal{P}(\gamma)$ is a surjective image of ω^ω in $L(A, \mathbb{R})$, which implies that Θ is a limit cardinal (under AD).

Corollary 4.2: *Assuming* AD, *if* $A \subseteq \omega^\omega$ *and* $\gamma < \Theta^{L(A,\mathbb{R})}$ *then* $\mathcal{P}(\gamma) \subseteq L(A, \mathbb{R})$ *and there is a surjection from* ω^ω *to* $\mathcal{P}(\gamma)$ *in* $L(A, \mathbb{R})$.

Suppose now that AD holds, $\leq$ is a prewellordering of ω^ω of length γ and $f\colon \gamma^\omega \to \omega^\omega$ is continuous and $A \subseteq \omega^\omega$. Strategies in the game $\mathcal{G}(f, A)$ can be coded by subsets of γ. It follows from the Moschovakis Coding Lemma then that any winning strategy for the game $\mathcal{G}(f, A)$ exists in $L(\leq, \mathbb{R})$. This gives the $<\Theta$-Determinacy part of Theorem 3.11.

5. AD$^+$ in $L(\mathbb{R})$

The fact that, in $L(\mathbb{R})$, AD implies AD$^+$ follows from several classical results from the Cabal Seminar. First, we have Kechris's theorem, giving $\mathsf{DC}_{\mathbb{R}}$.

Theorem 5.1: (Kechris [7]). $L(\mathbb{R}) \models (\mathsf{AD} \Rightarrow \mathsf{DC})$.

The following theorem of Martin and Steel characterizes the Suslin sets of $L(\mathbb{R})$ under AD.

Theorem 5.2: (Martin-Steel [14]). *Assuming* $\mathsf{AD} + V = L(\mathbb{R})$, *the Suslin subsets of* ω^ω *are exactly the* $\utilde{\Sigma}^2_1$ *sets.*

Finally, we have the Solovay Basis Theorem, which reflects potential counterexamples to AD$^+$ into the Suslin, co-Suslin sets (see Section 2.4 of [10] for further discussion).

Theorem 5.3: (Solovay). *In* $L(\mathbb{R})$, $\utilde{\Sigma}^2_1$ *facts have* $\utilde{\Delta}^2_1$ *witnesses.*

By the Moschovakis Coding Lemma, the assertion that there exist a pair (f, A) giving a counterexample to $<\Theta$-Determinacy is equivalent to a Σ^2_1-statement. If $<\Theta$-Determinacy fails in $L(\mathbb{R})$, then, the Solovay Basis Theorem implies that there is a counterexample where A is Suslin and co-Suslin. Theorem 3.19 however implies that this is impossible. We have then that AD implies $<\Theta$-Determinacy in $L(\mathbb{R})$. A similar argument applies to the statement that every subset of ω^ω is ∞-Borel, using the material in Section 6.2.

6. Reflecting ∞-Borel codes

In this section we introduce an alternate notion ∞-Borel code, and use it to prove that for every ∞-Borel set $A \subseteq \omega^\omega$ there is an ∞-Borel code for A in the inner model $L(A, \mathbb{R})$.

6.1. *Infinitary ∞-Borel codes*

In this section, we give an alternate (equivalent) definition of ∞-Borel set which corresponds to the classical notion of Borel set. This definition defines the ∞-Borel subsets of 2^ω using an infinitary language.

Definition 6.1: *For γ an infinite ordinal, $\mathcal{L}_{\gamma,0}$ is the language with* 0-*ary predicates $\{p_n : n \in \omega\}$ which is closed under the unary connective of negation and under wellordered disjunctions and conjunctions indexed by subsets of γ. We let $\mathcal{L}_{\infty,0}$ be the union over all ordinals γ of the languages $\mathcal{L}_{\gamma,0}$.*

An element $x \in 2^\omega$ may be viewed as an $\mathcal{L}_{\infty,0}$-structure, where $x \models p_n$ if and only if $x(n) = 1$.

Definition 6.2: *A set $A \subseteq 2^\omega$ is* γ-Borel *if there is a formula $\phi \in \mathcal{L}_{\gamma,0}$ such that $A = \{x \in 2^\omega : x \models \phi\}$.*

We let A_ϕ denote the set of $x \in 2^\omega$ such that $x \models \phi$. Then ϕ gives a description of a construction of A in terms of negations and wellordered unions and intersections, starting with the basic open sets in 2^ω. Using a definable paring function on the ordinals we can code elements of $\mathcal{L}_{\infty,0}$ by sets of ordinals, and for each suitable $S \subseteq \mathrm{Ord}$ we let ϕ_S be the sentence coded by S. When κ is an infinite cardinal each ϕ in $\mathcal{L}_{\kappa,0}$ is ϕ_S for some $S \subseteq \kappa$. Then there is a first-order formula ψ such that (S, ψ) is an ∞-Borel code for A_ϕ in the sense defined previously. Our two notions of ∞-Borel set are in fact equivalent in ZF.

Theorem 6.3: (Woodin). *For all $A \subseteq 2^\omega$ the following are equivalent:*

(1) There exist $S \subseteq \mathrm{Ord}$ and a first-order formula ϕ such that

$$A = \{x \in 2^\omega : L[S, x] \models \phi(S, x)\}.$$

(2) There is an $S \subseteq \mathrm{Ord}$ such that $A = \{x \in 2^\omega : x \models \phi_S\}$.

We have just sketched the reverse direction of this theorem. The proof of the forward direction is more involved, and proceeds by induction on γ (observing that if $A = \{x \in 2^\omega : L[S, x] \models \phi(S, x)\}$ then there is an ordinal γ such that $A = \{x \in 2^\omega : L_\gamma[S, x] \models \phi(S, x)\}$) and a second induction on the complexity of ϕ.

There are several ways to consider subsets of (finite powers of) ω^ω as ∞-Borel in our new sense. We skip the details here and treat both notions of ∞-Borel set as equivalent and meaningful for subsets of $(\omega^\omega)^n$ for any $n \in \omega \setminus \{0\}$.

The ∞-Borel sets are not always the same as the smallest class containing the open sets and closed under negations and wellordered unions. In particular, it is consistent with ZF that the collection of ∞-Borel sets is not closed under wellordered unions. In order to show that a given wellordered union of ∞-Borel sets is ∞-Borel one must be able to choose the codes, which is not possible in general.

In general ∞-Borel sets cannot be uniformized. In particular, models of the form $L(A, \mathbb{R})$ (where A is a subset of $L(\mathbb{R})$) can satisfy the statement that every subset of ω^ω is ∞-Borel, even though they cannot satisfy $\mathsf{Uniformization}$. Similarly (and again, unlike the case of Suslin representations), it can be that S is an ∞-Borel code for a nonempty set of reals which has no members in the model $L[S]$. For instance, the set of reals which are Cohen-generic over L has an ∞-Borel code in L.

In [9] and [13] the type of ∞-Borel code defined in Definition 3.4 is called an ∞-Borel* code; here we will not make a distinction, and hope that the intended meaning will be clear from context. In practice one needs both definitions for the ∞-Borel sets, both types of ∞-Borel code and the ability to translate between them.

6.2. *Local ∞-Borel codes*

In this section, we show that, assuming $\mathsf{AD}+\mathsf{DC}_\mathbb{R}$, each ∞-Borel set $A \subset \omega^\omega$ has an ∞-Borel code which is is projective in a parameter for A. This fact will complete the argument from Section 5 that AD implies AD^+ in $L(\mathbb{R})$.

We define the following ordinals, all of which are at most Θ:

- χ_{B}, the least χ such that all ∞-Borel sets are χ-Borel;
- κ_{B}, the supremum of the lengths of wellordered sequences of ∞-Borel codes for nonempty disjoint sets;
- λ_{B}, the supremum of the lengths of ∞-Borel prewellorderings.
- ρ_{B}, the supremum of the Wadge ranks of ∞-Borel sets.

Assuming $\mathsf{AD}+\mathsf{DC}_{\mathbb{R}}$ (or just Wadge Determinacy plus the wellfoundedness of the Wadge hierarchy), $\rho_{\mathrm{B}}=\Theta$ is equivalent to the statement that every subset of ω^ω is ∞-Borel. We will see below that $\mathsf{AD}+\mathsf{DC}_{\mathbb{R}}$ implies that all four ordinals are the same. Our proof of this fact will use Theorem 3.9, whose proof is not discussed in these notes. It may be that there is an easier proof.

Our goal in this section is to prove Theorem 6.7 below assuming only $\mathsf{AD}+\mathsf{DC}_{\mathbb{R}}$. We begin make a few preliminary observations on the four ordinals defined above. We first prove that $\chi_{\mathrm{B}}\leq\kappa_{\mathrm{B}}\leq\lambda_{\mathrm{B}}$, using a function $\phi\mapsto\phi^*$ on $\mathcal{L}_{\infty,0}$ sentences, defined recursively as follows:

- $p_n^*=p_n$;
- $(\neg\phi)^*=\neg\phi^*$;
- whenever Y is a set of ordinals,
$$\left(\bigvee_{\alpha\in Y}\phi_\alpha\right)^*=\bigvee_{\alpha\in X}\phi_\alpha^*,$$
where $X=\{\alpha\in Y: A_{\phi_\alpha}\not\subseteq\bigcup_{\beta\in Y\cap\alpha}A_{\phi_\beta}\}$, and
$$\left(\bigwedge_{\alpha\in Y}\phi_\alpha\right)^*=\bigwedge_{\alpha\in X}\phi_\alpha^*,$$
where $X=\{\alpha\in Y: A_{\phi_\alpha}\not\supseteq\bigcap_{\beta\in Y\cap\alpha}A_{\phi_\beta}\}$.

This function recursively removes from each conjunction and disjunction any terms which do not affect the final result. In particular, for any $\phi\in\mathcal{L}_{\kappa,0}$, $A_\phi=A_{\phi^*}$.

Parts of the proofs of Theorems 6.4 and 6.5 will be reused in the proof of Theorem 6.7.

Theorem 6.4: (Woodin). *Assuming* $\mathsf{AD}+\mathsf{DC}_{\mathbb{R}}$, $\chi_{\mathrm{B}}\leq\kappa_{\mathrm{B}}\leq\lambda_{\mathrm{B}}$.

Proof: To see that $\chi_{\mathrm{B}}\leq\kappa_{\mathrm{B}}$, let ϕ be any sentence in $\mathcal{L}_{\kappa,0}$. For any disjunction $\bigvee_{\alpha<\gamma}\psi_\alpha$ in ϕ^* the sentences $\psi_\alpha\wedge\neg(\bigvee_{\beta<\alpha}\psi_\beta)$ are ∞-Borel codes for nonempty disjoint sets. A similar remark applies to the conjunctions. It follows that ϕ^* is in $\mathcal{L}_{\kappa_{\mathrm{B}},0}$.

To see that $\kappa_{\mathrm{B}} \leq \lambda_{\mathrm{B}}$, note that one can convert a sequence of ∞-Borel codes for nonempty disjoint sets to an ∞-Borel code for a prewellordering in a uniform fashion by means of uniformly coding the product. Consider $\langle \phi_\alpha : \alpha < \gamma \rangle$ with the A_{ϕ_α}'s disjoint and nonempty. Then

$$\bigvee_{\alpha \leq \beta < \gamma} A_{\phi_\alpha} \times A_{\phi_\beta}$$

is a prewellordering of length γ. If $\xi \geq \omega$ is such that each ϕ_α is in $\mathcal{L}_{\xi,0}$, then the resulting code is in $\mathcal{L}_{\xi \cup |\gamma|,0}$. □

We observed above that $\rho_{\mathrm{B}} = \Theta$ if and only if all subsets of ω^ω are ∞-Borel. The same is true for λ_{B} (and, as well shall see, for all four ordinals). Given $A \subseteq \omega^\omega$, we let δ_A be the supremum of the lengths of the prewellorderings which are Wadge-below either A or $\omega^\omega \setminus A$.

Theorem 6.5: (Woodin). *Assuming* $\mathsf{AD} + \mathsf{DC}_{\mathbb{R}}$, $\lambda_{\mathrm{B}} = \Theta$ *if and only if every subset of* ω^ω *is* ∞*-Borel.*

Proof: By definition, Θ is the supremum of the lengths of the prewellorderings of ω^ω, so if every subset of ω^ω is ∞-Borel then $\lambda_{\mathrm{B}} = \Theta$. For the other direction, suppose that $A \subseteq \omega^\omega$ is not ∞-Borel and that P is an ∞-Borel prewellorder. Then $A \not\leq_W P$ as the ∞-Borel sets form an initial segment of the Wadge hierarchy. Thus either $P \leq_W A$ or $P \leq_W \omega^\omega \setminus A$, and in either case the length of P is less than δ_A, which is less than Θ. □

Putting together the results above, we have the following.

Theorem 6.6: (Woodin). *Assuming* $\mathsf{AD} + \mathsf{DC}_{\mathbb{R}}$, *the following statements are equivalent.*

- *At least one of* χ_{B}, κ_{B}, λ_{B} *and* ρ_{B} *is* Θ.
- *All subsets of* ω^ω *are* ∞*-Borel.*
- $\chi_{\mathrm{B}} = \kappa_{\mathrm{B}} = \lambda_{\mathrm{B}} = \rho_{\mathrm{B}} = \Theta$.

Proof: All that remains to be shown is that if all subsets of ω^ω are ∞-Borel then $\chi_{\mathrm{B}} = \Theta$. This follows from the Moschovakis Coding Lemma, which implies that for each $\chi < \Theta$ there is a surjection from ω^ω to $\mathcal{P}(\chi)$, and therefore one from ω^ω to the χ-Borel sets. □

Finally, we have the existence of local ∞-Borel codes.

Theorem 6.7: (Woodin). *Assume that* $\mathsf{AD} + \mathsf{DC}_{\mathbb{R}}$ *holds. If* $A \subseteq \omega^\omega$ *is* ∞*-Borel then it is* δ_A*-Borel.*

Proof: If $\delta_A \geq \chi_B$ we are done. Supposing otherwise, there is a $\phi \in \mathcal{L}_{\infty,0}$ such that ϕ^* contains a disjunction or conjunction of length δ_A. Take a minimal example and cut it off at length δ_A. Thus we get $\langle \phi_\alpha : \alpha < \delta_A \rangle$ where each ϕ_α is in $\mathcal{L}_{\delta_A,0}$ and the sets A_{ϕ_α} are nonempty and disjoint. As in the second half of the proof of Theorem 6.4, we obtain a δ_A-Borel prewellordering P of length δ_A. Then P is not Wadge-below A or $\omega^\omega \setminus A$. Hence $A \leq_{\mathrm{W}} P$, and since P is δ_A-Borel A is too. □

With the Coding Lemma and the fact that $\delta_A < \Theta^{L(A,\mathbb{R})}$ we have that $\mathsf{AD} + \mathsf{DC}_{\mathbb{R}}$ implies that if A is ∞-Borel then it is ∞-Borel in $L(A, \mathbb{R})$, and moreover that the assertion that $A \subseteq \omega^\omega$ is ∞-Borel is projective in A. In particular, the assertion that there is a subset of ω^ω which is not ∞-Borel is Σ^2_1. With Theorems 5.2 and 5.3, and the fact that Suslin sets are ∞-Borel, we get that AD implies that in $L(\mathbb{R})$ every subset of ω^ω is ∞-Borel.

We conclude this section by showing that the four ordinals defined above are all the same under $\mathsf{AD} + \mathsf{DC}_{\mathbb{R}}$. The proof uses Woodin's Theorem 3.9, whose prove we do not discuss in these notes.

Theorem 6.8: *Assume that* $\mathsf{AD} + \mathsf{DC}_{\mathbb{R}}$ *holds. Then*

$$\chi_{\mathrm{B}} = \kappa_{\mathrm{B}} = \lambda_{\mathrm{B}} = \rho_{\mathrm{B}} = \Theta^{L(\mathcal{P}(\chi_B))}.$$

Proof: By Theorem 6.6, the theorem holds in the case where all subsets of ω^ω are ∞-Borel. Suppose then that some $A \subseteq \omega^\omega$ is not ∞-Borel. Since $\mathsf{DC}_{\mathbb{R}}$ holds, $L(A, \mathbb{R})$ satisfies DC. Since the ∞-Borel sets form an initial segment of the Wadge hierarchy, they are all in $L(A, \mathbb{R})$. By Theorem 6.7, every ∞-Borel subset of ω^ω is ∞-Borel in $L(A, \mathbb{R})$. By the Moschovakis Coding Lemma, $L(A, \mathbb{R})$ contains all bounded subsets of its Θ. It follows that $\chi_{\mathrm{B}} = \chi_{\mathrm{B}}^{L(A,\mathbb{R})}$, $\lambda_{\mathrm{B}} = \lambda_{\mathrm{B}}^{L(A,\mathbb{R})}$ and $\rho_{\mathrm{B}} = \rho_{\mathrm{B}}^{L(A,\mathbb{R})}$.

By Theorem 3.9 (applied in $L(A, \mathbb{R})$), every subset of ω^ω in $L(\mathcal{P}(\chi_{\mathrm{B}}))$ is ∞-Borel. It follows then that the ∞-Borel sets are exactly the subsets of ω^ω in $L(\mathcal{P}(\chi_{\mathrm{B}}))$, all of which are ∞-Borel in $L(\mathcal{P}(\chi_{\mathrm{B}}))$. Moreover, $\chi_{\mathrm{B}} = \chi_{\mathrm{B}}^{L(\mathcal{P}(\chi_{\mathrm{B}}))}$, $\lambda_{\mathrm{B}} = \lambda_{\mathrm{B}}^{L(\mathcal{P}(\chi_{\mathrm{B}}))}$ and $\rho_{\mathrm{B}} = \rho_{\mathrm{B}}^{L(\mathcal{P}(\chi_{\mathrm{B}}))}$. By Theorem 6.6, applied in $L(\mathcal{P}(\chi_{\mathrm{B}}))$,

$$L(\mathcal{P}(\chi_{\mathrm{B}})) \models \chi_{\mathrm{B}} = \kappa_{\mathrm{B}} = \lambda_{\mathrm{B}} = \rho_{\mathrm{B}} = \Theta.$$

The theorem then follows from Theorem 6.4 (applied in V). □

7. Cone Measures

This section begins by presenting Martin's theorem that under AD the cone measure on the Turing degrees is an ultrafilter. This result lifts to larger degree notions, in particular those induced by constructibility relative to a fixed set of ordinals. This in turn induces the partial order $\leq_{\mathcal{D}}$ on degree notions, which turns out to be connected to Uniformization and the Suslin property (see Theorems 8.5 and 9.7).

7.1. *Turing reducibility*

Let HF denote the set of hereditarily finite sets. Given subsets x and y of HF, we say that x is *Turing-reducible* to y if x is Δ^0_1 definable over HF with a predicate for y. For x and y in ω^ω we write $x \leq_{\mathrm{Tu}} y$ to indicate that x is Turing-reducible to y, and $x \equiv_{\mathrm{Tu}} y$ to mean that $x \leq_{\mathrm{Tu}} y$ and $y \leq_{\mathrm{Tu}} x$.

Given $x \in \omega^\omega$, we let $[x]_{\mathrm{Tu}} = \{y \in \omega^\omega : x \equiv_{\mathrm{Tu}} y\}$. Sets of the form $[x]_{\mathrm{Tu}}$ (for some $x \in \omega^\omega$) are called *Turing degrees.* We write $\mathcal{D}_{\mathrm{Tu}}$ for the set of Turing degrees. The sets $\{y \in \omega^\omega : y \geq_{\mathrm{Tu}} x\}$ or $\{[y]_{\mathrm{Tu}} : y \in \omega^\omega,\ y \geq_{\mathrm{Tu}} x\}$ are both referred to as the *Turing cone* above x (for any $x \in \omega^\omega$, which is called a *base* for the cone).

Given an equivalence relation E on a set X, we say that a set $A \subseteq X$ is *E-invariant* if it is a union of E-classes, that is, if, for all $x, y \in X$ such that xEy, $x \in A$ if and only if $y \in A$. A subset of ω^ω is said to be *Turing-invariant* if it is $\equiv_{\mathrm{Tu}}$-invariant.

The following theorem of Martin shows that, under AD, every Turing-invariant subset of ω^ω either contains or is disjoint from a Turing cone.

Theorem 7.1: (Martin). *Let $A \subseteq \omega^\omega$ be Turing-invariant, let σ be a strategy in $\mathcal{G}_\omega(A)$ and let $x \in \omega^\omega$ be such that σ is Turing-reducible to x. If σ is a winning strategy for player I then A contains the Turing cone above x. If σ is a winning strategy for player II then A is disjoint from the Turing cone above x.*

We let Turing Determinacy (TD) be the statement that each Turing-invariant subset of ω^ω either contains or is disjoint from a cone. A theorem of Peng and Yu [18] shows that $\mathsf{CC}_{\mathbb{R}}$ follows from TD. It is an open question whether TD implies AD. Woodin has shown that this implication does hold in $L(\mathbb{R})$ (see [2]).

We let μ_{Tu} be the set of $A \subseteq \mathcal{D}_{\mathrm{Tu}}$ which contain a Turing cone. Martin's theorem implies μ_{Tu} is an ultrafilter on the Turing degrees, and $\mathsf{CC}_{\mathbb{R}}$ implies this ultrafilter is countably complete. The following theorem shows that TD implies that $\aleph_1 \not\leq 2^{\aleph_0}$.

Theorem 7.2: *If TD holds then ω_1 is measurable.*

Proof: Let U be the set of $A \subseteq \omega_1$ for which the set

$$\{[x]_{\mathrm{Tu}} : x \in \omega^\omega,\, \omega_1^{L[x]} \in A\}$$

is in μ_{Tu}. Then U is a countably complete ultrafilter on ω_1. □

The results above apply to a larger class of equivalence relations, including those described by the following definition. We will use a nonstandard definition.

Definition 7.3: *An equivalence relation E on ω^ω is* good *if it contains $\equiv_{\mathrm{Tu}}$ and there exists a preorder $\leq_E$ on ω^ω such that*

- $E = \leq_E \cap \geq_E$*;*
- *for every $x \in \omega^\omega$ the set $\{y \in \omega^\omega : y \leq_E x\}$ is countable;*
- *for all $x, y \in \omega^\omega$, if $x \leq_E y$ then there is $z \geq_{\mathrm{Tu}} x$ such that $z \equiv_E y$.*

In particular, for $S \subseteq \mathrm{Ord}$ the following equivalence relations are good.

- $x \equiv_S y \Leftrightarrow L[S, x] = L[S, y]$ (induced by setting $x \leq_S y$ to be $x \in L[S, y]$);
- $x \equiv_{\mathrm{HOD}_S} y \Leftrightarrow \mathrm{HOD}_{\{S,x\}} = \mathrm{HOD}_{\{S,y\}}$ (induced by setting $x \leq_{\mathrm{HOD}_S} y$ to be $x \in \mathrm{HOD}_{\{S,y\}}$).

Under AD, Martin's theorem implies that if E is a good equivalence relation, then every set of E-degrees either contains or is disjoint from a $\leq_E$-cone, and that the corresponding cone measures for these relations are (countably complete) ultrafilters. We let $\mathcal{D}_S$ be the set of $\equiv_S$-degrees, and μ_S be the corresponding cone measure. We call the corresponding cones S-cones. In the next section, we will consider ultraproducts by these measures.

7.2. *Cone measures and ultrapowers*

We have seen that AD implies TD and that the Turing measure μ_{Tu} is countably complete, and that the same holds for any good equivalence relation, including the relations $\equiv_S$ and $\equiv_{\mathrm{HOD}_S}$. We will be looking at ultraproducts of the form

$$\prod_{d \in \mathcal{D}_{\mathrm{Tu}}} M_d / \mu_{\mathrm{Tu}}$$

or

$$\prod_{d \in \mathcal{D}_S} M_d / \mu_S,$$

for S a set of ordinals.

If the ultraproduct by the Turing measure μ_{Tu} is wellfounded, then so is the ultraproduct by any good equivalence relation, since any ultraproduct for a good equivalence relation embeds into the corresponding ultraproduct by μ_{Tu}. By countable completeness, these ultraproducts are wellfounded provided $\mathsf{TD} + \mathsf{DC}$ holds. Recall that the implication $\mathsf{DC}_{\mathbb{R}} \Rightarrow \mathsf{DC}$ holds in models of the form $L(A, \mathbb{R})$ when A is a subset of $L(\mathbb{R})$. The following theorem of Solovay shows that $\mathsf{DC}_{\mathbb{R}}$ follows from the wellfoundedness of the ultrapower of the ordinals by the cone measure for a good equivalence relation. The theorem is essentially proved in [19]. A proof is also given in [13].

Theorem 7.4: (Solovay). *If $(E, \leq_E)$ is a good equivalence relation, μ_E is the corresponding cone measure and the ultrapower of the ordinals by μ_E is wellfounded, then $\mathsf{DC}_{\mathbb{R}}$ holds.*

The assertion that the ultrapower of the ordinals by μ_{Tu} is wellfounded was originally one of the axioms of AD^+. Woodin later showed however that this ultrapower is wellfounded if $V = L(\mathcal{P}(\mathbb{R}))$ and AD^+ (as currently stated) holds. The proof uses his reversal of the Derived Model Theorem.

Theorem 7.5: (Woodin). *If $V = L(\mathcal{P}(\mathbb{R}))$ and AD^+ holds then the ultrapower of the ordinals by μ_{Tu} is wellfounded*

We will eventually (in Theorem 9.5) be interested in the ordinal represented by the function $x \mapsto \omega_2^{L[S,x]}$ in the μ_S-ultrapower, for a fixed set $S \subseteq \mathrm{Ord}$. First, we look at functions of the form $x \mapsto \omega_1^{L[S,x]}$. Given a set S of ordinals, and function f on ω^ω, we say that f is *S-invariant* if $f(x) = f(y)$ whenever $x \equiv_S y$.

Let WO be the set of wellorderings of ω. The proof of Theorem 7.6 uses the following classical fact, known as $\underset{\sim}{\Sigma}^1_1$ boundedness: for every analytic $A \subseteq \mathrm{WO}$ there is $\beta < \omega_1$ such that for every $x \in A$, the ordertype of x is less than β. We write $\mathrm{otp}(x)$ for the ordertype of a wellordering x.

Theorem 7.6: (Woodin). *Assume that AD holds. Let S be a set of ordinals, and define the function $f\colon \omega^\omega \to \omega_1$ by setting $f(x)$ to be $\omega_1^{L[S,x]}$. Then f represents ω_1 in the ultrapower $\prod \mathrm{Ord}/\mu_S$.*

Proof: Let $[f]_S$ denote the element of $\prod \mathrm{Ord}/\mu_S$ represented by f. The countable completeness of μ_S implies that ω_1^V is (isomorphic to) an initial segment of the ultrapower $\prod \mathrm{Ord}/\mu_S$. For each $\alpha \in \omega_1$, $\{[x]_S : f(x) > \alpha\} \in \mu_S$, so $[f]_S \geq \alpha$.

To see that $[f]_S \leq \omega_1$ suppose that $g : \omega^\omega \to \omega_1$ is S-invariant and that for every $x \in \omega^\omega$ we have that $g(x) < \omega_1^{L[S,x]}$. We want to find an $\alpha < \omega_1$ such that $\{[x]_S : g(x) < \alpha\} \in \mu_S$.

Consider the game where

- player I plays $x(i)$ in round i, collectively defining $x \in \omega^\omega$;
- player II plays $(y(i), z(i)) \in \omega \times (\omega \times \omega)$ in round i, collectively defining $y \in \omega^\omega$ and $z \subseteq \omega \times \omega$;
- player II wins if and only if $y \geq_S x$, $z \in \mathrm{WO}$, and the ordertype of z is greater than $g(y)$.

If σ is a strategy for player I, player II can defeat it by playing a $y \in \omega^\omega$ such that σ is Turing reducible to y, and a $z \in \mathrm{WO} \cap L[S, y]$ with ordertype greater than $g(y)$ (such a z exists since $g(y) < \omega_1^{L[S,y]}$). Thus by AD we may fix a winning strategy σ for player II. For each $x \in \omega^\omega$, let $y_\sigma(x)$ denote the $y \in \omega^\omega$ that σ produces when player I plays x, and let $z_\sigma(x)$ denote the set $z \subseteq \omega \times \omega$ that σ produces when player I plays x. Then $\{z_\sigma(x) : x \in \omega^\omega\}$ is analytic and included in WO, so there is $\alpha < \omega_1$ such that for every $x \in \omega^\omega$, the ordertype of $z_\sigma(x)$ is less than α. The set of $x \in \omega^\omega$ such that σ is Turing-reducible to x contains an S-cone. For any such x, $y_\sigma(x) \equiv_S x$, and

$$\alpha > \mathrm{otp}(z_\sigma(x)) > g(y_\sigma(x)) = g(x). \qquad \square$$

The following application of cone measures is used in the proof of Theorem 9.5, which is the key theorem for producing Suslin representations from ∞-Borel codes. Theorem 7.7 was first proved by Steel assuming AD [20] and later by Woodin from TD by the proof given below. Almost the same proof gives the corresponding result for $\Diamond$. Whether one can also get $\Box$ or forms of condensation seems to be open.

Theorem 7.7: *Suppose that $\aleph_1 \not\leq 2^{\aleph_0}$ and let S be a set of ordinals such that μ_S is an ultrafilter. Then*

$$\{[y]_S : L[S, y] \cap V_{\omega_1} \models \mathsf{GCH}\} \in \mu_S.$$

Proof: We first prove the theorem with CH in place of GCH. It suffices to show that for each $y \in \omega^\omega$ there is an $x \in \omega^\omega$ such that $L[S, y][x] \models \mathsf{CH}$. Since $\aleph_1 \not\leq 2^{\aleph_0}$, for each $y \in \omega^\omega$ there exist $L[S, y]$-generic filters for each partial order in $L[S, y] \cap V_{\omega_1}$. Fix $y \in \omega^\omega$ and force over $L[S, y]$ with the iteration $\mathrm{Col}^*(\omega_1, \omega^\omega) * \mathbb{Q}$ (as defined in $L[S, y]$), where Col^* adds a bijection

f from ω_1 to ω^ω by countable approximations (without adding reals) and $\mathbb{Q}$ is the almost disjoint coding forcing (see [5]) to make the relation

$$\{(a,b) \in (\omega^\omega)^2 : f^{-1}(x) \leq f^{-1}(y)\}$$

an F_σ subset of $(\omega^\omega)^2$. If $a \in \omega^\omega$ is a real coding this F_σ set (and thus the generic filter) in the corresponding generic extension, then f is in $L[S,y][a]$. Since $\mathbb{Q}$ is a c.c.c. partial order of cardinality $\aleph_1$ in $L[S,y][f]$, $L[S,y][a] \models \mathsf{CH}$.

For the general case of GCH, let $x_0 \in \omega^\omega$ be such that for every $y \geq_S x_0$, $L[S,y] \models \mathsf{CH}$. Fix $y \geq_S x_0$ and an $L[S,y]$-cardinal $\gamma < \omega_1^V$. Let $g \subseteq \mathrm{Col}(\omega,\gamma)$ be an $L[S,y]$-generic filter. Then $L[S,y][g] \models \mathsf{CH}$ so $L[S,y] \models 2^\gamma = \gamma^+$. □

7.3. *A degree notion for sets of ordinals*

The proof of Theorem 9.5 uses the following order on sets of ordinals, induced by their corresponding degree notions. Given $S, T \subseteq \mathrm{Ord}$, we write $S \leq_{\mathcal{D}} T$ to mean that

$$\{[x]_{\mathrm{Tu}} : \omega^\omega \cap L[S,x] \subseteq L[T,x]\} \in \mu_{\mathrm{Tu}}.$$

We write $S <_{\mathcal{D}} T$ for $S \leq_{\mathcal{D}} T \wedge \neg(T \leq_{\mathcal{D}} S)$. A reflection argument shows that for any set S of ordinals there is a bounded $T \subseteq \Theta$ with $S \leq_{\mathcal{D}} T$. Ultimately, we will be concerned only with the restriction of $\leq_{\mathcal{D}}$ to bounded subsets of Θ.

Theorem 7.8 shows that $\leq_{\mathcal{D}}$ is a total order when μ_{Tu} is a countably complete ultrafilter. As with Theorem 7.7, Theorem 7.8 was first proved by Steel in [20] assuming AD and later by Woodin from $\mathsf{TD} + \mathsf{CC}_{\mathbb{R}}$. Woodin's proof uses a variation of Mathias forcing.

Theorem 7.8: *Assume that* TD *holds. If* $\neg(T \leq_{\mathcal{D}} S)$ *then*

$$\{[x]_{\mathrm{Tu}} : \omega^\omega \cap L[S,x] \in H(\aleph_1)^{L[T,x]}\} \in \mu_{\mathrm{Tu}}.$$

The proof of Theorem 7.10 uses the following classical result of Solovay. In many of our applications we will have $M = M'$, but we will use the general version at the end of Section 7.4.

Theorem 7.9: (Solovay). *If $M \subseteq M'$ are both transitive models of* ZF, *$\mathbb{P} \in M$ is a partial order, $\mathcal{P}(\mathbb{P}) \cap M'$ is countable, and $x \in M[G]$ for all M'-generic $G \subseteq \mathbb{P}$, then x is in M.*

Theorem 7.10 lists some consequences of Theorem 7.8 which will be used in Section 7.4. Parts (1a) and (2a) follow from Theorems 7.8 and 7.9 by a collapsing argument as in the second half of the proof of Theorem 7.7. Note that both parts apply to all ordinals γ below ω_1^V, not just $\omega_1^{L[T,x]}$. Part (1b) follows from part (1a) and Theorem 7.9, by another collapsing argument. Part (2b) follows from part (2a). Note also that neither part of the theorem assumes (or implies) that S is in $L[T,y]$. A natural application of part (1b) is when S is an ∞-Borel code.

Theorem 7.10: *Assume that* TD *holds, and let S and T be sets of ordinals.*

(1) If x is a base of a cone witnessing that $S \leq_{\mathcal{D}} T$, then the following hold for all $y \geq_{\mathrm{Tu}} x$.

(a) For every $\gamma < \omega_1^V$, $\mathcal{P}(\gamma) \cap L[S,y] \subseteq L[T,y]$.

(b) For every formula ϕ,

$$\{z \in \omega^\omega : L[S,z] \models \phi(S,z)\} \cap L[T,y] \in L[T,y].$$

(2) If $x \in \omega^\omega$ is a base of a Turing cone witnessing that $S <_{\mathcal{D}} T$, then the following hold for all $y \geq_{\mathrm{Tu}} x$.

(a) For every $\gamma < \omega_1^V$,

$$\mathcal{P}(\gamma) \cap L[S,y] \in H(|\gamma|^+)^{L[T,y]}.$$

(b) $\omega_1^{L[T,y]}$ is a strongly inaccessible cardinal in $L[S,y]$.

7.4. *Forcing with sets of reals*

Given a set S of ordinals (and assuming that μ_S induces a wellfounded ultrapower), we let δ_S^∞ be the ordinal represented by the function $x \mapsto \omega_2^{L[S,x]}$. Theorem 7.13 (in conjunction with part (2b) of Theorem 7.10) shows that if S is not $\leq_{\mathcal{D}}$-maximal, then $\delta_S^\infty < \Theta$, which is one of the hypotheses of Theorem 9.6. The proof of Theorem 7.13 uses a result of Martin on pointed perfect trees.

Given a set S of ordinals, a set $A \subseteq \omega^\omega$ is *S-positive* if it intersects every S-cone. A tree is *perfect* if each node has an incompatible pair of extensions.

Definition 7.11: *For $S \subseteq$ Ord, an S-*pointed *(perfect) tree is a perfect tree $a \subseteq \omega^{<\omega}$ such that for every $x \in [a]$, $a \in L[S,x]$.*

Theorem 7.12: (Martin). AD *implies that for every $B \subseteq \omega^\omega$ and every $S \subseteq$ Ord, B is S-positive if and only if there is an S-pointed perfect tree a with $[a] \subseteq B$.*

Given a set T of ordinals, we define the relation $<^{\mathrm{c}}_T$ on $(\omega^\omega)^2$ by setting $(x,y) <^{\mathrm{c}}_T (z,w)$ if $x = z$, $y,w \in L[T,x]$ and y comes before w in the constructibility order on $L[T,x]$ using T and x as parameters. Theorem 7.13 says (in conjunction with part (2) of Theorem 7.10) that if $S \subseteq \mathrm{Ord}$ is not $\leq_{\mathcal{D}}$-maximal, and f is an S-invariant function on ω^ω sending each x to $H(\gamma)^{L[S,x]}$, for some γ below the least strongly inaccessible cardinal of $L[S,x]$, then the structure represented by f in the μ_S ultrapower is coded by a subset of ω^ω.

Theorem 7.13: (Woodin). *Suppose that* AD *holds, and let S and T be sets of ordinals. Suppose that f is an S-invariant function on ω^ω such that for every $x \in \omega^\omega$, $f(x)$ is a transitive set in $H(\aleph_1)^{L[T,x]}$. Then $\prod f(x)/\mu_S$ is isomorphic to a relation projective in $\leq_S \times <^{\mathrm{c}}_T$.*

Let S^∞ denote the set represented by the constant function $x \mapsto S$ in the ultrapower given by μ_S. The countable completeness of μ_S give that for all $x, y \in \omega^\omega$, $x \in L[S,y]$ if and only if $x \in L[S^\infty, y]$. It follows that a tree is S-pointed if and only if it is S^∞-pointed.

Let $\mathbb{P}_S$ be the partial order of (not necessarily S-invariant) S-positive sets under containment. Theorem 7.12 implies that $\mathbb{P}_S$ is forcing-equivalent to the partial order of S^∞-pointed perfect trees, also under containment. Therefore, any x_G produced by a V-generic filter $G \subseteq \mathbb{P}_S$ is also generic over $L(S^\infty, \mathbb{R})$ for this partial order (which is in $L(S^\infty, \mathbb{R})$ while $\mathbb{P}_S$ may not be).

Since μ_S is countably complete, whenever A is an S-positive set and n is in ω there exists an $s \in \omega^n$ such that $\{x \in A : x{\restriction}n = s\}$ is S-positive. It follows that if G is a V-generic filter for $\mathbb{P}_S$ then there is a unique $x \in \omega^\omega$ such that, for each $n \in \omega$ the set $\{y \in A : y{\restriction}n = x{\restriction}n\}$ is in G. We let x_G denote the unique such x. A standard genericity argument shows that x_G will be Turing-above every element of $\omega^\omega \cap V$. This shows that $\mathbb{P}_S$ collapses every cardinal below Θ^V.

A generic filter G for $\mathbb{P}_S$ induces an ultrapower $\prod_{x\in\omega^\omega} V/G$, whose elements are represented by functions with domain ω^ω. Although we will not prove it here, it turns out that, assuming AD plus the wellfoundednes of the μ_S-ultrapower, the models $\prod_{x\in\omega^\omega} L[S,x]/G$ and $L[S^\infty, x_G]$ are isomorphic.

Now suppose that $A \subseteq \omega^\omega$ is such that $A \cap L[S,x] \in L[S,x]$ for all $x \in \omega^\omega$. Then the function $A \mapsto A \cap L[S,x]$ represents in $\prod_{x\in\omega^\omega} L[S,x]/G$ a set $A^* \subseteq \omega^\omega$ in $V[G]$ such that $A^* \in L[S^\infty, x_G]$ and $A^* \cap (\omega^\omega)^V = A$. It follows that for any V-generic x_G we have $A \in L(S^\infty, \mathbb{R})[x_G]$, so $A \in L(S^\infty, \mathbb{R})$ by Theorem 7.9. Along with part (1b) of Theorem 7.10, this gives Theorem 7.14.

Theorem 7.14: (Woodin). *Assume that* AD *holds, S is a set of ordinals, and the μ_S-ultrapower of the ordinals is wellfounded. If S is a $\leq_{\mathcal{D}}$-maximal set of ordinals, then every ∞-Borel set is in $L(S^\infty, \mathbb{R})$.*

Theorem 9.7 below shows that, assuming AD^+, the existence of a $\leq_{\mathcal{D}}$-maximal set of ordinals is equivalent to the failure of Uniformization.

By Theorem 6.7, the ∞-Borel sets are all ∞-Borel in $L(S^\infty, \mathbb{R})$, assuming the hypotheses of Theorem 7.14. Wellfoundedness of the μ_S-ultrapower follows from $\mathsf{AD}+\mathsf{DC}$, and also from AD^+, by Theorem 7.5. If AD holds and $A \subseteq \omega^\omega$ is not ∞-Borel, then every ∞-Borel set is ∞-Borel in $L(A, \mathbb{R})$ by Theorem 6.7, and one can apply Theorem 7.14 in $L(S, A, \mathbb{R})$ assuming only $\mathsf{AD} + \mathsf{DC}_{\mathbb{R}}$. That is, under these assumptions the model $L(S, A, \mathbb{R})$ satisfies the statement that the μ_S-ultrapower is wellfounded. In this case, however, the S^∞ in the statement of the theorem is as computed in $L(S, A, \mathbb{R})$.

8. The Vopěnka Algebra

In this section, we present a version of Vopěka's theorem that every set of ordinals is generic over HOD, and Woodin's applications of this theorem in analyzing the ∞-Borel sets. We start with some technical formalities.

Let X be a set with $X \in \mathrm{OD}_X$, let z be an element of HOD_X and let λ be the least ordinal such that $\mathcal{P}(\mathcal{P}(z)) \cap \mathrm{OD}_X \subseteq \mathrm{OD}_X^{V_\lambda}$. We let $P_{X,z}$ be the set of triples $(n, \bar{x}, \bar{\alpha})$ such that

- n is the Gödel number of a formula ϕ_n of arity $|\bar{x}| + |\bar{\alpha}| + 1$,
- $\bar{x} \in X^{<\omega}$,
- $\bar{\alpha} \in \lambda^{<\omega}$ and
- the set $b_{n,\bar{x},\bar{\alpha}} = \{C \subseteq z : V_\lambda \models \phi_n(C, \bar{x}, \bar{\alpha})\}$ is nonempty.

Order $P_{X,z}$ by setting $(n, \bar{x}, \bar{\alpha}) \leq (m, \bar{y}, \bar{\beta})$ to hold if $b_{n,\bar{x},\bar{\alpha}} \subseteq b_{m,\bar{y},\bar{\beta}}$. We write $[(n, \bar{x}, \bar{\alpha})]$ for the corresponding equivalence class of $(n, \bar{x}, \bar{\alpha})$ and let $\mathcal{V}_{X,z}$ denote the corresponding partial order on the equivalence classes. Then $\mathcal{V}_{X,z} \in \mathrm{HOD}_X$. Furthermore, for each $a \in z$, the set $T_{X,z}$ consisting of the pairs $(a, [(n, \bar{x}, \bar{\alpha})])$ for which $b_{n,\bar{x},\bar{\alpha}} = \{C \subseteq z : a \in C\}$ is in HOD_X.

For each $C \subseteq z$, we let $G_C^{X,z}$ be the set of $[(n, \bar{x}, \bar{\alpha})]$ for which V_λ satisfies $\phi_n(C, \bar{x}, \bar{\alpha})$. Vopěnka's Theorem says that under these assumptions on X and z, each set $G_C^{X,z}$ is a HOD_X-generic filter for $\mathcal{V}_{X,z}$. The key point in the proof is that union of the sets $b_{n,\bar{x},\bar{\alpha}}$ corresponding to any maximal antichain in $\mathcal{V}_{X,z}$ is all of $\mathcal{P}(z)$.

Theorem 8.1: (Vopěnka). *Let X be a set with $X \in \mathrm{OD}_X$, and let z be an* *element of* HOD_X. *For each* $C \subseteq z$ *the set* $G_C^{X,z}$ *is a* HOD_X*-generic filter in* $\mathcal{V}_{X,\gamma}$*, and* C *is in* $\mathrm{HOD}_X[G_C^{X,z}]$.

For the rest of this section, we restrict to the case where $X = \{S\}$, for some set S of ordinals, and $z = \omega$, and consider the Vopěnka algebra in models of the form $L[S, x]$, where $x \in \omega^\omega$. Using the definability order in $\mathrm{HOD}_{\{S\}}^{L[S,x]}$ we can definably copy $\mathcal{V}_{\{S\},\omega}^{L[S,x]}$ and $T_{\{S\},\omega}^{L[S,x]}$ to a partial order Q_x^S on some ordinals and a corresponding set $K_x^S \subseteq \omega \times \mathrm{Ord}$, both in $\mathrm{HOD}_{\{S\}}^{L[S,x]}$. Thus for every $c \in \mathcal{P}(\omega) \cap L[S, x]$ there exists a $L[S, Q_x^S, K_x^S]$-generic filter $G \subseteq Q_x^S$ with $c \in L[S, Q_x^S, K_x^S][G]$.

It follows immediately from either definition that the collection of ∞-Borel sets of ω^ω (which generalizes naturally to subsets of $(\omega^\omega)^n$ for any $n \in \omega$) is closed under complements. We will now outline a proof (assuming $\mathsf{TD} + \mathsf{DC}_\mathbb{R}$) that it is also projectively closed. This amounts to showing that if $A \subseteq (\omega^\omega)^2$ is an ∞-Borel set, and $B = \{x \in \omega^\omega : \exists y\, (x, y) \in A\}$, then B is also ∞-Borel. We start with the following lemma.

Lemma 8.2: (Woodin). *Let* S *be a set of ordinals, and suppose that* $\aleph_1 \not\leq 2^{\aleph_0}$. *Let* ϕ *be a ternary formula. Then for all* $x \in \omega^\omega$,

$$\exists y \in \omega^\omega\, L[S, x, y] \models \phi(S, x, y)$$

if and only if, for a μ_S*-cone of* $z \in \omega^\omega$*, the model* $L[S, Q_z^S, K_z^S, x]$ *satisfies the statement that some partial order of cardinality at most* $2^{|Q_z^S|}$ *forces the existence of a* $y \in \omega^\omega$ *such that* $L[S, x, y] \models \phi(S, x, y)$.

Proof: The reverse direction follows from the fact that each partial order of cardinality at most $2^{|Q_z^S|}$ in $L[S, Q_z^S, K_z^S, x]$ is countable in V, so $L[S, Q_z^S, K_z^S, x]$-generic filters exist.

For the other direction, fix $x \in B$ and $y \in \omega^\omega$ such that $L[S, x, y] \models \varphi(S, x, y)$. Let $z \in \omega^\omega$ be $\leq_S$-above both x and y. There is an $L[S, Q_z^S, K_z^S]$-generic filter $G \subseteq Q_z^S$ such that z is in the model $L[S, Q_z^S, K_z^S][G]$. Then $L[S, Q_z^S, K_z^S][G]$ is a generic extension of $L[S, Q_z^S, K_z^S, x]$ by a quotient of the regular open algebra of Q_z^S. □

The μ_S-ultrapower now gives an ∞-Borel code for the projection of the set coded by the pair (S, ϕ).

Theorem 8.3: (Woodin). *Let* S *be a set of ordinals, and suppose that* $\mathsf{DC}_\mathbb{R}$ *holds and* μ_S *is an ultrafilter. Let* ϕ *be a ternary formula and let*

$$B = \{x \in \omega^\omega : \exists y \in \omega^\omega\, L[S, x, y] \models \phi(S, x, y)\}.$$

Then there exist a binary formula ψ and a set T of ordinals such that $T \in \mathrm{OD}_{\{S\}}$ and

$$B = \{x \in \omega^\omega : L[T, x] \models \psi(T, x)\}.$$

Proof: Work in $L(S, \mathbb{R})$, and let

$$S^\infty = \prod S/\mu_S,$$
$$Q_S^\infty = \prod Q_z^S/\mu_S$$

and

$$K_S^\infty = \prod K_z^S/\mu_S.$$

For each $x \in \omega^\omega$, the constant function $y \mapsto x$ represents x in the ultrapower by μ_S. By Lemma 8.2, then, we have that for every $x \in \omega^\omega$, $x \in B$ if and only if $L[S^\infty, Q_S^\infty, K_S^\infty, x]$ satisfies the statement that there exists a partial order of cardinality at most $2^{|Q_S^\infty|}$ forcing the existence of a $y \in \omega^\omega$ such that

$$L[S^\infty, x, y] \models \phi(S^\infty, x, y). \qquad \square$$

Restating Theorem 8.3, we have the following.

Theorem 8.4: *If* $\mathsf{TD} + \mathsf{DC}_{\mathbb{R}}$ *holds, then the set of* ∞*-Borel subsets of* $(\omega^\omega)^{<\omega}$ *is projectively closed.*

The proof of the reverse direction of the following theorem (which we do not give) uses material from this section.

Theorem 8.5: $(\mathsf{ZF} + \mathsf{TD} + \mathsf{DC}_{\mathbb{R}})$. *Suppose that every subset of* ω^ω *is* ∞*-Borel. Then* Uniformization *is equivalent to the assertion that* $\leq_{\mathcal{D}}$ *has no greatest element.*

The forward direction of the theorem requires only that $\aleph_1 \not\leq 2^{\aleph_0}$ and every subset of ω^ω is ∞-Borel. Let S be a set of ordinals, and let $f : \omega^\omega \to \omega^\omega$ uniformize the set $\{(x, y) \in (\omega^\omega)^2 : y \notin L[S, x]\}$. Let (T, ϕ) be an ∞-Borel code for $\{(x, i, j) : (i, j) \in f(x)\}$. Then for all (x, i, j), $(i, j) \in f(x)$ if and only if $L[T, x] \models \phi(T, x, i, j)$, which shows that $f(x) \in L[T, x] \setminus L[S, x]$ for all x.

9. Strong Generic Codes

In this section, we show how to combine ∞-Borel codes, the Vopenka algebra and $<\Theta$-Determinacy to produce Suslin representations. Theorem 9.5 is the key technical result. A summary of results so far accumulated is given in Theorem 9.7. Again, we begin with some technical preliminaries.

Given a $\mathbb{P}$-name τ for an element of ω^ω, we let, for each $n \in \omega$, $D_{\tau,n}$ be the set

$$\{p \in \mathbb{P} : \exists m \in \omega\, p \Vdash (n,m) \in \tau\}.$$

For any set B, we say that a filter G contained in a partial order $\mathbb{P}$ is B-generic if it intersects every dense open subset of $\mathbb{P}$ in B. When $\{D_{\tau,n} : n \in \omega\} \subseteq B$, we write $A_{\mathbb{P},B,\tau}$ for the set of values τ_g, where g ranges over the set of all B-generic filters contained in $\mathbb{P}$.

When $\mathbb{P}$ is a partial order, we write $\operatorname{dom}(\mathbb{P})$ for the underlying set. Similarly, if B is a sequence, we write $\operatorname{dom}(B)$ for the corresponding index set, and $\operatorname{range}(B)$ for the set indexed by B. A *nice* $\mathbb{P}$-name for a subset of ω^2 is a set of pairs of the form $(p, (\check{n}, \check{m}))$, where $p \in \mathbb{P}$ and $n, m \in \omega$.

Definition 9.1: *Given $A \subseteq \omega^\omega$, an ordinal α, a partial order $\mathbb{P}$, a sequence B of dense open subsets of $\mathbb{P}$ and a nice $\mathbb{P}$-name τ for an element of ω^ω, we say that $(\mathbb{P}, B, \tau)$ is a* generic code *for A if $\{D_{\tau,n} : n \in \omega\} \subseteq \operatorname{range}(B)$ and*

$$A_{\mathbb{P},\operatorname{range}(B),\tau} = A.$$

If we say that C is a generic code, then we mean that C is a triple of the form $(\mathbb{P}_C, B_C, \tau_C)$, and write A_C for $A_{\mathbb{P}_C,\operatorname{range}(B_C),\tau_C}$. If in addition α is an ordinal containing $\operatorname{dom}(\mathbb{P}) \cup \operatorname{dom}(B)$, then we say that $(\mathbb{P}, B, \tau)$ is a generic α-code *for A. We say that $(\mathbb{P}, B, \tau)$ is a* generic ∞-code *if it is a generic α-code for some ordinal α.*

Theorem 8.1 (with C as ω and X as $\{S\}$) shows that if (S, ψ) is an ∞-Borel code for a set $A \subseteq \omega^\omega$, then $(\mathcal{V}_{\{S\},\omega^2}\restriction p, B, \tau\restriction p)$ is a generic code for A, where, letting τ be the $\mathcal{V}_{\{S\},\omega^2}$-name for the associated generic real,

- p is the $\mathcal{V}_{\{S\},\omega}$-condition corresponding to the set
$$\{x \in \omega^\omega : L[S,x] \models \psi(S,x)\},$$
- $\mathcal{V}_{\{S\},\omega^2}\restriction p$ and $\tau\restriction p$ are the corresponding restrictions of $\mathcal{V}_{\{S\},\omega^2}$ and τ below p and
- B lists the set of dense open subsets of $\mathcal{V}_{\{S\},\omega^2}\restriction p$ in $\mathrm{HOD}_{\{S\}}$.

Using the definability order on $\mathrm{HOD}_{\{S\}}$, one can convert this generic code to a generic ∞-code. The proof of Theorem 9.5 below uses the fact that, by Theorem 7.7, for a Turing cone of $x \in \omega^\omega$ there exists in $L[S,x]$ a generic $\omega_2^{L[S,x]}$-code for $A \cap L[S,x]$.

Definition 9.3 below defines a game, relative to a generic code C for a set A, in which player I attempts to build a countable subcode for a subset of A. Theorem 9.4 shows that a winning strategy for player I in this game gives a Suslin representation for A. The following definition gives the relevant notion of subcode.

Definition 9.2: *Suppose that $(\mathbb{P}, B, \tau)$ is a generic ∞-code. For any set X of ordinals, we let $\mathbb{P}_X$ be $\mathbb{P}{\restriction}\mathrm{dom}(\mathbb{P}){\cap}X$, B_X be $\langle B(\beta){\cap}X : \beta \in X{\cap}\mathrm{dom}(B)\rangle$ and τ_X be $((\mathrm{dom}(\mathbb{P}) \cap X) \times \omega^2) \cap \tau$. If $C = (\mathbb{P}, B, \tau)$, we write C_X for $(\mathbb{P}_X, B_X, \tau_X)$.*

Definition 9.3: *Suppose that $C = (\mathbb{P}, B, \tau)$ is a generic α-code for a set A, for some ordinal α. We associate to $(\mathbb{P}, B, \tau)$ a game on* $\mathrm{dom}(\mathbb{P}) \cup \mathrm{dom}(B)$, *called $\mathcal{G}_{\mathbb{P},B,\tau}$ or $\mathcal{G}_C$, where I and II collaborate to build a countable subset $\sigma \subseteq \alpha$, and I wins if $C_\sigma = (\mathbb{P}_\sigma, B_\sigma, \tau_\sigma)$ is a generic code for a subset of A. We say that $(\mathbb{P}, B, \tau)$ is a* strong generic code *(or* strong α-generic code*) for A_ϕ if II does not have a winning strategy in $\mathcal{G}_{\mathbb{P},B,\tau}$.*

For any infinite cardinal κ and κ-generic code $(\mathbb{P}, B, \tau)$, κ-Determinacy implies the determinacy of the game $\mathcal{G}_{\mathbb{P},B,\tau}$, since a run of the game continuously builds a subset of ω coding a generic ω-code. Unlike ∞-Borel codes, strong generic codes witness Suslinity, in the context of $<\Theta$-Determinacy.

Theorem 9.4: *Let κ be an infinite cardinal and let C be a generic κ-code for a set $A \subseteq \omega^\omega$. If player I has a winning strategy in the game $\mathcal{G}_C$, then A is κ-Suslin. Moreover, if Σ is a winning strategy for player I in $\mathcal{G}_C$, then $A = \mathrm{p}[T]$, for some tree $T \subseteq (\omega \times \kappa \times 2)^{<\omega}$ which is definable from C and Σ.*

Proof: (Sketch) Let C be $(\mathbb{P}, B, \tau)$, and fix a winning strategy Σ for player I in $\mathcal{G}_C$. Then for all $x \in \omega^\omega$, $x \in A$ if and only if there exist a countable $\sigma \subseteq \mathrm{dom}(\mathbb{P}) \cup \mathrm{dom}(B)$ produced by a run of $\mathcal{G}_C$ where I plays according to Σ and a B_σ-generic filter $g \subseteq \mathbb{P}_\sigma$ such that $x = \tau_{\sigma,g}$.

The set of (x, y, z) such that

- $x \in \omega^\omega$,
- $y \in \kappa^\omega$ is a run of $\mathcal{G}_C$ where I plays according to Σ,
- $z \in 2^\omega$ and

- $y[z^{-1}[\{1\}]]$ is a filter on $y[\omega] \cap \mathbb{P}$ for which the realization of

$$\tau_{y[\omega],y[z^{-1}[\{1\}]]}$$

  is x

is the set of paths through a tree on $\omega \times \kappa \times 2$ whose projection is A. □

Theorem 9.5 converts ∞-Borel codes into strong generic codes. The terms μ_S and $\mathcal{D}_S$ were defined in Section 7.1, for S a set of ordinals. Recall (from Section 7.4) that δ_S^∞ denotes $\prod \omega_2^{L[S,x]}/\mu_S$.

Theorem 9.5: ($\aleph_1 \not\leq 2^{\aleph_0}$). *Let S be a set of ordinals. Suppose that the following hold.*

- *μ_S is an ultrafilter on $\mathcal{D}_S$.*
- *The ultrapower $\mathrm{Ord}^{\mathcal{D}_S}/\mu_S$ is wellfounded, and $\delta_S^\infty < \Theta$.*

Then if S is an ∞-Borel code for a set $A \subseteq \omega^\omega$, then A has a strong generic δ_S^∞-code which is definable from S.

Theorems 7.10, 7.13, 9.4 and 9.5 now combine to give the following.

Theorem 9.6: ($\mathsf{ZF} + \mathsf{DC}_{\mathbb{R}} + {<}\Theta$-$\mathsf{Determinacy}$). *Every subset of ω^ω which has an ∞-Borel code which is not $\leq_{\mathcal{D}}$-maximal is Suslin.*

Proof: Suppose that $A \subseteq \omega^\omega$ has an ∞-Borel code which is not $\leq_{\mathcal{D}}$-maximal. A reflection argument shows that there exist bounded subsets S and T of Θ such that $S <_{\mathcal{D}} T$ and S is an ∞-Borel code for A. Fix $B \subseteq \omega^\omega$ such that A, S and T are all in $L(B,\mathbb{R})$. The hypotheses of Theorem 9.5 (with respect to S) are then satisfied in $L(B,\mathbb{R})$, and ${<}\Theta$-$\mathsf{Determinacy}$ also holds in $L(B,\mathbb{R})$. Work in $L(B,\mathbb{R})$. By Theorem 7.13, and part (2b) of Theorem 7.10, $\delta_S^\infty < \Theta$. Letting C be a strong generic δ_S^∞-code for A as given by Theorem 9.5, δ_S^∞-$\mathsf{Determinacy}$ gives a winning strategy for player I in the game $\mathcal{G}_C$. It follows from Theorem 9.4 that A is Suslin. □

In addition, one gets the following equivalences under AD^+.

Theorem 9.7: ($\mathsf{ZF} + \mathsf{AD}^+$). *The following are equivalent.*

(1) *There is no $\leq_{\mathcal{D}}$-maximal set of ordinals.*
(2) *Every subset of ω^ω is Suslin.*
(3) Uniformization.

Proof: By Theorem 9.6, (1) implies (2). The implication from (2) to (3) was discussed in Section 3.1. The equivalence of (3) and (1) follows from Theorem 8.5. □

There is a natural join operation on wellordered sequences of generic ∞-codes such that the join is a generic ∞-code for the union of the sets coded by the members of the sequence. We briefly sketch the construction and omit the routine details.

Suppose that ζ is an ordinal and $\langle C_\alpha : \alpha < \zeta \rangle$ is a sequence of generic ∞-codes. Let C_α be $(\mathbb{P}_\alpha, B_\alpha, \tau_\alpha)$, for each $\alpha < \zeta$. The join of the sequence is a generic ∞-code $(\mathbb{P}, B, \tau)$ such that $\mathbb{P}$ is the disjoint union of the $\mathbb{P}_\alpha$'s, converted to a partial order on an ordinal via a paring function. The $\mathbb{P}$-name τ is formed from the union of the copies of the τ_α's in the respective $\mathbb{P}_\alpha$-parts of $\mathbb{P}$. Finally each member of each B_α is copied over to the $\mathbb{P}_\alpha$-part of $\mathbb{P}$, and then extended trivially the other parts of $\mathbb{P}$ to make a dense open set. Then B is the collection of all of these dense open sets.

Theorem 9.8 shows that the join operation preserves strong generic codes, under the appropriate form of ordinal determinacy. The proof is a standard strategy-blending argument. The theorem is used in the proof of Theorem 10.2, which shows that under AD^+ the set of Suslin cardinals is closed below Θ.

Theorem 9.8: ($\mathsf{ZF} + \mathsf{DC}_\mathbb{R}$). *Let κ be an infinite cardinal below Θ such that* κ-$\mathsf{Determinacy}$ *holds, and let $\bar{C} = \langle C_\alpha : \alpha < \kappa \rangle$ be a sequence of strong κ-codes. Then the join of $\bar{C}$ is a strong generic κ-code for $\bigcup_{\alpha<\kappa} A_{C_\alpha}$.*

In the remaining two sections we will briefly sketch some structural consequences of AD, AD^+ and $\mathsf{AD}_\mathbb{R}$.

10. Suslin Cardinals

A *Suslin cardinal* is an ordinal γ for which there exists an $A \subseteq \omega^\omega$ which is γ-Suslin but not η-Suslin for any $\eta < \gamma$. Any such γ must be a cardinal.

A reflection argument shows that if $A \subseteq \omega^\omega$ is Suslin, then A is γ-Suslin for some $\gamma < \Theta$. It follows that all Suslin cardinals are below Θ. The Moschovakis Coding Lemma implies that, for each $\gamma < \Theta$, there is a surjection from ω^ω onto the set of γ-Suslin sets. It follows that if every subset of ω^ω is Suslin, then the Suslin cardinals are cofinal in Θ. The converse is also true, although its proof is more involved.

Theorem 10.1: *Assuming* AD, *every subset of* ω^ω *is Suslin if and only if the Suslin cardinals are cofinal in* Θ.

A join argument shows that the supremum of a countable sequence of Suslin cardinals is Suslin. A theorem of Steel and Woodin shows that if AD holds then the set of Suslin cardinals is closed below its supremum, which by Theorem 10.1 is below Θ if and only if some subset of ω^ω is not Suslin. The following theorem of Woodin shows that under AD^+ the supremum of the Suslin cardinals is also Suslin if it is less than Θ.

Theorem 10.2: (Woodin). *Assuming* AD^+, *the set of Suslin cardinals is closed below* Θ.

Woodin has shown the converse as well. The proof is well beyond the scope of these notes.

Theorem 10.3: (Woodin). *If* $\mathsf{AD}+\mathsf{DC}_{\mathbb{R}}$ *holds, then* AD^+ *holds if and only if the supremum of the Suslin cardinals is either* Θ *or a Suslin cardinal.*

We will sketch the proof of Theorem 10.2, using Theorems 10.4 and 10.5 below. The proof of Theorem 10.4, which we will not give, uses a classical pointclass analysis for the first conclusion, and an analysis of ultrapowers by cone measures (using Theorem 7.13 and the first conclusion) for the second. We say that a subset of ω^ω is $<\kappa$-Borel if it is γ-Borel for some $\gamma < \kappa$, and $<\kappa$-Suslin if it is γ-Suslin for some $\gamma < \kappa$.

Theorem 10.4: *If* AD^+ *holds and* κ *is a limit of Suslin cardinals and the cofinality of* κ *is uncountable, then every* $<\kappa$*-Borel set is* $<\kappa$*-Suslin, and for all bounded* $S \subseteq \kappa$, $\delta^\infty_S < \kappa$.

The following (ZF) theorem of Kunen and Martin generalizes Σ^1_1-boundedness, and follows from applying Σ^1_1-boundedness in a forcing extension by $\mathrm{Col}(\omega, \gamma)$.

Theorem 10.5: (Kunen-Martin). *For any cardinal* γ, *every* γ*-Suslin prewellordering of* ω^ω *has length less than* γ^+.

Proof: [Proof of Theorem 10.2.] It suffices to show that if $\kappa < \Theta$ is a limit of Suslin cardinals of uncountable cofinality, then κ is a Suslin cardinal. Fixing such a κ, there is by the Moschovakis Coding Lemma a set $A \subseteq \omega^\omega$ which is not κ-Borel. Since A is ∞-Borel, the $*$-operation from Section 6.2 applied to an ∞-Borel code (in $\mathcal{L}_{\infty,0}$) for A gives a κ-sequence of $<\kappa$-Borel codes for disjoint nonempty subsets of ω^ω. The proof of the second half of Theorem 6.4 then produces a κ-sequence of $\mathcal{L}_{\kappa,0}$-sentences such that the union of the sets they define is a prewellordering of length κ. Theorem 9.5 applied to the members of this sequence (along with Theorem 10.4) gives a strong κ-generic code for each of these sets. Theorem 9.8 shows that the join operation on strong generic codes gives a strong κ-generic code for the union of these sets. Theorem 9.4 then implies that this prewellordering is κ-Suslin. Theorem 10.5 implies that the prewellordering cannot be γ-Suslin for any $\gamma < \kappa$. □

11. The Solovay Sequence

The *Solovay sequence* [19] is the unique continuous sequence $\langle \theta_\alpha : \alpha \leq \beta \rangle$ such that

- θ_0 is the least ordinal which is not the surjective image of ω^ω under an OD function;
- for every ordinal α such that $\alpha + 1 \leq \beta$, $\theta_{\alpha+1}$ is the least ordinal which is not the surjective image of ω^ω under an $\mathrm{OD}_{\{A\}}$ function for any $A \subseteq \omega^\omega$ of Wadge rank θ_α;
- $\theta_\beta = \Theta$.

We call β the *length* of the Solovay sequence. In $L(\mathbb{R})$, $\theta_0 = \Theta$, so $\beta = 0$. For each $\alpha \leq \beta$, if we let Γ_α be the set of subsets of ω^ω of Wadge rank less than θ_α, then $\mathrm{HOD}_{\Gamma_\alpha}$ is a model of ZF containing ω^ω whose subsets of ω^ω are exactly the members of Γ_α. Assuming AD^+, every successor member of the Solovay sequence below Θ has cofinality ω ([13] contains a proof).

The proof of Theorem 1.3 shows that if $\mathsf{AD} + \mathsf{Uniformization}$ holds then β is a limit ordinal. On the other hand, Theorems 7.14 and 9.6 (noting that every bounded subset of Θ is in $L(A, \mathbb{R})$ for some $A \subseteq \omega^\omega$, and that the μ_S ultrapower of a set S of ordinals ordinal definable from S) give the forward direction of the following theorem. Recall from Theorem 3.3 that if AD holds and every subset of ω^ω is Suslin then $\mathsf{AD}_{\mathbb{R}}$ holds.

Theorem 11.1: (Woodin). *If AD^+ holds then the Solovay sequence has limit length if and only if every subset of ω^ω is Suslin.*

If the length of the Solovay sequence is at least $\alpha + 1$, then there is a largest Suslin cardinal below $\theta_{\alpha+1}$. The largest Suslin cardinal below $\theta_{\alpha+1}$ is either θ_α or $\underset{\sim}{\delta}^2_1(A)$ for any $A \subseteq \omega^\omega$ of Wadge rank θ_α.

As a final note, we observe that the length of the Solovay sequence in a model of $\mathsf{AD} + \mathsf{DC}_{\mathbb{R}} + V = L(\mathcal{P}(\mathbb{R}))$ determines whether or not DC holds in the model. If Θ has countable cofinality, then DC fails. If the Solovay sequence has length 0, or successor length, then DC holds in $L(\mathcal{P}(\mathbb{R}))$ if $\mathsf{DC}_{\mathbb{R}}$ does. Solovay [19] showed that, assuming $\mathsf{AD}_{\mathbb{R}}$, DC holds in $L(\mathcal{P}(\mathbb{R}))$ if and only the cofinality of Θ is uncountable. In conjunction with Theorem 11.1 we get that under AD^+, DC fails in $L(\mathcal{P}(\mathbb{R}))$ if and only if, in $L(\mathcal{P}(\mathbb{R}))$, the length of the Solovay sequence is a limit ordinal of countable cofinality.

References

1. H. Becker, *A property equivalent to the existence of scales*, Transactions of the American Mathematical Society, Vol. 287, No. 2 (Feb. 1985), 591–612.
2. S. Cody, *Note on* $\mathsf{TD} + \mathsf{DC}_{\mathbb{R}}$ *implying* $\mathsf{AD}^{L(\mathbb{R})}$, J. Symb. Log. 89 (2024) 1, 211–217.
3. Henle, A. Mathias, W.H. Woodin, *A barren extension*, in: **Methods in mathematical logic** (Caracas, 1983), 195–07, Lecture Notes in Math., 1130, Springer, Berlin, 1985.
4. T. Jech, **Set Theory**, Springer-Verlag, 2003.
5. R.B. Jensen, R.M. Solovay, *Some applications of almost disjoint sets*, in: **Mathematical Logic and Foundations of Set Theory**, (Proc. Internat. Colloq., Jerusalem, 1968), 84–104, North-Holland, Amsterdam, 1970.
6. A. Kanamori, **The Higher Infinite**, Spring-Verlag, 2009.
7. A.S. Kechris, *The Axiom of Determinancy implies Dependent Choices in* $L(\mathbb{R})$, The Journal of Symbolic Logic, 49(1):161–173, 1984.
8. A.S. Kechris, Kleinberg, Moschovakis, W.H. Woodin, *The axiom of determinacy, strong partition properties and nonsingular measures*, in: **Cabal Seminar 77–79** (Proc. Caltech-UCLA Logic Sem., 1977–79), 75–99, Lecture Notes in Math., 839, Springer, Berlin-New York, 1981.
9. R. Ketchersid, *More structural consequences of* AD, in : **Set theory and its applications**, 71–105, Contemporary Math 533 (2011).
10. P. Koellner, W.H. Woodin, *Large cardinals from determinacy*, in: The Handbook of Set Theory. Vol. 3, 1951–2119, Springer, Dordrecht, 2010.
11. P.B. Larson, **The Stationary Tower**. Notes on a course by W. Hugh Woodin. American Mathematical Society University Lecture Notes, 2004.
12. P.B. Larson, *A brief history of determinacy*, in : **Sets and Extensions in the Twentieth Century**, 457–507, Handb. Hist. Log., 6, Elsevier/North-Holland, Amsterdam, 2012.
13. P.B. Larson, **Extensions of the Axiom of Determinacy**. Univ. Lecture Ser. 78, American Mathematical Society, 2023.
14. D.A. Martin, J.R. Steel, *The extent of scales in* $L(\mathbb{R})$, in: **Cabal seminar 79–81**, 86–96, Lecture Notes in Math., 1019, Springer, Berlin, 1983.
15. D.A. Martin, J.R. Steel, *A proof of projective determinacy*, J. Amer. Math. Soc. 2 (1989), no. 1, 71–125.

16. J. Mycielski, *On the axiom of determinateness*, Fundamenta Mathematicae 53 (1964), 205–224.
17. I. Neeman, *Games of length* ω_1, J. Math. Log. 7 (2007) 1, 83–124.
18. Y. Peng, L. Yu, *TD implies* $CC_{\mathbb{R}}$, Adv. Math. 384 (2021), paper no. 107755.
19. R.M. Solovay, *The independence of* DC *from* AD, in: **Cabal Seminar 76–77** (Proc. Caltech-UCLA Logic Sem., 1976–77), 171–183, Lecture Notes in Math., 689, Springer, Berlin, 1978.
20. J.R. Steel, *A classification of jump operators*, J. Symbolic Logic 47 (1982), no. 2, 347–358.
21. J.R. Steel, *The derived model theorem*, in: **Logic Colloquium 2006**, 280–327, Lect. Notes Log., 32, Assoc. Symbol. Logic, Chicago, IL, 2009.
22. R. Van Wesep, *Wadge degrees and descriptive set theory*, in: **Cabal Seminar 76–77** (Proc. Caltech-UCLA Logic Sem., 1976–77), 151–170, Lecture Notes in Math., 689, Springer, Berlin, 1978.
23. W.H. Woodin, **The Axiom of Determinacy, Forcing Axioms and the Nonstationary Ideal**, DeGruyter, 2010.

https://doi.org/10.1142/9789819806584_0009

On the Borelness of Upper Cones of Hyperdegrees

Benoit Monin[†] and Liang Yu[‡]
[†]*LACL, Département d'Informatique, Faculté des Sciences et Technologie*
61 avenue du Général de Gaulle, 94010 Créteil Cedex, France
benoit.monin@computability.fr
[‡]*Department of Mathematics, Nanjing University*
Jiangsu Province 210093, P. R. of China
yuliang.nju@gmail.com

We show that for any x, $\mathcal{U}_h(x) = \{y \mid y \geq_h x\}$ is Borel if and only if $x \in L$. Moreover, the Borel rank of $\mathcal{U}_h(x)$ when it exists is less than $(\omega_1)^L$. We also show using Steel forcing with tagged trees that Borel ranks of these sets range unboundedly below $(\omega_1)^L$.

1. Introduction

We assume the reader is familiar with descriptive set theory, Gödel's constructible hierarchy and higher recursion theory. The reader can refer to [3] and [8] for more details about these topics.

We use $\leq_h$ to denote hyperarithmetic reduction partial order.

Definition 1.1: For any real x, let

$$\mathcal{U}_h(x) = \{y \mid y \geq_h x\}.$$

The goal of this chapter is to show the two following theorems:

Theorem 1.2: *Let $x \in 2^\omega$. Then $\mathcal{U}_h(x)$ is Borel iff $x \in L$. Moreover, when $x \in L$ the Borel rank of $\mathcal{U}_h(x)$ is less than $(\omega_1)^L$.*

Theorem 1.3: *For any $\gamma < (\omega_1)^L$ there is $x \in L$, such that the Borel rank of $\mathcal{U}_h(x)$ is greater than γ.*

Yu was partially supported by NSF of China No. 11671196, Humboldt foundation for experienced fellows and the Fundamental Research Funds for the Central Universities No. 14380028.

The heart of Theorem 1.3 lies in the following result from Steel:

Theorem 1.4: (Steel [10]). *For any countable admissible ordinal* α*, the set*

$$\{x \in 2^\omega \mid \omega_1^x = \alpha\}$$

is properly $\mathbf{\Pi}^0_{\alpha+2}$.

Steel has developed in [10] a new forcing notion — forcing with tagged trees — in order to study countable Δ^1_1 subsets of ω^ω, as well as independence results for subsystems of analysis. Steel mentions in his paper that his forcing can be used to show Theorem 1.4, without formally writing the proof. Such a proof has later been given in a paper of Marker [6] studying the equivalence relation given by $X \sim Y$ if $\omega_1^X = \omega_1^Y$. The proof of [6] is however rather condensed, and we propose here a more detailed version of it, with a development of Steel forcing similar to the one given in [2], another descriptive set theoretic paper, about Borel uniformizations. We hope the reader who doesn't know Steel forcing will appreciate this nice notion, whose core essence lies in the *rettaging tool* (Lemma 2.5) and in the *rettaging lemma* (Lemma 2.6).

In the next section, we give a detailed explanation of the forcing. In the section after that we show Theorem 1.4 and Theorem 1.3. Finally, we show Theorem 1.2 in the last section.

2. Steel Forcing

2.1. *The trees*

Let $\mathcal{T}$ be the set of trees of the Baire space, both finite and infinite. Let us fix a computable bijection $b : \omega \to \omega^{<\omega} - \{\epsilon\}$ (where ϵ is the empty word). We say that an element $x \in 2^\omega$ represents a tree T if $n \in x$ iff $b(n) \in T$. We easily observe that the set of sequences representing elements of $\mathcal{T}$ is a closed subset of 2^ω. Indeed, the condition for x to represent a tree is Π^0_1: "For every n, if $x(n) = 1$ then for every prefix τ of $b(n)$ we should have $x(b^{-1}(\tau)) = 1$".

It is clear that any tree is uniquely represented by a sequence this way. Also sometimes we will blur the distinction between an element of $\mathcal{T}$ and its representation in the Cantor space. We use on $\mathcal{T}$ the topology of the Cantor space induced on the set of representations of elements of $\mathcal{T}$. We easily verify that the set of representations of elements in $\mathcal{T}$ has no isolated point, therefore its elements are the paths of a perfect subtree of $2^{<\omega}$. It follows that $\mathcal{T}$ is topologically the same space as the Cantor space.[1]

We will denote by F the set of 'finite trees' corresponding to a cylinder in the set of representation of elements of $\mathcal{T}$. In particular, an element $p \in F$ specifies a set of nodes that are in the tree, and also a set of nodes that are not in the tree. Given an element $p \in F$, we denote by $[p]$ the set of all trees of $\mathcal{T}$ that extend p. If $T \in \mathcal{T}$ extends p, we write $p \prec T$, and if another finite tree q extends p, we write $p \preceq q$. It is clear that for any cylinder $[p]$, there are two finite sets of strings $\{\sigma_1, \dots, \sigma_n, \tau_1 \dots, \tau_m\}$ such that any tree T is in $[p]$ iff for $i \leq n$ we have $\sigma_i \in T$ and for $i \leq m$ we have $\tau_i \notin T$.

For a well-founded tree T, we write $|T|$ to denote the ordinal coded by T : for every node $\sigma \in T$ we inductively define $|\sigma| = \sup_i(|\sigma^\frown n_i|+1))$ where $\langle \sigma^\frown n_i \rangle_{i \in \omega}$ are all the children of σ. We then have $|T| = |\sigma|$ where σ is the root of T. For an ill-founded tree T we write $|T| = \infty$. If $\sigma \in T$ is such that the tree of nodes compatible with σ is ill-founded we write $|\sigma| = \infty$. For every countable ordinal α, we denote by $\mathcal{T}_\alpha$ the set of trees T in $\mathcal{T}$ such that for every node $\sigma \in T$ of length 1 (direct child of the root), either $|\sigma| < \alpha$ or $|\sigma| = \infty$. In particular, for any tree $T \in \mathcal{T}_\alpha$ we have either $|T| = \infty$ or $|T| \leq \alpha$.

2.2. *The tagging*

We now define P to be the set of elements $p \in F$, paired with a valid tagging function h which assigns to each node of p a countable ordinal, or the value ∞. A tagging is said to be *valid* if for any $\sigma_1 \prec \sigma_2 \in p$, we have $h(\sigma_1) > h(\sigma_2)$. By convention, ∞ is considered greater than any countable ordinal, and also greater than itself.

So an element of P is given by a pair (p, h) where $p \in F$ and where h is a valid tagging of p. Given $(p, h) \in P$ we write $[(p, h)]$ to denote the set of trees $T \in [p]$ such that for every node $\sigma \in p$, we have $|\sigma| = h(\sigma)$ (where $|\sigma|$ is performed within T). For $(p, h) \in P$ and $(q, g) \in P$, we say that $(p, h) \preceq (q, g)$ if $p \preceq q$ and $h \preceq g$ (the taggings g and h coincide on elements of p).

For any countable α, we then let P_α denotes the set of elements $(p, h) \in P$ such that for nodes $\sigma \in p$ of length 1 we have $h(\sigma) < \alpha$ or $h(\sigma) = \infty$. Given $(p, h) \in P_\alpha$, we write $[(p, h)]^\alpha$ to denote the set $[(p, h)] \cap \mathcal{T}_\alpha$. In

[1]One can easily prove it directly by constructing the homeomorphism, or use Brouwer's theorem, saying that any compact, metrisable, perfect, 0-dimensional space is homeomorphic to the Cantor space, see [5] for details.

particular, if $T \in [(p,h)]^\alpha$ then for every node $\sigma \in T$ distinct from the root, we have either $|\sigma| < \alpha$ or $|\sigma| = \infty$.

2.3. *The forcing relation*

For any countable α, β, we now define the forcing relation between $\mathbf{\Sigma}^0_\alpha$ or $\mathbf{\Pi}^0_\alpha$ subsets of $\mathcal{T}$ and elements of P_β.

For $(p,h) \in P_\beta$ and $(q,g) \in P$, we say that $(p,h) \preceq_\beta (q,g)$ if $(p,h) \preceq (q,g)$ and if in addition we have $(q,g) \in P_\beta$. Let $(p,h) \in P_\beta$ and let us define the relation $\Vdash_\beta$ by induction on the Borel complexity of sets.

- If $\mathcal{A}$ is $\mathbf{\Delta}^0_1$ (a finite union of cylinders) we say that $(p,h) \Vdash_\beta \mathcal{A}$ iff $[p] \subseteq \mathcal{A}$.
- If $\mathcal{A}$ is $\mathbf{\Sigma}^0_\alpha$ with $\mathcal{A} = \bigcup_n \mathcal{A}_n$, we say that $(p,h) \Vdash_\beta \mathcal{A}$ iff $\exists n\ (p,h) \Vdash_\beta \mathcal{A}_n$.
- If $\mathcal{A}$ is $\mathbf{\Pi}^0_\alpha$, we say that $(p,h) \Vdash_\beta \mathcal{A}$ iff $\forall (q,g) \succeq_\beta (p,h)\ (q,g) \nVdash_\beta \mathcal{A}^c$.

Note that the forcing relation that we gave might depend on the presentation of a given Borel set. Also for two different ways to write $\mathcal{A} = \bigcup_n \mathcal{A}_n$ or $\mathcal{A} = \bigcup_n \mathcal{A}'_n$, we might have that some $(p,h) \Vdash_\beta \bigcup_n \mathcal{A}_n$ but $(p,h) \nVdash_\beta \bigcup_n \mathcal{A}'_n$. We can show however that if $(p,h) \Vdash_\beta \bigcup_n \mathcal{A}_n$ then there must be some $(q,g) \succeq_\beta (p,h)$ such that $(q,g) \Vdash_\beta \bigcup_n \mathcal{A}'_n$, which will be good enough for us. Furthermore one can prove by induction that as long as our unions are increasing, the forcing relation then does not depend anymore on the presentation of a given Borel set.

To simplify the reading, instead of writing (p,h) for elements of P, we sometimes simply write p, the tagging function being implicit. When we do so, we will always specify it, so that there is no ambiguity. This slight abuse of notation starts with the next lemma, for which the tagging function is implicit:

Lemma 2.1: *For a $\mathbf{\Pi}^0_\alpha$ set $\mathcal{A} = \bigcap_n \mathcal{A}_n$, any countable β and any $(p,h) \in P_\beta$, we have*

$$p \Vdash_\beta \mathcal{A} \text{ iff } \forall n\ \forall q \succeq_\beta p\ \exists r \succeq_\beta q\ r \Vdash_\beta \mathcal{A}_n.$$

Proof: Suppose $p \Vdash_\beta \mathcal{A}$, then by definition, $\forall q \succeq_\beta p\ q \nVdash_\beta \bigcup_n \mathcal{A}^c_n$. Still following the definition of forcing we then have $\forall q \succeq_\beta p\ \forall n\ q \nVdash_\beta \mathcal{A}^c_n$ with $\mathcal{A}^c_n$ a $\mathbf{\Pi}^0_\gamma$ set for some $\gamma < \alpha$, and then $\forall n\ \forall q \succeq_\beta p\ \exists r \succeq_\beta q\ r \Vdash_\beta \mathcal{A}_n$.

Suppose $p \nVdash_\beta \mathcal{A}$, then by definition, $\exists q \succeq_\beta p\ q \Vdash_\beta \bigcup_n \mathcal{A}^c_n$. Still following the definition of forcing we have $\exists q \succeq_\beta p\ \exists n\ q \Vdash_\beta \mathcal{A}^c_n$ with $\mathcal{A}^c_n$ a $\mathbf{\Pi}^0_\gamma$ set for some $\gamma < \alpha$, and then $\exists n\ \exists q \succeq_\beta p\ \forall r \succeq_\beta q\ r \nVdash_\beta \mathcal{A}_n$. □

2.4. *The β-topology*

For any ordinal β, we call β-topology, the topology on $\mathcal{T}_\beta$ generated by the basis $[(p,h)]^\beta$ for any $(p,h) \in P_\beta$. Recall that the topology on $\mathcal{T}$ is such that for any $(p_1,h_1),(p_2,h_2) \in P_\beta$ we must have $[(p_1,h_1)] \cap [(p_2,h_2)] = \emptyset$ or $(p_1,h_1) \preceq (p_2,h_2)$ or $(p_2,h_2) \preceq (p_1,h_1)$. We would like to study genericity with respect to the β-topology, that is, elements of $\mathcal{T}_\beta$ which are in 'sufficiently many' dense open sets of this topological space.

This study can make sense only after we prove that generic elements actually exist, that is, we should make sure that $\mathcal{T}_\beta$ endowed with the β-topology is a Baire space:

Proposition 2.2: *For any β, the set $\mathcal{T}_\beta$, together with the β-topology is a Baire space.*

Proof: Recall that $\mathcal{T}$ is homeomorphic to the Cantor space.

Suppose that we have a sequence $\{\mathcal{U}_n\}_{n\in\omega}$ of subsets of $\mathcal{T}_\beta$ which are open in the β-topology. Each of them is a union of cylinders, so that for any n and any $(p,h) \in P_\beta$, there is some cylinder $[(q,g)]^\beta \subseteq \mathcal{U}_n$ so that $[(q,g)]^\beta \subseteq [(p,h)]^\beta$. Consider any condition $(p,h) \in P_\beta$. Let us show that $\bigcap_n \mathcal{U}_n \cap [(p,h)]^\beta$ is not empty by building an element inside it.

There must exist some $[(p_0,h_0)]^\beta \subseteq \mathcal{U}_0$ which is such that $[(p_0,h_0)]^\beta \subseteq [(p,h)]^\beta$. Then inductively for any n, assuming (p_n,q_n) is defined, we define (p_{n+1},q_{n+1}). We first define a pair (q,g) extending (p_n,q_n) the following way: we start by putting in (q,g) all tagged nodes of (p_n,q_n). Then for any leaf σ in p_n with tagging $\alpha+1$, we add $(\sigma^\frown k,\alpha)$ in (q,g) for some $\sigma^\frown k$ so that no string $\tau \succeq \sigma^\frown k$ is mentioned in p_n or in q so far.

For any node σ in p_n with tagging α limit, if no sequence $\{\alpha_m\}_{m\in\omega}$ is assigned to σ yet, we assign one so that $\alpha = \sup_m \alpha_m$. If a sequence $\{\alpha_m\}_{m\in\omega}$ is assigned to σ, we put $(\sigma^\frown k,\alpha_n)$ in (q,g) for some $\sigma^\frown k$ so that no string $\tau \succeq \sigma^\frown k$ is mentioned in p_n or in q so far.

Finally, for every node σ in p_n with tagging ∞, we add $(\sigma^\frown k,\infty)$ in (q,g) for some $\sigma^\frown k$ so that no string $\tau \succeq \sigma^\frown k$ is mentioned in p_n or in q so far. Then as q should correspond to a cylinder in the set of representations of trees, we might need to actively specify that some nodes are not in q (and then not in any extension of q). If needed we do so.

Now (q,g) is a valid extension of (p_n,h_n) and then there must exists a cylinder $[(p_{n+1},h_{n+1})]^\beta \subseteq \mathcal{U}_{n+1}$ such that $[(p_{n+1},h_{n+1})]^\beta \subseteq [(q,g)]^\beta$.

It is clear by construction that $\bigcap_n [(p_n,q_n)]^\beta \subseteq \bigcap_n \mathcal{U}_n$. We should now prove that $\bigcap_n [(p_n,q_n)]^\beta$ is not empty. Because $p_0 \preceq p_1 \preceq p_2 \preceq \ldots$ and

$h_0 \preceq h_1 \preceq h_2 \preceq \dots$ we have that $\bigcap_n [p_n]$ contains a unique element T and $\bigcap_n [h_n]$ contains a unique element H tagging every node in T (and saying nothing on nodes which are not in T). One easily show by induction on α that for any node $\sigma \in T$ such that $H(\sigma) = \alpha$ we have $H(\sigma) = |\sigma| = \alpha$. One also easily show that $H(\sigma) = \infty$ iff $|\sigma| = \infty$. □

We shall now prove that if $(p, h) \Vdash_\beta \mathcal{A} \subseteq \mathcal{T}$, then $\mathcal{A} \cap \mathcal{T}_\beta$ is co-meager in $[(p, h)]^\beta$ for the β-topology (we will simply say that $\mathcal{A}$ is co-meager in $[(p, h)]^\beta$). In particular, whenever $(p, h) \Vdash_\beta \mathcal{A}$, any generic enough element of $[p, h]^\beta$ belongs to $\mathcal{A}$.

Lemma 2.3: *Let $\mathcal{A}$ be any $\mathbf{\Sigma}^0_{\boldsymbol{\alpha}}$ or $\mathbf{\Pi}^0_{\boldsymbol{\alpha}}$ set and let $(p, h) \in P_\beta$. If $(p, h) \Vdash_\beta \mathcal{A}$ then $\mathcal{A} \cap \mathcal{T}_\beta$ is co-meager in $[(p, h)]^\beta$ for the β-topology.*

Proof: Consider $\mathcal{A}$ a $\mathbf{\Delta}^0_{\mathbf{1}}$ set and suppose that for some β and $(p, h) \in P_\beta$ we have $(p, h) \Vdash_\beta \mathcal{A}$. Then $[p] \subseteq \mathcal{A}$ and then also $[(p, h)]^\beta \subseteq \mathcal{A}$, so clearly $\mathcal{A}$ is co-meager in $[(p, h)]^\beta$.

The tagging function is now implicit. Consider $\mathcal{A} = \bigcap_n \mathcal{A}_n$ a $\mathbf{\Pi}^0_{\boldsymbol{\alpha}}$ set and suppose that for some β and $p \in P_\beta$ we have $p \Vdash_\beta \mathcal{A} = \bigcap_n \mathcal{A}_n$. Then $\forall n\ \forall q \geq_\beta p\ \exists r \geq_\beta q\ r \Vdash_\beta \mathcal{A}_n$. Therefore, for all n, by induction hypothesis, the set $\mathcal{A}_n$ is co-meager in a dense open subset of $[p]^\beta$. Therefore it is also co-meager in $[p]^\beta$. Also as every $\mathcal{A}_n$ is co-meager in $[p]^\beta$, then $\bigcap_n \mathcal{A}_n$ is co-meager in $[p]^\beta$.

Consider $\mathcal{A} = \bigcup_n \mathcal{A}_n$ a $\mathbf{\Sigma}^0_{\boldsymbol{\alpha}}$ set and suppose that for some β and $p \in P_\beta$ we have $p \Vdash_\beta \mathcal{A}$. Then $p \Vdash_\beta \mathcal{A}_n$ for some n. By induction hypothesis we have that $\mathcal{A}_n$ is co-meager in $[p]^\beta$ and then that $\bigcup_n \mathcal{A}_n$ is co-meager in $[p]^\beta$. □

Just a small step now remains to prove the Baire property of any Borel $\mathcal{A}$, for the β-topology, that is, any Borel set $\mathcal{A}$ is equal to an open set, up to a meager set. The tagging function is implicit in the following lemma.

Lemma 2.4: *For any $\mathbf{\Sigma}^0_{\boldsymbol{\alpha}}$ or $\mathbf{\Pi}^0_{\boldsymbol{\alpha}}$ set $\mathcal{A}$ and any β, the set $\{[p]^\beta : p \in P_\beta \wedge (p \Vdash_\beta \mathcal{A} \vee p \Vdash_\beta \mathcal{A}^c)\}$ is dense in $\mathcal{T}_\beta$, for the β-topology.*

Proof: Let $\mathcal{A}$ be $\mathbf{\Sigma}^0_{\boldsymbol{\alpha}}$. Consider any $p \in P_\beta$. Then either $p \Vdash_\beta \mathcal{A}^c$ or $p \nVdash_\beta \mathcal{A}^c$, in which case by definition $\exists q \succeq_\beta p\ q \Vdash_\beta \mathcal{A}$. □

For a fixed β, the more dense open sets (for the β-topology) T belongs to, the more generic it is. We argue that for any β and any countably many

Borel sets $\{\mathcal{A}_n\}_{n\in\omega}$, if a tree $T \in \mathcal{T}_\beta$ is generic enough, we have for any n that $T \in \mathcal{A}_n$ iff there is a prefix p of T such that $(p, |T|\restriction_p) \Vdash_\beta \mathcal{A}_n$. In what follows, the tagging function $|T|\restriction_p$ is implicit.

Pick some n and suppose that for some prefix p of T we have $p \Vdash_\beta \mathcal{A}_n$. Then using Lemma 2.3 we have that $\mathcal{A}_n$ is co-meager in $[p]^\beta$ and then if T is generic enough it belongs to $\mathcal{A}_n$. Suppose now that $T \in \mathcal{A}_n$. In particular, if T is generic enough, it is in the dense open set $\{[p]^\beta \ : \ p \in P_\beta \ \wedge p \Vdash_\beta \mathcal{A}_n \vee p \Vdash_\beta \mathcal{A}_n^c\}$. Also we cannot have that $p \Vdash_\beta \mathcal{A}^c$ for some $p \prec T$, as we just proved that in this case $T \in \mathcal{A}^c$ for T generic enough. Therefore, for some prefix p of T we have $p \Vdash_\beta \mathcal{A}_n$.

2.5. *The retagging lemma*

We now prove the main lemma of Steel forcing. For any ordinal α, any two ordinals $\beta_1, \beta_2 \geq \omega\alpha$, and $(p, h_1) \in P_{\beta_1}$, $(p, h_2) \in P_{\beta_2}$, we write $(p, h_1) \sim_{\omega\alpha} (p, h_2)$ if for every node σ in p we have $h_1(\sigma) < \omega\alpha$ iff $h_2(\sigma) < \omega\alpha$ iff $h_1(\sigma) = h_2(\sigma)$. In the following, given a condition (p, g), we call *retagging* of p any another function g' assigning nodes of p to ordinals.

Lemma 2.5: (The retagging tool). *Let β, α be countable ordinals with $\beta < \alpha$. Let $\beta_1, \beta_2 \geq \omega\alpha$ and $p \in F$ with $(p, h_1) \in P_{\beta_1}$, $(p, h_2) \in P_{\beta_2}$ and suppose $(p, h_1) \sim_{\omega\alpha} (p, h_2)$. Then for any $(q, g_1) \succeq_{\beta_1} (p, h_1)$, there exists a retagging g_2 of q such that $(q, g_2) \succeq_{\beta_2} (p, h_2)$ and with $(q, g_1) \sim_{\omega\beta} (q, g_2)$.*

Proof: We simply build g_2. On nodes σ of p, we set $g_2(\sigma) = h_2(\sigma)$, so the tagging g_2 will extend the tagging h_2. As $(p, h_1) \sim_{\omega\alpha} (p, h_2)$ then also $(p, g_1\restriction_p) \sim_{\omega\alpha} (p, g_2\restriction_p)$.

Also because $(p, h_1) \sim_{\omega\alpha} (p, h_2)$ and because $\omega\beta + \omega \leq \omega\alpha$, for every other node σ of q that is not in p and such that $g_1(\sigma) < \omega\beta + \omega$, we can set $g_2(\sigma) = g_1(\sigma)$ and have that g_2 is still a valid tagging so far.

Let M be the largest integer such that every node $\sigma \in q$ tagged by something strictly smaller than $\omega\beta + \omega$ is tagged by something strictly smaller than $\omega\beta + M$. So every node tagged in q so far by g_2 is such that $g_2(\sigma) < \omega\beta + M$ or $g_2(\sigma) \geq \omega\beta + \omega$. We have infinitely many values between $\omega\beta + M$ and $\omega\beta + \omega$ that we can use to extend g_2 in a valid tagging. It is then easy to check that $(q, g_1) \sim_{\omega\beta} (q, g_2)$ and that $(q, g_2) \succeq_{\beta_2} (p, h_2)$. □

Lemma 2.6: (The retagging lemma). *For any $\mathbf{\Pi}^0_\alpha$ or $\mathbf{\Sigma}^0_\alpha$ set $\mathcal{A}$, any countable ordinal $\beta_1, \beta_2 \geq \omega\alpha$ and any $p \in F$ with $(p, h_1) \in P_{\beta_1}$ and $(p, h_2) \in P_{\beta_2}$, if $(p, h_1) \sim_{\omega\alpha} (p, h_2)$, then $(p, h_1) \Vdash_{\beta_1} \mathcal{A}$ iff $(p, h_2) \Vdash_{\beta_2} \mathcal{A}$.*

Proof: Let $\alpha = 1$. Let $\beta_1, \beta_2 \geq \omega$. Let $p \in F$ with $(p, h_1) \in P_{\beta_1}$ and $(p, h_2) \in P_{\beta_2}$. Suppose $\mathcal{A}$ is a $\mathbf{\Pi^0_1}$. Let us suppose that $(p, h_1) \not\Vdash_{\beta_1} \mathcal{A}$. Then $\exists (q, g_1) \succeq_{\beta_1} (p, h_1)$ $(q, g_1) \Vdash_{\beta_1} \mathcal{A}^c$. Also $\mathcal{A}^c$ is given by a union of clopen set $\bigcup_n \mathcal{A}_n$ and we have by definition that $[q] \subseteq \mathcal{A}_n$ for some n. By the retagging tool there is a tagging g_2 of q such that $(q, g_2) \succeq_{\beta_2} (p, h_2)$. We also have $(q, g_2) \Vdash_{\beta_2} \mathcal{A}^c$ and then $(p, h_2) \not\Vdash_{\beta_2} \mathcal{A}$. Similarly, one shows that $(p, h_2) \not\Vdash_{\beta_2} \mathcal{A}$ implies $(p, h_1) \not\Vdash_{\beta_1} \mathcal{A}$.

Suppose now that $\mathcal{A}$ is a $\mathbf{\Pi^0_\alpha}$ set. Let us suppose that $(p, h_1) \not\Vdash_{\beta_1} \mathcal{A}$. Then we have an extension $(q, g_1) \succeq_{\beta_1} (p, h_1)$ such that $(q, g_1) \Vdash_{\beta_1} \mathcal{A}^c$. Let $\bigcup_n \mathcal{A}_n$ be the complement of $\mathcal{A}$. Then for some n we have $(q, g_1) \Vdash_{\beta_1} \mathcal{A}_n$. Let $\beta < \alpha$ be such that $\mathcal{A}_n$ is Π^0_β. By the retagging tool, we have a tagging g_2 with $(q, g_2) \sim_{\omega\beta} (q, g_1)$ and such that $(q, g_2) \succeq_{\beta_2} (p, h_2)$. By induction hypothesis, we have $(q, g_2) \Vdash_{\beta_2} \mathcal{A}_n$. As $(q, g_2) \succeq_{\beta_2} (p, h_2)$, it follows that $(p, h_2) \not\Vdash_{\beta_2} \mathcal{A}$. Similarly, one shows that $(p, h_1) \not\Vdash_{\beta_1} \mathcal{A}$ implies $(p, h_2) \not\Vdash_{\beta_2} \mathcal{A}$.

Suppose now that the lemma is true for any $\mathbf{\Pi^0_\alpha}$ set (for α countable). For any $\mathbf{\Sigma^0_\alpha}$ set $\mathcal{A} = \bigcup_n \mathcal{A}_n$ where each $\mathcal{A}_n$ is $\mathbf{\Pi^0_\beta}$ for some $\beta < \alpha$. Let $\beta_1, \beta_2 \geq \omega\alpha$. Let $(p, h_1) \in P_{\beta_1}$ and $(p, h_2) \in P_{\beta_2}$. We have $(p, h_1) \Vdash_{\beta_1} \mathcal{A}$ iff $(p, h_1) \Vdash_{\beta_1} \mathcal{A}_n$ for some n iff $(p, h_2) \Vdash_{\beta_2} \mathcal{A}_n$ iff $(p, h_2) \Vdash_{\beta_2} \mathcal{A}$. □

3. On the Borel Rank of $\mathcal{U}_h(x)$

The goal of this section is to show Theorem 1.3:

Theorem (1.3). For any $\gamma < (\omega_1)^L$ there is $x \in L$ such that the Borel rank of $\mathcal{U}_h(x)$ is greater than γ.

To do so, we show using Steel's forcing that for any countable admissible α, the set $\{x \in 2^\omega \mid \omega_1^x = \alpha\}$ is properly $\mathbf{\Pi}^0_{\alpha+2}$ and in particular not $\mathbf{\Sigma}^0_{\alpha+2}$. Let us first show that the set $\{x \in 2^\omega \mid \omega_1^x = \alpha\}$ is $\mathbf{\Pi}^0_{\alpha+2}$.

In what follows, given L_α admissible with α countable, we allow Borel codes of L_α to have their unions and/or intersections to be indexed by ordinals $\beta < \alpha$ and not necessarily indexed by ω. This is less restrictive when in particular L_α is not a model of "everything is countable" or when α is countable but not constructibly countable.

Definition 3.1: For $x \in 2^\omega$ let R^x_e be the c.e. subset of $\omega \times \omega$ of code e. We write $\mathcal{W}^x$ for the set of codes e such that $R^x_e \subseteq \omega \times \omega$ enumerates a well-order. For $e \in \mathcal{W}^x$ we write $|e|_x$ for the order-type of the order enumerated by R^x_e. We write $\mathcal{W}^x_{<\alpha}$ for the elements $e \in \mathcal{W}^x_{<\alpha}$, such that $|e|_x < \alpha$.

In this chapter, given L_α admissible with α countable, we allow Borel codes of L_α to have their unions and/or intersections to be indexed by ordinals $\beta < \alpha$ and not necessarily indexed by ω. This is less restrictive when in particular L_α is not a model of "everything is countable" or if α is countable but not constructibly countable.

Proposition 3.2: *Let α be admissible. Let $e \in \omega$. Let $\omega \leq \beta < \alpha$. The set $\{x \in 2^\omega \ : \ e \in \mathcal{W}^x_{<\beta}\}$ has a Σ^0_β Borel code in L_α. A function which to e, β assigns a Σ^0_β Borel code for $\{x \in 2^\omega \ : \ e \in \mathcal{W}^x_{<\beta}\}$ is $\Delta_1^{L_\alpha}$-definable uniformly in α admissible.*

Proof: We have $e \in \mathcal{W}^x_{<\omega}$ iff R^x_e describe a linear order (a $\Pi^0_2(x)$ statement), with a finite support (a $\Sigma^0_2(x)$ statement). The set $\{x \in 2^\omega \ : \ e \in \mathcal{W}^x_{<\omega}\}$ is then Σ^0_ω.

Then inductively we have $e \in \mathcal{W}^x_{<\beta+1}$ if R^x_e enumerates a linear order (a Π^0_2 statement), if there exists m such that m is maximal in R^x_e and such that $f(e,m) \in \mathcal{W}^x_{<\beta}$ where $R^x_{f(e,m)}$ is R^x_e restricted to elements strictly smaller than m. By induction $\{x \in 2^\omega \ : \ e \in \mathcal{W}^x_{<\beta+1}\}$ is Σ^0_β and then $\Sigma^0_{\beta+1}$. Finally for β limit we have $\{x \in 2^\omega \ : \ e \in \mathcal{W}^x_{<\beta}\} = \bigcup_{\gamma<\beta}\{x \ : \ e \in \mathcal{W}^x_{<\gamma}\}$. It is then a Σ^0_β set.

The construction of each Σ^0_β Borel code for the set $\{x \in 2^\omega \ : \ e \in \mathcal{W}^x_{<\beta}\}$ is clearly absolute within any admissible L_α. □

Proposition 3.3: *Let α be limit. The set $\{x \in 2^\omega \ : \ \omega_1^x = \alpha\}$ is $\boldsymbol{\Pi}^0_{\boldsymbol{\alpha}+\mathbf{2}}$.*

Proof: Let us first show that the set $\{x \in 2^\omega \ : \ \omega_1^x \leq \alpha\}$ is $\boldsymbol{\Pi}^0_{\boldsymbol{\alpha}+\mathbf{2}}$. For a given e and a given n, the set:

$$\mathcal{A}_{e,n} = \{x \in 2^\omega \ : \ \forall \beta < \alpha \ \Phi_e(x,n) \notin \mathcal{W}^x_{<\beta}\}$$

is $\boldsymbol{\Pi}^0_{\boldsymbol{\alpha}}$. Also for a given e the set:

$$\mathcal{B}_e = \{x \in 2^\omega \ : \ \exists \beta < \alpha \ \forall n \ \Phi_e(x,n) \in \mathcal{W}^x_{<\beta}\}$$

is $\boldsymbol{\Sigma}^0_{\boldsymbol{\alpha}}$. Then the set $\{x \in 2^\omega \ : \ \omega_1^x \leq \alpha\}$ is equal to $\bigcap_e((\bigcup_n \mathcal{A}_{e,n}) \cup \mathcal{B}_e))$ which is clearly a $\boldsymbol{\Pi}^0_{\boldsymbol{\alpha}+\mathbf{2}}$ set. Now the set $\{x \in 2^\omega \ : \ \omega_1^x = \alpha\}$ is the intersection of $\{x \in 2^\omega \ : \ \omega_1^x \leq \alpha\}$ together with the set $\bigcap_{\beta<\alpha \text{ with } \beta \text{ limit}} \{x \in 2^\omega \ : \ \omega_1^x > \beta\}$. By induction this set is $\boldsymbol{\Pi}^0_{\boldsymbol{\alpha}}$ and thus the set $\{x \in 2^\omega \ : \ \omega_1^x = \alpha\}$ is $\boldsymbol{\Pi}^0_{\boldsymbol{\alpha}+\mathbf{2}}$. □

We shall now show that the set $\{x \in 2^\omega \mid \omega_1^x = \alpha\}$ is not $\boldsymbol{\Sigma}^0_{\alpha+2}$. To do so, we need to use Steel's proof of Sacks' result saying that α is admissible and countable iff it equals ω_1^y for some $y \in 2^\omega$. Sacks showed something even stronger: for any countable admissible α and any $x \notin L_\alpha$ we can find $y \in 2^\omega$

such that $x \notin L_{\omega_1^y}[y]$ with $\omega_1^y = \alpha$. This full version of Sack's theorem will be needed for the last section of this chapter, which is why we start by proving it here.

3.1. *Sacks's theorem*

It is clear that for α admissible and $\beta < \alpha$ we have $P_\beta \in L_\alpha$ and P_β is uniformly Δ_1-definable over L_α. Furthermore

Lemma 3.4: *The relation $p \Vdash_\beta \mathcal{A}$ is Δ_1-definable over L_α uniformly in any admissible α and in any set $\mathcal{A}$ with a Borel code in L_α.*

Proof: It is clear by definition of the forcing. □

There are many different proofs of the following theorem. The first one by Sacks uses Sacks forcing. We present here the proof of Steel, using Steel's forcing. It can also be proved using the Friedman-Jensen's application of Barwise compactness (see [4]) together with omitting type theorem [1].

Theorem 3.5: *Let α be admissible. Let $T \in \mathcal{T}_\alpha$ be generic enough, then $\omega_1^T = \alpha$. Furthermore, if $x \notin L_\alpha$ then for T generic enough we have $x \notin L_{\omega_1^T}[T]$.*

Proof: We start by showing that if $T \in \mathcal{T}_\alpha$ is generic enough then $\omega_1^T = \alpha$. We will then show that if $T \in \mathcal{T}_\alpha$ is generic enough then for any $\beta < \alpha$ we have $x \notin L_\beta[T]$. Consider a functional $\Phi : \mathcal{T} \times \omega \to \omega$ and the set

$$\mathcal{A} = \{T \ : \ \forall n \ \exists \beta < \alpha \ \Phi(T, n) \in \mathcal{W}^T_{<\beta}\}.$$

Let $\mathcal{A}_n = \{T \ : \ \exists \beta < \alpha \ \Phi(T, n) \in \mathcal{W}^T_{<\beta}\}$ and $\mathcal{A}_{n,\beta} = \{T \ : \ \Phi(T, n) \in \mathcal{W}^T_{<\beta}\}$. Note that from Proposition 3.2, for each $\beta < \alpha$ and each n the set $\mathcal{A}_{n,\beta}$ is $\boldsymbol{\Sigma}^0_\beta$ uniformly in n and β, with a Borel code in L_α.

Suppose that for some $T \in \mathcal{T}_\alpha$ we have $T \in \mathcal{A}$. Suppose also that T is generic enough, so that T belongs to some $[(p, h)]^\alpha$ such that $(p, h) \Vdash_\alpha \mathcal{A}$. In particular, there is a smallest $\beta_0 < \alpha$ such that $(p, h) \in P_{\beta_0}$. In what follows the tagging is implicit.

Consider the $\Sigma_1^{L_\alpha}$ function $f : \alpha \to \alpha$ which to each β associates the smallest ordinal $\gamma \geq \omega\beta$ such that:

$$\forall n \ \forall q \succeq_{\omega\beta} p \ \exists r \succeq_\gamma q \ r \Vdash_\gamma \bigcup_{\zeta<\gamma} \mathcal{A}_{n,\zeta}.$$

The fact that f is Σ_1-definable over L_α follows from Lemma 3.4. Let us show that f is defined on every ordinal $\beta \geq \beta_0$. As we have $p \Vdash_\alpha \bigcap_n \bigcup_{\zeta<\alpha} \mathcal{A}_{n,\zeta}$, then also we have:

$$\forall n\ \forall q \succeq_\alpha p\ \exists r \succeq_\alpha q\ r \Vdash_\alpha \bigcup_{\zeta<\alpha} \mathcal{A}_{n,\zeta}.$$

So consider any n and any $q \succeq_{\omega\beta} p$. In particular, there must exist some $r \succeq_\alpha q$ such that $r \Vdash_\alpha \bigcup_{\zeta<\alpha} \mathcal{A}_{n,\zeta}$. Therefore, by the definition of the forcing relation we must have $r \Vdash_\alpha \mathcal{A}_{n,\zeta}$ already for some $\zeta < \alpha$. Also let γ be the smallest ordinal bigger than $\max(\omega\zeta, \omega\beta)$ such that $r \in P_\gamma$. Then by the retagging lemma, as $\mathcal{A}_{n,\zeta}$ is a $\mathbf{\Sigma}^0_\zeta$ set, we must have $r \Vdash_\gamma \mathcal{A}_{n,\zeta}$ and then $r \Vdash_\gamma \bigcup_{\zeta<\gamma} \mathcal{A}_{n,\zeta}$. As we can find such a γ for any n and any $q \succeq_{\omega\beta} p$, then by admissibility of L_α, the supremum of all those γ is still smaller than α. So the function is f is defined everywhere.

It is straightforward to check that the function f is continuous, that is, $f(\sup_n \beta_n) = \sup_n f(\beta_n)$. Therefore, if we define $\beta_{n+1} = f(\beta_n)$ for each n, we then have that $\beta_\omega = \sup_n \beta_n$ is a fixed point of f. Note also that as $\beta_{n+1} \geq \omega\beta_n$ we have $\omega\beta_\omega = \beta_\omega$. It follows that:

$$\forall n\ \forall q \succeq_{\beta_\omega} p\ \exists r \succeq_{\beta_\omega} q\ r \Vdash_{\beta_\omega} \bigcup_{\zeta<\beta_\omega} \mathcal{A}_{n,\zeta}.$$

Then we have $p \Vdash_{\beta_\omega} \bigcap_n \bigcup_{\zeta<\beta_\omega} \mathcal{A}_{n,\zeta}$. We now have to prove that $p \Vdash_\alpha \bigcap_n \bigcup_{\zeta<\beta_\omega} \mathcal{A}_{n,\zeta}$. Note that we cannot apply the tagging lemma directly because $\bigcap_n \bigcup_{\zeta<\beta_\omega} \mathcal{A}_{n,\zeta}$ is only a $\mathbf{\Pi}^0_{\beta_\omega+1}$ set. This is here that we exploit the possibility for a tagging to be ∞. Using this, we shall argue that we actually already have:

$$\forall n\ \forall q \succeq_\alpha p\ \exists r \succeq_\alpha q\ r \Vdash_\alpha \bigcup_{\zeta<\beta_\omega} \mathcal{A}_{n,\zeta}. \tag{*}$$

Consider any n and any $q \succeq_\alpha p$, and let q^* be a retagging of q, so that every node in q that is tagged by something bigger than or equal to β_ω is retagged by ∞ in q^*. Then we have some $r^* \succeq_{\beta_\omega} q^*$ with $r^* \Vdash_{\beta_\omega} \bigcup_{\zeta<\beta_\omega} \mathcal{A}_{n,\zeta}$. In particular, for some $\gamma < \beta_\omega$ we have $r^* \Vdash_{\beta_\omega} \mathcal{A}_{n,\gamma}$. Also by the retagging tool, as $q \sim_{\omega\beta_\omega} q^*$, we have some $r \succeq_\alpha q$ with $r \sim_{\omega\gamma} r^*$ and then, by the retagging lemma, we have $r \Vdash_\alpha \mathcal{A}_{n,\gamma}$, as $\mathcal{A}_{n,\gamma}$ is a $\mathbf{\Sigma}^0_\gamma$ set. It follows that $r \Vdash_\alpha \bigcup_{\zeta<\beta_\omega} \mathcal{A}_{n,\zeta}$ and then that (*) is actually true. Then we have $p \Vdash_\alpha \bigcap_n \bigcup_{\zeta<\beta_\omega} \mathcal{A}_{n,\zeta}$.

It follows that $\sup_n |\Phi(T,n)|_T \leq \beta_\omega < \alpha$. As we have this for every functional Φ for T generic enough, we then have $\omega_1^T = \alpha$.

Let us now show that if $T \in \mathcal{T}_\alpha$ is generic enough then for any $\beta < \alpha$ we have $x \notin L_\beta[T]$. Consider a name $\dot{x}$ for $x \in 2^\omega$. Then for any $n \in \omega$ and any β the sets $A_{\beta,n} = \{T : L_\beta[T] \models n \in \dot{x}\}$ have a Borel code in L_α uniformly in n and β. Let us show that the set $\{p \in P_\alpha : p \Vdash_\alpha A_{\beta,n}$ for $n \notin x$ or $p \Vdash_\alpha 2^\omega - A_{\beta,n}$ for $n \in x\}$ is dense. For $\beta < \alpha$ let $\beta^* < \alpha$ be such that each $A_{\beta,n}$ is $\mathbf{\Sigma^0_{\beta^*}}$. Consider a forcing condition $p \in P_\alpha$. Let p^* be a rettaging of p where all the nodes tagged by something greater than $\omega(\beta^*+1)$ is rettaged by ∞. Note that we have $p \sim_{\omega(\beta^*+1)} p^*$.

Let $A = \{n \in \omega : \exists q^* \succeq_{\omega(\beta^*+1)} p^* : q^* \Vdash_{\omega(\beta^*+1)} A_{\beta,n}\}$. Note that by Lemma 3.4 $A \in L_\alpha$. Suppose $x \nsubseteq A$. Then there is $q^* \succeq_{\omega(\beta^*+1)} p^*$ and $n \notin x$ such that $q^* \Vdash_{\omega(\beta^*+1)} A_{\beta,n}$. By the rettaging tool there is a rettaging $q \succeq_\alpha p$ such that $q \sim_{\omega\beta^*} q^*$. By the rettaging lemma we have $q \Vdash_\alpha A_{\beta,n}$. Suppose now $x \subseteq A$. As $x \notin L_\alpha$ and $A \in L_\alpha$ there must be $n \in x$ and $n \notin A$. By Lemma 2.4 there must be $q^* \succeq_{\omega(\beta^*+1)} p^*$ such that $q^* \Vdash_{\omega(\beta^*+1)} 2^\omega - A_{\beta,n}$. By the rettaging tool there is a rettaging $q \succeq_\alpha p$ such that $q \sim_{\omega\beta^*} q^*$. By the rettaging lemma we have $q \Vdash_\alpha 2^\omega - A_{\beta,n}$.

In any case, the set $\{p \in P_\alpha : p \Vdash_\alpha A_{\beta,n}$ for $n \notin x$ or $p \Vdash_\alpha 2^\omega - A_{\beta,n}$ for $n \in x\}$ is dense. It follows that if $T \in \mathcal{T}_\alpha$ is generic enough then $x \notin L_\beta[T]$ for any $\beta < \alpha$. □

3.2. *The Borel complexity of* $\{X \in 2^\omega : \omega_1^X = \alpha\}$

Theorem 3.6: *The set* $\{X \in 2^\omega : \omega_1^X = \alpha\}$ *is not* $\mathbf{\Sigma^0_{\alpha+2}}$.

Proof: We shall prove that the set of representations of elements of $\mathcal{T}$ which preserve α is not $\mathbf{\Sigma^0_{\alpha+2}}$. As this set is a closed subset of the Cantor space, it follows that also the set $\{X \in 2^\omega : \omega_1^X = \alpha\}$ is not $\mathbf{\Sigma^0_{\alpha+2}}$. In what follows, the tagging functions are implicit.

Suppose that $\{T \in \mathcal{T} : \omega_1^T = \alpha\} = \bigcup_m \bigcap_n \mathcal{A}_{n,m}$ where each $\mathcal{A}_{n,m}$ is a $\mathbf{\Sigma^0_\alpha}$ set. Then using Theorem 3.5, there must be some m such that the set $\bigcap_n \mathcal{A}_{n,m}$ contains some tree T which is generic enough for Steel forcing over P_α, so that $\omega_1^T = \alpha$. In particular, we have $p \Vdash_\alpha \bigcap_n \mathcal{A}_{n,m}$ for some $p \prec T$ with $(p, |T|\restriction_p) \in P_\alpha$.

So also we have $\forall n\, \forall q \succeq_\alpha p\, \exists r \succeq_\alpha q\; r \Vdash_\alpha \mathcal{A}_{n,m}$. Let α^+ be the smallest admissible ordinal bigger than α. We should now prove that we actually have:

$$\forall n\, \forall q \succeq_{\alpha^+} p\, \exists r \succeq_{\alpha^+} q\; r \Vdash_{\alpha^+} \mathcal{A}_{n,m}. \qquad (*)$$

Consider now any n and any $q \succeq_{\alpha^+} p$ and let q^* be a retagged version of q where each ordinal bigger than or equal to α in q is retagged by ∞

in q^*. Then $q^* \succeq_\alpha p$ and in particular, we have some $r^* \succeq_\alpha q^*$ such that $r^* \Vdash_\alpha \mathcal{A}_{n,m}$. Let $\mathcal{A}_{n,m} = \bigcup_{k\in\omega} \mathcal{A}_{n,m,k}$ with each $\mathcal{A}_{n,m,k}$ is $\mathbf{\Pi}^0_{\boldsymbol{\beta}}$ set for some $\beta < \alpha$. In particular, for some $\beta < \alpha$ we have $r^* \Vdash_\alpha \mathcal{A}_{n,m,k}$ where $\mathcal{A}_{n,m,k}$ is a $\mathbf{\Pi}^0_{\boldsymbol{\beta}}$ set. Also by the retagging tool, as $q \sim_{\omega\alpha} q^*$, we have some $r \succeq_{\alpha^+} q$ with $r \sim_{\omega\beta} r^*$ and then, by the retagging lemma, we have $r \Vdash_{\alpha^+} \mathcal{A}_{n,m,k}$, as $\mathcal{A}_{n,m,k}$ is a $\mathbf{\Pi}^0_{\boldsymbol{\beta}}$ set. It follows that $r \Vdash_{\alpha^+} \mathcal{A}_{n,m}$ and then that (*) is actually true. Then we have $p \Vdash_{\alpha^+} \bigcap_n \mathcal{A}_{n,m}$.

It follows that any $q \succeq_{\alpha^+} p$ also forces $\bigcap_n \mathcal{A}_{n,m}$. Take such an extension q with a node tagged by the ordinal α. For any $T \in [q]^{\alpha^+}$ we have some node of T starting a well-founded tree coding for α. Thus α is computable in T and $\omega_1^T > \alpha$. Also as $q \Vdash_{\alpha^+} \bigcap_n \mathcal{A}_{n,m}$, the set $\bigcap_n \mathcal{A}_{n,m}$ contains some generic tree that is in $[q]^{\alpha^+}$. Then $\bigcup_m \bigcap_n \mathcal{A}_{n,m}$ contains an element making α computable, which is a contradiction. □

We are now ready to show Theorem 1.3.

Theorem (1.3). *For any $\gamma < (\omega_1)^L$ there is $x \in L$ such that the Borel rank of $\mathcal{U}_h(x)$ is greater than γ.*

Proof: We show that for any $\gamma < (\omega_1)^L$ there is $x \in L$ such that the Borel rank of $\mathcal{U}_h(x)$ is not $\mathbf{\Sigma}^0_\gamma$. Fix an ordinal $\gamma < (\omega_1)^L$. Let $\nu < \alpha_0 < \alpha_1 < (\omega_1)^L$ be three contiguous admissible ordinals greater than $\gamma + \omega$ so that there is a real in $L_{\alpha_1} \setminus L_{\alpha_0}$. So we may assume that there is a real $x \in L_{\alpha_1} \setminus L_{\alpha_0}$ which is a master code. Thus $\omega_1^x = \alpha_1$. Then

$$\begin{aligned}\mathcal{U}_h(x) &= \{y \mid \omega_1^y \geq \omega_1^x\} = \{y \mid \omega_1^y \geq \alpha_1\} = \{y \mid \omega_1^y > \alpha_0\} \\ &= 2^\omega \setminus \bigcup_{\beta \leq \alpha_0} \{y \mid \omega_1^y = \beta\}.\end{aligned}$$

Suppose for contradiction that $\mathcal{U}_h(x)$ is $\mathbf{\Sigma}^0_\gamma$. Then $\{y \mid \omega_1^y = \alpha_0\} = (2^\omega \setminus \mathcal{U}_h(x)) \setminus \bigcup_{\beta \leq \nu} \{y \mid \omega_1^y = \beta\}$ would be $\mathbf{\Pi}^0_{\nu+3}$. Since $\alpha_0 > \nu + \omega$, we have a contradiction to Theorem 1.4. □

4. On the Borelness of $\mathcal{U}_h(x)$

The goal of this section is to show Theorem 1.2:

Theorem (1.2). *Let $x \in 2^\omega$. Then $\mathcal{U}_h(x)$ is Borel iff $x \in L$. Moreover when $x \in L$ the Borel rank of $\mathcal{U}_h(x)$ is less than $(\omega_1)^L$.*

We start with the following lemma:

Lemma 4.1: *If $x \in L$, then the set $\mathcal{U}_h(x)$ is a $\Delta^1_1(z)$ set for some real $z \in L$. In particular, the rank of z must be smaller than $(\omega_1)^L$ and so must be the rank of the Borel code of $\mathcal{U}_h(x)$.*

Proof: For any x we have $y \geq_h x$ if and only if there is an ordinal β computable in y such that $x \in L_\beta[y]$. Suppose now $x \in L$, then there is an ordinal $\alpha < (\omega_1)^L$ so that $x \in L_\alpha$. Suppose $y \geq_h x$. If $\omega_1^y \leq \alpha$ then we must have $x \in L_\beta[y]$ for some $\beta \leq \alpha$ which is computable in y. If $\omega_1^y > \alpha$, as $x \in L_\alpha$ then we must have $x \in L_\beta[y]$ for some $\beta \leq \alpha$ which is computable in y. In any case, the set $\{y \in 2^\omega : y \geq_h x\}$ equals the set $\{y \in 2^\omega : x \in L_\beta[y]$ for some ordinal $\beta \leq \alpha$ computable in $y\}$.

Now fix a real $z \in L$ coding a well order of ω with order type α. Then $\mathcal{U}_h(x)$ is $\Delta^1_1(z)$. □

To show the last theorem we use the two following results:

Theorem 4.2: (Simpson [9]). *Let $r \in 2^\omega$. There is an uncountable $\Sigma^1_1(r)$ closed set A so that for any reals $y_0 \neq y_1 \in A$, $L_{\omega_1^r}[y_0 \oplus r] \cap L_{\omega_1^r}[y_1 \oplus r] = L_{\omega_1^r}[r]$.*

Theorem 4.3: (Martin [7]). *Let $r \in 2^\omega$. If A is an uncountable $\Sigma^1_1(r)$-closed set, then for any y, there is a real $x \in A$ so that $y \in L_{\omega_1^{x \oplus r}}[x \oplus r]$.*

We use $\langle *, * \rangle$ to denote the Gödel paring function.

Lemma 4.4: *Suppose that $x \notin L$, then $\mathcal{U}_h(x)$ is not Borel.*

Proof: For a contradiction, suppose that $\mathcal{U}_h(x)$ is Borel. So there is a real $z \geq_T x$ so that $\mathcal{U}_h(x)$ is $\Delta^1_1(z)$. We claim that there is a countable ordinal α so that

$$\forall y (y \geq_h x \rightarrow x \in L_\alpha[y]).$$

Since $\mathcal{U}_h(x)$ is $\Delta^1_1(z)$, by Spector-Gandy's theorem (see [3]), there is a Σ_1-formula φ so that

$$\forall y (y \not\geq_h x \leftrightarrow L_{\omega_1^{y \oplus z}}[y \oplus z] \models \varphi).$$

So there is a total $\Pi^1_1(z)$-function $f : 2^\omega \to \omega$ uniformizing the following Π^1_1-relation

$$R(y, m) = \begin{cases} m = \langle 0, n \rangle, & y \geq_h x \wedge n \in \mathcal{W}^y \wedge x \in L_{|n|_y}[y]; \\ m = \langle 1, n \rangle, & y \not\geq_h x \wedge n \in \mathcal{W}^{y \oplus z} \wedge L_{|n|_{y \oplus z}}[y \oplus z] \models \varphi. \end{cases}$$

So f is also a $\Delta^1_1(z)$-function. Thus the range of f on $\mathcal{U}_h(x)$ is a $\Sigma^1_1(z)$ set and so there must be an ordinal $\alpha < \omega_1^z$ so that for any $y \geq_h x$ and ordinal notation $n \in \mathcal{W}^y$ with $f(y) = \langle 0, n \rangle$, we have that $|n|_y \leq \alpha$. The α is exactly what we want.

We may let α be an admissible ordinal. By Lemma 3.4, there is a real r so that $\omega_1^r = \alpha$ and $x \not\leq_h r$. Now by Theorem 4.2, there is an uncountable $\Sigma^1_1(r)$ closed set A so that for any reals $y_0 \neq y_1 \in A$, $L_{\omega_1^r}[y_0 \oplus r] \cap L_{\omega_1^r}[y_1 \oplus r] = L_{\omega_1^r}[r]$. By Theorem 4.3, there must be a two reals $y_1 \neq y_2 \in A$ so that $x \in L_{\omega_1^{y_1 \oplus r}}[y_1 \oplus r]$ and $x \in L_{\omega_1^{y_2 \oplus r}}[y_2 \oplus r]$. By the choice of A, we must have $x \notin L_{\omega_1^r}[y_i \oplus r]$ for some $i \in \{0, 1\}$. This contradicts the above claim saying that as $x \in L_{\omega_1^{y_i \oplus r}}[y_i \oplus r]$ we must have $x \in L_\alpha[y_i \oplus r]$ where $\omega_1^r = \alpha$. Thus $\mathcal{U}_h(x)$ is not Borel. □

References

1. Jon Barwise. *Admissible sets and structures.* Springer-Verlag, Berlin, 1975.
2. Howard Becker and Randall Dougherty. On disjoint borel uniformizations. *Advances in mathematics*, 146(2):167–174, 1999.
3. Chi Tat Chong and Liang Yu. *Recursion theory*, volume 8 of *De Gruyter Series in Logic and its Applications.* De Gruyter, Berlin, 2015. Computational aspects of definability, With an interview with Gerald E. Sacks.
4. Harvey Friedman and Ronald Jensen. Note on admissible ordinals. In Jon Barwise, editor, *The Syntax and Semantics of Infinitary Languages*, pp. 77–79, Berlin, Heidelberg, 1968. Springer Berlin Heidelberg.
5. Alexander S. Kechris. *Classical descriptive set theory*, volume 156 of *Graduate Texts in Mathematics.* Springer-Verlag, New York, 1995.
6. David Marker. An analytic equivalence relation not arising from a polish group action. *Fundamenta Mathematicae*, 130(3):225–228, 1988.
7. Donald A. Martin. Proof of a conjecture of Friedman. *Proc. Amer. Math. Soc.*, 55(1):129, 1976.
8. Yiannis N. Moschovakis. *Descriptive set theory*, volume 155 of *Mathematical Surveys and Monographs.* American Mathematical Society, Providence, RI, second edition, 2009.
9. Stephen G. Simpson. Minimal covers and hyperdegrees. *Trans. Amer. Math. Soc.*, 209:45–64, 1975.
10. John R. Steel. Forcing with tagged trees. *Ann. Math. Logic*, 15(1):55–74, 1978.

https://doi.org/10.1142/9789819806584_0010

A New Game Metatheorem for Ash–Knight Style Priority Constructions

Antonio Montalbán
Department of Mathematics, University of California, Berkeley, USA
www.math.berkeley.edu/~antonio/index.html
antonio@math.berkeley.edu

We develop a new metatheorem for $0^{(\eta)}$-priority constructions for transfinite ordinals η. Our metatheorem simplifies previous metatheorems by Ash, Ash–Knight and the author. A drawback of this simplification is that its applications are more limited: It can only be used when the object being built is an ω-presentation of a structure. However, in practical terms, this is not a huge limitation and the majority of proofs in the literature that use the Ash–Knight metatheorem can be done using our metatheorem.

Often in computability theory we need to build a computable object using non-computable information — sometimes even non-computable information about the very same object we are building. The main tool for such constructions is the priority method, which has become increasingly more involved and sophisticated since the 1950s. Priority arguments are classified in terms of how much non-computable information is needed to verify the construction. The most common priority constructions are the finite-injury ones [Fri57, Muc56]. They are used when the information needed is $0'$-computable. Infinite-injury priority constructions [Sho61, Sac63] are used when $0''$-computable guesses are needed. There are various zero-triple priority constructions [Lac76] in the literature, but they are very complicated, and far less common. Beyond that point, humans cannot keep track of the combinatorics anymore. Well, that is unless the level-by-level combinatorics of the proof is uniform and one can describe the work done at all the levels by a single procedure. There have been various proposals for general

The author was partially supported by NSF grant DMS-1363310.

$0^{(n)}$-injury constructions: Harrington's worker's method [Har76], Lempp and Lerman's trees of strategies [LL95, Ler10], Ash's [Ash86] and Ash–Knight's [AK00] η-systems, and Montalbán's systems of true stages [Mon14]. Among these methods, the Ash–Knight's η-system is the only one that contains a clear metatheorem, where if a certain combinatorial machinery can be put in place, one can then apply the metatheorem as a black box and produce the desired computable object out of Δ^0_η-information. The proof of the metatheorem is complicated. But the whole point of the metatheorem is that one does not need to know its proof to use it. A disadvantage for the metatheorem is that it is applicable in a limited number of situations. It has only been used in computable structure theory, where the combinatorial features needed for the metatheorem often occur naturally. Let me emphasize, though, that despite its limitations, it has been extremely useful in computable structure theory, and has been applied to a wide range of situations.

In this chapter, we propose a new metatheorem. The advantage of the new metatheorem is that it is much easier to use than the Ash–Knight's metatheorem. The disadvantage is that it is more limited. However, in most of the constructions from the computable-structure-theory literature where Ash–Knight's metatheorem is useful, our metatheorem is too. So the restriction in applicability does not seem to be too limiting in practice, while the simplification is considerable. At a meta-level, we believe that our metatheorem exhibits the interplay between the η-back-and-forth relations and Δ^0_η-information in the clearest possible way.

Montalbán's η-true stages method provides a different angle to view Ash's construction, giving it a more hands-on feeling, while adding some flexibility. For instance, Montalbán used his method in [Mon14] to prove the *tree of structures theorem*, a version of the pair-of-structures theorem which cannot be proved with Ash and Knight's system. It was also used by Greenberg and Turetsky [GT], and Day and Marks (in preparation).

The metatheorem we introduce here is not more but less flexible: it can only be used to build ω-presentations of structures. There are many proofs in computable structure theory that require exactly this: to build a ω-presentations with a particular property. After presenting the metatheorem in Sec. 2, we showcase its applicability by presenting proofs of various results from the literature using our metatheorem. We claim that these proofs are much simpler than the original ones, though we will let the reader be the judge of that. 'Simpler' might not be the right word. Instead, may we should say that it black boxes the computational part of the construction, leaving only the key combinatorial steps that are pertinent to the situation. The proofs we will go through are: Ash and Knight's pair of structures theorem (Sec. 3); Watnick's and Ash-Jockusch-Knight's theorem that a linear ordering $\mathcal{L}$ has a $\Delta^0_{2\alpha+1}$ copy if and only if $\omega^\alpha \cdot \mathcal{L}$ has a computable copy (Sec. 4); and Ash's theorem that, under enough effectiveness conditions, Δ^0_α-categoricity implies relative Δ^0_α-categoricity (Sec. 5). We will also discuss how the copy-vs-diagonalizer constructions introduced by the author in [Mon13] to prove the low$_n$ property can be viewed as a particular case of the metatheorem (Sec. 6).

We leave the proof of the metatheorem to the end of the chapter. We use Sec. 7 to review η-true stage systems, and then use them to prove the metatheorem in Sec. 8.

1. Background and Notation

Before stating the metatheorem in the next section, we need to lay down some basic background and notation.

1.1. *Diagrams*

Throughout this chapter, we assume we have a computable vocabulary τ and all our structures are τ-structures. For simplicity, and without loss of generality, we may assume τ is relational. An ω-presentation is just a copy of a structure with domain ω. Given an ω-presentation $\mathcal{A}$, we use $D(\mathcal{A}) \in 2^{\mathbb{N}}$ to denote its atomic diagram, where $D(\mathcal{A})(i)$ gives the truth value of ith atomic formula. That is, $D(\mathcal{A})(i) = 1$ if and only if $\mathcal{A} \models \varphi_i^{at}[x_j \mapsto j]$, where $\{\varphi_i^{at} : i \in \mathbb{N}\}$ is some effective enumeration of the atomic formulas on the variables $x_0, x_1, \ldots$. Given a finite tuple $\bar{a} = \langle a_0, \ldots, a_{|\bar{a}|-1}\rangle \in \mathcal{A}^{<\mathbb{N}}$, we define the *atomic diagram of* $\bar{a}$ *in* $\mathcal{A}$ to be the finite binary string $D_{\mathcal{A}}(\bar{a})$ of length $\ell_{|\bar{a}|}$, where, for $i < \ell_{|\bar{a}|}$,

$$D_{\mathcal{A}}(\bar{a})(i) = 1 \iff \mathcal{A} \models \varphi_i^{at}[x_j \mapsto a_j : j < |\bar{a}|],$$

and where ℓ_s is the number of atomic formulas using the variables $x_0, \ldots, x_{s-1}$ and the first s symbols from τ. We assume that in our enumeration $\{\varphi_i^{at} : i \in \mathbb{N}\}$ of the atomic formulas, those atomic formulas come first. It follows that $D(\mathcal{A}) = \bigcup_{s\in\mathbb{N}} D_{\mathcal{A}}(\langle 0, \ldots, s-1\rangle)$, where the union is a union of strings. Also, if $g\colon \omega \to \mathcal{A}$ is a bijection and $\mathcal{B}$ is the pull-back of $\mathcal{A}$ through g (i.e., g is an isomorphism $\mathcal{B} \to \mathcal{A}$), then $D(\mathcal{B}) = \bigcup_{s\in\mathbb{N}} D_{\mathcal{A}}(g \restriction s)$.

1.2. *Back-and-forth relations*

The back-and-forth relations measure how hard it is to distinguish two structures, or two tuples within a structure, or even two tuples from different structures. The idea is that two tuples are n-back-and-forth equivalent if we cannot differentiate them using only n Turing jumps. More background can be found in the author's book [MonP2].

Definition 1.1: Given structures $\mathcal{A}$ and $\mathcal{B}$, tuples $\bar{a} \in A^{<\mathbb{N}}$, $\bar{b} \in B^{<\mathbb{N}}$, and an ordinal $\eta > 0$, we say that $(\mathcal{A}, \bar{a})$ is η-back-and-forth below $(\mathcal{B}, \bar{b})$, and write $(\mathcal{A}, \bar{a}) \leq_\eta (\mathcal{B}, \bar{b})$, if every $\Pi^{\mathtt{in}}_\eta$ formula true about $\bar{a}$ in $\mathcal{A}$ is also true about $\bar{b}$ in $\mathcal{B}$. I.e.,

$$(\mathcal{A}, \bar{a}) \leq_\eta (\mathcal{B}, \bar{b}) \quad \iff \quad \Pi^{\mathtt{in}}_\eta\text{-}tp_{\mathcal{A}}(\bar{a}) \subseteq \Pi^{\mathtt{in}}_\eta\text{-}tp_{\mathcal{B}}(\bar{b}).$$

Here, $\Pi^{\mathtt{in}}_\eta$ denotes the class of infinitary Π_η formulas, and $\Pi^{\mathtt{in}}_\eta\text{-}tp_{\mathcal{A}}(\bar{a})$ denotes the $\Pi^{\mathtt{in}}_\eta$-type of $\bar{a}$ in $\mathcal{A}$. See [AK00, Chapter 6] and [MonP2] for more background on infinitary formulas.

For $\eta = 0$ we only look at finite segments of the quantifier-free types and we let $(\mathcal{A}, \bar{a}) \leq_0 (\mathcal{B}, \bar{b})$ if $D_{\mathcal{A}}(\bar{a}) \subseteq D_{\mathcal{B}}(\bar{b})$.

An equivalent definition for the back-and-forth relations is given in the following theorem.

Theorem 1.2: (Karp [Kar65]). *Consider structures $\mathcal{A}$ and $\mathcal{B}$, tuples $\bar{a} \in A^{<\mathbb{N}}$, $\bar{b} \in B^{<\mathbb{N}}$, and an ordinal $\eta > 0$. Then $(\mathcal{A}, \bar{a}) \leq_\eta (\mathcal{B}, \bar{b})$ if and only if, for every $\beta < \eta$ and every $\bar{d} \in B^{<\mathbb{N}}$, there exists $\bar{c} \in \mathcal{A}^{<\mathbb{N}}$ such that*

$$(\mathcal{A}, \bar{a}\bar{c}) \geq_\beta (\mathcal{B}, \bar{b}\bar{d}).$$

See [AK00, Secs. 15.1, 18.6] for a proof.

Given a computable list of computable ω-presentations $\mathbb{A} = (\mathcal{A}_i : i \in \mathbb{N})$ and a computable ordinal η, we say that the *back-and-forth relations in $\mathbb{A}$ are computable up to η* if the relations $\leq_\beta$ for $\beta \leq \eta$ are uniformly computable, that is, if the set of five-tuples

$$\{\langle \beta, i, \bar{a}, j, \bar{b} \rangle : \beta \in \eta + 1, (\mathcal{A}_i, \bar{a}) \leq_\beta (\mathcal{A}_j, \bar{b})\}$$

is computable.

This is essentially Ash and Knight's notion of η-friendliness, except that they required the set above to be only c.e.

Note that this notion is meaningful even when $\mathbb{A}$ consists of a single ω-presentation. That ω-presentation has to be extremely nice to allow us to compute all the back-and-forth relations among its tuples. When we understand a structure well, we expect to be able to understand its back-and-forth relations too. For example, every computable ordinal has an ω-presentation where the back-and-forth relations are computable up to η. Furthermore, for every computable ordinal α, if we represent ω^α in the standard way using Cantor normal forms, we get an ω-presentation of ω^α where the back-and-forth relations are computable up to α. A calculation of the back-and-forth relations on ordinals was done by Ash in [Ash86], see [AK00, Lemma 15.10]. The same applies if we take linear orderings of the form $\mathbb{Z}^\alpha$, where α is a computable ordinal. Ash also calculated the back-and-forth relations on super-atomic Boolean algebras in [Ash87, Lemma 5].

In [Mon10, Proposition 2.10], the author shows that if a class of structures is Σ_α-small, meaning that there are only countably many $\equiv_\alpha$-equivalence classes among the tuples within its structures, and it is Σ_α-small in an effective way, meaning that we can effectively understand how the back-and-forth relations work among these $\equiv_\alpha$-equivalence classes, then we can produce a computable list of structures in the class containing representatives of all $\equiv_\alpha$-equivalence classes, and where the back-and-forth relations are computable up to α.

Another point to make is that for every countable family of structures there is an oracle, namely the 2ηth jump of the list of ω-presentations, relative to which the back-and-forth relations up to η are computable. For many of the results below, we still get a meaningful result if we relativize to this oracle to be able to assume the back-and-forth relations up to η are computable.

1.3. $\Delta^0_{\eta+1}(X)$ *questions*

We will deal with finite and transfinite iterations of the Turing jumps. The standard notation for these iterates has been set, unfortunately, so that for finite n, $0^{(n)}$ is Σ^0_n-complete, while for infinite α, it is $0^{(\alpha+1)}$ who is Σ_α-complete. One can encapsulate both statements by saying that $0^{(\eta+1)}$ is $\Sigma^0_{1+\eta}$ complete for all η — though this is rather cumbersome. Instead, in this chapter we avoid using the $0^{(\alpha)}$ notation, and we only talk about Σ^0_η complete sets or of Δ^0_γ Turing complete sets.

Definition 1.3: Given $X \in 2^{\mathbb{N}}$, $n, e \in \mathbb{N}$ and a computable ordinal η, we say that n is *the answer to the eth* $\Delta^0_{\eta+1}(X)$ *question* if

$$n = \Phi_e^{S^X_\eta}(0),$$

where S_η^X is a $\Sigma_\eta^0(X)$ complete set, namely $X^{(\eta)}$ when η is finite and $X^{(\eta+1)}$ when η is infinite, and where Φ_e is the eth Turing functional.

Let us remark that any finite number of questions of the form $\Phi_{e_0}^{S_\eta^X}(k_0)$, $\Phi_{e_1}^{S_\eta^X}(k_1), ..., \Phi_{e_\ell}^{S_\eta^X}(e_k)$ can be encoded into a single question using an index e such that $\Phi_e^{S_\eta^X}(0)$ outputs a number encoding the tuple $\langle \Phi_{e_0}^{S_\eta^X}(k_0)$, $\Phi_{e_1}^{S_\eta^X}(k_1), ..., \Phi_{e_\ell}^{S_\eta^X}(k_\ell)\rangle$.

2. Game Constructions

Let η be a computable ω-presentation of an ordinal. The reader may assume $\eta = 1$ on a first read of this section — the case $\eta = 1$ is already interesting, and the metatheorem is still quite useful and meaningful. Suppose we have a list of structures

$$\mathbb{A} = \{\mathcal{A}_0, \mathcal{A}_1, \mathcal{A}_2, ...\},$$

where the back-and-forth relations are computable up to η. Let $\mathfrak{P}$ be a property of ω-presentations of structures, or equivalently, let $\mathfrak{P}$ be a subset of the set Mod_τ of all τ-ω-presentations.

We will now describe a type of construction that we will call an η-$\mathbb{A}$-*game*. This game involves three characters, the *engineer*, the *extender*, and the *oracle*. Together, when the game ends, they build an ω-presentation $\mathcal{L}$, which we call the *limit structure*. The goal of the engineer is for the limit structure $\mathcal{L}$ to satisfy property $\mathfrak{P}$. The extender is in charge of making $\mathcal{L}$ computable — he will not, in any way, coordinate his work with the engineer. The job of the oracle is to answer $\Delta_{\eta+1}^0(\mathcal{L})$ questions posed by the engineer. The game has infinitely many stages. At each stage $j \in \mathbb{N}$, first, the engineer plays a triple $\langle i_j, \bar{a}_j, e_j\rangle$ where $i_j, e_j \in \mathbb{N}$ and $\bar{a}_j \in A_{i_j}^{<\mathbb{N}}$, and second, the extender plays a tuple $\bar{b}_j \in A_{i_j}^{<\mathbb{N}}$ extending $\bar{a}_j$ and the oracle plays a number n_j which must be the answer to the e_jth $\Delta_{\eta+1}^0(D(\mathcal{L}))$ question.

engineer	$i_0, \bar{a}_0, e_0$		$i_1, \bar{a}_1, e_1$		$i_2, \bar{a}_2, e_2$		$\cdots$
extender		$\bar{b}_0$		$\bar{b}_1$		$\bar{b}_2$	$\cdots$
oracle		n_0		n_1		n_2	$\cdots$

At each stage $j > 0$, the tuple $\bar{a}_j$ played by the engineer must satisfy:

$$(\mathcal{A}_{i_j}, \bar{a}_j) \geq_\eta (\mathcal{A}_{i_{j-1}}, \bar{b}_{j-1}).$$

The tuple $\bar{b}_j$ played by the extender must be in the same structure just played by the engineer and must satisfy:

$$\bar{b}_j \supseteq \bar{a}_j.$$

After ω many moves, we get

$$D_{\mathcal{A}_{i_0}}(\bar{a}_0) \subseteq D_{\mathcal{A}_{i_0}}(\bar{b}_0) \subseteq D_{\mathcal{A}_{i_1}}(\bar{a}_1) \subseteq D_{\mathcal{A}_{i_1}}(\bar{b}_1) \subseteq D_{\mathcal{A}_{i_2}}(\bar{a}_2) \subseteq \cdots$$

and hence we get a limit ω-presentation $\mathcal{L}$ whose atomic diagram is the union of the diagrams of the tuples played:

$$D(\mathcal{L}) = \bigcup_{j \in \mathbb{N}} D_{\mathcal{A}_{i_j}}(\bar{a}_j).$$

The numbers e_j represent $\Delta^0_{\eta+1}(\mathcal{L})$ questions as in Definition 1.3.* The engineer is responsible of ensuring that all these questions converge, as we will see below. The numbers n_j played by the oracle must be the answers to these questions. Notice the self-reference in the $\Delta^0_{\eta+1}(\mathcal{L})$ questions, that the oracle will be able to answer even before having built $\mathcal{L}$. The reader may be smelling that the Recursion Theorem will come into play here.

A *strategy* for the engineer is a function that tells the engineer what to play next given the previous moves by the extender and the oracle. We say that a strategy is *valid* if, on all possible plays by the extender, all the $\Delta^0_{\eta+1}(\mathcal{L})$ questions e_j converge.

Theorem 2.1: *Given $\mathbb{A}$ and η as described above, for every computable valid strategy for the engineer in the η-$\mathbb{A}$-game, there is a $\Delta^0_{\eta+1}$ sequence of moves by the extender so that, if the engineer follows her strategy, the limit ω-presentation $\mathcal{L}$ is computable.*

Furthermore, we will prove that there is a uniform effective procedure that, given the strategy for the engineer, produces the ω-presentation $\mathcal{L}$ which comes from some sequence of moves by the extender.

To apply this theorem, one needs to describe a computable valid strategy for the engineer that, with the help of the oracle who is answering our $\Delta^0_{\eta+1}$ questions, will build an ω-presentation with property $\mathfrak{P}$ independently of what the extender does. One can then cite the theorem to conclude that, even if the construction relied on the $\Delta^0_{\eta+1}$ information provided by the oracle, the resulting ω-presentation is computable.

*From now on we write $\Delta^0_{\eta+1}(\mathcal{L})$ instead of $\Delta^0_{\eta+1}(D(\mathcal{L}))$.

3. Pairs of Structures

The pair of structures theorem of Ash and Knight [AK90] is one of the most useful applications of Ash and Knight's metatheorem. It gives a complete answer to the question of when two structures are hard to distinguish in a Σ^0_α way — under enough effectiveness conditions. Two of its main applications are the construction of the αth jump inversion [GHK+05] and of a structure whose degree spectrum is exactly the non-hyperarithmetic degrees [GMS13].

Theorem 3.1: *Let η be a computable ordinal and let $\mathcal{A}_0$ and $\mathcal{A}_1$ be a pair of ω-presentations with their back-and-forth relations computable up to η.*

If $\mathcal{A}_0 \geq_{\eta+1} \mathcal{A}_1$, then, for every $\Sigma^0_{\eta+1}$ set $S \subseteq \omega$, there is a computable list $\{\mathcal{C}_i : i \in \omega\}$ of computable structures such that

$$\mathcal{C}_i \cong \begin{cases} \mathcal{A}_0 & \text{if } i \notin S \\ \mathcal{A}_1 & \text{if } i \in S. \end{cases}$$

Proof: Let S be a $\Sigma^0_{\eta+1}$ subset of $\mathbb{N}$. Uniformly computably in $i \in \mathbb{N}$, we need to define a strategy for the engineer to build a structure that is isomorphic to $\mathcal{A}_1$ if $i \in S$, and to $\mathcal{A}_0$ if $i \notin S$. Theorem 2.1 will then guarantee that, also uniformly computably in i, we can choose a $\Delta^0_{\eta+1}$ sequence of moves by the extender so that the limit structure $\mathcal{C}_i$ is uniformly computably in i. The strategy must be computable, but the engineer is allowed to ask $\Delta^0_{\eta+1}$ questions along the way, and it will be the extender who, through Theorem 2.1, will guarantee that $\mathcal{C}_i$ is computable.

The question of whether $i \in S$ or not is not a $\Delta^0_{\eta+1}$ question, so we cannot ask the oracle directly about it. Instead, we use a computable list of indices e_0, e_1,... of $\Delta^0_{\eta+1}$ questions whose answers n_0, n_1,... are either all zeros if $i \notin S$ or start with zeros and then change to all ones if $i \in S$.† Notice that she can do this in a way that guarantees that the answers always converge.

Fix $i \in \mathbb{N}$. We define a strategy for the engineer for the η-$\mathbb{A}$-game where $\mathbb{A} = \{\mathcal{A}_0, \mathcal{A}_1\}$ as follows: In its first move, play the empty tuple in $\mathcal{A}_0$, and ask about e_0 — i.e., play the triple $\langle 0, \langle\rangle, e_0\rangle$. In the $j+1$st move, if $n_{j+1} = n_j$, then play any tuple $\bar{a}_{j+1}$ in $\mathcal{A}_{n_j}$ of length at least j extending $\bar{b}_j$,

†Let W be a c.e. operator such that $i \in S \iff i \in W^{S_\eta}$ where S_η is a Σ^0_η-complete set. Then let e_j be an index so that $\Phi^{S_\eta}_{e_j}(0) = 1$ if i is enumerated in W^{S_η} in less than j steps.

and ask about e_{j+1} — i.e., play the triple $\langle n_{j+1}, \bar{a}_{j+1}, e_{j+1}\rangle$. If $n_{j+1} \neq n_j$, we must have $n_j = 0$ and $n_{j+1} = 1$. In this case, play a tuple $\bar{a}_{j+1} \in \mathcal{A}_1^{<\mathbb{N}}$ such that $(\mathcal{A}_1, \bar{a}_{j+1}) \geq_\eta (\mathcal{A}_0, \bar{b}_j)$. The existence of such $\bar{a}_{j+1}$ follows from the hypothesis that $\mathcal{A}_0 \geq_{\eta+1} \mathcal{A}_1$.

At the end of the game we get that, if $i \notin S$, then $\{\bar{a}_j : j \in \mathbb{N}\}$ is an increasing sequence of tuples in $\mathcal{A}_0$, and hence the limit structure is isomorphic to $\mathcal{A}_0$. If $i \in S$ and s_0 is the first stage with $n_{s_0} = 1$, then $\{\bar{a}_j : j \in \mathbb{N}, j \geq s_0\}$ is an increasing sequence of tuples in $\mathcal{A}_1$, and hence the limit structure is isomorphic to $\mathcal{A}_1$. □

The theorem above is only the successor-ordinal version of the Ash–Knight pairs-of-structures theorem. The way we developed our metatheorem in this chapter only allows us to prove this case. To prove the limit-ordinal case, we would need to develop $<\eta$-$\mathbb{A}$-game constructions.

4. Linear Ordering Presentations

Here is another classical application of Ash's metatheorem. The theorem below was proved by Watnick [Wat84] for the case $\eta = 1$ and then extended to all η by Ash, Jockusch, and Knight [AJK90] using workers, and by Ash alone [Ash91] using 2η-systems.

Theorem 4.1: *Let $\mathcal{A}$ be a linear ordering with a first element. Then $\mathcal{A}$ has a $\Delta^0_{2\eta+1}$ copy if and only if $\omega^\eta \cdot \mathcal{A}$ has a computable copy.*

Proof: The right-to-left direction is the easy one. To prove it one has to consider the quotient over the convex equivalence relation on $\omega^\eta \cdot \mathcal{A}$, where two elements are equivalent if the segment between them embeds in an initial segment of ω^η. We concentrate on the other direction which is the one that uses the game metatheorem.

Assume that the least element of $\mathcal{A}$ is the 0 of its ω-presentation. We use $\mathcal{A}$ to denote the $\Delta^0_{2\eta+1}$ ω-presentation of $\mathcal{A}$.

The pool $\mathbb{A}$ of structures that we use for our game will consists of all the structures of the form $\omega^\eta \cdot \mathcal{F}$, where $\mathcal{F}$ is a finite linear ordering whose domain is an initial segment of $\mathbb{N}$. The back-and-forth relations between these structures are computable up to $2\eta + 1$. Precise calculations of the back-and-forth relations among ordinals are calculated in [AK00, Lemma 15.10]. These calculations are based on two main lemmas: One, that a tuple is β-back-and-forth below another if the ordering of the elements within the tuple are the same, and the intervals in between the elements of the first tuple are β-back-and-forth below the corresponding intervals for the other tuple [AK00, Lemma 15.8]; and, two, that for linear orderings $\mathcal{B}$ and $\mathcal{C}$, $\omega^\beta \cdot \mathcal{B} \leq_{2\beta+1} \omega^\beta \cdot \mathcal{C}$ if and only if $|\mathcal{B}| \geq |\mathcal{A}|$.

Another observation we need is that if $\mathcal{F}_0 \subseteq \mathcal{F}_1$ are linear orderings with the same first element 0, then $\omega^\eta \cdot \mathcal{F}_0$ is a $\Sigma^{\mathtt{in}}_{2\eta+1}$-elementary substructure of $\omega^\eta \cdot \mathcal{F}_1$, meaning that, for every tuple $\bar{b} \in \omega^\eta \cdot \mathcal{F}_0$, we have that $(\omega^\eta \cdot \mathcal{F}_0, \bar{b}) \geq_{2\eta+1} (\omega^\eta \cdot \mathcal{F}_1, \bar{b})$. The reason is that if an interval (b_i, b_j) of $\omega^\eta \cdot \mathcal{F}_0$ changes in $\omega^\eta \cdot \mathcal{F}_1$, it is because we add a few intervals of the form ω^η in between, that is, it changes from $\omega^\eta \cdot k_0 + \beta$ to $\omega^\eta \cdot k_1 + \beta$ for $0 < k_0 < k_1 \in \mathbb{N}$ and $\beta < \omega^\eta$, and we know from the lemma mentioned above, that $\omega^\eta \cdot k_1 \leq_{2\eta+1} \omega^\eta \cdot k_0$.

We describe a computable strategy for the engineer in the 2η-$\mathbb{A}$-game. At stage $j-1$, the engineer asks the oracle for a full description of $\mathcal{A} \restriction j$, i.e., the ordering $\mathcal{A}$ on the first j natural numbers. At the following stage, stage j, she chooses the structure $\omega^\eta \cdot \mathcal{F}_j$ in $\mathbb{A}$ where $\mathcal{F}_j = \mathcal{A} \restriction j$. Note that $\mathcal{F}_j$ naturally extends $\mathcal{F}_{j-1}$. By our observation above that $\omega^\eta \cdot \mathcal{F}_{j-1}$ is a $\Sigma^{\mathtt{in}}_{2\eta+1}$-elementary substructure of $\omega^\eta \cdot \mathcal{F}_j$, we know that $(\omega^\eta \cdot \mathcal{F}_{j-1}, \bar{b}_{j-1}) \leq_\eta (\omega^\eta \cdot \mathcal{F}_j, \bar{b}_{j-1})$, and hence that the engineer can play any tuple $\bar{a}_j$ extending $\bar{b}_{j-1}$. All she needs to do is make sure that she ends up including all members of $\omega^\eta \cdot \mathcal{A}$ eventually. The limit structure will then be isomorphic to $\omega^\eta \cdot \mathcal{A}$. □

The theorem is still true if $\mathcal{A}$ has no least element. In that case, instead of using the finite linear ordering $\mathcal{A} \restriction j$ to approximate $\mathcal{A}$, we need to use ω^* as follows. One needs to show that there is an $\mathcal{A}$-computable sequence of computable embeddings $f_j : \omega^* \to \omega^*$ whose direct limit is $\mathcal{A}$. For this, consider the isomorphism between ω^* and $\omega^* + \mathcal{A} \restriction j$, and embed $\omega^* + \mathcal{A} \restriction j$ into $\omega^* + \mathcal{A} \restriction j+1$ using the inclusion of $\mathcal{A}_j \subseteq \mathcal{A}_{j+1}$ and use the ω^* part for elements of $\mathcal{A}_{j+1}$ that are below the least element of $\mathcal{A}_j$.

A similar proof would show that $\mathcal{A}$ has a $\Delta^0_{2\eta+1}$ copy if and only if $\mathbb{Z}^\eta \cdot \mathcal{A}$ has a computable copy.

5. Δ^0_η-categoricity

A computable structure $\mathcal{A}$ is Δ^0_α-*categorical* if, for every computable copy $\mathcal{B}$ of $\mathcal{A}$, there is a Δ^0_α isomorphism between $\mathcal{A}$ and $\mathcal{B}$. Downey, Kach, Lempp, Lewis, Montalbán, and Turetsky [DKL+15] proved that this property cannot be characterized structurally. However, a variant of it, namely the on-a-cone version, can [AKMS89]. Given $X \in 2^{\mathbb{N}}$, $\mathcal{A}$ is $\Delta^0_\alpha(X)$-*categorical* if, for every X-computable copy $\mathcal{B}$ of $\mathcal{A}$, there is a $\Delta^0_\alpha(X)$ isomorphism between

$\mathcal{A}$ and $\mathcal{B}$. $\mathcal{A}$ is *relatively* Δ^0_α*-categorical* if it is $\Delta^0_\alpha(X)$-categorical for all X. $\mathcal{A}$ is Δ^0_α*-categorical on a cone* if there is a $C \in 2^{\mathbb{N}}$ such that $\mathcal{A}$ is $\Delta^0_\alpha(X)$-categorical for all $X \geq_T C$. There are several measures for different aspects of the complexity of a structure that turned out to be equivalent to $\boldsymbol{\Delta}^0_\alpha$-categoricity on a cone, as for example, the structure having a $\Sigma^{\mathtt{in}}_{\alpha+2}$ Scott sentence [Mon15]. Because of this robustness, we have proposed that the rank of $\mathcal{A}$ should be defined to be the least α such that $\mathcal{A}$ is $\boldsymbol{\Delta}^0_\alpha$-categorical on a cone.

Unfortunately, the three notions of plain, relative, and on-a-cone Δ^0_α-categoricity are not equivalent. Examples of this non-equivalence were built by Goncharov, Harizanov, Knight, McCoy, R. Miller and Solomon [GHK+05]. However, they are equivalent for most natural structures one encounters. Ash and Knight proved that these notions are equivalent if we have enough structural information about $\mathcal{A}$. To understand that result, we need to consider the notion of α-freeness:

Definition 5.1: (Ash [Ash87], [AK00, Section 17.4]). We say that the $\Pi^{\mathtt{in}}_\alpha$-type of a tuple $\bar{a} \in \mathcal{A}^{<\mathbb{N}}$ is $\Sigma^{\mathtt{in}}_\alpha$*-supported* if there is a $\Sigma^{\mathtt{in}}_\alpha$-formula $\varphi(\bar{x})$ true of $\bar{a}$ which implies all $\Pi^{\mathtt{in}}_\alpha$ formulas $\psi(\bar{x})$ true about $\bar{a}$. I.e.,

$$\mathcal{A} \models \varphi(\bar{a}) \quad \& \ \forall \bar{x}\big(\varphi(\bar{x}) \rightarrow \bigwedge\!\!\!\bigwedge_{\psi \in \Pi^{\mathtt{in}}_\alpha\text{-}tp_{\mathcal{A}}(\bar{a})} \psi(\bar{x})\big).$$

If the $\Pi^{\mathtt{in}}_\alpha$-type of $\bar{a}$ is not supported by a $\Sigma^{\mathtt{in}}_\alpha$ formula, we say that $\bar{a}$ is α*-free*. We say that $\bar{a}$ is α*-free over* $\bar{p}$ if it is α-free in the structure $(\mathcal{A}, \bar{p})$.

Using ideas from [AKMS89, AK00], one can show that a structure $\mathcal{A}$ is $\boldsymbol{\Delta}^0_\alpha$-categorical on a cone if and only if there is a tuple $\bar{p} \in \mathcal{A}^{<\mathbb{N}}$ such that every tuple $\bar{a} \in \mathcal{A}^{<\mathbb{N}}$ is α-free over $\bar{p}$ (see [Mon15]). It can be shown using [Mon10, Lemma 2.2] that a tuple $\bar{a}$ is $(\eta+1)$-free if and only if, for every $\bar{b} \supseteq \bar{a}$, there exist tuples $\bar{a}' \subseteq \bar{b}'$ such that

$$\bar{b} \leq_\eta \bar{b}' \quad \text{but} \quad \bar{a} \not\leq_{\eta+1} \bar{a}'.$$

In practice, when we understand the back-and-forth relations on a structure very well we can effectively decide which tuples are α-free over which tuples, and we can effectively find witness for the tuples that are not α-free. When that is the case we say that *α-freeness is computable* in $\mathcal{A}$. Under this assumptions, and the computability of the back-and-forth relations, Ash [Ash87] proved that plain, relative, and on-a-cone Δ^0_η categoricity coincide. That proof has a structural part and a computably-theoretic part.

The computably-theoretic part can be captured in the next lemma, which we state only for the successor case.

Lemma 5.2: *Let $\mathcal{A}$ be a computable ω-presentation where the back-and-forth relations up to $\eta+1$ and $\eta+1$-freeness are computable. If over every tuple $\bar{p}$ there is a tuple that is $\eta+1$-free, then $\mathcal{A}$ is not $\Delta^0_{\eta+1}$-categorical.*

This is essentially the η-$\mathcal{A}$-game version of the proof that computable categoricity implies relative computable categoricity for 2-decidable structures (i.e., the case $\eta = 0$ due to Goncharov following ideas of Nurtazin; see for instance [MonP1, Section VIII.4]). We recommend the reader to read that proof before reading the following argument.

Proof: We build $\mathcal{A}$ by defining a computable strategy for the engineer in an η-$\mathcal{A}$-game construction (i.e. $\mathbb{A} = \{\mathcal{A}\}$). We want to end up defining a copy of $\mathcal{A}$, so we will make sure that the tuples $\bar{a}_j$ stabilize in the limit. i.e., that for each $n \in \mathbb{N}$, $\lim_{j\to\infty} \bar{a}_j(n)$ exists — call it $g(n)$. We will then end up with a function $g\colon \omega \to \mathcal{A}$ and we will get that the limit structure $\mathcal{L}$ is exactly the pull-back of $\mathcal{A}$ through g (see [MonP1, Section I.1.7]). The objective is to build $\mathcal{L}$ so that it is not $\Delta^0_{\eta+1}$ isomorphic to $\mathcal{A}$.

We use a finite-injury priority construction with infinitely many requirements R_e for $e \in \mathbb{N}$. Requirement R_e ensures $\Phi_e^{S^0_\eta}$ is not an isomorphism from $\mathcal{L}$ to $\mathcal{A}$, where S^0_η is $0^{(\eta)}$ if η is finite and is $0^{(\eta+1)}$ if η is infinite, or in other words, S^0_η is a Σ^0_η-complete set. In practical terms, R_e will try to guarantee that $\Phi_e^{S^0_\eta} \circ g^{-1}$ is not an automorphism of $\mathcal{A}$ by finding a tuple $\bar{c}'_e$ so that its image $\tilde{c}_e$ through $\Phi_e^{S^0_\eta} \circ g^{-1}$ is not $\eta+1$-back-and-forth equivalent to $\bar{c}'_e$, and in particular, not automorphic to $\bar{c}'_e$. Since we do not know when or where $\Phi_e^{S^0_\eta}$ converges, we cannot ask the $\Delta^0_{\eta+1}$ oracle directly about its values. All we can ask is, given a tuple $\bar{n}$ and a number s, whether $\Phi_e^{S^0_\eta}$ converges on the numbers in $\bar{n}$ within s steps.

At each step j, an initial segment $R_0, ..., R_{k_j}$ of the list of requirements are active. The engineer goes through them one at the time checking if they *require attention* (defined below). Each requirement R_{e-1} outputs a tuple $\bar{p}_e[j] \subseteq \bar{b}_{j-1}$, which lower priority requirements are not allowed to modify. The tuple $\bar{p}_e[j]$ is then given to R_e as input, and R_e's output, namely $\bar{p}_{e+1}[j]$ must extend it. (We will usually omit the $[j]$ from the notation when the stage is understood from context.) When a stronger priority requirement acts, the weaker requirement is deactivated and its $\bar{p}_e$ becomes undefined to be re-defined later. We will see, however, that for each e, $\bar{p}_e[j]$ will stabilize as $j \to \infty$ and hence we will end up with a limit function $g\colon \omega \to \mathcal{A}$, where $g(n) = \lim_{j\to\infty} \bar{p}_e[j](n)$.

If none of the requirements R_e for $e \le k_j$ requires attention, the engineer initializes the first inactive requirement, namely, R_{k_j+1}: That means, for $e = k_j + 1$, the engineer looks for a tuple $\bar{c}_e$ that is η-free over $\bar{p}_e$ and adds it to the tuple, say on position $\bar{n}_e \in \mathbb{N}^{<\mathbb{N}}$. That is, she plays the tuple $\bar{a}_j = \bar{b}_{j-1}{}^\frown\bar{c}_e{}^\frown d$ and asks the oracle whether $\Phi_e^{S^0_\eta}(\bar{n}_e)$ converges within j steps, where d is the least element in $\mathcal{A}$ not yet played, and $\bar{n}_e = \langle|\bar{b}_{j-1}|, |\bar{b}_{j-1} + 1|, \ldots, |\bar{b}_{j-1}{}^\frown\bar{c}_e| - 1\rangle$. (As we will see, we will have $\bar{p}_e \subseteq \bar{b}_{j-1}$.) She will keep on asking about this convergence, but with larger time bounds, at every later stage $j' > j$ until she gets an answer. Actually, it is for all $e \le k_j$ simultaneously that she asks whether $\Phi_e^{S^0_\eta}(\bar{n}_e)$ converges within j steps, encapsulating all questions into one question as in the last paragraph of Sec. 1.3.

What do we mean by requiring attention, and what does the engineer do then? If the oracle answered that for some $e < k_j$, $\Phi_e^{S^0_\eta}(\bar{n}_e)$ converges within $j-1$ steps, we say that the least such e *requires attention*. Suppose that $\Phi_e^{S^0_\eta}(\bar{n}_e) = \tilde{c}_e$. So we have that $\bar{p}_{e+1}$ maps $\bar{n}_e$ to $\bar{c}_e$ while $\Phi_e^{S^0_\eta}$ maps $\bar{n}_e$ to $\tilde{c}_e$. Then, she checks if $\bar{c}_e \le_\eta \tilde{c}_e$. If not, she does not need to do anything, as we then know that $\bar{c}_e$ and $\tilde{c}_e$ are not non-automorphic. She plays $\bar{a}_j = \bar{b}_{j-1}{}^\frown d$ where d is the least element in $\mathcal{A}$ not yet played, declares R_e satisfied (unless it is later deactivated), deactivates lower priority requirements, and lets $k_{j+1} = e$. If yes, that is if $\bar{c}_e \le_\eta \tilde{c}_e$, she will switch $\bar{c}_e$ by a tuple $\bar{c}'_e \not\ge_{\eta+1} \bar{c}_e$, getting that $\bar{c}'_e$ and $\tilde{c}_e$ are non-automorphic. To find such $\bar{c}'_e$, we use the $(\eta+1)$-freeness of $\bar{c}_e$ over $\bar{p}_e$, applied to the tuple $\bar{b}_{j-1} \supseteq \bar{p}_e\bar{c}_e$. By $(\eta+1)$-freeness, we then know there is a tuple $\bar{c}'_e$ and a tuple $\bar{a}'_j \supseteq \bar{p}_e\bar{c}'_e$ such that $\bar{b}_{j-1} \le_\eta \bar{a}'_j$ but $\bar{p}_e\bar{c}_e \not\le_{\eta+1} \bar{p}_e\bar{c}'_e$. The engineer now plays $\bar{a}_j = \bar{a}'_j{}^\frown d$ where d is the least element in $\mathcal{A}$ not yet played, declares R_e satisfied (unless it is later deactivated), deactivates lower priority requirements, and lets $k_{j+1} = e$.

If R_e is never deactivated, we will end up with $\bar{p}_{e+1}[j] \subseteq \bar{a}_{j'}$ for all $j' \ge j$, and hence with $\bar{p}_{e+1}[j] \subseteq g$. We would have then satisfied R_e as either $\Phi_e^{S^0_\eta}$ is not total or $\Phi_e^{S^0_\eta} \circ g^{-1}$ maps either $\bar{c}_e$ or $\bar{c}'_e$ to $\tilde{c}_e$, and in either case these tuples are not $\eta + 1$-back-and-forth equivalent. $\square$

6. The Low$_n$ Property

Downey and Jockusch [DJ94] proved that Boolean algebras have the *low property*, that is, that every low Boolean algebra has a computable copy. Jockusch and Soare [JS91] had already shown this is not the case for linear orderings. A year later, Thurber [Thu95] showed that Boolean algebras have the low$_2$ property, that is, that every low$_2$ Boolean algebra has a computable copy. A few years later, Knight and Stob showed that Boolean algebras have the low$_4$ property. We still don't know whether Boolean algebras have the low$_n$ property for all $n \in \mathbb{N}$, or even if they have the low$_5$ property. Harris and Montalbán [HM14] showed that the low$_5$ problem for Boolean algebras is qualitatively more difficult than the previous ones: While for $n = 1, 2, 3, 4$, every low$_n$ Boolean algebra is $0^{(n+2)}$-isomorphic to a computable one, they built a low$_5$ Boolean algebra not $0^{(7)}$-isomorphic to any computable one.

One of the most interesting examples of a class with the low property is differentially closed fields of characteristic zero, as recently proved by Marker and Miller [MM17]. This allowed them to give a full description of the degree spectra of differentially closed fields of characteristic zero.

In [Mon13], we provided a framework for low-property proofs that separates the combinatorial/algebraic part of the proof from the computational part of the proof, hiding the computational part inside the framework, and letting the prover concentrate only on the combinatorial/algebraic part. We called that framework the *copy-vs-diagonalizer game*. In this section we just want to observe that the copy-vs-diagonalizer game is, essentially, a particular case of our new metatheorem.

We proved in [MonP1] that if a class of structures has the low$_1$ property, it must be Σ-small, meaning that there are countably many 1-back-and-forth equivalence classes among all tuples among all structures in the class. This can be easily extended to show that if a class of structures has the low$_n$ property, it must be Σ_n-small. Being Σ_n-small is not a sufficient condition, but a necessary one that helps setting up the constructions. All natural Σ_n-small classes are *effectively* Σ_n*-small* (see [Mon10, Definition 2.3], [Mon13, Definition 4.1], [MonP1, Definition X.24]) meaning that we have a complete effective understanding of the interactions between the m-back-and-forth equivalence classes for $m \leq n$. We showed in [Mon10] that we can then build a computable list of structure representing all n-back-and-forth equivalence classes where the back-and-forth relations are computable up to n. Call it $\mathbb{K}_n$. Actually, having such a list of structures is equivalent to having a computable representation of the n-back-and-forth structure of the class.

We thus have a setting to perform an n-$\mathbb{K}_n$-game construction. When the engineer is playing tuples in $\mathbb{K}_n$, all that matters is the n-back-and-forth type of that tuple. So, if we know the structure of the n-back-and-forth types and we are given a way to represent them nicely, we may think that instead of playing structures in $\mathbb{K}_n$, we are playing n-back-and-forth types.

If we are given a low$_n$ structure $\mathcal{A}$ in the class, the Δ_{n+1} oracle has access to its n-jump, and hence it knows how its tuples are m-back-and-forth related to the tuples of the structures in $\mathbb{K}_n$ — well, kind of: Given a tuple $\bar{k}$ in a structure $\mathcal{K}_i$ in $\mathbb{K}_n$, and a tuple $\bar{a} \in \mathcal{A}^{<\mathbb{N}}$, it is Π_n in the structure to tell if $(\mathcal{K}_i, \bar{k}) \leq_n (\mathcal{A}, \bar{a})$ (see [Mon10, Lemma 2.2]), but it is harder to tell if $(\mathcal{K}_i, \bar{k}) \geq_n (\mathcal{A}, \bar{a})$.[‡] What we showed in [Mon13] is that if we are given the nth jump of a structure we can produce an n-approximation to it. In the setting of this chapter, an *n-approximation* to a structure is a sequence of pairs $\langle i_0, \bar{k}_0\rangle, \langle i_1, \bar{k}_1\rangle, \langle i_2, \bar{k}_2\rangle, \ldots$ where $\bar{k}_j \in \mathbb{K}_{i_j}^{<\mathbb{N}}$ and

$$(\mathcal{K}_{i_0}, \bar{k}_0) \leq_n (\mathcal{K}_{i_1}, \bar{k}_1) \leq_n (\mathcal{K}_{i_2}, \bar{k}_2) \leq_n \cdots,$$

and the given structure $\mathcal{A}$ is the limit of this sequence in the sense that $D(\mathcal{A}) = \bigcup_j D_{\mathcal{K}_{i_j}}(\bar{k}_j)$.

The engineer can thus, at stage j, ask the oracle for the pair $\langle i_j, \bar{k}_j\rangle$. The engineer does not need to play in $\mathcal{K}_{i_j}$ in the next move, tough that may be a good idea. The engineer would like to copy $\mathcal{A}$. Her obstacle is that the extender is extending her tuples, and the fact that $(\mathcal{K}_{i_{j-1}}, \bar{k}_{j-1}) \leq_n (\mathcal{K}_{i_j}, \bar{k}_j)$ does not mean that every tuple $\bar{b}_{j-1}$ extending $\bar{a}_{j-1}$ has a matching tuple in $\mathcal{K}_{i_j}$. If she can overcome that obstacle, she will be able to produce a proof of the low$_n$-property.

7. η-true-stage Systems

In this section we review η-true-stage systems, introduced by the author in [Mon14], so we can use them in the proof of the metatheorem in the next section.

Let η be a computable ω-presentation of an ordinal where we can decide whether an ordinal is a successor or a limit, and we can calculate successors.

An *η-true-stage system* is a computable family $\{\leq_\xi : \xi \leq \eta\}$ of partial orderings on $\mathbb{N}$ that satisfies the following properties:

(TS0) $\leq_0$ is just the standard ordering on $\mathbb{N}$.

[‡]The asymmetry comes from the assumption that the back-and-forth relation in $\mathcal{K}_i$ are computable, while in $\mathcal{A}$ they probably aren't.

(TS1) The sequence of relations is *nested* (i.e., if $\gamma \leq \xi$ and $s \leq_\xi t$, then $s \leq_\gamma t$).

(TS2) For every ξ, there exist an infinite $\leq_\xi$-increasing sequence $t_0 <_\xi t_1 <_\xi \cdots$.

(TS3) The sequence of relations is *continuous* (i.e., if λ is a limit ordinal, then $\leq_\lambda = \bigcap_{\xi<\lambda} \leq_\xi$).

(♣) For every ξ, and every $r < s < t$, if $r \leq_{\xi+1} t$ and $s \leq_\xi t$, then $r \leq_{\xi+1} s$.

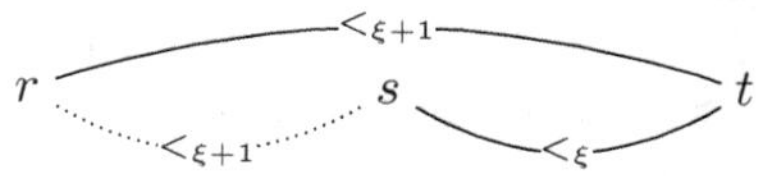

For each $\xi \leq \eta$, we say that t is a *ξ-true stage* if it belongs to an infinite $\leq_\xi$-increasing sequence. We define $\mathfrak{T}^\xi \in \mathbb{N}^\mathbb{N}$ to be the sequence of ξ-true stages listed in increasing order. One could thus define $\mathfrak{T}^\xi$ as the union of all infinite $\leq_\xi$-increasing sequences. One can prove using (♣) that $\mathfrak{T}^\xi$ is itself a $\leq_\xi$-increasing sequence. $\mathfrak{T}^\xi$ is thus the maximal $\leq_\xi$-increasing sequence. For successor ordinals, one can show using (♣) that

$$t \text{ is a } \xi+1\text{-true stage} \iff t \text{ is } \xi\text{-true and } (\forall s \geq_\xi t \text{ with } s\ \xi\text{-true})\ s \geq_{\xi+1} t.$$

For limit ordinals λ, t is a λ-true stage if and only if it is a ξ-true stage for all $\xi < \lambda$. This implies that the set of ξ-true stages is Π^0_ξ. $\mathfrak{T}^\xi$ does not have the same m-degree as the complete Π^0_ξ set, but the theorem below states that it may have the same Turing degree.

Definition 7.1: We say that an η-true-stage system is *complete* if $\mathfrak{T}^\xi$ is $\Delta^0_{\xi+1}$-Turing-complete for all $\xi \leq \eta$, uniformly in ξ.

Theorem 7.2: *There exists a complete η-true-stage system.*

This theorem was proved in [Mon14, Lemma 7.8]. A different construction was later given by Greenberg and Turetsky in [GT]. In [MonP2], we present yet a different construction which incorporates ideas from [GT]. In any case, for applications of the η-true-stage method, all that matters is that such a system exists. Let us now describe the key idea behind these systems.

We say that s is an *apparent ξ-true stage at t* if $s \leq_\xi t$. Given ξ and t, we define the *stage-t approximation to $\mathfrak{T}^\xi$*, denoted $\mathfrak{T}^\xi_t$, as the tuple enumerating the apparent ξ-true stages at t:

$$\mathfrak{T}^\xi_t = \langle s : s \leq_\xi t \rangle.$$

Note that

$$s \leq_\xi t \iff \mathfrak{T}^\xi_s \subseteq \mathfrak{T}^\xi_t,$$

where the inclusion is as strings, and that

$$t \text{ is } \xi\text{-true} \iff \mathfrak{T}^\xi_t \subseteq \mathfrak{T}^\xi.$$

An η-system thus provides a combinatorial framework to approximate the iterations of the jumps. The properties (TS1)–(TS3) and (♣) describe the behavior of these approximations.

8. The Proof of the Game Metatheorem

This section is dedicated to prove Theorem 2.1, which we re-state below.

Theorem 8.1: *Given $\mathbb{A}$ and η as described above, for every computable valid strategy for the engineer in the η-$\mathbb{A}$-game, there is a $\Delta^0_{\eta+1}$ sequence of moves by the extender so that, if the engineer follows her strategy, the limit ω-presentation $\mathcal{L}$ is computable.*

We split the proof in two independent parts by first considering a simplification of the game. In this simplification, the engineer does not get to choose which $\Delta^0_{\eta+1}$ question to ask, and the oracle just plays $\mathfrak{T}^\eta \restriction j$ at stage j, namely the string consisting of the first j many η-true stages. Recall that $\mathfrak{T}^\eta$ is $\Delta^0_{\eta+1}$ Turing complete, so if the engineer had a $\Delta^0_{\eta+1}$ question in mind, she will eventually be able to figure out the answer. Let us call this version of the game, the *simplified game*. Let us start seeing how to transform the general version of the game into an instance of the simplified version.

Proof: [Proof of Theorem 2.1 from the simplified version of Theorem 2.1] Let σ be a computable valid strategy for the engineer in the η-$\mathbb{A}$-game from Theorem 2.1. We will build a computable ω-presentation for a limit-structure obtained from a certain sequence by the extender where the engineer follows her strategy σ. This construction will use a number ℓ as a parameter — so we are actually building a different ω-presentation $\mathcal{L}_\ell$ for each $\ell \in \mathbb{N}$. We will then use the recursion theorem to find a computable index ℓ_0 for the diagram of the limit-structure computed using ℓ_0 as a parameter. I.e.,

$$D(\mathcal{L}_{\ell_0}) = \Phi_{\ell_0}.$$

Thus, we may assume the parameter ℓ is an index for the computable diagram we are building. For this to work, we must produce a computable ω-presentation $\mathcal{L}_\ell$ even if the ℓth computable function Φ_ℓ is not total.

We will build a computable valid strategy $\hat{\sigma}$ for engineer in the simplified η-$\mathbb{A}$-game, and we will do it uniformly in ℓ. We will do it in a way that, for every run of the simplified game following $\hat{\sigma}$, there is a run of the original game following σ that produces the same limit structure. The simplified version of Theorem 2.1 (which we prove below) will give a sequence of moves by the extender such that, when we follow $\hat{\sigma}$, produces a computable ω-presentation $\mathcal{L}_\ell$. We omit the subindex ℓ in what follows.

Here is how we define $\hat{\sigma}$: Let Γ be a computable operator such that $\Gamma^{\mathfrak{T}^\eta}(e)$ is the answer to the eth $\Delta^0_{\eta+1}(\Phi_\ell)$ question. Let $\hat{\sigma}$'s first move be the same as σ's; that is let $\hat{\sigma}(\langle\rangle) = \sigma(\langle\rangle)$. At each following stage, $\hat{\sigma}$ may either *pass* or *emulate* σ, depending on whether on not the oracle has given her enough information to answer the last $\Delta^0_{\eta+1}$-question she asked. At a stage $j+1$, whether she passes or emulates σ gets decided as follows: Suppose the extender has just played $\bar{b}_j$, and the oracle played the string $\mathfrak{T}^\eta \upharpoonright j$. Let j_0 be the last stage at which $\hat{\sigma}$ emulated σ — suppose it was the kth time where $\hat{\sigma}$ emulated σ. At that stage, σ asked a $\Delta^0_{\eta+1}(\mathcal{L})$-question, say e_k. If $\Gamma^{\mathfrak{T}^\eta \upharpoonright j}(e_k)\downarrow$ let $\hat{\sigma}$ emulate σ and play the string σ would play if the extender had played $\bar{b}_j$ and the oracle has played $n_k = \Gamma^{\mathfrak{T}^\eta \upharpoonright j}(e_k)$. If $\Gamma^{\mathfrak{T}^\eta \upharpoonright j}(e_k)\uparrow$, let $\hat{\sigma}$ pass, that is, let it play $i_{j+1} = i_j$ and $\bar{a}_{j+1}$ be any proper extension of $\bar{b}_j$ in $\mathcal{A}_{i_j}$.

If we apply the simplified version of Theorem 2.1 to this strategy $\hat{\sigma}$, we end up building a computable ω-presentation $\mathcal{L}$. The diagram of $\mathcal{L}$ is total independently of whether Φ_ℓ is total and of whether we ever get answers to the $\Delta^0_{\eta+1}(\mathcal{L})$-questions e_k. This is because if $\hat{\sigma}$ ends up passing from some point j_0 onwards, then the limit structure will end up isomorphic to $\mathcal{A}_{i_{j_0}}$. Thus, when ℓ_0 is given to us by the recursion theorem as above, we get that Φ_{ℓ_0} is total and is equal to the diagram of the limit structure $\mathcal{L}_{\ell_0}$ we just obtained. Since σ is a valid strategy, all the $\Delta^0_{\eta+1}(\mathcal{L})$-questions it asks converge, and hence, for all k, $\Gamma^{\mathfrak{T}^\eta \upharpoonright j}(e_k)$ eventually converges. This means that there are infinitely many stages at which $\hat{\sigma}$ emulates σ, and we thus get that for every sequence of moves by the extender in the simplified game, there is a sequence of moves in the original game which give us the same limit structure. We then get a run of the original game where the limit structure is computable. □

The following lemma is the key property about the back-and-forth relations that allows us to simplify the metatheorems of Ash and Knight [AK00] and Montalbán [Mon14] in the particular case where the object we are building is an ω-presentation of a structure. Those metatheorems are more general than the one of this chapter, and when they are applied in the literature the lemma below is used in the middle of each of the constructions. One of the ways in which our metatheorem simplifies the previous ones is by absorbing the lemma inside the proof of the metatheorem, so that the user of the metatheorem does not have to deal with it.

Lemma 8.2: *(Ash [Ash87, AK00]) Suppose we have a finite sequence of τ-structures $\mathcal{A}_0, ..., \mathcal{A}_k$, ordinals $\xi_{k-1} > \cdots > \xi_1 > \xi_0$, and tuples $\bar{a}_i \in A_i^{<\mathbb{N}}$ for $i \leq k$, such that*

$$(\mathcal{A}_k, \bar{a}_k) \leq_{\xi_{k-1}+1} (\mathcal{A}_{k-1}, \bar{a}_{k-1}) \leq_{\xi_{k-2}+1} \cdots \leq_{\xi_1+1} (\mathcal{A}_1, \bar{a}_1) \leq_{\xi_0+1} (\mathcal{A}_0, \bar{a}_0).$$

There exist a tuple $\bar{b} \in A_k^{<\mathbb{N}}$ extending $\bar{a}_k$ such that $(\mathcal{A}_j, \bar{a}_j) \leq_{\xi_j} (\mathcal{A}_k, \bar{b})$ for all $j < k$.

Proof: We will define a sequence of tuples $\bar{b}_j \in A_j^{<\mathbb{N}}$ extending $\bar{a}_j$ by induction on $j \leq k$, ending with $\bar{b} = \bar{b}_k$. We will make sure by induction that $(\mathcal{A}_i, \bar{a}_i) \leq_{\xi_i} (\mathcal{A}_j, \bar{b}_j)$ for all $i < j$.

$$\begin{array}{ccccccccc}
(\mathcal{A}_k, \bar{a}_k) & \leq_{\xi_{k-1}+1} & (\mathcal{A}_{k-1}, \bar{a}_{k-1}) & \leq_{\xi_{k-2}+1} & \cdots & \leq_{\xi_1+1} & (\mathcal{A}_1, \bar{a}_1) & \leq_{\xi_0+1} & (\mathcal{A}_0, \bar{a}_0) \\
\cap\!\!\!| & \leq_{\xi_{k-1}+1} & \cap\!\!\!| & \leq_{\xi_{k-2}+1} & \ddots & \leq_{\xi_1+1} & \cap\!\!\!| & \leq_{\xi_0+1} & \| \\
(\mathcal{A}_k, \bar{b}_k) & \geq_{\xi_{k-1}} & (\mathcal{A}_{k-1}, \bar{b}_{k-1}) & \geq_{\xi_{k-2}} & \cdots & \geq_{\xi_1} & (\mathcal{A}_1, \bar{b}_1) & \geq_{\xi_0} & (\mathcal{A}_0, \bar{b}_0)
\end{array}$$

Let $\bar{b}_0 = \bar{a}_0$. Given $\bar{b}_j$, since $(\mathcal{A}_{j+1}, \bar{a}_{j+1}) \leq_{\xi_j+1} (\mathcal{A}_j, \bar{a}_j)$, and $\bar{a}_j \subseteq \bar{b}_j$, there exists $\bar{b}_{j+1} \supseteq \bar{a}_{j+1} \in A_{j+1}^{<\mathbb{N}}$ such that $(\mathcal{A}_{j+1}, \bar{b}_{j+1}) \geq_{\xi_j} (\mathcal{A}_j, \bar{b}_j)$. We end up with $(\mathcal{A}_k, \bar{b}_k) \geq_{\xi_j} (\mathcal{A}_j, \bar{b}_j) \supseteq (\mathcal{A}_j, \bar{a}_j)$. □

Proof: [Proof of Theorem 2.1 for the simplified game] We build a computable sequence of pairs $\langle i_s, \bar{a}_s \rangle$ with $\bar{a}_s \in \mathcal{A}_{i_s}$ which satisfies that, for all $\xi \leq \eta$, and $r < s \in \mathbb{N}$,

$$r \leq_\xi s \quad \text{implies} \quad (\mathcal{A}_{i_r}, \bar{a}_r) \leq_\xi (\mathcal{A}_{i_s}, \bar{a}_s). \tag{8.1}$$

In particular, we get that if $s \leq t$, then $(\mathcal{A}_{i_s}, \bar{a}_s) \leq_0 (\mathcal{A}_{i_t}, \bar{a}_t)$, which means that $D_{\mathcal{A}_{i_s}}(\bar{a}_s) \subseteq D_{\mathcal{A}_{i_t}}(\bar{a}_t)$, and hence the limit sequence with diagram $\bigcup_s D_{\mathcal{A}_{i_s}}(\bar{a}_s)$ is computable. To show that this structure is the limit structure under some run of the game following the engineer's strategy σ, we will show that if we restrict ourselves to the sequence of η-true stages $t_0 \leq_\eta t_1 \leq_\eta t_2 \leq_\eta \cdots$, then the sequence $\langle i_{t_0}, \bar{a}_{t_0}\rangle, \langle i_{t_1}, \bar{a}_{t_1}\rangle, \langle i_{t_2}, \bar{a}_{t_2}\rangle, \ldots$ can be seen as the sequence of moves by the engineer following σ for some particular sequence of moves by the extender.

For each s, will also define a tuple $\bar{b}_s$ that belongs to A_{i_t} for the largest $t < s$ with $t \leq_\eta s+1$. (This is the tuple we will use as the move by the extender later.) The tuple $\bar{b}_s$ will satisfy that for every $\xi \leq \eta$ and every $r < s$,

$$r \leq_\xi s+1 \quad \text{implies} \quad (\mathcal{A}_{i_r}, \bar{a}_r) \leq_\xi (\mathcal{A}_{i_t}, \bar{b}_s). \tag{8.2}$$

Thus, if we then define i_{s+1} and $\bar{a}_{s+1}$ satisfying $(\mathcal{A}_{i_t}, \bar{b}_s) \leq_\eta (\mathcal{A}_{i_{s+1}}, \bar{a}_{s+1})$, we will immediately get property (8.1). We will define $\bar{b}_s$ using the previous lemma, so we first need to find the right setting to apply it. For each $\xi \leq \eta$, let $t_\xi < s+1$ be the largest t such that $t \leq_\xi s+1$. (It does not hurt to assume that $0 \leq_\xi s+1$ for all ξ and s, so such a t_ξ always exists.) Notice that if $r <_\xi s+1$, then $r \leq t_\xi$ and then by (♣) $r \leq_\xi t_\xi \leq_\xi s+1$. So, to satisfy property (8.2), it is enough to get $\bar{b}_s$ so that $(\mathcal{A}_{i_{t_\xi}}, \bar{a}_{t_\xi}) \leq_\xi (\mathcal{A}_{i_t}, \bar{b}_s)$ for all $\xi \leq \eta$.

There are infinitely many ξ's, but only finitely many possible values for $t_\xi < s+1$, so they must repeat a lot. Since the relations $(\leq_\xi)_{\xi\leq\eta}$ are nested, if $\xi \leq \zeta \leq \eta_0$ then $t_\zeta \leq t_\xi$. We now want to define stages $s_k < \ldots < s_0 < s+1$ so that $\{s_k, \ldots, s_0\} = \{t_\xi : \xi \leq \eta_0\}$ as sets, but we need to define them effectively. Let $s_0 = t_0 = s$. Given s_j, let $\xi_j \leq \eta_0$ be the greatest such that $s_j = t_{\xi_j}$, i.e., the greatest ξ such that $s_j \leq_\xi s+1$. We notice that such a greatest ordinal exists by the continuity of $(\leq_\xi)_{\xi\leq\eta}$. If $\xi_j = \eta$, then we let $k = j$ and that finishes the definition of $s_0, \ldots, s_k$. Otherwise, let $s_{j+1} = t_{\xi_j+1}$. Since we know $s_j \not\leq_{\xi_j+1} s$, we must have $s_{j+1} < s_j$. By (♣) we then have $s_{j+1} \leq_{\xi_j+1} s_j$, and hence $(\mathcal{A}_{i_{s_{j+1}}}, \bar{a}_{s_{j+1}}) \leq_{\xi_j+1} (\mathcal{A}_{i_{s_j}}, \bar{a}_{s_j})$. We now apply Lemma 8.2 to the sequence

$$\begin{aligned}(\mathcal{A}_{i_{s_k}}, \bar{a}_{s_k}) &\leq_{\xi_{k-1}+1} (\mathcal{A}_{i_{s_{k-1}}}, \bar{a}_{s_{k-1}}) \leq_{\xi_{k-2}+1} \cdots \leq_{\xi_1+1} (\mathcal{A}_{i_{s_1}}, \bar{a}_{s_1}) \\ &\leq_{\xi_0+1} (\mathcal{A}_{i_{s_0}}, \bar{a}_{s_0}),\end{aligned}$$

to get $\bar{b}_s$ satisfying (8.2).

The last step is to define $\bar{a}_{s+1}$ using the strategy σ for the engineer in the simplified game. Let $0 = t_0, \ldots, t_j$ are the apparent η-true stages below $s+1$. Note that $\mathfrak{T}^\eta_{t_i} = \mathfrak{T}^\eta_{s+1} \restriction i$. We then let

$$\langle i_{s+1}, \bar{a}_{s+1}\rangle = \sigma(\bar{b}_{t_1-1}, \mathfrak{T}^\xi_{t_1}, \bar{b}_{t_2-1}, \mathfrak{T}^\xi_{t_2}, \ldots, \bar{b}_{t_j-1}, \mathfrak{T}^\xi_{t_j}, \bar{b}_s, \mathfrak{T}^\xi_{s+1}).$$

That is, $\langle i_{s+1}, \bar{a}_{s+1}\rangle$ is what the engineer would play in her $j+1$st move if she was following σ and the previous moves by the extender where $\bar{b}_{t_1-1}$, $\bar{b}_{t_2-1}, \ldots, \bar{b}_{t_j-1}, \bar{b}_s$ and the previous plays by the oracle $\mathfrak{T}^{\xi}_{s+1} \upharpoonright 1$, $\mathfrak{T}^{\xi}_{s+1} \upharpoonright 2, \ldots, \mathfrak{T}^{\xi}_{s+1}$.

Now, consider the sequence $t_1 <_\eta t_2 <_\eta \cdots$ of η-true stages. We get that the following is a run of the simplified game following σ:

engineer	$i_0, \bar{a}_0$		$i_{t_1}, \bar{a}_{t_1}$		$i_{t_2}, \bar{a}_{t_2}$		$\cdots$
extender		$\bar{b}_{t_1-1}$		$\bar{b}_{t_2-1}$		$\bar{b}_{t_3-1}$	$\cdots$
oracle		$\mathfrak{T}^{\xi} \upharpoonright 1$		$\mathfrak{T}^{\xi} \upharpoonright 2$		$\mathfrak{T}^{\xi} \upharpoonright 3$	$\cdots$

It follows that the limit structure of this run of the game is the computable structure with diagram $\bigcup_s D_{\mathcal{A}_{i_s}}(\bar{a}_s)$ that we mentioned above. $\square$

References

AJK90. C. J. Ash, C. G. Jockusch, Jr. and J. F. Knight. Jumps of orderings. *Trans. Amer. Math. Soc.*, 319(2):573–599, 1990.

AK90. C. J. Ash and J. F. Knight. Pairs of recursive structures. *Ann. Pure Appl. Logic*, 46(3):211–234, 1990.

AK00. C. J. Ash and J. Knight. *Computable Structures and the Hyperarithmetical Hierarchy*. Elsevier Science, 2000.

AKMS89. C. J. Ash, J. Knight, M. Manasse and T. Slaman. Generic copies of countable structures. *Ann. Pure Appl. Logic*, 42(3):195–205, 1989.

Ash86. C. J. Ash. Stability of recursive structures in arithmetical degrees. *Ann. Pure Appl.] Logic*, 32(2):113–135, 1986.

Ash87. C. J. Ash. Categoricity in hyperarithmetical degrees. *Ann. Pure Appl. Logic*, 34(1):1–14, 1987.

Ash91. C. J. Ash. A construction for recursive linear orderings. *J. Symbolic Logic*, 56(2):673–683, 1991.

DJ94. R. Downey and C. G. Jockusch. Every low Boolean algebra is isomorphic to a recursive one. *Proc. Amer. Math. Soc.*, 122(3):871–880, 1994.

DKL+15. R. G. Downey, A. M. Kach, S. Lempp, A. E. M. Lewis-Pye, A. Montalbán and D. D. Turetsky. The complexity of computable categoricity. *Advances in Mathematics*, 268:423–466, 2015.

Fri57. R. M. Friedberg. Two recursively enumerable sets of incomparable degrees of unsolvability (solution of Post's problem, 1944). *Proc. Nat. Acad. Sci. U.S.A.*, 43:236–238, 1957.

GHK+05. S. Goncharov, V. Harizanov, J. Knight, C. McCoy, R. Miller and R. Solomon. Enumerations in computable structure theory. *Ann. Pure Appl. Logic*, 136(3):219–246, 2005.

GMS13. N. Greenberg, A. Montalbán and T. A. Slaman. Relative to any nonhyperarithmetic set. *Journal of Mathematical Logic*, 13(1), 2013.

GT. N. Greenberg and D. Turetsky. Completeness of the hyperaritmetic isomorphism equivalence relation. Submitted for publication.

Har76. L. Harrington. Mclaughlin's conjecture. Handrwitten notes, 11 pages, September 76.

HM14. K. Harris and A. Montalbán. Boolean algebra approximations. *Transactions of the AMS*, 366(10):5223–5256, 2014.

JS91. C. G. Jockusch, Jr. and R. I. Soare. Degrees of orderings not isomorphic to recursive linear orderings. *Ann. Pure Appl. Logic*, 52(1-2):39–64, 1991. International Symposium on Mathematical Logic and its Applications (Nagoya, 1988).

Kar65. C. R. Karp. Finite-quantifier equivalence. In *Theory of Models (Proc. 1963 Internat. Sympos. Berkeley)*, pp. 407–412. North-Holland, Amsterdam, 1965.

Lac76. A. H. Lachlan. A recursively enumerable degree which will not split over all lesser ones. *Ann. Math. Logic*, 9(4):307–365, 1976.

Ler10. M. Lerman. *A framework for priority arguments*, volume 34 of *Lecture Notes in Logic*. Association for Symbolic Logic, La Jolla, CA, 2010.

LL95. S. Lempp and M. Lerman. A general framework for priority arguments. *Bull. Symbolic Logic*, 1(2):189–201, 1995.

MM17. D. Marker and R. Miller. Turing degree spectra of differentially closed fields. *J. Symb. Log.*, 82(1):1–25, 2017.

Mon10. A. Montalbán. Counting the back-and-forth types. *Journal of Logic and Computability*, doi: 10.1093/logcom/exq048, 2010.

Mon13. A. Montalbán. Copyable structures. *Journal of Symbolic Logic*, 78(4):1025–1346, 2013.

Mon14. A. Montalbán. Priority arguments via true stages. *Journal of Symbolic Logic*, 79(4):1315–1335, 2014.

Mon15. A. Montalbán. A robuster Scott rank. *Proc. Amer. Math. Soc.*, 143(12):5427–5436, 2015.

MonP1. A. Montalbán. *Computable structure theory—within the arithmetic.* Perspectives in Logic. Cambridge University Press, Cambridge; Association for Symbolic Logic, Ithaca, NY, 2021.

MonP2. A. Montalbán. *Computable structure theory—beyond the arithmetic.* In preparation.

Muc56. A. A. Muchnik. On the unsolvability of the problem of reducibility in the theory of algorithms. *Dokl. Akad. Nauk SSSR, N.S.*, 108:194–197, 1956.

Sac63. G. E. Sacks. *Degrees of unsolvability.* Princeton University Press, Princeton, N.J., 1963.

Sho61. J. R. Shoenfield. Undecidable and creative theories. *Fund. Math.*, 49:171–179, 1960/61.

Thu95. J. J. Thurber. Every low_2 Boolean algebra has a recursive copy. *Proc. Amer. Math. Soc.*, 123(12):3859–3866, 1995.

Wat84. R. Watnick. A generalization of Tennenbaum's theorem on effectively finite recursive linear orderings. *J. Symbolic Logic*, 49(2):563–569, 1984.

https://doi.org/10.1142/9789819806584_0011

An I_0 Analogue of an AD Theorem

Xianghui Shi
School of Mathematical Sciences, Beijing Normal University
Key Laboratory of Mathematics and Complex Systems, Ministry of Education
Beijing 100875, P. R. China
shi.bnu@gmail.com

It is well-known that AD implies that there is no ω_1-sequence of distinct reals. A slightly stronger theorem says that AD implies there is no ω_1 sequence of distinct countable sets of reals. It is known that under $I_0(\lambda)$ the λ-version of the former statement holds in $L(V_{\lambda+1})$ (see Shi-Trang [1]). In this chapter, we prove the I_0 analogue of the latter: Assume ZFC + $I_0(\lambda)$, in $L(V_{\lambda+1})$ there is no λ^+-sequence of distinct subsets of $\mathscr{P}(\lambda)$ of size $\leq\lambda$. This answers a question of John Steel raised at the Arctic Set Theory Workshop 2019. We also analyze two other proofs of the AD theorem (for countable sets of reals), and discuss the possibility of as well as the obstacles in generalizing these arguments to I_0 context.

Keywords: Axiom of Determinacy, Axiom I_0.

1. Introduction

The Axiom of Determinacy (AD) is the assertion that every two players integer game is determined. $\mathsf{AD}^{L(\mathbb{R})}$ is AD restricted to sets of reals in $L(\mathbb{R})$. $I_0(\lambda)$ is the assertion that there is an elementary embedding j : $L(V_{\lambda+1}) \to L(V_{\lambda+1})$ with critical point $<\lambda$. There is a large degree of similarity between the structural properties of sets of reals obtained from AD (or more precisely, $\mathsf{AD}^{L(\mathbb{R})}$) and the structure properties of subsets of $V_{\lambda+1}$ that follows from axiom $I_0(\lambda)$ (see Woodin [2–4], Cramer [5]). In this chapter, we shall add one more example to this supporting pool.

It is a well-known AD-fact that there is no ω_1-sequence of distinct reals (see Mycielski [6]). In Shi-Trang [1], it was shown that the similar result holds in $L(V_{\lambda+1})$ under $I_0(\lambda)$, namely, there is no λ^+ sequence of distinct subsets of λ. Shi and Trang proved it using the $I_0(\lambda)$-fact that λ^+ is measurable in $L(V_{\lambda+1})$. One could also use Cramer's λ-Perfect Set Theorem (see pp. 226) to prove it. During the Arctic Set Theory workshop 2019, John Steel asked how about the I_0-analogue of the following AD fact:

Theorem 1.1: *Assume* $\mathsf{ZF}+\mathsf{AD}$. *There is no* ω_1*-sequence of distinct countable sets of reals.*

Clearly, the conclusion implies that there is no ω_1-sequence of distinct reals. For our purpose, the more appropriate version to generalize is the following:

Theorem 1.2: *Assume* $\mathsf{ZF}+\mathsf{AD}^{L(\mathbb{R})}$. *There is no* ω_1*-sequence of distinct countable sets of reals in* $L(\mathbb{R})$.

In this chapter, we answer Steel's question positively — the I_0 analogue of Theorem 1.2 also holds in $L(V_{\lambda+1})$ under $I_0(\lambda)$:

Main Theorem *Assume* $\mathsf{ZFC}+I_0(\lambda)$. *In* $L(V_{\lambda+1})$, *there is no* λ^+*-sequence of distinct subsets of* $\mathscr{P}(\lambda)$ *of size* $\leq\lambda$.

We have found three proofs of Theorem 1.2. We manage to adapt one of them to show the main theorem. We will first present this argument and then run it under $I_0(\lambda)$ to show its I_0 counterpart (see §2). Although we haven't found appropriate ways to generalize the other two arguments to prove the main theorem, the questions of whether (or to what extent) the relevant ingredients can be generalized to I_0-context raise their own interest.

2. Proof of Main Theorem

The proof of the main theorem in this section is due to W. Hugh Woodin, who sketched this approach in a conversation with Vincenzo Dimonte and the author happened the next day after John Steel raised the question. The presentation here is reconstructed by the author later in the summer of 2019 based on the recollection of that conversation. This section starts with a proof of Theorem 1.2, which is then lifted to prove its $I_0(\lambda)$, the main theorem.

2.1. *A proof of Theorem 1.2*

This proof of Theorem 1.2 needs the following ingredients.

Fact 1: *Assume* $\mathsf{ZF}+\mathsf{AD}$. *The following statements hold in* $L(\mathbb{R})$:

(i) (*Perfect Set Theorem, Mycielski [6]*) *Every set of reals has the perfect set property, i.e. either it is countable or it contains a perfect subset.*
(ii) (*Mycielski [6]*) *There is no ω_1-sequence of distinct reals.*
(iii) (*Solovay, see Kanamori [7]*) *ω_1 and ω_2 are measurable cardinals.*[a]

In the argument, the HOD-type inner models play the central role. We write OD_X (HOD_X) for the class of sets that are (hereditarily, respectively) ordinal definable with parameters from X.

Another key ingredient is the following generalized Vopěnka Theorem.

Fact 2: (Vopěnka, see Jech [8] and Woodin et al. [9]) *Suppose α is an ordinal, $P \subset \mathscr{P}(\alpha)$, X is a parameter. Then for every $A \in P$, $\mathsf{HOD}_{\{X,P,A\}}$ is a generic extension of $\mathsf{HOD}_{\{X,P\}}$ and the partial order $\mathbb{B}$ is "given" by the Boolean algebra of all subsets of P which are* OD *in parameter $\{X, P\}$.*

Here by "given" we mean to identify the Boolean algebra $\mathsf{OD}_{\{X,P\}} \cap \mathscr{P}(P)$ is with its isomorphic copy in $\mathsf{HOD}_{\{X,P\}}$ via an isomorphism in $\mathsf{OD}_{\{X,P\}}$. The original Vopěnka Theorem is stated under the assumption that $V = L[A]$. The author learned the generalized version from Hugh Woodin, who observed that Vopěnka's argument essentially proves this stronger result. The statement in Fact 2 is a variation of Theorem 2.4 in Woodin et al. [9] Theorem 2.4 in Woodin et al. [9] is stated without parameter, its proof there can be easily parameterized and adapted to prove Fact 2.

Proof of Theorem 1.2, method I. Assume towards a contradiction that $\bar{\sigma} = \langle \sigma_\alpha : \alpha < \omega_1 \rangle$ is a sequence of countable sets of reals, and $\sigma_\alpha \neq \sigma_\beta$ for $\alpha \neq \beta$. Let $\Sigma = \bigcup\{\sigma_\alpha \mid \alpha < \omega_1\}$. Work in $L(\mathbb{R})$. There are two cases:

Case I. Σ is countable. Then given an enumeration of Σ, for each $\alpha < \omega_1$, there is a unique $a_\alpha \subset \omega$ such that a_α to ω is the same as σ_α to Σ. Then $\langle a_\alpha : \alpha < \omega_1 \rangle$ is an ω_1-sequence of distinct reals, contradicting Mycielski's theorem, Fact 1(ii).

Case II. Σ is not countable. By Mycielski's Perfect Set Theorem (Fact 1(i)) there is a perfect tree T such that $[T] \subset \Sigma$. Here $[T]$ denote the set of cofinal branches of T. Treat T as a subset of ω. As the conversions between any two of $\mathbb{R}$, ${}^\omega\omega$ and ${}^\omega 2$ are all definable, we shall not distinguish these three sets. Let $\mathsf{HOD}^* = \mathsf{HOD}_{\{T,\bar{\sigma}\}}$, and $\mathsf{HOD}^*_r = \mathsf{HOD}_{\{T,\bar{\sigma},r\}}$ for any real $r \in L(\mathbb{R})$ (all computed in $L(\mathbb{R})$) We claim that

[a]For a cardinal $\kappa > \omega$, κ is measurable if there is a κ-complete ultrafilter over κ.

Claim 1: HOD* *computes cofinalities* $\geq \omega_1$*, and hence cardinals* $\geq \omega_1$*, correctly.*

Now grant this claim, we complete the argument as follows. By Fact 1(iii), ω_2 is measurable in $L(\mathbb{R})$. There is an ordinal definable measure μ on ω_2 (see Kanamori [10], Theorem 28.6). Let $\mu^* = \mu \cap \mathsf{HOD}^*$. Then $\mu^* \in \mathsf{HOD}^*$ and μ^* is an ultrafilter on ω_2 in HOD^*. As μ is ω_2-complete, μ^* is ω_2-complete in HOD^*. So ω_2 is measurable in HOD^*, in particular, it is a limit cardinal in HOD^*. But by definition ω_2 is a successor cardinal in $L(\mathbb{R})$. Contradiction!

All that is left is to prove the claim.

Proof of Claim 1. Identify the perfect tree T as an tree isomorphism $\pi' : 2^{<\omega} \to T$, then its completion $\pi : 2^\omega \to [T] \subset \Sigma = \bigcup_\alpha \sigma_\alpha$. So every real x is definable from T and some real in σ_{α_x}, where α_x is the least α such that $\pi(x) \in \sigma_\alpha$. Note that for any real x, x and $\pi(x)$ are mutually definable via T, so $\mathsf{HOD}^*_x = \mathsf{HOD}^*_{\pi(x)}$. To prove the claim, it suffices to show that cofinalities $\geq \omega_1^{L(\mathbb{R})}$ are correctly computed in HOD^*. Suppose γ is an ordinal and C is a cofinal subset of γ in $L(\mathbb{R})$ such that its ordertype $\mathrm{otp}(C) = \mathrm{cf}^{L(\mathbb{R})}(\gamma) \geq \omega_1^{L(\mathbb{R})}$. It suffices to find a $C^* \in \mathsf{HOD}^*$ such that $\sup C^* = \sup C = \gamma$ and $\mathsf{HOD}^* \models$ "$|C^*| = \mathrm{cf}^{L(\mathbb{R})}(\gamma)$".

C is OD from some real, say r, then $C \in \mathsf{HOD}^*_r$. Apply the generalized Vopěnka Theorem to parameters $\{T, \bar{\sigma}\}$ and $\pi(r) \in \sigma_{\alpha_r}$, we have that HOD^*_r is a generic extension of HOD^* via $\mathbb{B}_{\alpha_r}$, where $\mathbb{B}_{\alpha_r}$ is the Boolean algebra of all subsets of σ_{α_r} that are OD from σ_{α_r}. Note that the club filter Club_{ω_1} on $\omega_1^{L(\mathbb{R})}$ witnesses that $\omega_1^{L(\mathbb{R})}$ is measurable in $L(\mathbb{R})$. It is easy to check that $\mathrm{Club}_{\omega_1} \cap \mathsf{HOD}^*$ is in HOD^* and witnesses that $\omega_1^{L(\mathbb{R})}$ is measurable and hence strongly inaccessible in HOD^*. In HOD^*, $|\mathbb{B}_{\alpha_r}| \leq 2^{|\sigma_{\alpha_r}|} < \omega_1^{L(\mathbb{R})}$, therefore $\mathbb{B}_{\alpha_r}$ satisfies $\omega_1^{L(\mathbb{R})}$-chain condition. There is a function $F \in \mathsf{HOD}^*$ such that for each $\beta \in \mathrm{dom}(F) = \mathrm{otp}(C)$, $C(\beta) \in F(\beta) \subset \gamma$, where $C(\beta)$ denotes the β-th element of C, and $F(\beta)$ has size $\leq |\mathbb{B}_{\alpha_r}| < \omega_1^{L(\mathbb{R})}$ (in HOD^*). Let $C^* = \bigcup_{\beta \in \mathrm{dom}(F)} F(\beta)$. Then $C^* \in \mathsf{HOD}^*$, $C \subset C^*$, $\sup C^* = \sup C$ and $\mathrm{otp}(C) \leq \mathrm{otp}(C^*) \leq \mathrm{otp}(C) \cdot \omega_1^{L(\mathbb{R})}$, so $|C^*| = \mathrm{cf}^{L(\mathbb{R})}(\gamma)$ in HOD^*. $\dashv_{\text{(Claim 1)}}$

This completes the first proof of Theorem 1.2.

2.2. *Preparing for proving main theorem*

Before moving on to the proof of the main theorem, let us lay down necessary background I_0 theory. Axiom I_0 is the assertion $\exists \lambda I_0(\lambda)$, where $I_0(\lambda)$

states that there is an elementary embedding $j : L(V_{\lambda+1}) \prec L(V_{\lambda+1})$ with critical point $\mathrm{crit}(j) < \lambda$. Write $\kappa_0 = \mathrm{crit}(j)$. Let $\kappa_{n+1} = j(\kappa_n)$. Then $\lambda = \sup_{n<\omega} \kappa_n$ and it is the first fixed point of j beyond its critical point. Note that λ is a limit of measurable cardinals, hence is a strong limit. Let α-DC denote the following statement: Any $< \alpha$-closed trees of height α with no terminal nodes have cofinal branches. Write $<\alpha$-DC for $(\forall\beta < \alpha)(\beta$-DC$)$ and α-DC_X for α-DC restricted to trees (as binary relation) on X. In our notation, the principle of Dependent Choice DC is ω-DC.

Fact 3: (Woodin [2]) *Assume* ZFC+$I_0(\lambda)$. *The following statements hold in* $L(V_{\lambda+1})$*:*

(i) $<\lambda^+$-DC.
(ii) λ^+ *is measurable.*

To run the argument for Theorem 1.2 in the $I_0(\lambda)$ context, it seems that one also needs λ^{++} to be measurable in $L(V_{\lambda+1})$. For that, let's see how λ^+ is shown to be measurable in $L(V_{\lambda+1})$.

For cardinal $\kappa > \omega$, let $S^\kappa_\omega =_{\mathrm{def}} \{\alpha < \delta \mid \mathrm{cf}(\alpha) = \omega\}$. The ultrafilter needed for Fact 3(ii) is obtained as follows:

Theorem 2.1: (Woodin [2]) *Assume* ZFC+$I_0(\lambda)$. *Let j ba an elementary embedding witnessing $I_0(\lambda)$. In $L(V_{\lambda+1})$, for every regular cardinal κ such that $\lambda < \kappa < \Theta^\lambda$ and $j(\kappa) = \kappa$, there is a $\delta < \mathrm{crit}(j)$ and a sequence $\langle S_\alpha : \alpha < \delta\rangle$ of mutually disjoint stationary subsets of S^κ_ω such that on each S_α, the club filter on κ can not be split into two disjoint stationary sets.*[b]

Therefore in $L(V_{\lambda+1})$, on each S_α, the club filter focusing on S_α, $\mathrm{Club}_\kappa(S_\alpha) =_{\mathrm{def}} \{A \subset \kappa \mid S_\alpha \setminus A \text{ is nonstationary}\}$, is an ultrafilter. On the other hand, Fact 3(i) provides enough choice to prove the λ^+-completeness of the filter $\mathrm{Club}_{\lambda^+}(S_\alpha)$. Therefore λ^+ is measurable in $L(V_{\lambda+1})$.

But it is not known whether one can have $<\lambda^{++}$-DC (in $L(V_{\lambda+1})$), without which we do not know how to show that λ^{++} is regular, do not know how to show the filter of the form $\mathrm{Club}_{\lambda^{++}}(S_\alpha)$ is an ultrafilter, let alone how to show the λ^{++}-completeness of these particular club filters on λ^{++}. Fortunately, as we shall see later, for this particular problem, there are ways to get around this.

Besides the measurability of λ^+, the analogue of perfect set theorem also holds in $L(V_{\lambda+1})$. To state it, we need the following definitions.

---

[b]The same result holds for every cofinality $< \kappa$.

Definition 2.2: *Let λ be a singular cardinal of countable cofinality.*

- *Let $\bar{\gamma} = \langle \gamma_n : n < \omega \rangle$ be nondecreasing and such that $\lambda = \sup_n \gamma_n$. A* $\bar{\gamma}$-splitting tree *is a subtree of $\lambda^{<\omega}$ such that every node t on the tree has $\gamma_{|t|}$ many immediate successors.*
- *A tree is* λ-splitting *if it is $\bar{\gamma}$-splitting for some $\bar{\gamma}$.*
- *A set $P \subset \lambda^\omega$ is λ-perfect if $P = [T]$ for some λ-splitting tree.*

Theorem 2.3: (λ-Perfect Set Theorem) (Cramer [5], see also Shi [11]) *Assume $\mathsf{ZFC} + I_0(\lambda)$. In $L(V_{\lambda+1})$, every subset of $\mathscr{P}(\lambda)$ has λ-perfect set property, i.e. it is either of size $\leq\lambda$ or contains a λ-perfect subset.*

As mentioned in Introduction (§1), using either Fact 3(ii) or Theorem 2.3, one can prove that

Theorem 2.4: (Shi-Trang [1]) *Assume $\mathsf{ZFC} + I_0(\lambda)$. In $L(V_{\lambda+1})$ there is no λ^+-sequence of distinct subsets of λ.*

We proceed to prove the main theorem.

2.3. *Proof of main theorem*

Work in $L(V_{\lambda+1})$. Assume towards a contradiction that $\bar{\sigma} = \langle \sigma_\alpha : \alpha < \lambda^+ \rangle$ is such a sequence. Let $\Sigma = \bigcup\{\sigma_\alpha \mid \alpha < \lambda^+\}$. Argue as before, case I leads to a contradiction with Theorem 2.4. We argue for case II.

Fix a sequence of measurable cardinals $\langle \kappa_n : n < \omega \rangle$ such that $\lambda = \sup_n \kappa_n$. By Cramer's λ-Perfect Set Theorem, there is a λ-perfect tree T such that $[T] \subset \Sigma$. Identify T as a tree embedding map from $\bigcup_{n<\omega} T_n$ into ${}^{<\omega}\lambda$, where $T_0 = \{\varnothing\}$ and for $n > 0$, $T_n = \prod_{i<n} \kappa_i$, and $\bigcup_{n<\omega} T_n$ and ${}^{<\omega}\lambda$ are ordered by end-extension. The completion of this function gives an embedding of $\prod_{n<\omega} \kappa_n$, the set of all cofinal branches of $\bigcup_n T_n$, into $[{}^{\omega}\lambda]$. This T can be coded by a subset of λ.

Since λ is a strong limit cardinal, and AC holds below λ, there is a wellordering Ξ of V_λ (viewed as a mapping $\Xi : V_\lambda \to \mathsf{Ord}$) such that $\{\Xi(u) \mid u \subset V_{\kappa_i}\} \subset \kappa_{i+1}$ for $i < \omega$. This Ξ has ordertype λ, it can also be coded by a subset of λ. Using such an wellordering of V_λ, one can define bijections between any two of $V_{\lambda+1}$, $\mathscr{P}(\lambda)$, ${}^{\omega}\lambda$ and $\prod_{n<\omega} \kappa_n$.[c] For the notational convenience, we shall not distinguish them.

---

[c]One can first define injections from each of $V_{\lambda+1}$, $\mathscr{P}(\lambda)$, ${}^{\omega}\lambda$ into $\prod_{n<\omega}[\kappa_n, \kappa_{n+1})$, and hence into $\prod_{n<\omega} \kappa_n$: for instance, given $x \in V_{\lambda+1}$, define $\pi(x) = \langle \kappa_i + \Xi(x \cap V_{\kappa_i}) : i < \omega \rangle$, where the addition is the ordinal addition. And then use the injection of the reverse direction to construct the bijection.

Let R be a stationary subset of $S^{\lambda^+}_\omega$ such that $\text{Club}_{\lambda^+}(R)$ witnesses λ^+ is measurable in $L(V_{\lambda+1})$. Let $r \in V_{\lambda+1}$ be such that $R \in \mathsf{OD}_{\{\Xi,r\}}$. Thus $\text{Club}_{\lambda^+}(R) \in \mathsf{OD}_{\{\Xi,r\}}$. Write HOD^* for $\mathsf{HOD}^{L(V_{\lambda+1})}_{\{\Xi,T,\bar\sigma,r\}}$ and HOD^*_p for $\mathsf{HOD}_{\{\Xi,T,\bar\sigma,r,p\}}$, with p as a parameter.

The completion of T induces a copy of $V_{\lambda+1}$ inside Σ, thus every member of $V_{\lambda+1}$ is (ordinal) definable from Ξ, T and a member of σ_α for some $\alpha < \lambda^+$. By Fact 2, for each $A \in \mathscr{P}(\lambda)$, $\mathsf{HOD}^*_A = \mathsf{HOD}^*[G]$, where G is a generic filter for the Boolean algebra $\mathbb{B}_{\alpha_A}$ of all subsets of σ_{α_A} that are definable from ordinals and parameters in $\{\Xi, T, \bar\sigma, r\}$. Argue as in the proof of Claim 1, then as every $\mathbb{B}_\alpha$ ($\alpha < \lambda^+$) has size $< \lambda^+$ in HOD^*, we have

Claim 2: HOD^* *computes cofinality* $\geq \lambda^+$, *and hence cardinals* $\geq \lambda^+$ *correctly.*

And moreover,

Claim 3: *For every regular cardinal* $\kappa \geq \lambda^+$, *every closed unbounded subset of* κ *contains a closed unbounded subset in* HOD^*.

Proof of Claim 3. Let C be a closed unbounded subset of κ in $L(V_{\lambda+1})$. Write $C(\alpha)$, $\alpha < \kappa$, to denote its α-th elements in increasing order. Argue as in the cofinality-preserving argument for Claim 2 (see the proof of Claim 1), via a λ^+-cc Boolean algebra over HOD^*, there is a function $F \in \mathsf{HOD}^*$ such that for each $\alpha < \kappa$, $C(\alpha) \in F(\alpha) \subset \mathsf{Ord}$ and $F(\alpha)$ has size $< \lambda^+$ in HOD^*. The C^* for Claim 2 is $\bigcup_\alpha F(\alpha)$. Now we use the regularity of κ to get a $C^\dagger \in \mathsf{HOD}^*$ such that $C^\dagger \subset C$. Use the regularity of κ to define a function $h : \kappa \to \mathsf{Ord}$ as follows: $h(0) = 0$, and for $\alpha < \kappa$, $h(\alpha) = \sup \bigcup_{\beta<\alpha} F(h(\beta))$. Let $C^\dagger$ be the set of limit points of h, i.e. $C^\dagger = \{\sup_{\beta<\alpha} h(\beta) \mid \alpha < \kappa,\ \alpha \text{ is a limit ordinal}\}$. Clearly $C^\dagger$ is closed unbounded and is in HOD^*. We check that $C^\dagger \subset C$. Let $\alpha < \kappa$ be a limit ordinal. For $\beta < \alpha$, $C(h(\beta)) \in F(h(\beta))$ so $h(\beta) \leq C(h(\beta)) \leq h(\beta+1) \leq C(h(\beta+1))$, so $\sup_{\beta<\alpha} h(\beta) = \sup_{\beta<\alpha} C(h(\beta)) \in C$. $\dashv_{(\text{Claim 3})}$

Claim 2 implies that λ^{++} is regular in $L(V_{\lambda+1})$. By Theorem 2.1, there is a stationary $S \subset S^{\lambda^{++}}_\omega$ such that $\text{Club}_{\lambda^{++}}(S)$ is an ultrafilter in $L(V_{\lambda+1})$. Claim 3 implies that the club filter on λ^{++} is λ^{++}-complete, therefore λ^{++} is measurable in $L(V_{\lambda+1})$. S is in $\mathsf{OD}_{\{\Xi,t\}}$ for some $t \in \mathscr{P}(\lambda)$, so $S \in \mathsf{HOD}^*_t$. As a model intermediate between HOD^* and $L(V_{\lambda+1})$, both Claim 2 and Claim 3 hold for HOD^*_t. Let $\mu^* = \text{Club}_{\lambda^{++}}(S) \cap \mathsf{HOD}^*_t$. Then $\mu^* = (\text{Club}_{\lambda^{++}}(S))^{\mathsf{HOD}^*_t}$ is an ultrafilter in HOD^*_t, and as $\text{Club}_{\lambda^{++}}(S)$ is λ^{++}-complete, μ^* is λ^{++}-complete in HOD^*_t. Thus λ^{++} is measurable, hence a limit cardinal, in HOD^*_t. Contradiction!

3. Two Other Proofs of Theorem 1.2

We have two other ways to prove Theorem 1.2. In this section, we discuss these two approaches, although we haven't managed to adapt them to prove the main theorem, the question how much the relevant ingredients occurred in these two arguments can be generalized to $I_0(\lambda)$-context raises independent interest. The analyses of these approaches can help the readers have a better understanding of the subtleties of the proof in the previous section, and at the meantime can shed some light on future studies of the analogy/difference between the theory of $L(\mathbb{R})$ under $\mathsf{AD}^{L(\mathbb{R})}$ and that of $L(V_{\lambda+1})$ under $I_0(\lambda)$.

3.1. *Proof of Theorem 1.2, method II*

The second proof of Theorem 1.2 (in fact, of Theorem 1.1) is a folklore. Here we sketch the argument so that we can discuss the obstacles to generalize it to $I_0(\lambda)$ context.

Let $\bar{\sigma}$ and Σ be as in the first proof of Theorem 1.2. The case Σ is countable is the same. Consider the case Σ is not countable. Let T be the perfect tree with $[T] \subset \Sigma$, and $\pi : 2^\omega \to [T]$ be a bijection. For each $r \in 2^\omega$, define α_r = least α such that $\pi(r) \in \sigma_\alpha$. This defines a prewellordering $\preccurlyeq$ of 2^ω of length ω_1 and it has the property that every initial segment $\{r \mid \alpha_r \leq \alpha\}$, $\alpha < \omega_1$, is countable. By the *Baire Property Theorem*,[d] $\preccurlyeq$ has the Baire property. The contradiction follows from *Kuratowski-Ulam Theorem*,[e]

- $(\forall^* y)\ \{x \mid x \preccurlyeq y\}$ is countable $\implies$ $\preccurlyeq$ is meager;
- $(\forall^* x)\ \{y \mid x \preccurlyeq y\}$ is co-countable $\implies$ $\preccurlyeq$ is comeager.

Here $\forall^* y$ abbreviates "for comeager many y". This concludes the second proof.

This argument proves both Theorem 1.1 and Theorem 1.2. One can also use the Lebesgue Measurability Theorem and Fubini Theorem in place of the Baire Property Theorem and Kuratowski-Ulam Theorem to run the same arguments. Notice that the Perfect Set Theorem, Baire Property Theorem and Lebesgue Measurability Theorem (as well as DC) all hold in the

[d] Every set of reals has the property of Baire.

[e] For any $E \subset \mathbb{R} \times \mathbb{R}$ with the property of Baire, the following are equivalent:

(1) E is meager (or comeager) in the product topology.
(2) $\forall^* x \in \mathbb{R}$, $\{y \mid \langle x, y\rangle \in E\}$ is meager (resp. comeager).
(3) $\forall^* y \in \mathbb{R}$, $\{x \mid \langle x, y\rangle \in E\}$ is meager (resp. comeager).

classical Solovay's model (see Solovay [12]), by this argument the conclusion is also true in Solovay's model.

Now let us discuss the generalization. The key elements of this argument are the Perfect Set Theorem, the Baire Property Theorem and Kuratowski Theorem. As mentioned before, we have the λ-version of Perfect Set Theorem under $I_0(\lambda)$. The real obstacle to lift the second argument to λ is the two theorems regarding sets with Baire property. In fact, we are still looking for a suitable version of Baire Category Theorem for subsets of $V_{\lambda+1}$, which is necessary for defining an appropriate notion of the property of Baire for subsets of $V_{\lambda+1}$. From forcing point of view, the classical Baire Category Theorem says that for any countable collection $\mathcal{D}$ of dense open subsets of the Cohen poset (over $\mathbb{R}$), there is a real that is $\mathcal{D}$-generic. The generic absoluteness theorem below gives some sort of Baire category theorem for subsets of $V_{\lambda+1}$.

Theorem 3.1: (Woodin [4], Cramer [13]) *Assume* ZFC+$I_0(\lambda)$. *Suppose j witness $I_0(\lambda)$ and M_ω is the ω-th j-iterate of V_λ. Suppose $\mathbb{P}$ is a λ-good poset in M_ω. Then there is a $G \in V$ that is $\mathbb{P}$-generic over M_ω. And furthermore, if $\mathrm{cf}(\lambda)^{M_\omega[G]} = \omega$ and $\alpha < \Theta^\lambda$ is good, then for some $\alpha < \lambda$, $L_{\bar{\alpha}}(V_{\lambda+1} \cap M_\omega[G]) \prec L_\alpha(V_{\lambda+1})$.*

But the genericity only involves dense open sets that are in M_ω (or in $M_\omega[g]$, where g is $\mathbb{P}$-generic over M_ω and $\mathbb{P}$ is some λ-good poset). This seems still far way from getting a λ-version Baire Property Theorem.

3.2. *Proof of Theorem 1.2, method III*

For the third method, we would like to thank William Chan for sharing his observation (Fact 4). A straightforward approach to prove Theorem 1.2 is to reduce it to Myscielski's Theorem (see Fact 1(ii)), namely, turning an ω_1-sequence of distinct countable sets of reals into an ω_1-sequence of distinct reals. In a choice context, with a wellordering of reals, one can uniformly code every countable set of reals in the sequence by a real to achieve this. In the AD context, although the reals are not wellorderable, one can still achieve that. But the argument is highly nontrivial.

In the 90's, Woodin proved the following dichotomy theorem for equivalence relations on reals.

Theorem 3.2: (Woodin, see Fact 6.2 in Chan [14]) *Assume* ZF + AD + $\mathsf{DC}_{\mathbb{R}}$ *and every set of reals has an ∞-Borel code. Let E be an* *equivalence relation on* $\mathbb{R}$*, with an* ∞*-Borel code* (S, φ)*. Then exactly one of the following holds.*

(1) $\mathbb{R}/E$ *is wellorderable.*
(2) *There is an injection* $\mathbb{R}$ *into* $\mathbb{R}/E$*.*

Moreover, if $\mathbb{R}/E$ *is wellorderable, then there is a uniform procedure to find a* $\mathsf{OD}_{\{S\}}^{L(S,\mathbb{R})}$ *wellordering of* $\mathbb{R}/E$*.*

For a set X, let $\mathscr{P}_{\mathrm{WO}}(X) =_{\mathrm{def}} \{A \subset X \mid A \text{ is wellorderable}\}$. Based on the uniformity in the theorem above, Chan made the following observation:

Fact 4: (Fact 6.4 in Chan [14]) *Assume* $\mathsf{ZF} + \mathsf{AD} + \mathsf{DC}_{\mathbb{R}}$ *and every set of reals has an* ∞*-Borel code. Let* $\kappa < \Theta$ *and* E *be an equivalence relation on* $\mathbb{R}$*. Suppose* $\Phi : \kappa \to \mathscr{P}_{\mathrm{WO}}(\mathbb{R}/E)$ *is a function. Then there is a sequence* $\langle \prec_\alpha : \alpha < \kappa \rangle$ *such that each* $\prec_\alpha$ *is a wellordering of* $\Phi(\alpha)$*.*

Now let $\bar{\sigma} = \langle \sigma_\alpha : \alpha < \omega_1 \rangle$ be a sequence of distinct wellorderable sets of reals. (Every countable set of reals is wellorderable.) Apply Chan's observation to the case E is the identity relation. ($\mathbb{R}/E$ is essentially $\mathbb{R}$.) There is a sequence $\Gamma = \langle \prec_\alpha : \alpha < \omega_1 \rangle$ such that each $\prec_\alpha$ wellorders σ_α. Via this Γ, one immediately has the following version of Theorem 1.2.

Theorem 1.2* *Assume* $\mathsf{ZF} + \mathsf{AD} + \mathsf{DC}_{\mathbb{R}}$ *and every set of reals has an* ∞*-Borel code. Then there is no* ω_1*-sequence of distinct wellorderable sets of reals.*

[The case $\Sigma = \bigcup_\alpha \sigma_\alpha$ is countable goes as before. For the case that Σ is not countable, order elements in Σ via Γ as follows, for each $x \in \Sigma$, let α_x be the least α such that $x \in \sigma_\alpha$, and β_x be the index of x in the $\Gamma(\alpha) = \prec_{\alpha_x}$ enumeration of σ_{α_x}. Comparing elements of Σ via $\langle \alpha_x, \beta_x \rangle$ lexicographically gives a wellordering of Σ. This reduces $\bar{\sigma}$ to an uncountable sequence of distinct reals. Contradiction!]

By Woodin's theorem that $\mathsf{AD}^{L(\mathbb{R})}$ implies that every set of reals in $L(\mathbb{R})$ has an ∞-Borel code, Theorem 1.2 follows as an immediate corollary.

Now let us examine the argument to see if it can be adapted to show the I_0-analogue. Chan's observation is a straightforward ordinal definability computation based on the uniform procedure mentioned in the dichotomy theorem. It is routine to generalize this part to obtain its λ-counterpart. However to run the third method for the main theorem, there are two more issues one has to deal with:

(1) The availability of the λ-version of ∞-Borel codes for subsets of $V_{\lambda+1}$ in $L(V_{\lambda+1})$ under $I_0(\lambda)$.
(2) Generalization of the dichotomy in Theorem 3.2 to $L(V_{\lambda+1})$ under $I_0(\lambda)$.

These two are in fact much more fundamental problems for the structure of $L(V_{\lambda+1})$.

In the search of the λ-analogue of ∞-Borel code for subsets of $V_{\lambda+1}$, there is a step which seems to be a serious obstruction for running the whole argument. To prove Woodin's theorem that $\mathsf{AD}^{L(\mathbb{R})}$ implies that every set of reals in $L(\mathbb{R})$ has an ∞-Borel code, an important step is to take ultrapowers of local HODs with respect to Martin's measure of Turing degrees (or its variations). Unfortunately, its λ-counterpart is not available under $I_0(\lambda)$. Let $\leq_T^\lambda$ denote the λ-analogue of Turing reduction. A cone of λ-Turing degrees is a subset of $V_{\lambda+1}$ of the form $\{X \in V_{\lambda+1} \mid B \leq_T^\lambda X\}$ for some $B \in V_{\lambda+1}$. Let λ-TD be the assertion that every degree invariant subset of $V_{\lambda+1}$ either contains or is disjoint from a cone of λ-Turing degrees. According to Theorem 6.2 in Shi [11], it is relatively consistent with ZFC + $I_0(\lambda)$ that λ-TD is false in $L(V_{\lambda+1})$. In fact this holds for λ-analogue of any reasonable definability degree notion. So one can not expect to find a λ-analogue of Martin's measure from $I_0(\lambda)$. It is not clear whether there is an alternative measure over $V_{\lambda+1}$ that plays the rule of Martin's measure, or whether there is a different argument to get around this obstacle.

It might be a long shot to prove the main theorem via either the λ-Baire property or the λ-dichotomy property. But the two problems — to what extents these two properties can be generalized — each leads to a much bigger project than the main result in this chapter.

Acknowledgement

The credit of the main result goes to W. Hugh Woodin, who sketched the argument to the author during the 4th Arctic Set Theory Workshop. The author also likes to thank John Steel for raising the question and William Chan for inspiring conversations. He is also indebted to the anonymous reviewer, whose report helps improve the quality of this chapter. Part of this notes was written during the author's participation in the Higher Recursion Theory and Set Theory Conference, at the Institute of Mathematical Sciences, National University of Singapore, in July 2019. He believes that the inclusion of this notes in this proceedings fits well with the theme of the conference — to honor Slaman and Woodin's contributions to recursion theory and set theory. He would like to thank IMS for the support of his visits. The author is also partially supported by NSFC (No. 11771050).

References

1. X. Shi and N. Trang, I_0 and combinatorics at λ^+, *Arch. Math. Logic* **56**, 131 (2017).
2. W. H. Woodin, *Notes on an AD-like axiom* (July 6, 1990). Seminar notes taken by George Kafkoulis.
3. W. H. Woodin, Suitable extender models I, *Journal of Mathematical Logic* **10**, 101 (2010).
4. W. H. Woodin, Suitable extender models II: beyond ω-huge, *Journal of Mathematical Logic* **11**, 115 (2011).
5. S. S. Cramer, Inverse limit reflection and the structure of $L(V_{\lambda+1})$, *J. Math. Log.* **15**, p. 1550001 (38 pages) (2015).
6. J. Mycielski, On the axiom of determinateness, *Fund. Math.* **53**, 205 (1963/1964).
7. A. Kanamori, *The higher infinite*, Springer Monographs in Mathematics, second edn. (Springer-Verlag, Berlin, 2009), Large cardinals in set theory from their beginnings, Paperback reprint of the 2003 edition.
8. T. Jech, *Set theory*, Springer Monographs in Mathematics (Springer-Verlag, Berlin, 2003), The third millennium edition, revised and expanded.
9. W. H. Woodin, J. Davis and D. Rodríguez, The HOD dichotomy, in *Appalachian set theory 2006–2012*, London Math. Soc. Lecture Note Ser. Vol. 406 (Cambridge Univ. Press, Cambridge, 2013) pp. 397–418.
10. A. Kanamori, *The higher infinite*, Springer Monographs in Mathematics, second edn. (Springer-Verlag, Berlin, 2003), Large cardinals in set theory from their beginnings.
11. X. Shi, Axiom I_0 and higher degree theory, *J. Symb. Log.* **80**, 970 (2015).
12. R. M. Solovay, A model of set-theory in which every set of reals is Lebesgue measurable, *Ann. of Math.* **92**, 1 (1970).
13. S. S. Cramer, Generic absoluteness from tree representations in $L(V_{\lambda+1})$ cardinals, Preprint (2015).
14. W. Chan and S. C. Jackson, Applications of infinity-borel codes to definability and definable cardinals, Preprint, URL: `https://williamchan-math.github.io/papers/applications_infinity_borel_code.pdf` (October 12, 2019).

https://doi.org/10.1142/9789819806584_0012

Type Omitting Theorems for Fragments of Second Order Logics: Classical and Modal

Richard A. Shore

Department of Mathematics, Cornell University
Ithaca NY 14853, USA
shore@math.cornell.edu

Type omitting arguments provide conditions under which one can find a model $\mathcal{M}$ of a given theory T which does not have elements $\bar{a}$ which satisfy all the formulas $\varphi(\bar{x})$ in a given type Γ, i.e. a specified collection of formulas. Such theorems are an important feature of first order logic and have been proven for a few more general logics. Harrington, Shore and Slaman [2017] proved a theorem about Σ_1^1 sets which implied such results for omega logic (a classical result) and computable infinitary logic as well as other classical theorems of hyperarithmetic theory. In this chapter, we show that the same theorem implies analogous type omitting theorems for many other logics including a wide range of fragments of second order logic both classical and modal. That is, for the modal logics we are also allowed to use any one of these fragments of second order logic at each world.

1. Introduction

This chapter grew out of research that began with Ted Slaman during a meeting at the IMS in Singapore. That work later involved Leo Harrington and led to the joint paper Harrington, Shore and Slaman [2017] (hereafter HSS). We spoke about many of the applications of that work presented here at a later meeting at the IMS (Higher Recursion Theory and Set Theory) held during May and June of 2019 in celebration of the work of both Ted

Partially supported by NSF Grant DMS-1161175. This chapter grew out of research begun at meetings held at, and supported by, the Institute of Mathematical Sciences at the National University of Singapore.

Slaman and Hugh Woodin on the occasion of their sixty-fifth birthdays. We are pleased to present this chapter in their honor.

In classical first order logic we begin with a language $\mathcal{L}$ consisting of (logical) symbols common to all first order languages and then a specified set of (nonlogical) relation, function and constant symbols for this particular language. As usual, one next inductively defines the formulas and sentences of $\mathcal{L}$. Then one defines a semantics for $\mathcal{L}$ by first specifying the structures $\mathcal{M}$ for it as sets M with interpretations of its nonlogical symbols and then defining satisfaction, $\mathcal{M} \models \varphi$ for sentences of $\mathcal{L}$ (or, as we prefer, for sentences in a language extending $\mathcal{L}$ by adding constant symbols, say c_a, for every $a \in M$). We will here consider only countable languages $\mathcal{L}$ (whether first order or more general). For our purposes, we may as well also assume that the basic language and so formulas, etc. are recursive as otherwise we can just relativize our results to the language. Note that, for first order languages, the relation $\mathcal{M} \models \varphi$ for countable structures is arithmetic in $\mathcal{M}$, i.e. in the list of atomic sentences (with names for its elements) true in $\mathcal{M}$ uniformly in $\mathcal{M}$ and φ. (Of course, the complexity of the arithmetic definition varies with that of φ.) A set T of sentences of $\mathcal{L}$ is a theory if it has a model, i.e. an $\mathcal{L}$ structure $\mathcal{M}$ such that $\mathcal{M} \models \varphi$ for every $\varphi \in T$. (For first order logic, one also has a proof theory and a completeness theorem establishing the equivalence of having a model with being consistent.)

For a theory T in a language $\mathcal{L}$, classically one defines an *n-type of $\mathcal{L}$* over T to be a set p of formulas of $\mathcal{L}$ each having free variables precisely $x_1, \ldots, x_n$ which is consistent with T (i.e. realized in some model of T in the sense about to be defined) and complete, i.e. maximal consistent. We say p is *realized in a model $\mathcal{M}$ of T* if there is an $\bar{a} = a_1, \ldots, a_n$ with each $a_i \in M$ such that $\mathcal{M} \models \varphi(\bar{a})$ for every $\varphi \in p$. Otherwise, we say that p is *omitted in $\mathcal{M}$*. By a type omitting theorem we mean one that gives conditions on T and p that guarantee that there is a model $\mathcal{M}$ of T that omits p.

The classical type omitting theorem of first order logic says (in one semantically oriented terminology) that if, for every formula $\varphi(\bar{x})$ of $\mathcal{L}$ consistent with T (i.e. there is a model of T satisfying $\exists \bar{x} \varphi(\bar{x})$), there is a $\gamma \in p$ such that T such that $\varphi \& \neg \gamma$ is consistent with T, then there is a model $\mathcal{M}$ of T which omits p. There are a number of standard variations and generalizations. The classical result (for countable first order languages) follows from compactness and can also be proven by direct applications of the constructions such as that of Henkin used to prove completeness. (For the classical versions see e.g. Chang and Keisler [1990] or Hodges [1993].)

We are interested in type omitting theorems for logics more general than first order that may not have compactness or completeness theorems. The classical example is the Gandy-Kreisel-Tait theorem for ω-logic (Gandy, Kreisel and Tait [1960] (hereafter GKT), Theorem 1). Other known examples involve admissible sets, (computable) infinitary logic or the like as in Barwise [1975] and Montalbán [2021]. These examples often have notions of proof (with infinitary rules) and appropriate completeness theorems but not always the usual compactness theorems. While some of these infinitary languages are also treated in HSS, we ignore them here and begin our analysis by recalling the classical theorem for ω-logic.

The definition of ω-logic begins with two (or many, even infinitely many) sorted logic. Here, in addition to the usual syntactic apparatus of first order logic, we have additional kinds of variables x^i in the language for each sort i. In addition, $\mathcal{L}$ may have function and relation symbols ($f^{i,n}$ and $R^{i,n}$) of any arity n whose domain and range are restricted to the sort i. The structures $\mathcal{M}$ for $\mathcal{L}$ have designated subsets N_i for the sorts i which, together with the interpretations of $f^{i,n}$ and $R^{i,n}$ as functions and relations on N_i, form structures $\mathcal{N}_i$. We may write this as $\langle \mathcal{M}, \mathcal{N}_i \rangle$ or the like. Of course, from one point of view this is really just a notational extension of standard first order logic in which one has distinguished unary predicates whose interpretation are taken to be the N_i and one does the appropriate translation to express that the domain or range of a function (relation) may be an N_i. Nonetheless, this approach often has the benefit of providing a more natural presentation of classes of structures such as vector spaces in which one sort is the underlying field and the other the vectors.

The true power and interest of these logics arise when one requires, for example, (some of) the $\mathcal{N}_i$ to be specific first order structures. We call this an $\mathcal{N}$-logic. The classical example of ω-logic has a unary function on N_1 and requires that the associated $\mathcal{N}_1$ is isomorphic to ω with successor. Alternatively, one can include the usual functions and relations of first order arithmetic on $\mathcal{N}_1$ and require of the structures that it is isomorphic to the standard natural numbers $\mathbb{N}$. The structures for such a sorted language in which $\mathcal{N}$ is isomorphic to the standard structure $\mathbb{N}$ are also called ω-models (although $\mathbb{N}$-model might be better in the latter version). One may then consider a language in which there is an $\in$ relation $n \in A$ between members n of N_1 and A of $M - N_1$ (the sets). With some axioms including extensionality for the sets, this gives structures for (varieties of) second order arithmetic. One can extend further and also have other types $\mathcal{N}_i$

($i > 1$) with membership relations $\in$ between members of N_i and N_{i+1} ($\mathcal{N}_1 \cong \mathbb{N}$). This last version of models of arithmetic of all finite orders with the first bottom sort assumed to be isomorphic to $\mathbb{N}$ is the setting for the primary result of GKT. Another natural example continuing on with vector spaces is requiring that the field be $\mathbb{Q}$.

What GKT explicitly state as their Theorem 1 is that if one has any Π^1_1 set $\mathfrak{A}$ of axioms in the language of ω-logic, then any set $B \subseteq \mathbb{N}$ appearing in every ω-model of $\mathfrak{A}$ is hyperarithmetic (i.e. Δ^1_1). (By $\mathfrak{A}$ being Π^1_1 we mean that the set of number codes for the sentences in $\mathfrak{A}$ is a Π^1_1 set of numbers, i.e. of the form $\{n|(\forall F)\exists x R(F, x, n)\}$ where R is a recursive (or equivalently arithmetic) relation on $\mathbb{N}^{\mathbb{N}} \times \mathbb{N} \times \mathbb{N}$. Remember that we are assuming our language is recursive and so then is the set of Gödel numbers for the formulas etc.) We can see this result as a very specific form of a type omitting argument where the type is one of a variable X over the sort of sets of natural numbers of the form $\{n \in X | n \in B\} \cup \{\neg(n \in X) | n \notin B\}$ for some $B \subseteq \mathbb{N}$. This type is omitted in some ω-model of $\mathfrak{A}$ as long as B is not hyperarithmetic and $\mathfrak{A}$ is a Π^1_1 set of sentences in the extended language. Many strengthenings and extensions of GKT can be found in Grilliot [1972].

We now want to state similar but stronger type omitting theorems in settings as described above that will include also many common (and some uncommon) subsystems of (weak) second order logic as well as logics like ω-logic.

2. $\mathcal{N}$-Logics and Subsystems of Weak Second Order Logic

Many common subsystems of second order logic are (or can be) defined by adding on variables and quantifiers that range over finite subsets of, or relations on, the domain of a structure. Examples include the following:

(1) Weak monadic second order logic has new variables X_i ranging over finite subsets of the domain and a syntax that views them as unary predicates and a semantics that interprets $X_i(t)$ as saying that the interpretation of a term t is a member of X_i.

(2) Weak second order logic has new variables $R^{i,n}$ that range over finite relations of arity n on the domain and a syntax that allows us to write $R^{i,n}(t_1, \ldots, t_n)$ for terms of the language and interprets this formula as saying the relation holds of the interpretation of the terms.

(3) Cardinality logic (for $\aleph_0$) has a new quantifier Q_0 where $Q_0x\varphi(x)$ in interpreted to mean there are infinitely many a such that $\varphi(a)$ is true in the structure.
(4) Ancestral logic allows one to talk about the transitive closure of any definable relation by adding a new operator (quantifier) TC and making, for each formula $\varphi(x,y)$, $TC_{x,y}\varphi(x,y)(u,v)$ into a formula with free variables u and v (and x and y bound) where $TC_{x,y}\varphi(x,y)(a,b)$ holds if there is a finite sequence $a = c_0 = \cdots = c_k = b$ such that $\varphi(c_i, c_{i+1})$ holds for every $i < k$.

Clearly, we can combine these fragments of second order logic with multisorted logics by allowing restrictions of the new types of variables to the sorts in the natural way. (One could also interpret them as multisorted logics in which some sorts are required to be, e.g. the collection of finite subsets of, or relations on, some other sort.)

To the extent that formalizations are desired for the syntax and semantics of these languages, they are straightforward and left to the reader. (One can also describe proof systems and completeness theorems for these systems but in general, like ω-logic, they require infinitely branching proofs and determining if the proof tree is well founded. Thus provability is often at the Π^1_1 level rather than Σ^0_1.) We do, however, want to point out two crucial facts. The first is, we think, fairly well-known to those who work with such logics: They all have the downward Skolem-Löwenheim property. In particular, if $\mathcal{M}$ is a structure for any of them (in the appropriate sense) then it has a countable elementary substructure (in the same sense). Note that this relies on the assumption that the only sorts that we specify up to isomorphism are required to be countable. Otherwise, the proof is of the same kind as for first order logic. We use this fact below just to equate a theory having a model with its having a countable model. The second even more crucial fact, is that the property of a countable sequence of countable sets with countable sequences of functions and relations (on some of them) being a structure for one of these logics is Σ^1_1 in $\mathcal{M}$ and the list of countable structures $\mathcal{N}_j$ specified as being the sort $\mathcal{M}_{i_j}$ with its specific sublanguage. The only requirement that pushes the definition beyond recursive conjunctions of arithmetic conditions are those that each $\mathcal{M}_{i_j}$ (with designated functions and relations from the list) is isomorphic to a given $\mathcal{N}_j$. This is clearly a Σ^1_1 predicate. Everything else, such as some function or relation being a k-ary one on some domain or some variables ranging over the finite subsets of, or relations on, a particular domain, is clearly arithmetic. Given that $\mathcal{M}$ is a countable structure for one of these languages in the appropriate sense, the satisfaction relation $\mathcal{M} \models \varphi(a_1, \ldots, a_n)$ (for $a_i \in M$) is, uniformly in $\mathcal{M}$ and φ, clearly arithmetic in $\mathcal{M}$ and so the full satisfaction relation is recursive in $\mathcal{M}^{(\omega)}$ the effective join of the nth Turing jumps of $\mathcal{M}$ for $n \in \mathbb{N}$.

For the rest of this section, $\mathcal{L}$ will be a (recursive) multisorted language as above and we will fix a semantics for it extending the usual first order one by interpreting the new variables and quantifiers in the intended manner and imposing restrictions on the acceptable interpretations of the sorts as in the examples. To be definite we use the sorts $\mathcal{N}_{\langle k \rangle}$ as the ones that are required to be isomorphic to some given sequence $\hat{\mathcal{N}_k}$ of countable first order structures. We could as well require of our structures, for example, that for various i and j, $\mathcal{N}_i$ is elementary equivalent or isomorphic to, or an elementary substructure of, $\mathcal{N}_j$. We call any of these an $\mathcal{N}$-logic for $\mathcal{N} = \left\langle \hat{\mathcal{N}_k} \right\rangle$. An *$\mathcal{N}$-theory* in $\mathcal{L}$ is a set of sentences of $\mathcal{L}$ which has an *$\mathcal{N}$-model,* i.e. there is an $\mathcal{N}$-structure which satisfies every $\varphi \in T$. We could generalize further by allowing, e.g. quantification over the sorts, their variables and even formulas with the satisfaction relation. Such versions of language and logic seem perhaps unnatural in this setting. They, will however, be both natural and crucial in the presentation of modal logics that are built on some $\mathcal{N}$-logic in each world and so we leave such examples to the next section.

The only properties of $\mathcal{N}$-logic that we need to prove our type omitting theorem are the following ones that we have already noted:

Property of $\mathcal{N}$-Logic 1: For countable $\mathcal{M}$, being an $\mathcal{N}$-structure is a Σ^1_1 property of $\mathcal{M}$ and $\mathcal{N}$.

Property of $\mathcal{N}$-Logic 2: For sentences φ of $\mathcal{L}$ and countable $\mathcal{N}$-structures $\mathcal{M}$ for $\mathcal{L}$ (even with constants for the elements a of M), the set $\{(\mathcal{M}, \varphi) | \mathcal{M} \models \varphi\}$ is Σ^1_1 in $\mathcal{N}$.

Property of $\mathcal{N}$-Logic 3: Any $\mathcal{N}$-theory T has a countable $\mathcal{N}$-model.

We can generalize $\mathcal{N}$-logics in various ways that preserve these three properties. For example, instead of requiring that $\mathcal{N}_{\langle k \rangle}$ is isomorphic to a fixed structure $\hat{\mathcal{N}_k}$ we could require only that it is isomorphic to one of a countable list of countable structures, $\hat{\mathcal{N}}_{k,l}$. A natural example here is for vector spaces requiring that the field be finite. Alternatively, we can require that a sort be infinite (so infinite fields). At a more complex level we could require that some binary relation (definable) in one of the sorts not be well-founded. The only constraint on conditions on the structures is Property 1. Our type omitting theorems rely only on these three principles.

Before stating our type omitting theorem we generalize the notion of a type in a language $\mathcal{L}$ over an $\mathcal{N}$-theory T.

Definition 2.1: Let T be a $\mathcal{N}$-theory in a language $\mathcal{L}$ for some version of $\mathcal{N}$-logic. Let Γ be any set of formulas of $\mathcal{L}$ and $n \in \mathbb{N}$. A Γ-n-type is a subset p of Γ with each element having free variables precisely $x_1, \ldots, x_n$. (If Γ is the set of all formulas of $\mathcal{L}$ we do not mention it. Note that requiring the variables to always range over the entire structure does not involve any loss of generality as one can add formulas of the form $\exists x^j(x^j = x_i)$ to the types as long as one includes them in Γ.) The type p is *realized* in a structure $\mathcal{M}$ for $\mathcal{L}$ if there are elements $a_1, \ldots, a_n$ of M such that $p = \{\varphi \in \Gamma | \mathcal{M} \models \varphi(a_1, \ldots, a_n)\}$. If p is not realized in $\mathcal{M}$ we say it is *omitted* in $\mathcal{M}$.

We can now state our type omitting theorem for logics satisfying Properties 1–3 above. Note that we can omit Property 3 if we require that all theories considered have countable models. All the models we construct to prove any of our type omitting theorems will also be countable

Theorem 2.2: *With the conditions just described on our $\mathcal{N}$-logic, if T is a Π^1_1 (in $\mathcal{N}$) $\mathcal{N}$-theory (i.e. the set of Gödel numbers for the sentences of T is a Π^1_1 (in $\mathcal{N}$) set of numbers); Γ is a Σ^1_1 in $\mathcal{N}$ set of formulas of $\mathcal{L}$ and p is a Γ-n-type which is not Σ^1_1 in $\mathcal{N}$, then there is a countable $\mathcal{N}$-model of T not realizing p.*

This theorem is an immediate corollary of the main theorem of HSS relativized to $\mathcal{N}$.

Theorem 2.3: (HSS Theorem 2.1) *If a real X is Σ^1_1 in every member G of a nonempty Σ^1_1 class $\mathcal{K}$ of reals then X is itself Σ^1_1.*

Here reals are just members of Cantor space, i.e. subsets of $\mathbb{N}$ or functions from $\mathbb{N}$ to $\{0, 1\} = 2$. We note that the proof of this theorem is quite elementary and takes about a page. So the full proofs of our type omitting results are quite short and simple.

Proof: [Proof of Theorem 2.2] By Properties 1 and 2 above, the class of countable $\mathcal{N}$-models of T (coded as reals in any reasonable way) is a Σ^1_1 in $\mathcal{N}$ class of reals. The Σ^1_1 in $\mathcal{N}$ class of such reals is nonempty by

Property 3. As Γ is a Σ^1_1 set of formulas, for any $a_1, \ldots, a_n$ in M, the set $\{\varphi \in \Gamma | \mathcal{M} \models \varphi(a_1, \ldots, a_n)\}$ is $\Sigma^1_1(\mathcal{M}, \mathcal{N})$ by Property 2. Thus by Definition 2.1, were p realized in every countable $\mathcal{N}$-model, it would be Σ^1_1 in $\mathcal{N}$ for a contradiction. □

We mention some easy extensions of [HSS Theorem 2.1] and the corresponding ones of Theorem 2.2 that include the ones in Grilliot [1972] for the appropriate logics in his settings.

First, as usual, if one can omit a type, one can omit a countable sequence of types:

Theorem 2.4: *If T is a Π^1_1 (in $\mathcal{N}$) $\mathcal{N}$-theory; Γ_i a uniformly Σ^1_1 in $\mathcal{N}$ sequence of sets of formulas of $\mathcal{L}$ and p_i are Γ_i-n-types none of which is Σ^1_1 in $\mathcal{N}$, then there is a countable $\mathcal{N}$-model of T not realizing any p_i.*

This theorem follows from the following result of HSS the same way Theorem 2.2 followed from HSS Theorem 2.1.

Theorem 2.5: (HSS Theorem 2.6) *If $\mathcal{K}$ is a nonempty Σ^1_1 class reals and X_n a countable sequence of reals none of which is Σ^1_1, then there is a $G \in \mathcal{K}$ such that no X_n is Σ^1_1 in G.*

As suggested by Grilliot's variations on his omitting types theorems, one can strengthen these results by proving "minimal pair" versions. We do a bit more.

Theorem 2.6: *If $\mathcal{K}_0$ and $\mathcal{K}_1$ are nonempty Σ^1_1 classes of reals then there are $G_0 \in \mathcal{K}_0$ and $G_1 \in \mathcal{K}_1$ such that any X which is Σ^1_1 in both G_0 and G_1 is itself Σ^1_1. In fact, one can prove more: For any $G_0 \in \mathcal{K}_0$ there is a $G_1 \in \mathcal{K}_1$ such that any X which is Σ^1_1 in both G_0 and G_1 is itself Σ^1_1.*

Proof: A construction of a pair of generics satisfying minimal pair type conditions provides a minimal pair G_0 and G_1. A suggestion we received for a direct proof leads to the stronger result: If every Y which is Σ^1_1 in G_0 is Σ^1_1, any $G_1 \in \mathcal{K}_1$ works. If not, let X_n list all the sets which are Σ^1_1 in G_0 but not Σ^1_1. By [HSS, Theorem 2.6], there is a $G_1 \in \mathcal{K}_1$ such that no X_n is Σ^1_1 in G_1. This G_1 is as required. □

The same argument as for Theorem 2.2 now gives the type omitting version.

Theorem 2.7: *If T_0 and T_1 are Π^1_1 $\mathcal{N}$-theories in $\mathcal{L}$ then there are $\mathcal{N}$-models $\mathcal{M}_0$ and $\mathcal{M}_1$ of T_0 and T_1, respectively, such that any Γ-n-type p (with Γ Σ^1_1 in $\mathcal{N}$) realized in both models is Σ^1_1 in $\mathcal{N}$. Indeed, for any model $\mathcal{M}_0$ of T_0 there is an $\mathcal{M}_1 \models T_1$ such that any type realized in both models is Σ^1_1.*

Similarly, one can prove countable sequence versions of both minimal pair theorems first getting, for a countable sequence $\mathcal{K}_i$ of nonempty Σ^1_1 classes of reals, a sequence $G_i \in \mathcal{K}_i$ (starting with any $G_0 \in \mathcal{K}_0$) such that any X which is Σ^1_1 in both G_j and G_k for any $j \neq k$ is itself Σ^1_1. Then, on the type omitting side, this tells us that if T_i are Π^1_1 $\mathcal{N}$-theories in $\mathcal{L}$ then there are $\mathcal{N}$-models $\mathcal{M}_i$ of T_i (starting with any $\mathcal{M}_0 \models T_0$) such that any Γ-n-type p with Γ Σ^1_1 in $\mathcal{N}$ realized in both $\mathcal{M}_j$ and $\mathcal{M}_k$ for any $j \neq k$ is Σ^1_1 in $\mathcal{N}$.

Finally, to bound the complexity of the desired models and reproduce the relevant applications in Grilliot [1972], we note that the proof of Theorem 2.8 of HSS, shows that in all of these cases the desired sets G and so models $\mathcal{M}$ can be constructed hyperarithmetically (recursively) in Kleene's $\mathcal{O}^{\mathcal{N}}$, the complete Π^1_1 in $\mathcal{N}$ set, if the sets X_i (types p_i) are uniformly hyperarithmetic (recursive) in $\mathcal{O}^{\mathcal{N}}$. Moreover, they can be taken to have strictly smaller hyperdegree than $\mathcal{O}^{\mathcal{N}}$.

We mention all of these variations partially to make a comment on a comment of Grilliot [1972] to his Application 2 of his analogous result. Grilliot points out that his theorems give basis results for Σ^1_1 classes. The strongest of them being the countable "minimal pair" version: If S_m is the mth nonempty Σ^1_1 set of functions $(\mathbb{N} \to \mathbb{N})$ in a recursive in $\mathcal{O}$ list, then there are (uniformly) recursive in $\mathcal{O}$ functions $f_{m,n}$ such that $f_{m,n} \in S_m$ for all $m, n \in \mathbb{N}$ and any function hyperarithmetical in more than one $f_{m,n}$ is hyperarithmetical. He says this seems to be about the strongest (basis theorem) that can be obtained while still insisting that the solutions be recursive in $\mathcal{O}$. We point out that our approach strengthens these results not by imposing stronger conditions on the solutions but by moving everything to the realm of sets rather than functions. For functions being Σ^1_1 is the same as being Δ^1_1 (hyperarithmetic). Of course, this is not true for sets. Our approach shows that if one replaces functions $(\mathbb{N} \to \mathbb{N})$ by sets (contained in $\mathbb{N}$) in any of the basis results he proves including the last, one can replace "hyperarithmetical" by "Σ^1_1" in the conclusions.

We conclude this section with a fragment of second order logic stronger than all of the ones considered so far and some natural settings for such a "logic". (We do not know if this particular fragment, which is not really a logic under most definitions, has been previously studied.) The mathematical settings it is intended to capture are extension problems. Given a (countable) model $\mathcal{M}$ of a theory T in one language $\mathcal{L}$ and a language $\mathcal{L}'$ extending $\mathcal{L}$ and a theory T' in $\mathcal{L}'$ extending T, what can we say about the (possible) extensions of $\mathcal{M}$ to an $\mathcal{M}' \models T'$? For example, starting with a countable group what can one say about the (possible) extensions to a ordered groups. A similar question studied in logic starts with a model of some fragment of arithmetic of some or all finite types or of set theory and asks what can we say about the (possible) extensions to a model of some stronger version of arithmetic or set theory. Many such questions are studied for subsystems of second order arithmetic such as when can one extend models of RCA_0 (or some other theory) to models of stronger theories perhaps without extending the first order part of the original model. One more unusual example here is the investigation of extensions of models of ZF to ones of KM as in Zygmunt [1981].

We extend any of the logics considered so far by including sentences that are existential second order. That is for a given language $\mathcal{L}$ a theory or type is allowed to be contain sentences of the form $\psi = \exists R_1^{n_1} \dots \exists R_k^{n_k} \varphi$ where φ is a sentence in the language extending $\mathcal{L}$ by adding on new n_i-ary relation symbols R_i $(1 \le i \le k)$. (Note that, up to logical equivalence, we can close the set of formulas we are allowing under conjunction and disjunction but not negation.) We call this set of formulas $\mathcal{L}^{Ext}$. The semantics is the natural one. A structure $\mathcal{M}$ for $\mathcal{L}$ satisfies $\psi = \exists R_1^{n_1} \dots \exists R_k^{n_k} \varphi$ if there are n_i-ary relations $R_i^{n_i}$ on M such that extending $\mathcal{M}$ to a structure $\mathcal{M}'$ by adding on these relations satisfies φ. It is clear that if $\mathcal{M}$ is a countable structure for $\mathcal{L}$, satisfaction for formulas ψ of the required form is Σ_1^1 in $\mathcal{M}$ (and $\mathcal{N}$ if this is an $\mathcal{N}$-language). Moreover, these structures still have the downward Skolem-Löwenheim property. So all of the Properties 1–3 required above hold for $\mathcal{L}^{Ext}$. Thus we have a type omitting theorem for theories and types in this language as well.

Theorem 2.8: *If T is a Π_1^1 set of formulas of $\mathcal{L}^{Ext}$ with a model, Γ a Σ_1^1 set of formulas of $\mathcal{L}^{Ext}$ and $\{p_i\}$ is a set of Γ-n_i-types of $\mathcal{L}^{Ext}$ none of which is Σ_1^1 then there is a model $\mathcal{M}$ of T in which no p_i is realized.*

Essentially, this theorem gives our usual conditions under which we can say that there are models (of a set of sentences in $\mathcal{L}$) which can be extended to relations satisfying additional axioms (as expressed by a theory T in

$\mathcal{L}^{Ext}$) involving the new relations which cannot be further extended to one satisfying some one of a collection of sentences p_i in $\mathcal{L}^{Ext}$ and so which cannot be extended with yet additional relations to satisfy a type in the larger extended language.

Sample known results which are immediate consequences (and can be phrased in terms of $\mathcal{N}$-logic alone) are that if some countable model of a theory of arithmetic or set theory can be extended to be one of a stronger theory (with, for example, more types) then no set (or higher type object of a level in the language) not already Σ_1^1 over the model can be realized in every extension to the stronger theory. If the original theory has enough comprehension this condition typically means that no new sets (or higher type object) can be in every model of the stronger theory. There are classical such results for models of arithmetic as in GKT. Less well-known generalizations of some of these results for extending models of ZF to ones of KM appear in Ratajczyk [1979]. Our versions say more than this not only because they allow a wider variety of types but they also allow one to consider two step extensions of a given model and to talk about all models of a given theory which have extensions satisfying another theory and give conditions under which one of the models has no further extensions to any of a countable list of theories with associated types realized.

3. Modal Logics: Classical and Beyond

To set the stage for a wide variety of (as far as we know) generally new modal logics, we begin with a description of basic modal first order logic. A language $\mathcal{L}^{\Box,\Diamond}$ here has a first order language $\mathcal{L}$ (which for simplicity we may (but need not) take to be relational except for some constant symbols as well as modal operators $\Box$ and $\Diamond$. The role of structures are played by Kripke frames $\mathcal{F} = \langle W, S, \{\mathcal{F}(p) | p \in W\}\rangle$ (or more simply $\langle W, S, \mathcal{F}(p)\rangle\rangle$ where S is a binary relation on a set W and for each $p \in W$, $\mathcal{F}(p)$ is a (first order) structure on $F(p)$ for $\mathcal{L}$. One then defines the forcing relation $p \Vdash \varphi$ between $p \in W$ and sentences of $\mathcal{L}$, φ being forced in $\mathcal{F}$ in the usual inductive fashion on sentences. (Some sample steps are given below.) There are however, some choices to be made in the definition of the semantics. One important class of variations involves the relationships among the domains $F(p)$. (See for example, Brainer and Ghilardi [2007].) At the core, for ordinary first order modal logic, we adopt one common one, monotonic domains $((\forall p, q)(S(p, q) \rightarrow F(p) \subseteq F(q))$ as in Nerode and Shore [1997]. This choice makes many things more straightforward (than say variable domains where the basic structures do not even fit our conventions about classical first order logic). For our purposes, we will see that we can use this setting to get our results for some other choices including, for example, requiring constant domains ($F(p) = F(q)$ for all $p, q \in W$). We can, however, easily allow constants and function symbols with the provisos inherent in monotonicity: If $S(p, q)$ and c is a constant symbol of the language then its interpretation remains the same, i.e. $c^{\mathcal{F}(p)} = c^{\mathcal{F}(q)}$. Similarly, if $\bar{a} \in F(p)^n$ and f is an n-ary function symbol of the language then $f^{\mathcal{F}(q)}(\bar{a}) = f^{\mathcal{F}(p)}(\bar{a})$.

We know of only three examples of type omitting theorems for modal logic. The first two (Mortimer [1974] and Bowen [1979, §15]) are quite tied to first order modal logic and have very restrictive definitions of types. A recent one (Litak et al. [2018]) has been brought to our attention and moves in a much more general direction toward coalgebraic logics. Our goal here is to introduce a wide range of modal logics stronger than first order ones in a variety of ways for which the basic theorems of HHS also supply a number of type omitting theorems.

We begin our generalizations in a well-known way by allowing many (pairs of) modal operators $\Box_i$ and $\Diamond_i$ each with their own accessibility relation S_i. (Think that we are representing knowledge for multiple agents.) We also allow the common logic and language used in the possible worlds to be any of the $\mathcal{N}$ ones considered in the previous section such as ω-logic or weak second order logic, etc. (We do not know of any investigations into such modal logics with worlds equipped with these fragments of second order logic and specified sorts. There are however, versions of full second order modal logic presented and studied in Cocchiarella and Freund [2008].) We can also allow the apparatus of common knowledge (Fagin, Halpern, Moses and Vardi [1995]. We can also allow any types of restrictions of the frames $\mathcal{F}$ being considered (say to a class $\mathcal{C}$) as long as the relation of satisfying these restrictions ($\mathcal{F} \in \mathcal{C}$) is a Σ^1_1 in $\mathcal{N}$ property for countable frames $\mathcal{F}$.

As examples of conditions that can define such classes $\mathcal{C}$ of frames $\mathcal{F}$, we can allow any restrictions on the S_i that are Σ^1_1 properties of the W_i if countable, e.g. for all i in some Π^1_1 set the S_i are transitive, reflexive, symmetric or the like. More unusually, one can require that they be isomorphic to any fixed countable binary relations such as $\mathbb{N}, \mathbb{Z}$ or $\mathbb{Q}$ or to any one of a countable collection of binary relations such as all finite ones or infinite ones or ones with a descending chain. Modal logics with some of these restrictions on the allowed frames are often called hybrid logics

(Areces and ten Cate [2007]). Some of these restrictions are well-known to be equivalent to first order axioms on the binary relation. Others are clearly not. We can impose additional restrictions on the domains such as being constant: $\forall p, q, (F(p) = F(q))$. We can also impose restrictions on the frame as a whole that may involve the forcing relationship. For example, once we see that p forcing φ in a frame $\mathcal{F}$, $p \Vdash_{\mathcal{F}} \varphi$, is at worst a Σ_1^1 in $\mathcal{N}$ relation, we can require that some sentence be forced at some world or at every world or many things in between. Our only constraint is that the analog of Property 1 continues to hold: A countable set $\mathcal{F}$ being an allowed frame for $\mathcal{L}$ (which we take to include any relevant $\mathcal{N}$) is a Σ_1^1 in property of $\mathcal{F}$ and $\mathcal{L}$.

We note that we also want the basic relation of a sentence φ being forced at a world p to be Σ_1^1 (in a countable frame $\mathcal{F}$ and any $\mathcal{N}$ used in the logic). This is easy to see for the usual inductive definitions of forcing extended to our more general logics. The typical definition for classical first order logic defines $p \Vdash_{\mathcal{F}} \varphi$ for p in a frame $\mathcal{F}$ inductively for sentences φ in the (classical) language $\mathcal{L}(p)$ gotten by expanding the basic language of the frame $\mathcal{L}$ by adding at least constant symbols c_a for every $a \in F(p)$. If, for example, we want to have the logic be weak second order logic as in the previous section, then we want $\mathcal{L}(p)$ to also include new n-ary relation symbols R_e^n for each finite n-ary relation e on $F(p)$ (we write this as $e \in F(n,p) = [F(p)^n]^{<\omega}$) with the natural interpretations. If we have an $\mathcal{N}$ logic then we require the appropriate type in each $\mathcal{F}(p)$ to be isomorphic to $\mathcal{N}$. The definition then proceeds as would be expected by induction. For example, $p \Vdash \varphi$ for atomic sentences φ of $\mathcal{L}(p)$ if and only if $\mathcal{F}(p) \models \varphi$; $p \Vdash \varphi \wedge \psi$ iff $p \Vdash \varphi$ and $p \Vdash \psi$; $p \Vdash \exists x \varphi(x)$ iff $p \Vdash \varphi(c_a)$ for some $a \in F(p)$ where if, for example, the variable x is typed to be say in $\mathcal{N}$ then a must be in the isomorphic copy of $\mathcal{N}$ in $\mathcal{F}(p)$; $p \Vdash \exists R^n \varphi(R^n)$ iff $\exists e \in F(n,p)(p \Vdash \varphi(R_e^n))$; $p \Vdash \Box_i \varphi$ iff $\forall q(S_i(p,q) \to q \Vdash \varphi)$ and so as expected for $\vee, \to, \neg, \forall x, \forall R^n$ and $\Diamond_i$ or the operators for common knowledge. (Note that by our assumption of monotonicity, $S_i(p,q) \to F(p) \subseteq F(q)$, implies that if φ is a formula of $\mathcal{L}(p)$ then it is also one of $\mathcal{L}(q)$ so the induction makes sense.) It is straight-forward to see that such definitions make $p \Vdash \varphi$ uniformly arithmetic in $\mathcal{F}$ (and the $\mathcal{N}$) where the quantifier complexity of the arithmetic definition depends uniformly on the syntactic form of φ. Thus over all $p \in W$ and sentences φ of $\mathcal{L}(p)$, the relation $p \Vdash \varphi$ is clearly Σ_1^1 in $\mathcal{F}$ and $\mathcal{N}$. All of this remains true if we allow as well the apparatus of common knowledge as in Fagin, Halpern, Moses and Vardi [1995] where we view the sequence of groups of agents as part of the logic as well and so can quantify over them to get the operators E_G and iterate the E_G to get the C_G as and keep the relation $p \Vdash \varphi$ arithmetic as before. (A version of common knowledge with infinitely many agents and groups as we would allow here is considered in Halpern and Shore [2004].)

Thus the analog of Property 2 holds for all these modal logics: for sentences φ of $\mathcal{L}$ and countable frames $\mathcal{F}$ for $\mathcal{L}$ the relation $p \Vdash \varphi$ is Σ_1^1 in $\mathcal{L}$ (which, remember, we take to include the relevant $\mathcal{N}$) and $\mathcal{F}$. As usual, a Skolem-Löwenheim type argument (involving W as well as the S_i and the $\mathcal{F}(p)$) shows that we maintain Property 3 as well for frames and theories about them in the expanded languages and associated generalized modal logics.

We next give definitions of theories and types for these modal logics analogous to those for classical logics in §2. We fix one of our generalized $\mathcal{N}$ logics and languages $\mathcal{L}$ as well as a Σ_1^1 (in $\mathcal{L}$ and $\mathcal{N}$) class $\mathcal{C}$ of acceptable frames for $\mathcal{L}$.

Definition 3.1: A $\mathcal{C}$-theory is a set T of sentences φ of $\mathcal{L}$ such that there is a $\mathcal{C}$-frame $\mathcal{F}$ such that $p \Vdash \varphi$ for every $\varphi \in T$ and $p \in W$. (We write this as $\mathcal{F} \Vdash T$.) For Γ a set of formulas of $\mathcal{L}$ and $n \in \mathbb{N}$, a Γ-n-type is a subset P of Γ with each element having free variables precisely $x_1, \ldots, x_n$. The type P is *realized* in a $\mathcal{C}$-frame $\mathcal{F}$ for $\mathcal{L}$ if there are elements $a_1, \ldots, a_n$ of some $\mathcal{F}(p)$ such that $P = \{\varphi \in \Gamma | p \Vdash \varphi(a_1, \ldots, a_n)\}$. In this case, we also say that $\mathcal{F}$ realizes P. If P is not realized in $\mathcal{F}$ we say it is *omitted* in $\mathcal{F}$.

We now have a type omitting theorem for our modal $\mathcal{L}$ analogous to Theorem 2.2 for classical logic.

Theorem 3.2: *For a modal $\mathcal{N}$-language and logic $\mathcal{L}$ with a class $\mathcal{C}$ of frames with the properties described above analogous to those of §2, a Π_1^1 (in $\mathcal{L}$) $\mathcal{C}$-theory T, a Σ_1^1 in $\mathcal{L}$ set Γ of formulas and P a Γ-n-type which is not Σ_1^1 in $\mathcal{L}$, there is a countable $\mathcal{C}$-frame $\mathcal{F}$ such that $\mathcal{F} \Vdash T$ which does not realize P.*

Proof: As before we apply Theorem 2.6 of HSS. The only things to note is that once again the $\mathcal{C}$-frames $\mathcal{F}$ such that $\mathcal{F} \Vdash T$ is a Σ_1^1 in $\mathcal{L}$ class and if P is realized in some $\mathcal{C}$-frame $\mathcal{F}$ then it is Σ_1^1 in $\mathcal{L}$ and $\mathcal{F}$ by definition and the assumed properties of $\mathcal{L}$ and $\mathcal{C}$. □

In the same style we can prove the analogs of Theorems 2.4 and 2.7 and their common generalization for types and sequences of types in the modal setting.

We close with some remarks and conjectures about expected proof systems and corresponding soundness and completeness results for the generalized modal logics we have considered. We believe that one can modify standard tableaux style proof systems for classical modal logics (as in, for example, Nerode and Shore [1997]) to get ones for these logics as well. We assume that we are dealing with systems with equality with the associated axioms/tableaux rules. As the intended semantics uses true equality, one expects the usual moding out by an equivalence relation for completeness proofs. Of course, these proof systems cannot be any simpler than those for the underlying generalized classical logics. Thus we expect to define tableaux and systematic tableaux which are infinitely branching trees. A proof of φ should then be as usual a tableau with root labeled $F\varphi$ on which every path is contradictory. Thus being a proof is a Π^1_1 property. (We can also think of terminating paths when they become contradictory. Then proofs are the well founded tableaux.)

To give an indication of how such systems should work we mention a few examples of new types of atomic tableaux. Our basic setting here is that for modal logics (with equality) in Nerode and Shore [1997]. If we have quantification over finite sets S, then we would have one with root $Tp \Vdash \exists S\varphi$ and immediate successors for each $n \in \mathbb{N}$. The successor for n would introduce n many new constants $c_1, \dots, c_n$ and have label $Tp \Vdash \varphi'$ where each instance in φ of $t \in S$ for terms t (or $S(t)$ depending on the syntax) is replaced by the disjunction over $i \in [1, n]$ of $t = c_i$. If we quantify over finite relations then we would have one with root $Tp \Vdash \exists R^n\varphi$ and infinitely many successors, one for each finite n-ary relations R^n_e with domain $d_1, \dots, d_m$ of size m which we label with $p \Vdash T\varphi'$ where we introduce m many new constants $c_1, \dots, c_m$ and φ' is gotten by replacing each occurrence in φ of $R^n(t_1, \dots, t_n)$ for terms t_i by $\bigvee_{j_i, j_2, \dots, j_n \in [1,m] \ \& \ R^n_e(d_{j_1}, \dots, d_{j_n})} \bigwedge_{i \in [1,n]} (t_i = c_{j_i})$. If the base logic is an $\mathcal{N}$ logic requiring, for example, some sort to be isomorphic to a countable structure $\mathcal{N}$ as in ω-logic, then we list the elements c_i of N intending them to be elements of the appropriate sort in every $\mathcal{F}(p)$ and put them in along every path of the tableau as well as $Tp \Vdash \theta$ for every θ in the atomic diagram of $\mathcal{N}$. We also have, for example, a new atomic tableau with root $Tp \Vdash \exists x \in N\varphi(x)$ which has infinitely many immediate successors each of the form $Tp \Vdash \varphi(c_i)$ for $i \in \mathbb{N}$.

Of course, we also have the expected variations for universal quantifications and the usual definition for $\Box_i\varphi$ and $\Diamond_i\varphi$. Similar moves should work for all the fragments of second order $\mathcal{N}$-logic discussed in §2. We can also handle many restrictions on the class $\mathcal{C}$ of acceptable frames in similar ways. For example, if S_i is required to be ω with the usual ordering $<$, then we have worlds p_i in W for $i \in \omega$ on which we specify S_i accordingly. Then, for example, we would introduce a new atomic tableau with root $Tp_i \Vdash \Diamond_i\varphi$ whose immediate successors are $Tp_j \Vdash \varphi$ for $j > i$ and when we get a node α labeled $Tp_i \Vdash \Box\varphi$ we promise to put a node labeled $Tp_j \Vdash \varphi$ for every $j > i$ on every path below α.

Admittedly, this is all quite sketchy but we hope that the project of producing proof systems generating (recursively for recursive $\mathcal{L}$ and theories T) trees whose well founded members can be taken as proofs from T should not be too difficult to carry out. The goal then would be to prove soundness and completeness theorems as one does for classical first order modal logics for as many of the generalized modal logics as possible.

References

1. Areces, C. and ten Cate, B. [2007], Hybrid Logics, Chapter 14 in Blackburn, Van Benthem and Wolter [2007], pp. 821–868.
2. Barwise, Jon [1975], *Admissible sets and structures. An approach to definability theory*, Perspectives in Mathematical Logic, Springer-Verlag, Berlin; New York.
3. Blackburn, P., Van Benthem, J. and Wolter, F. eds. [2007], *Handbook of Modal Logic*, Studies in Logic and Practical Reasoning vol. 3, Elsevier, Amsterdam; Boston.
4. Brainer, T. and Ghilardi, S., First Order Modal Logic, Chapter 9 in Blackburn, Van Benthem and Wolter [2007], pp. 549–620.
5. Bowen, Kenneth A. [1979], *Model Theory for Modal Logic*, Synthese Library **127**, Springer, Berlin.
6. Chang, C. C. and Keisler, H. J. [1990], *Model Theory* 3rd edition, Studies in Logic and the Foundations of Mathematics **73**, North-Holland Publishing Co., Amsterdam.
7. Cocchiarella, N. B. and Freund, M. A. [2008], *Modal Logic: An Introduction to Its Syntax and Semantics*, Oxford University Press, Oxford; New York.
8. Fagin, R., Halpern, J. Y., Moses, Y. and Vardi, M. Y. [1995], *Reasoning about Knowledge*, M.I.T. Press, Cambridge MA.
9. Gandy, R. O., Kreisel, G. and Tait, W. W. [1960] and [1961], Set Existence I and II, *Bul. Acad. Polon. Sci. Ser. Sci. Math. Astronom. Phys.* **8**, 577–582 and **9**, 881–882.
10. Grilliot, T. J. [1972], Omitting types: applications to recursion theory, *J. Symb. Logic* **37**, 81–89.

11. Halpern, J. and Shore, R. A. [2004], Reasoning about common knowledge with infinitely many agents, *Information and Computation* **191**, 1–40.
12. Harrington, L., Shore, R. A. and Slaman, T. A. [2017], Σ_1^1 in every real in a Σ_1^1 class of reals is Σ_1^1, in *Computability and Complexity, Essays Dedicated to Rodney Downey on the Occasion of His 60th Birthday*, A. Day, M. Fellows, N. Greenberg, B. Khoussainov, A. Melnikov and F. Rosamond eds., LNCS 10010, Springer, 455–468.
13. Hodges, W. [1993], *Model Theory*, Encyclopedia of Mathematics and its Applications **42**, Cambridge University Press, Cambridge UK.
14. Litak, T., Pattinson, D., Sano, K. and Schröder, L. [2018], Model Theory and Proof Theory of Coalgebraic Predicate Logic, *Logical Methods in Computer Science* **12**, 1–52.
15. Montalbán, A. [2021], *Computable Structure Theory: Within Arithmetic*, Perspectives in Logic, Association for Symbolic Logic and Cambridge University Press.
16. Mortimer, Michael [1974], Some results in modal model theory, *J. Symb. Logic* **39**, 496–508.
17. Nerode, A. and Shore, R. A. [1997], *Logic for Applications* 2nd ed., Graduate Texts in Computer Science, Springer-Verlag, New York.
18. Ratajczyk, Zygmunt [1981], A characterization of expandability of models for ZF to models for KM, *Fund. Math.* **113**, 9–19.

https://doi.org/10.1142/9789819806584_0013

Mouse Pairs and Suslin Cardinals

John R. Steel
Professor of Mathematics
Department of Mathematics, UC Berkeley
970 Evans Hall, MC 3840, Berkeley, CA 94720-3840, USA
steel@math.berkeley.edu

Building on the results of [19] and [18], we obtain optimal Suslin representations for mouse pairs. This leads us to an analysis of HOD^M, for any model M of $\mathsf{AD}_{\mathbb{R}}$ in which there are Wadge-cofinally many mouse pairs. We show also that if there is a least branch hod pair with a Woodin limit of Woodin cardinals, then there is a model of AD^+ in which not every set of reals is ordinal definable from a countable sequence of ordinals.

1. Introduction

The book [19] introduces *mouse pairs*, and develops their basic theory. It then uses one variety of mouse pair, the *least branch hod pairs*, to analyze HOD in certain models of the Axiom of Determinacy. Here we shall carry that work further, by obtaining optimal Suslin representations for mouse pairs. This enables us to show that the collection of models M of $\mathsf{AD}_{\mathbb{R}}$ to which the HOD-analysis of [19] applies is closed downward under Wadge reducibility. It also leads to a characterization of the Solovay sequence in such models M of $\mathsf{AD}_{\mathbb{R}}$ in terms of the Woodin cardinals of HOD^M.[1]

Unless otherwise stated, we shall assume AD^+ throughout this chapter.

1.1. *Mouse pairs*

Let us recall, in outline, some of the definitions and results of [19].

[1][19] is a substantial revision and expansion of *Normalizing iteration trees and comparing iteration strategies*, which first appeared on the author's webpage in 2015. The revisions and expansions to the 2015 draft that were completed before August 2020 are currently available on the author's webpage. The remainder will appear in [19].

A *pure extender premouse* is a transitive structure $M = (J_\alpha^{\vec{E}}, \in, \vec{E}, F)$, where $\vec{E}^\frown F$ is a coherent sequence of extenders. ($F = \emptyset$ is allowed.) We use Jensen indexing: if $E = E_\alpha^M$, then $\alpha = i_E(\text{crit}(E)^{+,M})$. We also use the *projectum-free spaces* fine structure of [19], for reasons that are explained in [19].[2] So in this chapter, *premouse* means *pfs premouse of type 1* in the sense of [19]. The new elements of pfs fine structure will rarely come up in this chapter, however, so familiarity with the standard fine structure should suffice to follow all but a few details.

A *least branch premouse* (lpm) is a transitive structure M that is like a pure extender premouse, except that at certain "branch active" stages, information about an iteration strategy for M is fed into M. When working in the AD^+ context, we deal primarily with countable premice, and iteration strategies for them that are defined on countable trees.

A *mouse pair* is a pair (P, Σ) such that P is a premouse of one of the two types, and Σ is a *complete iteration strategy* for P that normalizes well, is internally lift consistent, and has strong hull condensation. A complete strategy is one that is defined on countable stacks of countable[3] *plus trees.* In this chapter, it will suffice to consider the restriction of Σ to stacks of plus trees that are λ-*tight*, in the sense of [19], and these are just what are called *normal* iteration trees in the standard terminology. That is, the extenders used have increasing lengths, and they are applied to the largest possible initial segment of the earliest possible model.[4]

Normalizing well, internal lift consistency, and strong hull condensation are explained by

Definition 1.1: Suppose Σ is an iteration strategy for a premouse P.

(1) (Tail strategy) If s is a stack by Σ with last model Q, then Σ_s is the strategy for Q given by: $\Sigma_s(t) = \Sigma(s^\frown t)$.

[2]The new element is that $\rho_{k+1}(M)$ and $\rho_k(M)$ are always put into the hull collapsing to the $k+1$-st core of M. One must do this, in one form or another, in order to obtain iteration strategies that normalize well.

[3]In [19], the requirement is that the strategy be defined on finite stacks. One can show that any such strategy extends to countable stacks, however. See [16] or [18].

[4]For E an extender on the M-sequence, E^+ is the extender representing E-then-D, where D is the order zero measure of $\text{Ult}(M, E)$ on $\lambda(E)$. Plus trees are allowed to use extenders of the form E^+, where E is on the sequence of the current last model. $\mathcal{T}$ is λ-tight if it never makes use of this option. $\mathcal{T}$ is λ-*separated* if it always makes use of this option. [19] makes use of the good behavior of Σ of λ-separated trees to deduce that it behaves well on arbitrary plus trees, and hence on the λ-tight plus trees. We are taking [19] as our starting point, so we can sweep the non-λ-tight plus trees under the rug, and we are left with the usual normal trees. [19] uses "normal" so as to allow extenders of the form E^+ to be used, but we shall not do that here.

(2) (Pullback strategy) If $\pi\colon N \to P$ is elementary, then Σ^π is the strategy for N given by: $\Sigma^\pi(s) = \Sigma(\pi s)$, where πs is the lift of s by π to a stack on P.

(3) (Normalizes well) Σ *normalizes well* iff whenever s is a stack of plus trees by Σ with last model Q, and $\mathcal{V} = V(s)$ is the quasi-normalization of s, with associated map $\pi\colon Q \to R$, then
   (i) $\mathcal{V}$ is by Σ,
   (ii) $\Sigma_s = (\Sigma_{\langle \mathcal{W} \rangle})^\pi$, and
   (iii) for any plus tree $\mathcal{T}$ on Q, $\mathcal{T}$ is by Σ_s iff its normal companion is by Σ_s.

(4) (Strong hull condensation) Σ has *strong hull condensation* iff whenever $\mathcal{U}$ is a plus tree by Σ, and $\Phi\colon \mathcal{T} \to \mathcal{U}$ is a tree embedding, then
   (i) $\mathcal{T}$ is by Σ, and
   (ii) if $\pi\colon \mathcal{M}^{\mathcal{T}}_\alpha \to \mathcal{M}^{\mathcal{U}}_{u(\alpha)}$ is a map of Φ, then $\Sigma_{\mathcal{T}\restriction\alpha+1} = \Sigma^\pi_{U\restriction u(\alpha+1)}$.

(5) (Internal lift consistency) Σ is *internally lift consistent* iff whenever (Q, Ω) is an iterate of (P, Σ) via $\mathcal{T}$, $R \lhd Q$, and $\mathcal{U}$ is a plus tree on R, then $\mathcal{U}$ is by $\Sigma_{\mathcal{T},R}$ iff $\mathcal{U}^+$ is by $\Sigma_{\mathcal{T},Q}$, where $\mathcal{U}^+$ is the lift of $\mathcal{U}$ to a tree on Q.

The reader should see [19] for more complete definitions. Item (3) implies that Σ behaves in the same way with respect to embedding normalizations $s \mapsto W(s)$, because $W(s)$ is the normal companion of $V(s)$. We don't need to go into quasi-normalization in this chapter.

We understand elementarity fine structurally, of course. Our convention is that every premouse P has a degree of soundness $k(P)$ attached to it, and elementarity means elementarity at that quantifier level. We shall say that $\mathcal{T}$ *is a pseudo-hull of* $\mathcal{U}$ iff there is a tree embedding of $\mathcal{T}$ into $\mathcal{U}$.[5] Internal lift consistency is related to strong hull condensation, and the two can be combined naturally into one property.

There is one additional property demanded of least branch hod pairs (lbr hod pairs) (P, Σ): if s is a stack on P by Σ with last model Q, then $\dot{\Sigma}^Q \subseteq \Sigma_s$. That is, the strategy predicate of Q must be consistent with the tail strategy for Q determined by Σ. This property is called *pushforward consistency* in [19].

[5]Tree embeddings, and much of the general machinery around embedding normalization, were developed by Farmer Schlutzenberg and the author. See [16] and [19]. See also [4]. Strong hull condensation is a stronger version of the hull condensation property isolated by Sargsyan in [14].

If (M, Σ) is a mouse pair, and s is a stack by Σ with last model N, then we call (N, Σ_s) an *iterate of* (M, Σ). If the branch M-to-N of s does not drop, we call it a non-dropping iterate. In that case, we have an iteration map $i_s\colon M \to N$.

Definition 1.2: Let M be a premouse, then M is *projectum stable* iff for $k = k(M)$, the $r\Sigma_k$ cofinality of $\rho_k(M)$ is not measurable by the M-sequence. A pair (M, Σ) is projectum stable iff M is projectum stable.

The projectum-free spaces fine structure requires that $\rho_{k(M)}(M)$ not be measurable by the M-sequence, but Definition 1.2 is a stronger requirement. Iterations of a projectum stable premouse M preserve all projecta $\rho_i(M)$, where $i \leq k(M)$. This implies that they preserve the projectum-free spaces notion of soundness, in a way that iterations of arbitrary premice may not. We shall be able to restrict our attention to projectum stable mouse pairs.[6]

Notice that if $k(M) = 0$, then M is projectum stable.

Theorem 1.3: (Comparison theorem, [19]). *Assume* AD^+, *and let* (P, Σ) *and* (Q, Ψ) *be projectum stable mouse pairs of same type; then they have a common iterate* (R, Ω) *such that on at least one of the two sides, the iteration does not drop.*

The iteration trees on P and Q witnessing Theorem 1.3 can be taken to be normal, but they cannot always be obtained by iterating away least extender disagreements, as in the standard comparison process.[7,8]

Even for pure extender pairs, Theorem 1.3 goes beyond the usual comparison theorem for premice, because of the agreement between tail strategies it requires. In fact, it is no easier to prove the theorem for pure extender pairs than it is to prove it for least branch hod pairs. The proof in both cases is the same, and it makes use of the properties of the iteration strategies isolated in the definition of mouse pair.

Working in the category of mouse pairs enables one to state a general Dodd-Jensen lemma. Let us say $\pi\colon (P, \Sigma) \to (Q, \Psi)$ is elementary iff π is elementary from P to Q, and $\Sigma = \Psi^\pi$. The iteration maps associated to non-dropping iterations of a mouse pair are elementary.[9]

---

[6][19] uses "strongly stable" instead of "projectum stable".

[7][19] first proves that comparison via λ-separated trees is possible, then uses this to prove that comparison via λ-tight trees is possible.

[8]There is a comparison theorem for pairs that are not projectum stable, but its statement is more awkward.

[9]This is actually not obvious; it is a property of the iteration strategy known as pullback consistency. It follows from strong hull condensation.

Theorem 1.4: (Dodd-Jensen lemma, [19]). *Let* (P, Σ) *be a mouse pair, and* (Q, Ψ) *be an iterate of* (P, Σ) *via the stack* s. *Suppose* $\pi \colon (P, \Sigma) \to (Q, \Psi)$ *is elementary; then* s *does not drop, and for all ordinals* $\eta \in P$, $i_s(\eta) \leq \pi(\eta)$.

The proof is just the usual Dodd-Jensen proof; the point is just that the language of mouse pairs enables us to formulate the theorem in its proper generality. There is no need to restrict to mice with unique iteration strategies, as is usually done.

Similarly, we can define the mouse order in its proper generality, without restricting to mice with unique iteration strategies. If (P, Σ) and (Q, Ψ) are projectum stable pairs of the same type, then $(P, \Sigma) \leq^* (Q, \Psi)$ iff (P, Σ) can be elementarily embedded into an iterate of (Q, Ψ). The Comparison and Dodd-Jensen theorems imply that $\leq^*$ is a prewellorder on each type.

If (P, Σ) is a mouse pair, then $\Sigma \subseteq \mathrm{HC}$. We shall often want to think of Σ as a set of reals, so let us fix a natural coding. For $x \in \mathbb{R} = \omega^\omega$, we say $\mathrm{Code}(x)$ iff $E_x =_{\mathrm{df}} \{\langle n, m\rangle \mid x(2^n 3^m) = 0\}$ is a wellfounded, extensional relation on ω. If $\mathrm{Code}(x)$, then

$$\pi_x \colon (\omega, E_x) \cong (M, \in)$$

is the transitive collapse map, and

$$\mathrm{set}(x) = M \text{ and } \mathrm{set}_0(x) = \pi_x(0).$$

So Code is Π^1_1, and set_0 maps Code onto HC. For $A \subseteq \mathrm{HC}$, we let

$$\mathrm{Code}(A) = \{x \in \mathrm{Code} \mid \mathrm{set}_0(x) \in A\}.$$

Definition 1.5: Let $A \subset \mathrm{HC}$ and $\kappa \in \mathrm{OR}$; then A is *κ-Suslin* iff $\mathrm{Code}(A)$ is κ-Suslin.

1.2. *Hod pair capturing*

We shall show that least branch hod pairs can be used to analyze HOD in models of $\mathsf{AD}_{\mathbb{R}}$, provided that there are enough such pairs.

Definition 1.6: (AD^+)

(a) *Hod Pair Capturing* (HPC) is the assertion: for every Suslin-co-Suslin set A, there is a least branch hod pair (P, Σ) such that A is definable from parameters over $(\mathrm{HC}, \in, \Sigma)$.
(b) $L[E]$ *capturing* (LEC) is the assertion: for every Suslin-co-Suslin set A, there is a pure extender pair (P, Σ) such that A is definable from parameters over $(\mathrm{HC}, \in, \Sigma)$.

An equivalent (under AD^+) formulation would be that the sets of reals coding strategies of the type in question, under some natural map of the reals onto HC, are Wadge cofinal in the Suslin-co-Suslin sets of reals. The restriction to Suslin-co-Suslin sets A is necessary, for AD^+ implies that if (P, Σ) is a pair of one of the two types, then Σ is Suslin and co-Suslin. This is one of the main results of this chapter. It is easy to show that HPC implies the capturing pairs (P, Σ) can be taken to satisfy $k(P) = 0$, and hence to be projectum stable. Similarly for LEC.

Another equivalent (under AD^+) to LEC or HPC, as the case may be, is the assertion that the mouse order on pairs of the given type has order type equal to the sup of the Suslin cardinals.

Remark 1.7: HPC is a cousin of Sargsyan's *Generation of Full Pointclasses*. See [13] and [14], §6.1.

Assuming AD^+, LEC follows from[10] the well-known Mouse Capturing: for reals x and y, x is ordinal definable from y iff x is in a pure extender mouse over y. This equivalence is shown in [24]. (See especially Theorem 16.6.) [19] shows that under AD^+, LEC implies HPC. (See Theorem 5.58.) We do not know whether HPC implies LEC, or whether LEC implies Mouse Capturing.

The natural conjecture is that LEC and HPC hold in all models of AD^+ that have not reached an iteration strategy for a premouse with a long extender. Because our capturing mice have only short extenders on their sequences, LEC and HPC cannot hold in larger models of AD^+.

Definition 1.8: NLE ("No long extenders") is the assertion: there is no countable, $\omega_1 + 1$-iterable pure extender premouse M such that there is a long extender on the M-sequence.

Conjecture 1.8.1: Assume AD^+ and NLE; then LEC.

Conjecture 1.8.2: Assume AD^+ and NLE; then HPC.

As we remarked above, Conjecture 1.8.1 implies 1.8.2. Conjecture 1.8.1 follows from[11] a slight strengthening of the usual Mouse Set Conjecture MSC. (The hypothesis of MSC is that there is no iteration strategy for a

[10]Modulo the intertranslatability of the projectum-free spaces hierarchy with the standard one. This should be relatively straightforward, but it has not been carried out as yet.

[11]Modulo the translation just mentioned.

pure extender premouse with a superstrong, which is slightly stronger than NLE.) MSC has been a central target for inner model theorists for a long time.[12]

We shall prove here that both types of capturing localize to initial segments of the Wadge hierarchy, in the following sense.

Theorem 1.9: *Assume AD^+, and let $M \models \mathsf{AD}^+$ be such that $\mathbb{R} \cup OR \subseteq M$; then*

(i) if there is a pure extender pair (P, Σ) such that $\Sigma \notin M$, then LEC holds in M, and
(ii) if there is an lbr hod pair (P, Σ) such that $\Sigma \notin M$, then HPC holds in M.

For LEC, this was done in [24], but our proof here is different even in that case. We shall also sketch a proof of

Theorem 1.10: *Assume AD^+, and that there is a countable ω_1+1-iterable pure extender mouse with a long extender on its sequence. Let $M \models \mathsf{AD}^+ + \mathsf{NLE}$ be such that $\mathbb{R} \cup OR \subseteq M$; then $M \models$ LEC, and hence $M \models$ HPC.*

So, stepping out to the full universe, where ZFC holds and, it seems quite likely, there is a Hom_∞ iteration strategy for a mouse with a long extender, we see that Conjectures 1.8.1 and 1.8.2 hold in all Wadge cuts in Hom_∞. This makes it likely that the theory $\mathsf{AD}^+ + \mathsf{NLE}$ proves LEC and HPC. A counter-model would have to be "nonstandard".

1.3. *HOD in models of* $\mathsf{AD}_\mathbb{R}$

We shall show that, assuming HPC and $\mathsf{AD}_\mathbb{R}$, and letting θ be the least ordinal that is not the surjective image of $\mathbb{R}$, $V_\theta \cap \mathrm{HOD}$ is a direct limit of lbr hod mice under the iteration maps given by comparisons. This yields

Theorem 1.11: *Assume $\mathsf{AD}_\mathbb{R}$ and HPC; then $V_\theta \cap HOD$ is the universe of a least branch premouse.*

Under $\mathsf{AD}_\mathbb{R} + V = L(P(\mathbb{R})) + \mathsf{HPC}$, the full HOD is $L[M_\infty]$, where M_∞ is a direct limit of least branch hod mice. So HOD $\models$ GCH, and more generally, can be analyzed fine-structurally. In particular, by [28] and [29], the Schimmerling-Zeman characterization of $\Box$ holds in HOD.

[12] We retract the claim to have proved LEC implies MSC made in [19].

We believe Theorem 1.11 remains true if $\mathsf{AD}_{\mathbb{R}}$ is weakened to AD^+ in its hypothesis, but we do not have a proof.

Our proof of Theorem 1.11 gives a characterization of the Solovay sequence in terms of the Woodin cardinals in HOD.

Definition 1.12: (AD^+.) For $A \subseteq \mathbb{R}$, $\theta(A)$ is the least ordinal α such that there is no surjection of $\mathbb{R}$ onto α which is ordinal definable from A and a real. We set

$$\begin{aligned}\theta_0 &= \theta(\emptyset),\\ \theta_{\alpha+1} &= \theta(A), \text{ for any (all) } A \text{ of Wadge rank } \theta_\alpha,\\ \theta_\lambda &= \bigcup_{\alpha<\lambda} \theta_\alpha, \text{ for } \lambda \text{ a limit ordinal.}\end{aligned}$$

$\theta_{\alpha+1}$ is defined iff $\theta_\alpha < \Theta$.[13] Note $\theta(A) < \Theta$ iff there is some $B \subseteq \mathbb{R}$ such that $B \notin \mathrm{OD}(\mathbb{R} \cup \{A\})$. In this case, $\theta(A)$ is the least Wadge rank of such a B. The sequence of θ_α's is called the *Solovay sequence.* It is an important feature of any model of AD^+, one tied to the pattern of pointclasses with the scale property.

Theorem 1.13: *Assume $\mathsf{AD}_{\mathbb{R}} + V = L(P(\mathbb{R})) + \mathsf{HPC}$; then the following are equivalent:*

(1) HOD $\models$ "δ is Woodin, and there is no extender E on the HOD-sequence such that $\mathrm{crit}(E) < \delta$ *and* $V_\delta \subseteq \mathrm{Ult}(V, E)$ ",[14]
(2) $\delta = \theta_0$, or $\delta = \theta_{\alpha+1}$ for some α.

In particular, θ_0 is the least Woodin cardinal in HOD.

That θ_0 and the $\theta_{\alpha+1}$ are Woodin in HOD is due to Woodin, cf. [9]. Woodin also proved an approximation to the statement that they are cutpoints in HOD (unpublished). The rest of (2) $\to$ (1), and all of (1) $\to$ (2), will be proved here.

We shall also prove a conjecture of G. Sargsyan that provides a similar characterization of arbitrary successor Woodin cardinals in HOD.

Definition 1.14: For any κ, we let $\theta(\kappa^\omega)$ be the least ordinal α such that there is no ordinal definable surjection from $[\kappa]^\omega$ onto α.

---

[13] Θ is the least ordinal which is not the surjective image of $\mathbb{R}$.
[14] That is, δ is a cutpoint Woodin of HOD. See Definition 3.18.

Theorem 1.15: *Assume $AD_{\mathbb{R}} + V = L(P(\mathbb{R})) + \mathsf{HPC}$; then for any $\kappa < \theta$ such that κ is a successor cardinal of HOD, $\theta(\kappa^\omega)$ is the least $\delta > \kappa$ such that HOD $\models \delta$ is Woodin.*

We do not have any neat descriptive set theoretic characterization of those limits of Woodins in HOD that are themselves Woodin.

1.4. *Organization*

In §2, we review some material from [19] that did not find its way into this section, and we describe briefly the results on full normalization and very strong hull condensation for mouse pairs that are proved in [18].[15] In §3, we use very strong hull condensation and full normalization to obtain optimal Suslin representations for mouse pairs. This is the main technical tool used in the rest of this chapter. §4 concerns those Suslin representations in the special case that our mouse pair is mouse-minimal among pairs not in some strongly closed pointclass. We prove Theorems 1.9 and 1.10 in this section. In §5, we prove Theorem 1.11, and in §6, we prove Theorems 1.13 and 1.15. Finally, in §7, we show how to obtain models of some strong determinacy theories from hypotheses on the existence of mouse pairs.

2. Preliminaries

Let us review some material from [19] and [18].

2.1. *Background universes and constructions*

Assume AD^+, and let Γ, Γ_1 be good (i.e. closed under $\exists^{\mathbb{R}}$) lightface pointclasses with the scale property such that $\Gamma \subsetneq \Delta_1$. Let A be a universal Γ_1 set, and let $U \subset \mathbb{R}$ code $\{\langle \varphi, x \mid (V_{\omega+1}, \in, A) \models \varphi[x]\}$. Let S and T be trees on some $\omega \times \kappa$ that project to U and $\neg U$. Using his work in [9], Woodin has shown ([22, Lemma 3.13]) that there is a countable transitive N^*, a wellorder w of N^*, and an iteration strategy Σ such that for $\delta = o(N^*)$,

(a) (fullness) $N^* = V_\delta^{L(N^* \cup \{S,T,w\})}$,
(b) N^* is f-Woodin, for all $f \colon \delta \to \delta$ such that $f \in C_\Gamma(N^*, w)$,[16]

[15] Part of the author's work on normalization was done, either independently by or jointly with, Farmer Schlutzenberg. See [16].
[16] For a countable and transitive, $C_\Gamma(a)$ is the largest countable $\Gamma(a \cup \{a\})$ subset of $P(a)$. Its theory (under determinacy hypotheses) was first developed by Kechris and

(c) for all $\eta \leq \delta$, there is an $f\colon \eta \to \eta$ such that $f \in C_{\Gamma_1}(V_\eta^{N^*}, w \cap V_\eta^{N^*})$ and $V_\eta^{N^*}$ is not f-Woodin, and
(d) Σ is an (ω_1, ω_1)-iteration strategy for $L(N^*, S, T, w)$, with respect to nice[17] trees based on N^*.

Concerning item (d), recall that ω_1-iterability implies $\omega_1 + 1$-iterability, granted AD.

Definition 2.1: Assume AD^+, and let Γ be a good pointclass with the scale property, and let N^*, δ, S, T, w, and Σ be as in (a)–(d); then

(1) we call $\langle N^*, \delta, S, T, w, \Sigma\rangle$ a *coarse* Γ*-Woodin tuple*, and
(2) letting $M = (L[N^*, S, T, w], \in, S, T)$, we call (M, Σ) the associated *coarse* Γ*-Woodin pair.*

Of course, S and T determine U, and hence A and Γ_1. U is self-dual, so S is only there for notational convenience.

Let $M = L[N^*, S, T, w]$, where $\langle N^*, \delta, S, T, w, \Sigma\rangle$ is a coarse Γ-Woodin tuple. Let $p[T] = \mathrm{Th}(V_{\omega+1}, \in, A)$, and let Γ_1 be the good pointclass whose universal set is A. If P is a wellfounded iterate of M, and g is P-generic over $\mathrm{Col}(\omega, i(\delta))$, then $P[g]$ is projectively-in-A correct.[18] Thus the C_Γ and C_{Γ_1} operators are correctly defined over $P[g]$. It follows that M and its iterates are C_{Γ_1}-full, and Σ is guided at $\mathcal{T}$ by a Q-structure in $C_{\Gamma_1}(\mathcal{M}(\mathcal{T}))$, where

$$\mathcal{M}(\mathcal{T}) = \left(\bigcup_{\alpha < lh(\mathcal{T})} V^{M_\alpha}_{\mathrm{lh}(E_\alpha)}, \bigcup_{\alpha < \mathrm{lh}(\mathcal{T})} i_{0,\alpha}(w) \cap V^{M_\alpha}_{\mathrm{lh}(E_\alpha)} \right).$$

(We have omitted some superscript $\mathcal{T}$'s here.) That is,

Lemma 2.2: *Assume AD^+, and let (M, Σ) be a coarse Γ-Woodin pair. Let $\vec{T}, \mathcal{U}$ be a stack of nice trees played by Σ; then the following are equivalent*

(1) $\Sigma_{\vec{T}}(\mathcal{U}) = b$,
(2) $C_{\Gamma_1}(\mathcal{M}(\mathcal{U})) \subseteq \mathcal{M}^{\mathcal{U}}_b$,
(3) $\mathcal{M}^{\mathcal{U}}_b$ *is wellfounded.*

Moschovakis. See [5], [6], [7], and the survey [20]. Harrington and Kechris showed in [1] that $C_\Gamma(a) = P(a) \cap L[T, a]$, for any tree T of a Γ scale on a universal Γ set. This is probably the most useful characterization of $C_\Gamma(a)$ is our context.

[17] Roughly, $\mathcal{T}$ is nice iff the extenders it uses have inaccessible lengths and strengths.

[18] $p[T] \subseteq p[i(T)]$ and $p[S] \subseteq p[i(S)]$, while $p[i(T)] \cap p[i(S)] = \emptyset$ because P is wellfounded. So $p[T] = p[i(T)]$.

Since Σ is guided by C_{Γ_1} Q-structures, Code(Σ) is projective in A, hence Wadge reducible to U via a recursive function. On the other hand, letting T_0 be the tree of a Γ scale on a universal Γ set A_0, there are club many $\eta < \delta$ such that $L(V_\eta, w \cap V_\eta, T_0) \models \eta$ is Woodin, and this is preserved in iterations by Σ. Moreover, $p[i(T_0)] = A_0$ for such iterations, so letting τ be a $\text{Col}(\omega, \delta)$-term for $p[T_0]$, (M, Σ, τ) *captures* A_0.

Let (M, Σ) be the coarse Γ-Woodin pair associated to $\langle N^*, \delta, S, T, w, \Sigma\rangle$; then

$$\mathcal{F}^{N^*,w} = \{E \in V_\delta^M \mid M \models E \text{ is nice and } i_E(w) \cap V_{\text{lh}(E)+1}^{Ult(V,E)} = w \cap V_{\text{lh}(E)+1}^{\text{Ult}(V,E)}\}.$$

We call $(w, \mathcal{F})$ the *maximal coherent pair* associated to $\langle N^*, \delta, S, T, w, \Sigma\rangle$. There are two types of background construction associated to the tuple $\langle N^*, \delta, S, T, w, \Sigma\rangle$:

(a) the *maximal pure extender pair construction* $\mathbb{C} = \langle (M_{\nu,k}, \Omega_{\nu,k}) \mid \langle \nu, k\rangle \leq_{\text{lex}} \langle \delta, 0\rangle\rangle$, whose levels $(M_{\nu,k}, \Omega_{\nu,k})$ are pure extender pairs, and
(b) the *maximal hod pair construction* $\mathbb{C} = \langle (M_{\nu,k}, \Omega_{\nu,k}) \mid \langle \nu, k\rangle \leq_{\text{lex}} \langle \delta, 0\rangle\rangle$, whose levels $(M_{\nu,k}, \Omega_{\nu,k})$ are lbr hod pairs.

Both constructions add an extender F whenever doing so produces a premouse, and there is a background extender $F^* \in \mathcal{F}^{N^*,w}$ such that $F \subseteq F^*$. There is at most one such F at any stage. We choose a background extender for F by minimizing first in the Mitchell order, and then in w. This choice of background extenders enters into the definition of $\Omega_{\nu,k}$, which is the iteration strategy obtained by lifting trees on $M_{\nu,k}$ to trees on N^*, and then using Σ to pick branches. The hod pair construction adds information about the $\Omega_{\nu,k}$ as well.

One of the main results of [19] is

Theorem 2.3: *[Comparison II, [19]] Assume AD^+, let (P, Σ) be a projectum stable mouse pair, and let Γ be an good pointclass such that* Code(Σ) $\in \Gamma \cap \check{\Gamma}$. *Let $\langle N^*, \delta, S, T, w, \Sigma\rangle$ be a coarse Γ-Woodin tuple such that $P \in HC^{N^*}$, and let $\mathbb{C}$ be the associated maximal construction of the same mouse pair type as (P, Σ), with levels $(M_{\nu,k}, \Omega_{\nu,k})$; then there is a $\langle \nu, k\rangle$ such that*

(a) $\langle \nu, k\rangle <_{\text{lex}} \langle \delta, 0\rangle$,
(b) (P, Σ) iterates to $(M_{\nu,k}, \Omega_{\nu,k})$ via a normal tree, and
(c) for all $\langle \eta, j\rangle <_{\text{lex}} \langle \nu, k\rangle$, (P, Σ) iterates strictly past $(M_{\eta,j}, \Omega_{\eta,j})$ via a normal tree.

This easily implies Theorem 1.3. We do not know how to prove Theorem 1.3 without going through Theorem 2.3.

2.2. *Very strong hull condensation*

One of the main results of [18] is

Theorem 2.4: ([18]). *Assume AD^+, and let (P, Σ) be a projectum stable mouse pair. Let (Q, Ψ) be an iterate of (P, Σ); then there is a normal tree $\mathcal{T}$ such that (Q, Ψ) is an iterate of (P, Σ) via $\mathcal{T}$.*

The normal tree $\mathcal{T}$ in the theorem is uniquely determined by P, Q, and Σ. So we get

Corollary 2.5: *Let (P, Σ) be a projectum stable mouse pair; then Σ is positional. That is, if (Q, Ψ) and (Q, Λ) are iterates of (P, Σ), then $\Psi = \Lambda$.*

Definition 2.6: Let (P, Σ) be a projectum stable mouse pair, and let s be a normal stack by Σ with last model Q; then $X(s)$ is the unique normal tree by Σ with last model Q. If $s = \langle \mathcal{T}, \mathcal{U} \rangle$ is a stack of length 2, then we write $X(\mathcal{T}, \mathcal{U})$ for $X(s)$.

Remark 2.7: $X(s)$ depends on Σ as well as s. If t is another stack by Σ with last model Q, then $X(t) = X(s)$, so we could write $X(Q)$ or $X(Q, \Sigma)$ for the common value. By positionality, $\Sigma_s = \Sigma_t$, and we shall write Σ_Q for the common value.

The key to the proof of Theorem 2.4 is the definition of *weak tree embedding*, and a proof that if (P, Σ) is a projectum stable mouse pair, then Σ condenses to itself under weak tree embeddings. Let $\mathcal{T}$ and $\mathcal{U}$ be normal trees on the same premouse P. A weak tree embedding of $\mathcal{T}$ into $\mathcal{U}$ is a system of maps

$$\Phi = \langle u, \langle s_\xi \mid \xi < \mathrm{lh}(\mathcal{T}) \rangle, \langle t_\xi \mid \xi + 1 < \mathrm{lh}(\mathcal{T}) \rangle, p \rangle$$

having the properties of a tree embedding (see [19]), except that somewhat less is required of the map p sending exit extenders of $\mathcal{T}$ to exit extenders of $\mathcal{U}$. Namely, $p(E_\alpha^\mathcal{T})$ may not be $t_\alpha(E_\alpha^\mathcal{T})$, but instead connected to $t_\alpha(E_\alpha^\mathcal{T})$ inside $\mathcal{M}_{u(\alpha)}^\mathcal{U}$ via a sequence of fine structural hulls. The Condensation Theorem for $\mathcal{M}_{u(\alpha)}^\mathcal{U}$ implies that $p(E_\alpha^\mathcal{T})$ is on its extender sequence.

Definition 2.8: Let $\mathcal{T}$ and $\mathcal{U}$ be iteration trees on a premouse P; then $\mathcal{T}$ is a *weak hull of $\mathcal{U}$* iff there is a weak tree embedding of $\mathcal{T}$ into $\mathcal{U}$.

The definition of weak tree embedding is natural, but it would take some space to lay it out here in full, because of the fine structure involved. So we ask the reader to see [18] or [21]. Fortunately, we don't need to make use of the details of the definition here. One elementary fact we use is

Proposition 2.9: *If $\Phi\colon \mathcal{T} \to \mathcal{U}$ is a weak tree embedding, $\gamma \leq \mathrm{lh}(\mathcal{T})$, and $\mathcal{V}$ extends $\mathcal{U}$, then $\Phi \restriction \gamma$ is a weak tree embedding of $\mathcal{T} \restriction \gamma$ into $\mathcal{V}$.*

Here $\Phi \restriction \gamma$ consists of the maps of Φ that act on $\mathcal{T} \restriction \gamma$. The deeper facts we use are

Theorem 2.10: ([18], [21]). *Let (P, Σ) be a projectum stable mouse pair, and suppose that $\mathcal{U}$ is by Σ, and $\mathcal{T}$ is a weak hull of $\mathcal{U}$; then $\mathcal{T}$ is by Σ.*

This is a strengthening of the strong hull condensation property that is part of the definition of mouse pair. It is called *very strong hull condensation* in [18] and [21].

Theorem 2.11: ([18], [21]). *Let (P, Σ) be a projectum stable mouse pair, and let $\langle \mathcal{T}_\alpha \mid \alpha < \beta \rangle$ be a stack of normal trees by Σ; then*

(a) $\mathcal{T}_0$ is a weak hull of $X(\vec{\mathcal{T}})$, and
(b) $X(\vec{\mathcal{T}})$ is a weak hull of $W(\vec{\mathcal{T}})$.

(Recall that $W(\vec{\mathcal{T}})$ is the embedding normalization of $\vec{\mathcal{T}}$; see [19] or [16].) The weak tree embeddings asserted to exist in Theorem 2.11 are constructed in an inductive process that is woven into the proof that $\mathcal{T}$ can be fully normalized. The process gives a concrete inductive construction of $X(\vec{\mathcal{T}})$, in which these weak tree embeddings play a central role. Theorems 2.10 and 2.11 show that the process produces Σ-iterates of P.

3. Optimal Suslin Representations

Let (P, Σ) be a projectum stable mouse pair. By $\mathcal{F}(P, \Sigma)$ we mean the system of all non-dropping Σ-iterates (Q, Ψ) of (P, Σ), ordered by

$$(Q, \Psi) \prec (R, \Phi) \text{ iff } (R, \Phi) \text{ is a non-dropping iterate of } (Q, \Psi).$$

By the Comparison Theorem, $\mathcal{F}(P, \Sigma)$ is countably directed, and by Dodd-Jensen, the iteration map

$$\pi^{\Psi}_{Q,R}\colon (Q, \Psi) \to (R, \Phi)$$

is independent of Φ and the stack of trees used to get from (Q,Ψ) to (R,Φ). We let

$$M_\infty(P,\Sigma) = \text{ direct limit of } \mathcal{F}(P,\Sigma)$$

under the iteration maps. By countable directedness, $M_\infty(P,\Sigma)$ is wellfounded, so we assume it has been transitivized. Let

$$\pi^\Sigma_{P,\infty}\colon P \to M_\infty(P,\Sigma)$$

be the direct limit map.[19]

Proposition 3.1: *Let (P,Σ) be a projectum stable mouse pair; then*

(a) $M_\infty(P,\Sigma) \in HOD$,
(b) Σ is the unique Ψ such that $(P,\Psi) \equiv^* (P,\Sigma)$ and $\pi^\Sigma_{P,\infty} = \pi^\Psi_{P,\infty}$, and hence
(c) Σ is ordinal definable from $\pi^\Sigma_{P,\infty}$.

Proof: For (a): If (P,Σ) and (Q,Ψ) have a common non-dropping iterate, then $M_\infty(P,\Sigma) = M_\infty(Q,\Psi)$. Thus $M_\infty(P,\Sigma)$ is definable from the ordinal rank of (P,Σ) in the mouse order. But $M_\infty(P,\Sigma)$ has a definable wellorder, so $M_\infty(P,\Sigma) \in$ HOD.

For (b): Let (R,Φ) be a common nondropping iterate of (P,Σ) and (P,Ψ), with $i = \pi^\Sigma_{P,R}$ and $j = \pi^\Psi_{P,R}$. Then

$$\begin{aligned}\pi^\Phi_{R,\infty} \circ i &= \pi^\Sigma_{P,\infty}\\ &= \pi^\Psi_{P,\infty}\\ &= \pi^\Phi_{R,\infty} \circ j.\end{aligned}$$

Thus $i = j$. But $\Sigma = \Phi^i$ and $\Psi = \Phi^j$ by pullback consistency, and so $\Sigma = \Psi$.

(c) follows at once from (b). □

If $\theta_0 \le o(M_\infty(P,\Sigma))$, then Σ is not ordinal definable from a real. Thus under reasonable hypotheses, (c) cannot be improved. Similarly, if $\theta_0 \le o(M_\infty(P,\Sigma))$, then there is a $\Psi \ne \Sigma$ such that (P,Ψ) is mouse equivalent to (P,Σ). (By the Basis Theorem, there is also a $\boldsymbol{\Delta}^2_1$ strategy Λ such that (P,Λ) is a mouse pair, so P can have mouse-inequivalent expansions.)

Proposition 3.2: *Let (P,Σ) and (Q,Ψ) be projectum stable mouse pairs; then $(P,\Sigma) \equiv^* (Q,\Psi)$ if and only if $M_\infty(P,\Sigma) = M_\infty(Q,\Psi)$.*

---

[19]When $(Q,\Psi) \in \mathcal{F}(P,\Sigma)$, then $\Psi = \Sigma_Q$, and we may sometimes write $\pi^\Sigma_{Q,R}$ for $\pi^{\Sigma_Q}_{Q,R}$. There is no ambiguity here, because Σ determines P, so Q and Σ determine Σ_Q.

Proof: The "only if" part is obvious. Suppose now that $M_\infty(P, \Sigma) = M_\infty(Q, \Psi)$. Let A be a set of reals coding Σ and Ψ, let G be V-generic over $\mathrm{Col}(\omega_1, \mathbb{R})$, and let $N = L(A, \mathbb{R})[G]$, so that $N \models \mathsf{ZFC}$. In N, we have sequences $(P_\alpha, \Sigma_\alpha)$ and (Q_α, Ψ_α) that are cofinal in $\mathcal{F}(P, \Sigma)$ and $\mathcal{F}(Q, \Psi)$ respectively, such that if $\pi_{\alpha,\beta} = \pi^{\Sigma}_{P_\alpha, P_\beta}$, and $\tau_{\alpha,\beta} = \pi^{\Psi}_{Q_\alpha, Q_\beta}$, then

(a) for λ a limit ordinal, $P_\lambda = \bigcup_{\alpha<\lambda} \mathrm{ran}(\pi_{\alpha,\lambda})$ and $Q_\lambda = \bigcup_{\alpha<\lambda} \mathrm{ran}(\tau_{\alpha,\lambda})$, and
(b) $M_\infty(P, \Sigma) = \bigcup_{\alpha<\omega_1} \mathrm{ran}(\pi_{\alpha,\infty})$ and $M_\infty(Q, \Psi) = \bigcup_{\alpha<\omega_1} \mathrm{ran}(\tau_{\alpha,\infty})$.

Since $M_\infty(P, \Sigma) = M_\infty(Q, \Psi)$, we can find a $\lambda < \omega_1$ such that $\mathrm{ran}(\pi_{\lambda,\infty}) = \mathrm{ran}(\tau_{\lambda,\infty})$. Clearly, this implies $P_\lambda = Q_\lambda$ and $\pi_{\lambda,\infty} = \tau_{\lambda,\infty}$. By Proposition 3.1(b), $\Sigma_\lambda = \Psi_\lambda$. Thus (P, Σ) and (Q, Ψ) have a common iterate (in $V[G]$, but therefore in V), as desired. □

The key to our Suslin representation of Σ is the fact that $M_\infty(P, \Sigma)$ is an iterate of P via a normal tree all of whose countable weak hulls are by Σ. We show this now.

Lemma 3.3: *[Strategy extension] Let (P, Σ) be a projectum stable mouse pair; then there is a unique extension Ψ of Σ such that*

(a) Ψ is defined on all normal trees $\mathcal{T}$ by Ψ such that $\mathcal{T}$ is the surjective image of $\mathbb{R}$, and
(b) if $\mathcal{T}$ is by Ψ, then all countable weak hulls of $\mathcal{T}$ are by Σ.

Proof: For uniqueness, suppose b and c are cofinal branches of $\mathcal{T}$, and that all countable weak hulls of $\mathcal{T}^\frown b$ and $\mathcal{T}^\frown c$ are by Σ. If $b \neq c$, then taking a Skolem hull of the situation, we get countable weak hulls $\mathcal{S}^\frown d$ and $\mathcal{S}^\frown e$ of $\mathcal{T}^\frown b$ and $\mathcal{T}^\frown c$ such that $d \neq e$. But $d = \Sigma(\mathcal{S}) = e$, contradiction.

So it is enough to show that if $\mathcal{T}$ is a normal tree of limit length, $\mathcal{T}$ is the surjective image of $\mathbb{R}$, and all countable weak hulls of $\mathcal{T}$ are by Σ, then $\mathcal{T}$ has a cofinal branch b such that all countable weak hulls of $\mathcal{T}^\frown b$ are by Σ. Fix then such a $\mathcal{T}$.

Let μ be the supercompactness measure on $P_{\omega_1}(\mathcal{T})$. Since $\mathcal{T}$ is wellorderable in order type $< \theta$, μ exists. For μ-a.e. σ, let

$$\pi_\sigma \colon \mathcal{T}_\sigma \to \mathcal{T}$$

be the uncollapse map. It is clear that μ-a.e., π_σ generates a weak hull embedding of $\mathcal{T}_\sigma$ into $\mathcal{T}$. (In fact, it generates a full hull embedding.) So $\mathcal{T}_\sigma$

is by Σ, and we can set

$$b_\sigma = \Sigma(\mathcal{T}_\sigma).$$

Set

$$b = [\lambda\sigma.b_\sigma]_\mu.$$

We are taking the ultrapower of wellorderable structures, so there is enough elementarity that b is a cofinal branch of $\mathcal{T}$. Suppose that $\mathcal{S}^\frown d$ is a countable weak hull of $\mathcal{T}^\frown b$. It is easy to see that for μ a.e. σ, $\mathcal{S}^\frown d$ is a weak hull of $\mathcal{T}_\sigma{}^\frown b_\sigma$. Thus $\mathcal{S}^\frown d$ is by Σ, as desired. □

Definition 3.4: Let (P, Σ) be a projectum stable mouse pair; then for normal trees $\mathcal{T}$ on P such that $\mathcal{T}$ is the surjective image of $\mathbb{R}$, $\Sigma^+(\mathcal{T}) = b$ iff every countable weak hull of $\mathcal{T}^\frown b$ is by Σ.

Lemma 3.5: *Let (P, Σ) be a projectum stable mouse pair; then there is a normal tree $\mathcal{U}$ by Σ^+ such that*

(a) $\mathcal{U}$ has last model $\mathcal{M}^{\mathcal{U}}_\infty = M_\infty(P, \Sigma)$, and
(b) $i^{\mathcal{U}}_{0,\infty} = \pi^{\Sigma}_{P,\infty}$.

Proof: We build $\mathcal{U}$ by induction, iterating away extender disagreements with $M_\infty(P, \Sigma)$, and using Σ^+ to pick branches. Suppose that at some stage we have constructed $\mathcal{T}$ of length $\alpha + 1$.

Claim 1. If β is least such that $\mathcal{M}^{\mathcal{T}}_\alpha|\beta \neq M_\infty(P, \Sigma)|\beta$, then $\mathcal{M}^{\mathcal{T}}_\alpha|\beta$ is extender active, and $M_\infty(P, \Sigma)|\beta$ is passive.

Proof: Note that if P is an lpm, this says something about the agreement of the internal strategies. We aren't trying to line up external strategies, although it would be possible to say something sensible in that direction.

Let $A \subseteq \mathbb{R}$ be the codeset of Σ. It is easy to see that $\mathcal{T}$ and $M_\infty(P, \Sigma)$ are definable from A in $L_\xi(A, \mathbb{R})$, for ξ least such that ZF^- holds in $L_\xi(A, \mathbb{R})$. Using $\mathsf{DC}_\mathbb{R}$ we get a countable $\tau \subseteq \mathbb{R}$ and an embedding

$$\pi\colon L_\gamma(B, \tau) \to L_\xi(A, \mathbb{R}).$$

Let $\pi(\mathcal{S}) = \mathcal{T}$ and

$$\pi(N) = M_\infty(P, \Sigma).$$

$B = A \cap \tau$ codes $\Sigma \restriction L_\gamma(B, \tau)$, so

$$N = \text{ direct limit of } \mathcal{F}(P, \Sigma) \cap L_\gamma(B, \tau).$$

Working outside $L(B, \tau)$ we can pick a countable stack s cofinal in this direct limit system, and by Theorem 2.4, we have the normal tree

$$X = X(s)$$

by Σ which iterates (P, Σ) to (N, Σ_N). But note that $\mathcal{S}$ is also by Σ, because it is a weak hull of $\mathcal{T}$, and $\mathcal{T}$ is by Σ^+. Normal trees by a fixed strategy are determined by their last models, so $\mathcal{S}$ is an initial segment of X. Thus our claim is true about $\mathcal{S}$ in $L_\gamma(B, \mathbb{R})$. By elementarity, our claim is true. □

Claim 2. Neither of $\mathcal{M}_\alpha^{\mathcal{T}}$ and $M_\infty(P, \Sigma)$ is a proper initial segment of the other.

Proof: The reflection argument of Claim 1 shows this as well. □

Claims 1 and 2 yield Lemma 3.5 at once. □

Definition 3.6: Let (P, Σ) be a projectum stable mouse pair; then $\mathcal{U}(P, \Sigma)$ is the unique normal tree by Σ^+ whose last model is $M_\infty(P, \Sigma)$.

One might be tempted to think that $(P, \Sigma) <^* (Q, \Psi)$ iff $o(M_\infty(P, \Sigma)) < o(M_\infty(Q, \Psi))$, but this is trivially false. Let P be the least active mouse, and let Q be a mouse such that $P \lhd Q$ with $Q \models$ "all sets are countable". Let Σ and Ψ be the (in case, unique) strategies for the two. Then $o(M_\infty(P, \Sigma)) > \omega_1$, but $M_\infty(Q, \Psi) = Q$, because it is impossible to iterate Q without dropping.

Note also that in the example above, Ψ is not $o(M_\infty(Q, \Psi))$-Suslin. What we get directly is a Suslin representation for the $\mathcal{U}(Q, \Psi)$-certified part of Ψ.

Definition 3.7: Let (P, Σ) be a projectum stable mouse pair; then

(a) For $\mathcal{T}$ a countable normal iteration tree on P, $\mathcal{T}$ is $\mathcal{U}(P, \Sigma)$-*certified* iff $\mathcal{T}$ is a weak hull of $\mathcal{U}(P, \Sigma)$.
(b) For s a countable stack on P, s is $\mathcal{U}(P, \Sigma)$-certified iff $W(s)$ is a weak hull of $\mathcal{U}(P, \Sigma)$.
(c) Σ^{uc} is the set of all countable $\mathcal{U}(P, \Sigma)$-certified stacks.

Clearly, every $\mathcal{U}(P, \Sigma)$-certified normal tree or stack is by Σ. This includes all trees and stacks that do not drop on their main branches, by the following two lemmas.

Lemma 3.8: *Let (P, Σ) be a projectum stable mouse pair, and $\mathcal{T}$ a countable normal tree on P. Equivalent are*

(a) $\mathcal{T}$ is $\mathcal{U}(P, \Sigma)$-certified,
(b) there is a countable, normal tree $\mathcal{S}$ by Σ whose main branch does not drop, and such that $\mathcal{T}$ is a weak hull of $\mathcal{S}$.

Proof: That (a) implies (b) follows by a simple Skolem hull argument. Now let $\mathcal{S}$ be such that $\mathcal{S}$ is by Σ, and $\mathcal{S}$ has last model Q, P-to-Q does not drop, and $\mathcal{T}$ is a weak hull of $\mathcal{S}$. Let $\mathcal{V}$ be the normal tree by Σ_Q^+ on Q having last model $M_\infty(Q, \Sigma_Q) = M_\infty(P, \Sigma)$. A simple Skolem hull argument shows that

$$X(\mathcal{S}, \mathcal{V}) = \mathcal{U},$$

and by using Theorem 2.11, that $\mathcal{S}$ is a weak hull of $\mathcal{U}$. It follows that $\mathcal{T}$ is a weak hull of $\mathcal{U}$. □

For stacks, we have

Lemma 3.9: *Let (P, Σ) be a projectum stable mouse pair, and s a countable normal stack on P that is by Σ. Suppose that either*

(a) s has limit length, and does not drop along its unique branch, or
(b) s has a last tree $\mathcal{T}$, with base model $Q = \mathcal{M}_0^{\mathcal{T}}$ such that P-to-Q does not drop, and $\mathcal{T}$ is $\mathcal{U}(Q, \Sigma_Q)$-certified.

Then s is $\mathcal{U}(P, \Sigma)$-certified.

Proof: If (a) holds, then we can extend s by one model to t by Σ that has a last model, and does not drop going to it. But then $W(t)$ is by Σ, has a last model, and does not drop going to it. So $W(t)$ is $\mathcal{U}(P, \Sigma)$-certified by Lemma 3.8. But $W(s) = W(t)$.

So assume (b) holds, and let $s = t^\frown\langle\mathcal{T}\rangle$, where t is by Σ and doesn't drop, with last model Q, and $\mathcal{T}$ is $\mathcal{U}(Q, \Sigma_Q)$-certified. By Lemma 3.8, $\mathcal{T}$ is a weak hull of $\mathcal{V}$, for some $\mathcal{V}$ on Q by Σ_Q such that $\mathcal{V}$ has a last model, and does not drop going to it. Since $t^\frown\langle\mathcal{V}\rangle$ is by Σ, so is $W(t^\frown\langle\mathcal{V}\rangle)$, so by Lemma 3.8, $W(t^\frown\langle\mathcal{V}\rangle)$ is a weak hull of $\mathcal{U}(P, \Sigma)$. But by [18], $W(t^\frown\mathcal{T})$ is a weak hull of $W(t^\frown\mathcal{V})$. So $W(t^\frown\mathcal{T})$ is a weak hull of $\mathcal{U}(P, \Sigma)$, as desired. □

Definition 3.10: Let (P, Σ) be a projectum stable mouse pair; then

(a) Σ^{nd} is the collection of countable normal trees by Σ having a last model, and such that the branch to that model does not drop.
(b) A countable normal $\mathcal{T}$ is $M_\infty(P,\Sigma)$-*relevant* iff $\mathcal{T}$ has some (perhaps improper) extension $\mathcal{S}$ such that $\mathcal{S}$ is in Σ^{nd}.
(c) A countable stack s is $M_\infty(P,\Sigma)$-*relevant* iff s is by Σ, and either has limit length and does not drop along its unique branch, or has a last tree $\mathcal{T}$ with base model Q such that P-to-Q does not drop, and $\mathcal{T}$ is $M_\infty(Q,\Sigma_Q)$-relevant.
(d) Σ^{rl} is the restriction of Σ to all $M_\infty(P,\Sigma)$-relevant stacks.

One can give a more concrete characterization of the M_∞-relevant normal trees, but we don't need it here.

Given sets of reals A and B, we say that A is *positive*-$\boldsymbol{\Sigma}^1_{\mathbf{n}}(B)$ iff A is definable over $(V_{\omega+1},\in,B)$ from parameters by a Σ_n formula in which $\dot{B}$ has only positive occurrences. Note that if A is positive-$\boldsymbol{\Sigma}^1_1(B)$ and B is κ-Suslin, then A is κ-Suslin. Note also that each of the codesets of Σ^{nd}, Σ^{rl}, and Σ^{uc} is positive-$\boldsymbol{\Sigma}^1_1$ in each of the others, by Lemmas 3.8 and 3.9 and some simple computations. So we have

Proposition 3.11: *Let (P,Σ) be a projectum stable mouse pair; then for any κ, Σ^{uc} is κ-Suslin iff Σ^{nd} is κ-Suslin iff Σ^{rl} is κ-Suslin.*

Σ^{rl} is the fragment of Σ that gets used in forming $M_\infty(P,\Sigma)$. It is contained in Σ^{uc}, but the containment may be proper. By Lemma 3.8, Σ^{uc} consists of all weak hulls of trees by Σ^{rl}.

Theorem 3.12: *Let (P,Σ) be a projectum stable mouse pair; then*

(a) Σ^{uc} is $|M_\infty(P,\Sigma)|$-Suslin, and
(b) Σ^{uc} is not α-Suslin for any $\alpha < |M_\infty(P,\Sigma)|$.

Proof: Part (b) follows easily from the Kunen-Martin theorem. For let

$$(Q,\eta)R(S,\gamma) \Leftrightarrow [(Q,\Sigma_Q) \prec (S,\Sigma_S) \wedge \pi^{\Sigma}_{Q,\infty}(\eta) < \pi^{\Sigma}_{S,\infty}(\gamma)].$$

Code(R) is positive-$\boldsymbol{\Sigma}^1_1$ in Code(Σ^{uc}), so if Code(Σ^{uc}) is α Suslin, then so is Code(R). But R is a prewellorder of order type $o(M_\infty(P,\Sigma))$, so by Kunen-Martin, $\alpha \geq |M_\infty(P,\Sigma)|$.

For (a), let $\mathcal{U} = \mathcal{U}(P,\Sigma)$, and $\kappa = |M_\infty(P,\Sigma)| = |\mathcal{U}(P,\Sigma)|$. We may assume $\Sigma^{\text{uc}} \neq \emptyset$, as otherwise (a) is trivial. This implies $\kappa \geq \omega_1$. Let $C \subset \kappa$ code $\mathcal{U}$ and $M_\infty(P,\Sigma)$ in some simple way. (Note that they are both essentially sets of ordinals.) Let α be least such that $L_\alpha[C] \models \mathsf{ZFC}^-$.

Let T be the tree on $\omega \times L_\alpha[C]$ of all attempts to build an (x, π) such that $\pi\colon (\omega, E_x) \to L_\alpha[C]$ is elementary, and $\pi(0) = \mathcal{U}$. For $(x, \pi) \in [T]$, let

$$\sigma_x = \pi \circ \pi_x^{-1}$$

be the uncollapse map from $(\text{set}(x), \in)$ to $(\text{ran}(\pi), \in)$. Let $\mathcal{U}_x = \sigma_x^{-1}(\mathcal{U}) = \text{set}_0(x)$. So if $x \in p[T]$, then $\mathcal{U}_x$ is a pseudo-hull of $\mathcal{U}$, and every countable weak hull of $\mathcal{U}$ is a weak hull of some $\mathcal{U}_x$, where $x \in p[T]$. Let

$$\begin{aligned} A(y,x) \Leftrightarrow \text{Code}(y) \wedge \text{Code}(x) \wedge \\ \text{set}_0(y) \text{ is a stack of normal trees on } P \wedge \\ W(\text{set}_0(y)) \text{ is a weak hull of } \text{set}_0(x). \end{aligned}$$

Clearly, A is Σ^1_2 in P, hence κ-Suslin. But

$$y \in \text{Code}(\Sigma^{\text{uc}}) \Leftrightarrow \exists x (A(y,x) \wedge x \in p[T]).$$

Thus $\text{Code}(\Sigma^{\text{uc}})$ is κ-Suslin. □

In order to get a Suslin representation for the full Σ, and not just its M_∞-relevant part, we must amalgamate Suslin representations like that above.

Definition 3.13: H^e_α is the common value of $M_\infty(P, \Sigma)$, for all projectum stable pure extender pairs (P, Σ) of mouse rank α. H^h_α is the common value of $M_\infty(P, \Sigma)$, for all projectum stable lbr hod pairs (P, Σ).

When context permits, we shall simply write H_α, letting the superscript "e" or "h" be understood. Notice that H_α is essentially a set of ordinals, and thus $\langle H^e_\alpha \mid \alpha < \theta\rangle$ and $\langle H^h_\alpha \mid \alpha < \theta\rangle$ are in HOD. Assuming $\mathsf{AD}_\mathbb{R} + \mathsf{HPC}$, $\text{HOD} = L[\langle H^h_\alpha \mid \alpha < \theta\rangle]$, as we shall see.

The next lemma strengthens Lemma 3.5.

Lemma 3.14: *Let (P, Σ) be a projectum stable mouse pair of mouse rank α; then for all $\beta \leq \alpha$ there is a unique normal tree $\mathcal{W}_\beta$ on P such that $\mathcal{W}_\beta$ is by Σ^+, and has last model H_β.*

Remark 3.15: Of course, the branch P-to-H_β in $\mathcal{W}_\beta$ drops iff $\beta < \alpha$.

Proof: Fix β, and let $M_\infty(Q, \Psi) = H_\beta$. We may assume $\beta < \alpha$, as otherwise Lemma 3.5 applies. Comparing (P, Σ) with (Q, Ψ), we obtain a normal $\mathcal{T}$ on P by Σ with last model (R, Φ) such that (R, Φ) is a nondropping iterate of (Q, Ψ). Then $H_\beta = M_\infty(R, \Phi)$, so by Lemma 3.5 there is a normal tree $\mathcal{U}$ by Φ^+ on R whose last model is H_β. We set

$$\mathcal{W}_\beta = X(\mathcal{T}, \mathcal{U}).$$

Then $\mathcal{W}_\beta$ is a normal tree on P by Σ^+ that has last model H_β. (This can be shown by passing to the appropriate countable elementary submodel of V.) Uniqueness is clear. □

Corollary 3.16: *Let (P, Σ) be a mouse pair of mouse rank α, and let $\gamma = \sup_{\beta \leq \alpha} o(H_\beta)$; then*

(1) Σ is γ-Suslin, and
(2) either
  (a) Σ is not μ-Suslin for any $\mu < |\gamma|$, or
  (b) γ is singular, and $\gamma = \nu^+$ for some cardinal ν such that Σ is ν-Suslin, but not μ-Suslin for any $\mu < \nu$.

Proof: (*Sketch.*) Let $\langle \mathcal{W}_\beta \mid \beta \leq \alpha \rangle$ be the sequence of normal trees by Σ^+ given by the lemma. We verify that a countable stack s is by Σ by producing a $\beta \leq \alpha$ and a weak tree embedding from $W(s)$ into $\mathcal{W}_\beta$. The details can be arranged as they were in the proof of Theorem 3.12. The result is a tree

$$T \subseteq \omega \times (\alpha + 1) \times \gamma,$$

such that

$$p[T] = \text{Code}(\Sigma).$$

We claim that $|\alpha| \leq \gamma$. For if not, then since each H_β is essentially a subset of $o(H_\beta)$, we get a sequence of γ^+ distinct subsets of γ, contrary to the boldface GCH. Thus Σ is γ-Suslin.

Suppose now that $\nu \leq |\gamma|$ is least such that Code(Σ) is ν-Suslin. For each $\beta < \alpha$, there is a dropping iterate (Q, Ψ) of (P, Σ) such that $M_\infty(Q, \Psi) = H_\beta$. But Code$(\Psi)$ is positive-$\boldsymbol{\Sigma}^1_1$ in Code(Σ), so Code(Ψ) is ν-Suslin, so $|H_\beta| \leq \nu$ by Theorem 3.12. Thus $\gamma \leq \nu^+$. If $\gamma < \nu^+$ then we have conclusion (a).

Assume then $\gamma = \nu^+$. We have also $|\alpha| \leq \nu$, because the relation

$Q < R \Leftrightarrow$ there are normal trees $\mathcal{T}$ by Σ with last model R and
$\mathcal{S}$ by Σ_R with last model Q such that R-to-Q drops

is positive-$\boldsymbol{\Sigma}^1_1$ in Code(Σ), and hence ν-Suslin. It follows that the $o(H_\beta)$ for $\beta \leq \alpha$ witness that ν^+ is singular. Thus we have (b). □

Remark 3.17: We don't know whether clause (b) in the conclusion of Corollary 3.16 ever actually occurs.

We consider now a second way to get Suslin cardinals from mouse pairs.

Definition 3.18: If P is a premouse and $\kappa \in P$, then $o(\kappa)^P = \sup\{\eta \mid \mathrm{crit}(E_\eta^P) = \kappa\}$. (The sup is nonstrict.) δ is a *cutpoint* of P iff $\forall\alpha < \delta(o(\alpha)^P \leq \delta)$. δ is a *strong cutpoint* of P iff $\forall\alpha < \delta(o(\alpha)^P < \delta)$.

By this definition, there could be E on the P-sequence such that $\mathrm{lh}(E)$ is a cutpoint of P. However, we are usually interested in cutpoints that are cardinals of P, hence not of this form.

Definition 3.19: Let P be a premouse; then

(1) $\tau^P = \sup\{\kappa^{+,P} \mid \kappa < \rho_{k(P)}(P) \wedge \exists\eta > \kappa^{+,P}(\kappa = \mathrm{crit}(E_\eta^P))\}$.
(2) P has *limit type* iff τ^P is a limit of P-cardinals, and $\forall\alpha < \tau^P(o(\alpha)^P < \tau^P)$. *$P$ has a top block* iff P is not of limit type.
(3) If P has a top block, then β^P is the least $\alpha < \tau^P$ such that $o(\alpha)^P \geq \tau^P$. We say that β^P *begins the top block* of P.

By convention, $\tau^P = 0$ when there are no E on the P sequence with critical point $< \rho_{k(P)}(P)$. In that case, P has no non-dropping iterates, and we aren't interested in it at the moment.

Clearly, either τ^P is a cardinal of P, or $\tau^P = o(P)$ and P has no largest cardinal. If α is the largest P-cardinal $< \tau^P$, then $o(\alpha)^P > \tau^P$. Thus β^P does indeed exist whenever P is not of limit type. If P has limit type, then τ^P is a limit of successor cardinal strong cutpoints in P. If P has a top block, then β^P is a limit of successor cardinal strong cutpoints in P.[20]

Definition 3.20: Let (P, Σ) be a projectum stable mouse pair; then $\tau_\infty(P, \Sigma) = \pi_{P,\infty}^\Sigma(\tau^P)$. If P has a top block, then $\beta_\infty(P, \Sigma) = \pi_{P,\infty}^\Sigma(\beta^P)$.

Theorem 3.21: *Let (P, Σ) be a projectum stable mouse pair; then Σ^{uc} is $\tau_\infty(P, \Sigma)$-Suslin.*

Proof: Let $\tau_\infty = \tau_\infty(P, \Sigma)$, and let $\mathcal{U} = \mathcal{U}(P, \Sigma)$. We have $i_{0,\alpha}^{\mathcal{U}} = \pi_{(P,\Sigma),\infty}$, where $\alpha+1 = \mathrm{lh}(\mathcal{U})$. $[0, \alpha)_U$ uses no extenders overlapping τ_∞, since this is in the range of the branch embedding. But $[0, \alpha)_U$ uses no extenders with

[20]We prove the second statement. If $\alpha < \beta^P$, then $o(\alpha)^P < \beta^P$, as otherwise $o(\alpha)^P \geq \tau^P$ by coherence, contradiction. But β^P is a limit (in fact, measurable) cardinal of P, and it is easy to see by coherence that $o(\alpha)^{+,P}$ is a strong cutpoint of P.

critical point $\geq \tau_\infty$, because τ^P is defined so that any such extender would cause $[0,\alpha)_U$ to drop. Thus all extenders used in $[0,\alpha)_U$ have length $< \tau_\infty$. It follows that all extenders used in $\mathcal{U}$ have length $< \tau_\infty$.

Now we get a τ_∞ Suslin representation of Σ^{uc} from the proof of Theorem 3.12. □

It follows that $|o(M_\infty(P,\Sigma)| \leq \tau_\infty(P,\Sigma)$. That is also easy to see directly from the proof of Theorem 3.21.

It is possible that neither τ_∞ nor β_∞ are cardinals of V, where V is the model of AD^+ in which we are living. For example, let δ be the Woodin cardinal of M_1, κ the least strong to δ, and ν a cardinal of M_1 such that $\kappa < \nu < \delta$. Let $P = M_1|\nu$, and let Σ be its unique iteration strategy. Then $\omega_1 < \beta_\infty(P,\Sigma) < \tau_\infty(P,\Sigma) < \omega_2$. If we take $P = M_1|\delta$ instead, we still have $\omega_1 < \beta_\infty(P,\Sigma) < \omega_2$, but now $\tau_\infty(P,\Sigma)$ is a cardinal, namely ω_ω. (See [2].)[21]

Suppose (P,Σ) has a top block. Must Σ^{uc} be $\beta_\infty(P,\Sigma)$-Suslin? In general, no, as the example of M_1 and its strategy shows. What becomes Suslin at β_∞ is the fragment of Σ consisting of weak hulls of $\mathcal{U}(P,\Sigma) \restriction (\beta_\infty(P,\Sigma)+1)$.

Definition 3.22: Let (P,Σ) be a projectum stable mouse pair having a top block, let $\mathrm{lh}(\mathcal{U}(P,\Sigma)) = \gamma+1$, and $\nu \in [0,\gamma]_U$ be least such that $\mathrm{crit}(i^{\mathcal{T}}_{\nu,\gamma}) > i_{0,\nu}(\beta^P)$; then we set $\mathcal{U}_0(P,\Sigma) = \mathcal{U}(P,\Sigma) \restriction \nu+1$. We write $M^0_\infty(P,\Sigma)$ for the last model of $\mathcal{U}_0(P,\Sigma)$.

Another way of saying it is, the main branch extender of $\mathcal{U}_0(P,\Sigma)$ is $E \restriction \beta_\infty$, where E is the main branch extender of $\mathcal{U}(P,\Sigma)$. We have $|\mathcal{U}_0(P,\Sigma)| = |\beta_\infty(P,\Sigma)|$. In the cases we shall care about below, β_∞ is a cardinal of V, so $\mathrm{lh}(\mathcal{U}_0(P,\Sigma)) = \beta_\infty(P,\Sigma)+1$.

Definition 3.23: Let (P,Σ) be a projectum stable mouse pair; then

(a) For $\mathcal{T}$ a countable normal iteration tree on P, $\mathcal{T}$ is $\mathcal{U}_0(P,\Sigma)$-*certified* iff $\mathcal{T}$ is a weak hull of $\mathcal{U}_0(P,\Sigma)$.
(b) For s a countable stack on P, s is $\mathcal{U}_0(P,\Sigma)$-*certified* iff $W(s)$ is a weak hull of $\mathcal{U}_0(P,\Sigma)$.
(c) Σ^{ucb} is the set of all countable $\mathcal{U}_0(P,\Sigma)$-certified stacks.

[21]Note that τ_∞ is the next Suslin after β_∞ in this example. That is part of a more general pattern.

Definition 3.24: Let (P, Σ) be a projectum stable mouse pair, and $i\colon P \to Q$ an iteration map coming from a non-dropping iteration by Σ; then i is *ν-generated* iff for $k = k(P)$, $Q^k = \{i(f)(a) \mid f \in P^k \wedge a \in [\nu]^{<\omega}\}$.

Of course, when $\nu = \rho_{k(P)}(P)$, every nondropping iteration map i is $i(\nu)$-generated. We say a branch b of $\mathcal{T}$ is ν-generated iff b does not drop, and $i_b^{\mathcal{T}}$ is ν-generated. We are mainly interested below in the case that P has a top block, and b is $i_b(\beta^P)$-generated. In this case, $\mathcal{T}^\frown b$ is $\mathcal{U}_0(P, \Sigma)$-certified:

Lemma 3.25: *Let (P, Σ) be a projectum stable mouse pair having a top block, and $\mathcal{T}$ a countable normal tree on P. Equivalent are*

(a) $\mathcal{T}$ is $\mathcal{U}_0(P, \Sigma)$-certified,
(b) $\mathcal{T}$ is a weak hull of some countable, normal tree $\mathcal{S}$ by Σ whose main branch b does not drop, and is $i_b(\beta^P)$-generated.

Proof: That (a) implies (b) follows by a simple Skolem hull argument.

Now let $\mathcal{S}$ be as in (b), with last model Q. As in the proof of Lemma 3.8, letting $\mathcal{V}$ be the normal tree by Σ_Q^+ on Q having last model $M_\infty(Q, \Sigma_Q) = M_\infty(P, \Sigma)$, we have

$$X(\mathcal{S}, \mathcal{V}) = \mathcal{U},$$

where $\mathcal{U} = \mathcal{U}(P, \Sigma)$. So there is a weak tree embedding Φ of $\mathcal{S}$ into $\mathcal{U}$. Let $\alpha + 1 = \mathrm{lh}(\mathcal{S})$, and

$$\Phi = \langle u, \langle s_\xi \mid \xi \leq \alpha\rangle, \langle t_\xi \mid \xi < \alpha\rangle, p\rangle,$$

and let $v = v^\Phi$. We have that

$$s_\alpha\colon Q \to \mathcal{M}^{\mathcal{U}}_{v(\alpha)},$$

moreover $v(\alpha) \leq_U \gamma$ where $\gamma + 1 = \mathrm{lh}(\mathcal{U})$,

$$i^{\mathcal{U}}_{0,v(\alpha)} = s_\alpha \circ i^{\mathcal{S}}_{0,\alpha},$$

and

$$\mathcal{M}^{\mathcal{U}}_{v(\alpha)} = \{i^{\mathcal{U}}_{0,v(\alpha)}(f)(a) \mid a \in [\lambda]^{<\omega}\},$$

where $\lambda = \sup(\{s_\alpha(\lambda(E)) \mid E \text{ is used in } [0, \alpha)_S\})$. Since the branch extender of $[0, \alpha)_S$ has all measures concentrating on β^P,

$$\lambda \leq s_\alpha(i^{\mathcal{S}}_{0,\alpha}(\beta^P)) = i^{\mathcal{U}}_{0,v(\alpha)}(\beta^P).$$

But this implies $v(\alpha) + 1 \leq \text{lh}(\mathcal{U}_0)$, where $\mathcal{U}_0 = \mathcal{U}_0(P, \Sigma)$. (In fact, because $\mathcal{S}$ is countable, $v(\alpha)$ is somewhere on the main branch of $\mathcal{U}_0$ strictly below its last model.) Thus Φ is a weak tree embedding of $\mathcal{S}$ into $\mathcal{U}_0$. So $\Phi \upharpoonright \mathcal{T}^\frown b$ is a weak tree embedding of $\mathcal{T}^\frown b$ into $\mathcal{U}_0$. □

The next two lemmas show that $M^0_\infty(P, \Sigma)$ can be viewed as a direct limit of countable mouse pairs.

Lemma 3.26: *Let (P, Σ) be a projectum stable mouse pair, let $\mu < \rho_{k(P)}(P)$, and let s and t be nondropping stacks on P with last models Q and R that are $i_s(\mu)$ and $i_t(\mu)$-generated, respectively. Then there are normal trees $\mathcal{T}$ on Q by Σ_Q and $\mathcal{V}$ on R by Σ_R with a common last model S, and such that for $j = i^{\mathcal{T}} \circ i_s = i^{\mathcal{V}} \circ i_t$, j is $j(\mu)$-generated.*

Proof: We can compare (Q, Σ_Q) with (R, Σ_R) by iterating away least extender disagreements. Because Σ is positional, no strategy disagreements show up as we do so, and eventually we get $\mathcal{T}$ and $\mathcal{V}$ whose last models have no disagreement at all. By Dodd-Jensen, $\mathcal{T}$ and $\mathcal{V}$ have a common last model (S, Σ_S), neither side drops, and $i^{\mathcal{T}} \circ i_s = i^{\mathcal{V}} \circ i_t$. Let $j = i^{\mathcal{T}} \circ i_s$; we must see j is $j(\mu)$-generated.

Suppose not, and let $\gamma + 1 = \text{lh}(\mathcal{T})$. Since i_s is $i_s(\mu)$-generated, the generator of j above $j(\mu)$ must have been introduced by $i^{\mathcal{T}}$. Thus we have a least $\alpha <_T \gamma$ such that $i^{\mathcal{T}}_{0,\alpha} \circ i_s(\mu) = j(\mu)$. Similarly, for $\delta + 1 = \text{lh}(\mathcal{V})$, we have a least $\beta <_V \delta$ such that $i^{\mathcal{V}}_{0,\beta} \circ i_t(\mu) = j(\mu)$. But then

$$\begin{aligned} \mathcal{M}^{\mathcal{T}}_\alpha &= \text{Ult}(P, E_j \upharpoonright j(\mu)) \\ &= \mathcal{M}^{\mathcal{V}}_\beta. \end{aligned}$$

Since $\alpha+1 < \text{lh}(\mathcal{T})$ and $\beta+1 < \text{lh}(\mathcal{V})$, and we were iterating away extender disagreements, this is a contradiction. □

Remark 3.27: Lemma 3.26 describes another way in which the system of iterates of a mouse pair behaves like the system of iterates of a sound mouse projecting to ω.

By the lemma, for any $\mu \leq \rho_{k(P)}(P)$, we can form a directed system

$$\mathcal{F}_\mu(P, \Sigma) = \{(Q, \Sigma_s) \mid i_s \colon P \to Q \text{ is } i_s(\mu)\text{-generated}\}.$$

For $\mu = \rho_{k(P)}(P)$, this is just $\mathcal{F}(P, \Sigma)$. Let

$$M^\mu_\infty(P, \Sigma) = \text{ direct limit of } \mathcal{F}_\mu(P, \Sigma),$$

and $\pi^{\mu}_{P,\infty}\colon P \to M^{\mu}_{\infty}(P,\Sigma)$ be the canonical embedding. It is quite easy to see

Lemma 3.28: *Let (P,Σ) be a projectum stable mouse pair, and $\mathcal{U} = \mathcal{U}(P,\Sigma)$ have length $\gamma+1$. Let $\mu < \rho_{k(P)}(P)$ and let $\alpha \leq \gamma$ be largest such that $i^{\mathcal{U}}_{0,\alpha}$ is $i^{\mathcal{U}}_{0,\alpha}(\mu)$-generated; then $\mathcal{M}^{\mathcal{U}}_{\alpha} = M^{\mu}_{\infty}(P,\Sigma)$, and $i^{\mathcal{U}}_{0,\alpha} = \pi^{\mu}_{P,\infty}$.*

In the special case $\mu = \beta^P$, we write $M^0_{\infty}(P,\Sigma)$ for $M^{\mu}_{\infty}(P,\Sigma)$, etc. The normal tree from P to $M^0_{\infty}(P,\Sigma)$ is then the initial segment of $\mathcal{U}(P,\Sigma)$ that we have called $\mathcal{U}_0(P,\Sigma)$

Definition 3.29: Let (P,Σ) be a projectum stable mouse pair with a top block; then

(a) Σ^{ndb} is the collection of countable normal trees $\mathcal{T}$ by Σ having a last model, and such that the branch b to that model does not drop, and is $i_b(\beta^P)$-generated.[22]
(b) A countable normal $\mathcal{T}$ is $M^0_{\infty}(P,\Sigma)$*-relevant* iff $\mathcal{T}$ has some (perhaps improper) extension $\mathcal{S}$ such that $\mathcal{S}$ is in Σ^{ndb}.
(c) A countable stack s is $M^0_{\infty}(P,\Sigma)$*-relevant* iff s is by Σ, and either
   (i) s has limit length, does not drop along its unique branch b, and b is $i_b(\beta^P)$-generated, or
   (ii) $s = v^\frown\langle\mathcal{T}\rangle$, v does not drop, has last model Q, i_v is $i_v(\beta^Q)$-generated, and $\mathcal{T}$ is $M^0_{\infty}(Q,\Sigma_Q)$-relevant.
(d) Σ^{rlb} is the restriction of Σ to all $M^0_{\infty}(P,\Sigma)$-relevant stacks.

Σ^{rlb} is the fragment of Σ relevant to forming $M^0_{\infty}(P,\Sigma)$. It is contained in Σ^{ucb}, perhaps properly. By Lemma 3.25, Σ^{ucb} is the collection of weak hulls of trees by Σ^{rlb}. For any α, Σ^{ndb} is α-Suslin iff Σ^{rlb} is α-Suslin iff Σ^{ucb} is α-Suslin.

Theorem 3.30: *Let (P,Σ) be a projectum stable mouse pair having a top block; then Σ^{ucb} is $\beta_{\infty}(P,\Sigma)$-Suslin, but not α-Suslin for any $\alpha < |\beta_{\infty}(P,\Sigma)|$.*

Proof: We verify that a stack s is by Σ^{ucb} by producing a weak tree embedding from $W(s)$ into some countable Skolem hull of $\mathcal{U}_0(P,\Sigma)$. This implies that $\mathrm{Code}(\Sigma^{\mathrm{ucb}})$ is β_{∞}-Suslin. The details are the same as those in the proof of Theorem 3.12.

[22] The "b" in the superscript of Σ^{ndb}, etc., is for "beta" or "bottom".

Again, the second part follows easily from Kunen-Martin. Let $\alpha < |\beta_\infty(P,\Sigma)|$, and let

$$(Q,\eta)R(S,\gamma) \Leftrightarrow [(Q,\Sigma_Q) \prec (S,\Sigma_S) \wedge \pi^\Sigma_{Q,\infty}(\eta) < \pi^\Sigma_{(S,\infty}(\gamma)].$$

Code(R) is positive-$\mathbf{\Sigma}^1_1$ in Code(Σ^{rlb}), and R is a prewellorder of order type $o(M^0_\infty(P,\Sigma))$, so Code(Σ^{rlb}) cannot be α-Suslin. □

So if (P,Σ) has a top block, there are two Suslin cardinals associated to it, $|\beta_\infty|$ and $|\tau_\infty|$. It is possible that they are equal; for example, this is true if P is a cardinal initial segment M_1 strictly below its Woodin cardinal δ but above the least strong to δ, and Σ is the strategy induced by the unique strategy for M_1. If $P = M_1$ and Σ is its strategy, then τ_∞ is the next Suslin cardinal after β_∞. We conjecture that these are the only two possibilities.

Conjecture 3.30.1: Let (P,Σ) be a mouse pair having a top block, and suppose that $\beta_\infty(P,\Sigma) < |\tau_\infty(P,\Sigma)|$; then $\tau_\infty(P,\Sigma)$ is the least Suslin cardinal strictly greater than $\beta_\infty(P,\Sigma)$.

We shall prove some approximations to this conjecture now, and use them in the later sections. A full proof of the conjecture would require a finer analysis than we have done, and perhaps involve new results on pointclass envelopes and self justifying systems.

The proof of Theorem 3.30 shows that in certain cases, the full Σ^{rl} is $\beta_\infty(P,\Sigma)$-Suslin.

Definition 3.31: Let P be a premouse. We say that P is *top block stable* iff P is projectum stable, P has a top block, and

(a) $o(\beta^P) = \tau^P$, and
(b) whenever E is on the P-sequence and $\mathrm{crit}(E) < \rho_{k(P)}(P)$, then i^P_E is continuous at τ^P.

We say that a mouse pair (P,Σ) is top block stable iff P is top block stable.

β^P is the least strong to τ^P, so when $\tau^P < o(P)$, clause (a) is equivalent to τ^P being a cutpoint. Clause (a) and coherence imply that τ^P is not the critical point of any extender from the P-sequence, so it is a strong cutpoint. Clause (b) can be put somewhat more concretely:

Proposition 3.32: *Let P have a top block, and suppose $o(\beta^P) = \tau^P$; then P is top block stable iff P is projectum stable, and either $\tau^P = \rho_{k(P)}(P)$, or $\tau^P < \rho_{k(P)}(P)$ and* $\mathrm{cof}^P(\tau^P)$ *is not the critical point of a total extender from the P-sequence.*

Lemma 3.33: *Let (P, Σ) be a mouse pair having a top block, and suppose that P is not top block stable; then*

(a) whenever s is by Σ^{rl}, then $W(s)$ is $\mathcal{U}_0(P, \Sigma)$ certified,
(b) Σ^{rl} is $\beta_\infty(P, \Sigma)$-Suslin, and
(c) Σ is γ-Suslin, where $\gamma = \sup(\beta_\infty(P, \Sigma) \cup \{o(H_\beta) \mid \beta <$ mouse rank of $(P, \Sigma)\}$.

Proof: We prove (a). Part (b) then follows at once, and we get (c) using the proof of Theorem 3.16.

Let $\mathcal{U} = \mathcal{U}(P, \Sigma)$ and $\mathcal{U}_0 = \mathcal{U}_0(P, \Sigma)$. Let $s = t^\frown\langle \mathcal{V} \rangle$ be a stack with last tree $\mathcal{V}$ by Σ^{rl}; then we can extend $\mathcal{V}$ normally to $\mathcal{V}_1$ by Σ_t^{nd}. $W(s)$ is a weak hull of $\mathcal{T}$, where $\mathcal{T} = W(t^\frown\langle \mathcal{V}_1 \rangle)$. Thus it is enough to show that whenever $\mathcal{T}$ is a normal tree by Σ^{nd}, then $\mathcal{T}$ is a weak hull of $\mathcal{U}_0$.

So fix a normal $\mathcal{T}$ by Σ^{nd}, with last model $Q = \mathcal{M}_\gamma^{\mathcal{T}}$. Let

$$\begin{aligned}\delta &= \sup(\{\lambda(E_\xi^{\mathcal{T}}) \mid \xi + 1 < \mathrm{lh}(\mathcal{T})\}) \\ &= \sup(\{\lambda(E_\xi^{\mathcal{T}}) \mid \xi + 1 \leq_T \gamma)\}).\end{aligned}$$

Claim 1. $\delta \leq \tau^Q$.

Proof: Suppose not; then there must be an $\alpha <_T \gamma$ such that $i_{0,\alpha}^{\mathcal{T}}(\tau^P) \leq \mathrm{crit}(i_{\alpha,\gamma}^{\mathcal{T}}))$. Fix such an α, and let $R = \mathcal{M}_\alpha^{\mathcal{T}}$ and $E = E_\xi^{\mathcal{T}}$ where $\xi + 1 \leq_T \gamma$ and $\alpha = T\mathrm{pred}(\xi + 1)$. E is total over R and $\mathrm{crit}(E) < \rho_{k(R)}(R)$ because $[0, \gamma)_T$ does not drop in model or degree. By coherence and the initial segment condition, the normal measure of E is indexed on the R-sequence, so $\mathrm{crit}(E) < \tau^R$. But $\tau^R = i_{0,\alpha}^{\mathcal{T}}(\tau^P)$, contradiction. □

Claim 2. If $\delta < \tau^Q$, then $\mathcal{T}$ is $\mathcal{U}_0$-certified.

Proof: If $\beta^Q \geq \delta$, then $\mathcal{T}$ is $\mathcal{U}_0$ certified by Lemma 3.25. Otherwise, we have $\beta^Q < \delta < \tau^Q$, so we have E on the sequence of Q such that $\mathrm{crit}(E) = \beta^Q$ and $\delta < \mathrm{lh}(E)$. But then the normal extension $\mathcal{T}^\frown\langle E \rangle$ of $\mathcal{T}$ is $\mathcal{U}_0$-certified by Lemma 3.25, so $\mathcal{T}$ is $\mathcal{U}_0$-certified. □

So we assume $\delta = \tau^Q$. Since τ^Q is in $\mathrm{ran}(i_{0,\gamma}^{\mathcal{T}})$, it is not $\lambda(E)$ for any E used in $\mathcal{T}$, so γ is a limit ordinal. Let $b = [0, \gamma)_T$and $i_b = i_b^{\mathcal{T}}$. We have that b does not drop,

$$i_b(\tau^P) = \delta(\mathcal{T}),$$

$Q = \mathcal{M}_b^{\mathcal{T}}$, and

$$i_b(\langle \beta^P, \tau^P \rangle) = \langle \beta^Q, \tau^Q \rangle$$

by the elementarity of i_b.

If $o(\beta^P)^P > \tau^P$, then $o(\beta^Q)^Q > \tau^Q$, so there is an E on the Q sequence with $\text{crit}(E) = i_b(\beta^P)$ and $\text{lh}(E) > \delta(\mathcal{T})$. Let

$$\mathcal{S} = \mathcal{T}^\frown \langle E \rangle,$$

where E is applied to $\mathcal{M}_\nu^{\mathcal{T}}$, for $\nu \in b$ least such that $\text{crit}(i_{\nu,b}^{\mathcal{T}}) > i_{0,\nu}^{\mathcal{T}}(\beta^P)$. That is, $\mathcal{S}$ is normal. Then $\mathcal{S}$ is $\mathcal{U}_0$-certified by Lemma 3.25, so $\mathcal{T}$ is $\mathcal{U}_0$-certified.

So we may assume $o(\beta^P) = \tau^P$, and thus clause (b) of top block stability fails for P. By the proposition, letting η_0 be the $r\Sigma_k$ cofinality of τ^P we have that $\eta_0 < \tau^P$, and η_0 is measurable in P. Because P is projectum stable, there is an $f_0 \in P$ mapping η_0 cofinally into τ^P. Letting $\eta = i_b(\eta_0)$ and $f = i_b(f_0)$, f witnesses that τ^Q has measurable cofinality η in Q.

Now let D be a total normal measure on η from the Q sequence. Let

$$\mathcal{S}_0 = X(\mathcal{T}^\frown b, \langle D \rangle),$$

and

$$\Phi \colon \mathcal{T}^\frown b \to \mathcal{S}_0$$

be the associated weak hull embedding. The last model of $\mathcal{S}_0$ is $N = \text{Ult}(Q, D)$, and i_D^Q is discontinuous at τ^Q, so we have E on the N sequence with $\text{crit}(E) = \beta^N$ and $\text{lh}(E) > \sup(i_D^Q “\tau^Q)$. The extenders used on b have lengths $< \tau^Q$, so the extenders used on the main branch of $\mathcal{S}_0$ have lengths $< \sup(i_D^Q “\tau^Q) < \text{lh}(E)$. Thus $\mathcal{S}_1$ is normal, where

$$\mathcal{S}_1 = \mathcal{S}_0{}^\frown \langle E \rangle$$

is obtained by applying E to the appropriate earlier model in $\mathcal{S}_0$. Letting W be the last model of $\mathcal{S}_1$, the branch extender P-to-W of $\mathcal{S}_1$ has all component measures on β^P. It follows by our previous arguments that there is a weak tree embedding

$$\Psi \colon \mathcal{S}_1 \to \mathcal{U}_0.$$

But then $\Psi \circ \Phi$ is a weak tree embedding of $\mathcal{T}^\frown b$ into $\mathcal{U}_0$, as desired. □

The next lemma says that when dealing with top block stable (P, Σ), we may as well assume $\tau^P = o(P)$ and $k(P) = 0$.

Proposition 3.34: *Let (P, Σ) be top block stable, $\tau = \tau^P$, $\tau_\infty = \tau_\infty(P, \Sigma)$; then*

(1) $\Sigma^{rl} = \Sigma^{rl}_{P|\langle\tau,0\rangle}$, *and*
(2) $M_\infty(P,\Sigma)|\langle\tau_\infty, 0\rangle = M_\infty(P|\langle\tau,0\rangle, \Sigma_{P|\langle\tau,0\rangle})$.

Proof: Let $\Psi = \Sigma_{P|\langle\tau,0\rangle}$. Suppose that $\mathcal{T}$ is by Σ^{rl}, with models $M_\alpha = \mathcal{M}^{\mathcal{T}}_\alpha$. Let

$$N_\alpha = M_\alpha|\langle\hat{i}^{\mathcal{T}}_{0,\alpha}(\tau), 0\rangle$$

if $[0,\alpha)_T$ has not dropped so far that $\hat{i}^{\mathcal{T}}_{0,\alpha}(\tau)$ is undefined. Otherwise, let $N_\alpha = M_\alpha$. Thus $N_0 = P|\langle\tau,0\rangle$. Let $\mathcal{S}$ use the same extenders as $\mathcal{T}$, but have models N_α. We claim that $\mathcal{S}$ is an iteration tree by Ψ. (The main thing is that it is an iteration tree at all.)

For this, suppose $\mathcal{S} \restriction \alpha+1$ is a tree by Ψ. First, we must see $E = E^{\mathcal{T}}_\alpha$ is on the N_α sequence. If $\hat{i}^{\mathcal{T}}_{0,\alpha}(\tau)$ is undefined, this is clear. But if $\hat{i}^{\mathcal{T}}_{0,\alpha}(\tau)$ is defined, then $\mathrm{lh}(E) < \hat{i}^{\mathcal{T}}_{0,\alpha}(\tau^P)$, since otherwise, $\hat{i}^{\mathcal{T}}_{0,\alpha}(\tau^P) < \mathrm{crit}(E)$, so E is applied in $\mathcal{T}$ to M_α, forcing a drop that cannot be un-dropped, and $\mathcal{T}$ is irrelevant. So again, E is on the N_α sequence.

Second, we must see that $\mathcal{M}^{\mathcal{S}}_{\alpha+1} = N_{\alpha+1}$. Let $\beta = \mathrm{pd}_T(\alpha+1)$. Since the same extenders are used in $\mathcal{S}$ and $\mathcal{T}$, $\beta = \mathrm{pd}_S(\alpha+1)$. Let $Q = (\mathcal{M}^{\mathcal{T}}_{\alpha+1})^*$ be what E is applied to in $\mathcal{T}$. If $Q \lhd N_\beta$, then then $\mathcal{M}^{\mathcal{S}}_{\alpha+1} = \mathrm{Ult}(Q,E) = M_{\alpha+1} = N_{\alpha+1}$. If $N_\beta \lhd Q$, then $\mathrm{crit}(E) < o(N_\beta) = \hat{i}^{\mathcal{T}}_{0,\beta}(\tau)$, and $\hat{i}^{\mathcal{T}}_{\beta,\alpha+1}$ is continuous at $o(N_\beta)$ by our regularity assumption in stability. Thus $\mathcal{M}^{\mathcal{S}}_{\alpha+1} = i^Q_E(N_\beta) = N_{\alpha+1}$.

This proves (1) of Proposition 3.34. Part (2) follows easily. □

Definition 3.35: Let (P,Σ) be top block stable; then we say (P,Σ) is *minimal* iff $\tau^P = o(P)$ and $k(P) = 0$.

The next definition isolates an important class of $\mathcal{U}_0$-certified trees.

Definition 3.36: Let (P,Σ) be a mouse pair having a top block, and let $\mathcal{T}$ be a normal tree of limit length by Σ; then we say $\mathcal{T}$ is (P,Σ)-*maximal* iff for $b = \Sigma(\mathcal{T})$,

(i) b does not drop, and
(ii) if $\tau^P < o(P)$, then $i^{\mathcal{T}}_b(\tau^P) = \delta(\mathcal{T})$, and
(iii) if $\tau^P = o(P)$, then $\delta(\mathcal{T}) = o(\mathcal{M}^{\mathcal{T}}_b)$.

Otherwise, we say that $\mathcal{T}$ is (P,Σ)-*short*.

So if τ^P is a cutpoint of P, $\mathcal{T}$ is (P,Σ)-maximal, and $b = \Sigma(\mathcal{T})$, then $\mathcal{T}^\frown b$ has no proper, normal, $M_\infty(P,\Sigma)$-relevant extension.

Proposition 3.37: *Let (P, Σ) be top block stable and minimal, and let $\mathcal{T}$ be a countable, normal tree of limit length by Σ; then equivalent are*

(a) $\mathcal{T}$ is $M^0_\infty(P, \Sigma)$-relevant,
(b) $\mathcal{T}$ is $M_\infty(P, \Sigma)$-relevant, and short.

Proof: Let $\mathcal{T}$ be $M^0_\infty(P, \Sigma)$ relevant, as witnessed by a normal $\mathcal{S}$ extending $\mathcal{T}$ that has a last model Q, and is such that P-to-Q does not drop, and is β^Q-generated. Suppose $\mathcal{T}$ is maximal, and let $b = \Sigma(\mathcal{T})$. Then $\mathcal{T}^\frown b$ has no proper normal extensions, so $\mathcal{S} = \mathcal{T}^\frown b$. But b is not $i_b(\beta^P)$-generated, by maximality. This is a contradiction.

Conversely, let $\mathcal{T}$ be $M_\infty(P, \Sigma)$-relevant, and short. Let $b = \Sigma(\mathcal{T})$. We shall show that there is a normal $\mathcal{S}$ extending $\mathcal{T}^\frown b$ by at most one model such that $\mathcal{S}$ has a last model Q, the branch P-to-Q does not drop, and P-to-Q is β^Q-generated.

For suppose first that b does not drop. If $i^{\mathcal{T}}_b(\beta^P) \geq \delta(\mathcal{T})$, then we can take $\mathcal{S} = \mathcal{T}$. If $i^{\mathcal{T}}_b(\beta^P) < \delta(\mathcal{T})$, then since $i^{\mathcal{T}}_b(\tau^P) > \delta(\mathcal{T})$, there is an E on the sequence of $M^{\mathcal{T}}_b$ such that $\mathrm{lh}(E) > \delta(\mathcal{T})$ and $\mathrm{crit}(E) = i^{\mathcal{T}}_b(\beta^P)$. We can then take $\mathcal{S} = \mathcal{T}^\frown b^\frown \langle E \rangle$.

Now suppose that b does drop. Since $\mathcal{T}$ is $M_\infty(P, \Sigma)$-relevant, there is an E on the $\mathcal{M}^{\mathcal{T}}_b$ sequence such that $\mathrm{lh}(E) > \delta(\mathcal{T})$ and $\mathrm{crit}(E) < \delta(\mathcal{T})$. (Note $\mathrm{lh}(E) = \delta(\mathcal{T})$ is impossible.) Let κ be the least possible critical point of such an E. κ is a strong cutpoint of $\mathcal{M}^{\mathcal{T}}_b$. Let ν be least such that $\lambda(E^{\mathcal{T}}_\nu) > \kappa$.

Claim 1.

(i) For all $\eta \geq \nu$, $\mathrm{crit}(E^{\mathcal{T}}_\eta) > \kappa$.
(ii) For all $\eta \geq \nu + 1$, κ is a strong cutpoint in $\mathcal{M}^{\mathcal{T}}_\eta$, and $\mathrm{crit}(E^{\mathcal{T}}_\eta) > \kappa$.
(iii) $[0, \nu)_T \cap D^{\mathcal{T}} = \emptyset$.
(iv) $P(\kappa) \cap \mathcal{M}^{\mathcal{T}}_\nu = P(\kappa) \cap \mathcal{M}^{\mathcal{T}}_b$.

Proof: For (i): Let $\lambda = \lambda(E_\eta)$. Since λ is a cardinal in $\mathcal{M}^{\mathcal{T}}_b$ and $o(\kappa) > \lambda$ in $\mathcal{M}^{\mathcal{T}}_b$, $\mathcal{M}^{\mathcal{T}}_b|\lambda \models o(\kappa) = \infty$. Letting $\mu = \mathrm{crit}(E_\eta)$, we get that $\mathcal{M}^{\mathcal{T}}_\eta|\mu \models \exists\xi(o(\xi) = \infty)$. Fix such a ξ; then $\mathcal{M}^{\mathcal{T}}_{\eta+1}|\lambda \models o(\xi) = \infty$. This implies $o(\xi) \geq \kappa$ in $\mathcal{M}^{\mathcal{T}}_b$, and hence $\kappa \leq \xi$ because κ is a strong cutpoint in $\mathcal{M}^{\mathcal{T}}_b$. But $\xi < \mu$, so $\kappa < \mu$.

For (ii): Let $\lambda = \lambda(E_\nu)$. If $\xi < \kappa$ and $o(\xi) \geq \kappa$ in $\mathcal{M}^{\mathcal{T}}_\eta$, then since λ is a cardinal, then $\mathcal{M}^{\mathcal{T}}_\eta|\lambda \models o(\xi) = \infty$. But $\mathcal{M}^{\mathcal{T}}_\eta|\lambda = \mathcal{M}^{\mathcal{T}}_b|\lambda$, so κ is not a strong cutpoint in $\mathcal{M}^{\mathcal{T}}_b$, contradiction.[23]

[23]The proof does not show that κ is a strong cutpoint in $\mathcal{M}^{\mathcal{T}}_\nu$, because λ might not be a cardinal there.

For (iii) and (iv): Since $\mathcal{T}$ is $M_\infty(P,\Sigma)$-relevant, we can fix a normal extension $\mathcal{S}$ of $\mathcal{T}^\frown b$ such that $\mathrm{lh}(\mathcal{S}) = \eta + 1$ and $[0,\eta)_S \cap D^{\mathcal{S}} = \emptyset$. We may assume that $\eta = \xi + 1$ for some $\xi \geq \mathrm{lh}(\mathcal{T})$. By the proof of (ii), for all γ such that $\nu + 1 \leq \gamma$, κ is a strong cutpoint in $\mathcal{M}^{\mathcal{S}}_\gamma$, and $\kappa \leq \mathrm{crit}(E^{\mathcal{S}}_\gamma)$ if $\gamma \leq \xi$. Thus $\nu \leq_S \eta$, so if $[0,\nu)_T$ does not drop. Similarly, $\mathrm{crit}(i_{\nu,\eta}) \geq \kappa$, so $P(\kappa) \cap \mathcal{M}^{\mathcal{T}}_\nu = P(\kappa) \cap \mathcal{M}^{\mathcal{T}}_\eta$, which implies (iv). □

Now let E on the $M^{\mathcal{T}}_b$ sequence be such that $\mathrm{lh}(E) > \delta(\mathcal{T})$ and $\mathrm{crit}(E) = \kappa$. We claim that $\mathcal{S} = \mathcal{T}^\frown b^\frown \langle E \rangle$ witnesses that $\mathcal{T}$ is $M^0_\infty(P,\Sigma)$-relevant. It is enough to show that the measures of its main branch extender concentrate on β^P, and for that, it is enough to see that $\kappa \leq i^{\mathcal{T}}_{0,\nu}(\beta^P)$. But if not, then $i^{\mathcal{T}}_{0,\nu}(\beta^P) < \kappa < i^{\mathcal{T}}_{0,\nu}(\tau^P)$, so $i^{\mathcal{T}}_{0,\nu}(\beta^P)$ is strong to κ in $M^{\mathcal{T}}_\nu$, and hence in $M^{\mathcal{T}}_b$. By coherence then, $i^{\mathcal{T}}_{0,\nu}(\beta^P)$ is strong to $\delta(\mathcal{T})$ in $M^{\mathcal{T}}_b$, contrary to our choice of κ. □

Corollary 3.38: *Let (P,Σ) be top block stable and minimal; then all short, $M_\infty(P,\Sigma)$-relevant normal trees are $\mathcal{U}_0(P,\Sigma)$-certified.*

One could extend the notions to stacks of normal trees in a natural way, but it turns out then that a stack s is (P,Σ)-maximal if and only if $W(s)$ (equivalently $X(s)$) is (P,Σ)-maximal. So we shall just take that as our definition. Notice that maximality and shortness involve Σ, as well as $\mathcal{T}$ and P, and that Q-structures are irrelevant.

Definition 3.39: Let (P,Σ) be a mouse pair having a top block; then

(i) Σ^{sh} is the restriction of Σ to (P,Σ)-short trees.
(ii) $\Sigma^{\mathrm{sh,rl}}$ is the restriction of Σ to $M_\infty(P,\Sigma)$ relevant, (P,Σ)-short trees.

Notice that Σ^{sh} can be partial, that is, undefined at trees played according to it. Σ itself cannot be partial this way. This leads to the possibility that $\mathrm{Code}(\Sigma^{\mathrm{sh}})$ could be complete in some nicely closed nonselfdual pointclass. That could not happen with the full Σ.

Corollary 3.40: *Let (P,Σ) be a projectum stable mouse pair having a top block; then*

(a) for any stack s by $\Sigma^{sh,rl}$, $W(s)$ is a weak hull of $\mathcal{U}_0(P,\Sigma)$.
(b) $\Sigma^{sh,rl}$ is $\beta_\infty(P,\Sigma)$-Suslin, but not α-Suslin for any $\alpha < |\beta_\infty(P,\Sigma)|$.
(c) Σ^{sh} is γ-Suslin, where $\gamma = \sup(\beta_\infty(P,\Sigma) \cup \{o(H_\beta) \mid \beta <$ mouse rank of $(P,\Sigma)\}$.

Proof: (a) and (b) follow at once from Lemma 3.33, Proposition 3.34, and Corollary 3.38. The proof of Corollary 3.16 then gives (c). □

The following lemma implies that $\beta_\infty(P, \Sigma)$ is a limit of Suslin cardinals in the cases we shall be interested in later.

Lemma 3.41: *Let (P, Σ) be a projectum stable mouse pair, and γ a strong cutpoint of P. Suppose that for all $\alpha < \gamma$, there is a $\beta < \gamma$ such that β is a strong cutpoint cardinal of P, and $\Sigma^{rl}_{P|\beta}$ is not Σ_1-definable from parameters over $(HC, \in, \Sigma^{rl}_{P|\alpha})$. Then $\sup(\pi^{\Sigma}_{P,\infty}``\gamma)$ is a limit of Suslin cardinals.*

Proof: Let $\alpha < \beta < \gamma$ be strong cutpoint cardinals of P such that $\Sigma^{\text{rl}}_{P|\beta}$ is not Σ_1-definable from parameters over $(\text{HC}, \in, \Sigma^{\text{rl}}_{P|\alpha})$. We may assume α and β are successor cardinals, so that

$$\alpha_\infty =_{\text{df}} \pi^{\Sigma}_{P,\infty}(\alpha) = \tau_\infty(P|\alpha, \Sigma_{P|\alpha}),$$

and

$$\beta_\infty =_{\text{df}} \pi^{\Sigma}_{P,\infty}(\beta) = \beta_\infty(P|\alpha, \Sigma_{P|\beta}).$$

By Theorem 3.12, $|\alpha_\infty|$ and $|\beta_\infty|$ are Suslin cardinals, so it is enough to show $|\alpha_\infty| < |\beta_\infty|$. But if not, then $\Sigma^{\text{rl}}_{P|\beta}$ is $|\alpha_\infty|$-Suslin. Since $\Sigma^{\text{rl}}_{P|\alpha}$ is not κ-Suslin for any $\kappa < |\alpha_\infty|$, every $|\alpha_\infty|$-Suslin set is Σ_1-definable from parameters over $(\text{HC}, \in, \Sigma^{\text{rl}}_{P|\alpha})$. (See [3], §3.) Thus $\Sigma^{\text{rl}}_{P|\beta}$ is Σ_1-definable from parameters over $(\text{HC}, \in, \Sigma^{\text{rl}}_{P|\alpha})$, contradiction. □

Definition 3.42: Let (P, Σ) be a projectum stable mouse pair having a top block; then (P, Σ) is *projectively inaccessible* iff whenever (Q, Ψ) is a non-dropping iterate of (P, Σ), and $\alpha < \beta^Q$, there is a $\xi < \beta^Q$ such that ξ is a strong cutpoint cardinal of Q, and $\Psi^{\text{rl}}_{Q|\xi}$ is not Σ_1-definable from parameters over $(\text{HC}, \in, \Sigma^{\text{rl}}_{Q|\alpha})$.

Corollary 3.43: *Let (P, Σ) have a top block, and be projectively inaccessible; then $\beta_\infty(P, \Sigma)$ is a limit of Suslin cardinals, moreover $\text{cof}(\beta_\infty(P, \Sigma)) > \omega$.*

Proof: This follows at once from Lemma 3.41. Note that β^P is measurable in P and $< \rho_{k(P)}(P)$, so β_∞ has uncountable cofinality. □

The next theorem is our main result on the adjacency of $|\beta_\infty|$ and τ_∞ in the sequence of Suslin cardinals.

Theorem 3.44: *Let (P, Σ) be a projectum stable, projectively inaccessible mouse pair having a top block, let $\beta_\infty = \beta_\infty(P, \Sigma)$, $\tau_\infty = \tau_\infty(P, \Sigma)$, and suppose that Σ^{sh} is β_∞-Suslin. Let μ be the least Suslin cardinal $> \beta_\infty$, if one exists; then either*

(a) Σ^{rl} is β_∞-Suslin, or
(b) μ exists, and Σ^{rl} is μ-Suslin.

Remark 3.45: The hypothesis that Σ^{sh} is $\beta_\infty(P, \Sigma)$-Suslin is not redundant. We have shown that $\Sigma^{\text{sh,rl}}$ is β_∞-Suslin, but the full Σ^{sh} may not be β_∞-Suslin.

Proof: If P is not top block stable, then (a) holds by Lemma 3.33, so we may assume P is top block stable. By Proposition 3.34, we may assume $o(P) = \tau^P$, and $k(P) = 0$. If $|\tau_\infty| = \beta_\infty$, then (a) holds by Theorem 3.21, so we assume $\beta_\infty < |\tau_\infty|$. It follows that μ exists.

By Definition 3.42, β_∞ is a Suslin limit of Suslin cardinals, and of uncountable cofinality. By results of Kechris (see [3], §3), there is an ω-parametrized pointclass Γ_0 that is "Σ^1_2-like", in that it is closed under $\exists^{\mathbb{R}}$, number quantifiers, recursive substitution, and has the scale property, such that for any set of reals B, B is β_∞-Suslin iff B is $\Gamma_0(x)$ for some real x. We may assume that $\text{Code}((P, \Sigma^{\text{sh}}))$ is Γ_0. Let $A_0 \subseteq \omega \times \mathbb{R}$ be a universal Γ_0 set, and let T_0 be the tree of a Γ_0 scale on A_0.

Let $\langle N^*, \delta^*, w, S, T, \Sigma^* \rangle$ be a coarse Γ_0-Woodin tuple, and let $\mathbb{C} = \langle (M_{\nu,k}, \Omega_{\nu,k}) \mid \langle \nu, k \rangle \leq_{\text{lex}} \langle \delta^*, 0 \rangle \rangle$ be the maximal construction associated to $\langle N^*, \delta^*, w, S, T, \Sigma^* \rangle$ that has the same mouse pair type as (P, Σ).

For $\eta \leq \delta^*$, let

$$N^*|\eta = (V_\eta^{N^*}, w \cap V_\eta^{N^*}).$$

Let δ be the least Γ_0-Woodin of N^*, that is, δ is least such that

$$L[T_0, N^*|\delta] \models \delta \text{ is Woodin}.$$

The Woodinness is automatically witnessed by extenders in $\mathcal{F}^{N^*|\delta} = \mathcal{F} \cap V_\delta^{N^*}$. $\delta < \delta^*$ because a scale on $\neg A_0$ is coded into A_1. Let $\mathbb{C}$ be the maximal construction of N^* of the same mouse-type as P. The levels of $\mathbb{C}$ are mouse pairs $(M_{\nu,k}, \Omega_{\nu,k})$, and for $\langle \nu, k \rangle \leq_{\text{lex}} \langle \delta, 0 \rangle$, these are just the levels of the maximal construction of $L[T_0, N^*|\delta]$. By the Comparison Theorem 2.3 either

(i) (P, Σ) iterates to $(M_{\nu,k}, \Omega_{\nu,k})$ for some $\langle \nu, k \rangle <_{\text{lex}} \langle \delta, 0 \rangle$, or
(ii) (P, Σ) iterates past $(M_{\delta,0}, \Omega_{\delta,0})$ (perhaps not strictly).

Note here that Σ^* is defined on all countable trees in V, hence all the $\Omega_{\nu,k}$ are as well. N^* is sufficiently correct that whichever of (i) and (ii) holds in N^*, holds also in V of the extended strategies.

Suppose first that (i) holds, and let $(M, \Omega) = (M_{\nu,k}, \Omega_{\nu,k})$. Let $\eta < \delta$ be such that Ω is induced by Ψ, where Ψ is the restriction of Σ^* to trees based on $N^*|\eta$. Since δ is the least Γ_0 Woodin of (N^*, w), Ψ is guided by C_{Γ_0}, in that for $\mathcal{T}$ of limit length by Ψ,

$$\Psi(\mathcal{T}) = b \text{ iff } \exists f \in C_{\Gamma_0}(\mathcal{M}(\mathcal{T}))[\mathcal{M}(\mathcal{T}) \text{ is not } f\text{-Woodin}].$$

Thus Code(Ψ) is in $\Gamma_0(N^*|\eta)$, and hence Code(Ω) is in $\Gamma_0(N^*|\eta)$. But Σ is pullback consistent, so $\Sigma = \Sigma^i_M = \Omega^i$, where $i\colon P \to M$ is the iteration map. So Code(Σ) is $\Gamma_0(N^*|\eta, i)$, and therefore β_∞-Suslin.

Suppose next that (ii) holds, and let $(M, \Omega) = (M_{\delta,0}, \Omega_{\delta,0})$. Let $\mathcal{T}$ be the normal tree on P of limit length by which (P, Σ) iterates past (M, Ω), and let $b = \Sigma(\mathcal{T})$. Since $\mathcal{T}$ is normal, all its proper initial segments are short. $\mathcal{T}$ itself cannot be short, for otherwise $\mathcal{T}^\frown b \in L[T_0, N^*|\delta]$ because Σ^{sh} is coded into A_0, so $\Sigma^{\text{sh}} \cap L[T_0, N^*|\delta] \in L[T_0, N^*|\delta]$. But $\mathcal{T}^\frown b$ kills the Woodinness of δ. (Let $i_{\eta,b}(\gamma) = \delta$; then $i_{\eta,b}$ is continuous at γ, witnessing that δ is singular.)

So $\mathcal{T}$ is (P, Σ)-maximal. Looking from the point of view of the full N^*, we see then that (P, Σ) iterates to (M, Ω) via $\mathcal{T}^\frown b$. N^* is sufficiently correct that this is true in V. We have that $\Sigma = \Omega^{i_b}$, so Code(Σ) $= f^{-1}$(Code(Ω)), where f is a $\mathbf{\Delta}^1_2$ function. So it will suffice to show that Ω is μ-Suslin. But Ω is obtained from $\Sigma^*_{N^*|\delta}$ via a conversion system ([19]), so Code(Ω) $=$ g^{-1}(Code($\Sigma^*_{N^*|\delta}$)) for some $\mathbf{\Delta}^1_2$ function g. Thus it suffices to show that $\Sigma^*_{N^*|\delta}$ is μ-Suslin. For that, it is enough to consider the restriction of Σ^* to normal trees based in $N^*|\delta$.

Let S_μ be the collection of μ-Suslin sets. Second Periodicity, Martin's characterization of the next Suslin cardinal via pointclass envelopes, (see [3]), and results of Woodin on self-justifying systems (see [31]) show that there is a family $\mathcal{B}$ of sets of reals such that

(1) $\mathcal{B} \subseteq S_\mu \cap \breve{S}_\mu$,
(2) $A_0 \in \mathcal{B}$, $\mathcal{B}$ is closed under complements, and
(3) for all $B \in \mathcal{B}$, there is a scale on B all of whose associated prewellorders are in $\mathcal{B}$.

Such a $\mathcal{B}$ is called a self-justifying system. In the case that $\Gamma_0 = S_{\beta_\infty}$ is not closed under $\forall^{\mathbb{R}}$, the Second Periodicity Theorem yields $\mathcal{B}$ as above. In this case, μ is the prewellordering ordinal of $\forall^{\mathbb{R}}\Gamma_0$, and $\mu = \beta^+_\infty$. We can then take $\mathcal{B} \subseteq \forall^{\mathbb{R}}\Gamma_0$. $S_\mu = \exists^{\mathbb{R}}\forall^{R}\Gamma_0$, so (1) is not the best possible upper bound on $\mathcal{B}$. In the case that Γ_0 is closed under $\forall^{\mathbb{R}}$, we get a self-justifying system $\mathcal{B}$ as above such that $\mathcal{B}$ is contained in the boldface envelope $\boldsymbol{\Lambda}(\boldsymbol{\Gamma_0}, \beta_\infty)$ of Γ_0. (See [3] and [31].) Again, this is a better upper bound than given in (1).

We may assume that $\mathcal{B}$ has been chosen such that for all $B \in \mathcal{B}$ there is an $x \in N^*$ such that $B = (A_1)_x$. This then gives us standard $\mathrm{Col}(\omega, \delta)$ terms τ_B in N^* such that $(\tau_B)_g = B \cap N^*[g]$ for all g on $\mathrm{Col}(\omega, \delta)$. We then get that $\Sigma^*_{N^*|\delta}$ is determined by the fact that it moves these terms correctly. More precisely, if $\mathcal{T}$ is a normal tree on $N^*|\delta$ of limit length, then $\Sigma^*(\mathcal{T})$ is the unique cofinal wellfounded branch b of $\mathcal{T}$ such that

$$C_{\Gamma_0}(\mathcal{M}(\mathcal{T})) \subseteq \mathcal{M}_b^{\mathcal{T}},$$

and either

(i) there is an $f \colon \delta(\mathcal{T}) \to \delta(\mathcal{T})$ witnessing that $\delta(\mathcal{T})$ is not Woodin with respect to extenders in $\mathcal{M}(\mathcal{T})$, with $f \in C_{\Gamma_0}(\mathcal{M}(\mathcal{T}))$, or

(ii) $i_b^{\mathcal{T}}(\delta) = \delta(\mathcal{T})$, and for all $B \in \mathcal{B}$, for all g on $\mathrm{Col}(\omega, i_b(\delta))$, $i_b(\tau_B)_g = B \cap \mathcal{M}_b^{\mathcal{T}}$.

The reader should see [22] for a full explanation here. The relation $R(\mathcal{T}, b) \Leftrightarrow C_{\Gamma_0}(\mathcal{M}(\mathcal{T})) \subseteq \mathcal{M}_b^{\mathcal{T}}$ is in Γ_0-dual, hence in S_μ. Clause (i) above is a Γ_0 condition. Clause (ii) is also an S_μ condition, because $\mathcal{B} \subseteq S_\mu \cap \check{S}_\mu$, and the quantifier $\forall g$ can be replaced by "for comeager many g". Thus the restriction of Σ^* to normal trees based on $N^*|\delta$ is μ-Suslin, as desired. □

Corollary 3.46: *Under the hypotheses of Theorem 3.44, either $|\tau_\infty| = \beta_\infty$, or $|\tau_\infty|$ is the least Suslin cardinal strictly greater than β_∞.*

We would guess that the hypothesis that (P, Σ) is projectively inaccessible is not needed in Theorem 3.44 and its corollary. We used projective inaccessibility to conclude that S_{β_∞} has the scale property. Probably it would suffice to assume that $|\beta_\infty|$ has uncountable cofinality.

We can improve Corollary 3.46 in the case that S_{β_∞} is inductive-like.

Theorem 3.47: *Let (P, Σ) be a projectum stable mouse pair having a top block, $\beta_\infty = \beta_\infty(P, \Sigma)$, and $\tau_\infty = \tau_\infty(P, \Sigma)$. Suppose that S_{β_∞} is closed under $\forall^{\mathbb{R}}$, and that $\Sigma^{sh} \in S_{\beta_\infty}$; then τ_∞ is the least Suslin cardinal strictly greater than β_∞.*

Proof: Let Γ_0 be an ω-parametrized pointclass that is is closed under both real quantifiers, recursive substitution, has the scale property, and is such that for any set of reals B, B is β_∞-Suslin iff B is $\Gamma_0(x)$ for some real x. We may assume that Code$((P, \Sigma^{\mathrm{sh}}))$ is Γ_0. Let $A_0 \subseteq \omega \times \mathbb{R}$ be a universal Γ_0 set. The prewellordering ordinal of Γ_0 is β_∞.

If (Q, Ψ) is a nondropping iterate of (P, Σ), and $(R, \Phi) = (Q, \Psi)|\eta$ for some cardinal cutpoint $\eta < \beta^Q$, then both Φ and its complement are $\mathbf{\Gamma_0}$. (Φ is total, and coded into Σ^{sh}.) Such pairs (Q, Φ) are Wadge cofinal in $\mathbf{\Delta_0}$ because the ordinals of the form $o(M_\infty(Q, \Phi))$ are cofinal in β_∞. Thus (P, Σ) is projectively inaccessible.

Σ^{rl} cannot be in Γ_0, because it is (essentially) a total function, so it would then be in Δ_0. More precisely, a normal tree $\mathcal{T}$ is *not* by Σ^{rl} iff there is some limit ordinal $\lambda < \mathrm{lh}(\mathcal{T})$ such that $\mathcal{T} \upharpoonright \lambda$ is by Σ^{sh}, and a b such that $b \neq [0, \lambda)_T$, and either $\Sigma^{\mathrm{sh}}(\mathcal{T}) = b$ or $\Sigma^{\mathrm{rl}}(\mathcal{T}) = b$. So if Σ^{rl} is $\mathbf{\Gamma}_0$, it is $\mathbf{\Delta}_0$, contradiction.

Thus Σ^{rl} is not β_∞-Suslin. By Lemma 3.33, P is top block stable, and by Theorem 3.44, $|\tau_\infty| = \mu$, where μ is the least Suslin cardinal $> \beta_\infty$. We want to show $\tau_\infty = \mu$. By Proposition 3.34 we may assume that $\tau^P = o(P)$ and $k(P) = 0$.

Let Γ_1, N^*, δ, $\mathbb{C}$, and the self-justifying system $\mathcal{B}$ be as in the proof of the Theorem 3.44. We showed that if (P, Σ) iterates to some $(M_{\nu,k}, \Omega_{\nu,k})$ with $\nu < \delta$, then Σ is β_∞-Suslin, so we have that (P, Σ) iterates past $(M, \Omega) = (M_{\delta,0}, \Omega_{\delta,0})$. As we showed, this implies (P, Σ) iterates to (M, Ω), via some tree $\mathcal{T}^\frown b$ such that $i_b(\tau^P) = \delta$. So $M_\infty(P, \Sigma) = M_\infty(M, \Omega)$, and it is enough to show that for all $\gamma < \delta$, $i^\Omega_{M,\infty}(\gamma) < \mu$.

So fix $\gamma < \delta$. Because iterations of (M, Ω) convert to iterations of $N^*|\delta$ by Σ^* in the way they do, $\pi^\Omega_{M,\infty}(\gamma) \leq \pi^{\Sigma^*}_{N^*|\delta,\infty}(\gamma)$ so it would be enough to show $\pi_{(N^*|\delta,\Sigma^*),\infty}(\gamma) < \mu$. However, to do things this way, we need a comparison theorem for coarse iterates of $(N^*, \Sigma^*_{N^*|\delta})$. That can be proved, but we shall take a more direct route.

We recall briefly some well-known concepts. (See for example [30].) For any countable transitive N, let

$$P(N) = C_{\Gamma_0}(N) = L[T_0, N]|o(N)^+,$$

where $o(N)^+$ is computed in $L[T_0, N]$. Say that N is *suitable* iff $P(N) \models o(N)$ is Woodin, and N is a rank initial segment of $P(N)$. For $B \in \mathcal{B}$, and N suitable, let τ^N_B be the standard term capturing B over $L[T_0, N]$. (That is, $(\tau^N_B)_g \cap L[T_0, B][g] = B \cap L[T_0, N][g]$ for any g on $\mathrm{Col}(\omega, o(N))$, and τ^N_B is its own forcing relation.) Let

$$P_B(N) = (P(N), \tau^N_B),$$

and

$$\gamma_B(N) = \sup(\{\xi < o(N) \mid \xi \text{ is definable over } P(N)\{).$$

We assume that B has been chosen so that $\gamma < \gamma_B(N^*|\delta)$, and A_0 and $\neg A_0$ are Wadge reducible to B.

If N is suitable, and $\mathcal{T}$ is a normal tree on N, we say that $\mathcal{T}$ is Γ_0-*guided* iff for all limit $\lambda < \text{lh}(\mathcal{T})$, there is an $f\colon \delta(\mathcal{T}) \to \delta(\mathcal{T})$ witnessing that $\delta(\mathcal{T})$ is not Woodin with respect to extenders in $\mathcal{M}(\mathcal{T})$, with $f \in C_{\Gamma_0}(\mathcal{M}(\mathcal{T})) \cap \mathcal{M}_\lambda^{\mathcal{T}}$. Say that $\mathcal{T}$ is Γ_0 *maximal* iff $\mathcal{T}$ is Γ_0 guided, $\mathcal{T}$ has limit length, and $M(\mathcal{T})$ is suitable. In this case, if b is a cofinal branch of $\mathcal{T}$, then b *respects* B iff, regarding $\mathcal{T}$ as a tree on $P(N)$, $M_b^{\mathcal{T}} = P(\mathcal{M}(\mathcal{T}))$, $i_b(o(N)) = \delta(\mathcal{T})$, and $i_b(\tau_B^N) = \tau_B^{\mathcal{M}(\mathcal{T})}$.

Now consider the tree W on HC, whose nodes are finite sequences

$$\langle (N_0, \gamma_0, \mathcal{T}_0), ..., (N_k, \mathcal{T}_k, \gamma_k)\rangle$$

such that

(a) $N_0 = N^*|\delta$ and $\gamma_0 = \gamma$,
(b) for $m < k$,N_m and N_{m+1} are suitable, and $\mathcal{T}_m$ is a normal tree on N_m by having last model N_{m+1} such that $i_{0,\infty}^{\mathcal{T}_m}(\gamma_m) > \gamma_{m+1}$, and
(c) for $m < k$, either $\mathcal{T}_m$ is Γ_0-guided, or $\mathcal{T}_m = \mathcal{S}^\frown b$ where $\mathcal{S}$ is Γ_0-maximal, and b respects B.

The order on W is of course sequence extension. To see that W is wellfounded, note that if $\mathcal{S}$ on N is Γ_0-maximal, b respects B, and $c = \Sigma^*(\mathcal{S})$, then $\mathcal{M}_b^{\mathcal{S}} = \mathcal{M}_b^{\mathcal{S}}$ and $i_b^{\mathcal{S}}$ agrees with $i_c^{\mathcal{S}}$ on γ_B^N. Thus an infinite path through W would give rise to an infinite stack of trees by Σ^* with illfounded direct limit model. Finally, the conversion of trees on (M, Ω) to trees on $N^*|\delta$ by Σ^* shows that

$$\pi_{M,\infty}^{\Omega}(\gamma) \leq |W|.$$

(Here $|W|$ is the ordinal rank.) So it is enough to show that $|W| < \mu$.

Because Γ_0 is inductive like, its envelope is closed under real quantifiers. W is clearly projective in B. There is a prewellorder Y such that $|Y| = |W|$ and Y is projective in W, and hence in the envelope of Γ_0. Let V be a tree on $\omega \times \mu$ such that $p[V] = \neg A_0$. If $\mu \leq |Y|$, then by the Coding Lemma, V is $\mathbf{\Sigma}_1^1(Y)$ in the codes. This implies that every Γ_0-dual relation has a uniformization in the envelope of Γ_0. That is not true ([3, §3]). So $|W| < \mu$, as desired. □

Which of the Suslin cardinals are associated to mouse pairs? We would guess that all of them are.

Conjecture 3.47.1: Assume $\mathsf{AD}^+ + \mathsf{NLE}$, and let κ be a Suslin cardinal; then

(1) there is a pure extender pair (P, Σ) such that $\kappa = |\tau_\infty(P, \Sigma)|$, or (P, Σ) has a top block, and $\kappa = |\beta_\infty(P, \Sigma)|$, and
(2) there is a least branch pair (P, Σ) such that $\kappa = |\tau_\infty(P, \Sigma)|$, or (P, Σ) has a top block, and $\kappa = |\beta_\infty(P, \Sigma)|$.

Part (1) is known in the case κ is a projective ordinal, i.e., some δ^1_{2n+1} or its cardinal predecessor. We believe that part (2) also holds for projective κ, but do not have a full proof. We have stated the conjecture so that one would need to show NLE implies LEC in order to prove it, because the statement is prettier that way. The more accessible version would be: LEC implies (1), and HPC implies (2).[24]

4. Pointclass Generators

Here is some terminology related to local mouse capturing.

Definition 4.1: Let $\boldsymbol{\Delta}$ be a pointclass; then

(a) $\boldsymbol{\Delta}$ is *strongly closed* iff $\boldsymbol{\Delta}$ is closed under complements and real quantification, and closed downward under $\leq_w$,
(b) $\boldsymbol{\Delta} \models$ "A is Suslin" iff A admits a scale $\vec{\varphi}$ such that $\{(n, x, y) \mid \varphi_n(x) \leq \varphi_n(y)\} \in \boldsymbol{\Delta}$,
(c) $\boldsymbol{\Delta} \models$ HPC iff whenever $\boldsymbol{\Delta} \models$ "A is Suslin", then $A \leq_w \text{Code}(\Sigma)$ for some lbr hod pair (P, Σ) such that $\text{Code}(\Sigma) \in \boldsymbol{\Delta}$. We define $\boldsymbol{\Delta} \models$ LEC in parallel fashion.

If $\boldsymbol{\Delta}$ is strongly closed, then its Wadge ordinal, its prewellordering ordinal, and the sup of the lengths of $\boldsymbol{\Delta}$ wellfounded relations are all the same. (Cf. [8].) We write $o(\boldsymbol{\Delta})$ for this ordinal. Then $\boldsymbol{\Delta} \models$ "A is Suslin" iff A is κ-Suslin for some $\kappa < o(\boldsymbol{\Delta})$, and $\boldsymbol{\Delta} \models$ "all sets are Suslin" iff $o(\boldsymbol{\Delta})$ is a limit of Suslin cardinals.

The following is Theorem 1.9 from the introduction.

Theorem 4.2: *Let $\boldsymbol{\Delta}$ be strongly closed; then*

---
[24] Those versions have recently been proved. See [17].

(a) *if there is a pure extender pair* (P, Σ) *such that* $\mathrm{Code}(\Sigma) \notin \boldsymbol{\Delta}$, *then* $\boldsymbol{\Delta} \models$ *LEC, and*
(b) *if there is an lbr hod pair* (P, Σ) *such that* $\mathrm{Code}(\Sigma) \notin \boldsymbol{\Delta}$, *then* $\boldsymbol{\Delta} \models$ *HPC.*

Proof: We prove (a). The proof of (b) is the same, mutatis mutandis.

Let (P, Σ) be a pure extender pair of minimal mouse rank such that $\mathrm{Code}(\Sigma) \notin \boldsymbol{\Delta}$. We may as well assume $\boldsymbol{\Delta} \models$ "all sets are Suslin", since otherwise we can just replace $\boldsymbol{\Delta}$ with $\{A \mid \boldsymbol{\Delta} \models$ "A is Suslin and co-Suslin"$\}$.

Claim 1. If $\Sigma^{\mathrm{rl}} \in \boldsymbol{\Delta}$, then $\boldsymbol{\Delta} \models \mathsf{LEC}$.

Proof: Let α be the mouse rank of (P, Σ), and for $\beta \leq \alpha$, let $H_\beta = H^e_\beta$ be the common $M_\infty(Q, \Psi)$ for all (Q, Ψ) of mouse rank β. Since $\Sigma^{\mathrm{rl}} \in \boldsymbol{\Delta}$, H_α has a code in $\boldsymbol{\Delta}$. By the minimality of (P, Σ), each H_β for $\beta < \alpha$ has a code in $\boldsymbol{\Delta}$. The $o(H_\beta)$ for $\beta < \alpha$ must be unbounded in $o(\boldsymbol{\Delta})$, since otherwise Corollary 3.16 implies that Σ is γ-Suslin for some $\gamma < o(\boldsymbol{\Delta})$, so $\mathrm{Code}(\Sigma) \in \boldsymbol{\Delta}$. But then the Wadge ranks of mouse pairs $<^* (P, \Sigma)$ must be unbounded in $\boldsymbol{\Delta}$. □

Claim 2. If $\Sigma^{\mathrm{rl}} \notin \boldsymbol{\Delta}$ and P has limit type, then $\boldsymbol{\Delta} \models \mathsf{LEC}$.

Proof: We have that τ^P is a limit of successor cardinal cutpoints in P, and that $\tau_\infty(P, \Sigma) \geq o(\boldsymbol{\Delta})$. (Otherwise $\Sigma^{\mathrm{rl}} \in \boldsymbol{\Delta}$ by Theorem 3.21.) It follows that for each $\gamma < o(\boldsymbol{\Delta})$ there is a non-dropping iterate (Q, Σ_Q) of (P, Σ) and a strict cutpoint $(R, \Sigma_R) \lhd^* (Q, \Sigma_Q)$ such that $o(M_\infty(R, \Sigma_R)) \geq \gamma$. But then the $\mathrm{Code}(\Sigma_R)$ for such R must be Wadge cofinal in $\boldsymbol{\Delta}$. □

Claim 3. If $\Sigma^{\mathrm{rl}} \notin \boldsymbol{\Delta}$ and P has a top block, then $\boldsymbol{\Delta} \models \mathsf{LEC}$.

Proof: $o(\boldsymbol{\Delta})$ is a limit of Suslin cardinals, so if $\beta_\infty(P, \Sigma) < o(\boldsymbol{\Delta})$, then $\tau_\infty(P, \Sigma) < o(\boldsymbol{\Delta})$, by Corollary 3.46. This would imply $\Sigma^{\mathrm{rl}} \in \boldsymbol{\Delta}$. Thus $\beta_\infty(P, \Sigma) \geq o(\boldsymbol{\Delta})$. On the other hand,

$$\beta_\infty(P, \Sigma) = \sup(\{\tau_\infty(Q|\beta^Q, \Sigma_{Q|\beta^Q}) \mid (Q, \Sigma_Q) \text{ is an iterate of } (P, \Sigma)\}.$$

Thus $\beta_\infty(P, \Sigma) \leq o(\boldsymbol{\Delta})$ by the minimality of (P, Σ), and pairs of the form $(Q|\beta^Q, \Sigma_{Q|\beta^Q})$ are Wadge cofinal in $\boldsymbol{\Delta}$. □

Claims 1–3 complete the proof. □

Definition 4.3: Let $\mathbf{\Delta}$ be a selfdual pointclass closed downward under $\leq_w$, and (P, Σ) be a mouse pair; then we say (P, Σ) is a *generator of* $\mathbf{\Delta}$ iff $\Sigma \notin \mathbf{\Delta}$, and whenever $(Q, \Psi) <^* (P, \Sigma)$, then $\Psi \in \mathbf{\Delta}$. We say (P, Σ) is a *pointclass generator* iff it is a generator of some strongly closed $\mathbf{\Delta}$.

It would of course make sense to look at generators for pointclasses that are not strongly closed, but we shall not do that.

Clearly, (P, Σ) generates $\mathbf{\Delta}$ iff it generates $\{A \mid \mathbf{\Delta} \models A \text{ and } \neg A$ are Suslin$\}$. So (P, Σ) is a pointclass generator iff it is a generator for some strongly closed $\mathbf{\Delta}$ such that $\mathbf{\Delta} \models$ "all sets are Suslin". Clearly, if (P, Σ) and (Q, Ψ) are generators of $\mathbf{\Delta}$ of the same mouse type, then they are mouse-equivalent.

If $\mathbf{\Delta}$ is the collection of projective sets, then it has a unique generators (P, Σ) and (Q, Ψ) in each hierarchy. P and Q are least in their respective hierarchies such that $\mathbb{R} \cap P$ and $\mathbb{R} \cap Q$ are the lightface projective reals. In this case, Σ^{rl} and Ψ^{rl} are trivial, in fact, empty. For $\mathbf{\Delta}$ with more closure, this sort of thing does not happen.[25] For example, in the pure extender hierarchy, the generator for $P(\mathbb{R}) \cap L(\mathbb{R})$ (or equivalently, for $\mathbf{\Delta}_1^{2^{L(\mathbb{R})}}$) is (P, Σ), where $P = M_\omega | \delta$, for δ the least Woodin cardinal of M_ω, and Σ is the canonical strategy for P. In the hod pair hierarchy, the generator of $P(\mathbb{R}) \cap L(\mathbb{R})$ is something similar. In both cases, $\Sigma^{\mathrm{rl}} \notin L(\mathbb{R})$.

If (P, Σ) generates $\mathbf{\Delta}$, then $o(H_\beta) < o(\mathbf{\Delta})$ for all β strictly less than the mouse rank of (P, Σ). So our results on Suslin representations, when applied to pointclass generators, yield the following. Recall that whenever $\mathbf{\Delta}$ is strongly closed and $\mathrm{cof}(o(\mathbf{\Delta}) > \omega$, then $\mathbf{\Delta} = \mathbf{\Gamma} \cap \check{\mathbf{\Gamma}}$, for some nonselfdual $\mathbf{\Gamma}$ closed under $\forall^{\mathbb{R}}$. (See [27].)

Theorem 4.4: *Let $\mathbf{\Delta}$ be strongly closed, $\mathbf{\Delta} \models$ "all sets are Suslin", and $\delta = o(\mathbf{\Delta})$. Suppose $\mathrm{cof}(\delta) > \omega$, and let Γ be nonselfdual, closed under $\forall^{\mathbb{R}}$, and such that $\mathbf{\Delta} = \mathbf{\Gamma} \cap \check{\mathbf{\Gamma}}$. Let (P, Σ) be a pointclass generator for $\mathbf{\Delta}$, $\tau_\infty = \tau_\infty(P, \Sigma)$, and $\beta_\infty = \beta_\infty(P, \Sigma)$ if P has a top block; then:*

(1) (P, Σ) has mouse rank δ.
(2) If (P, Σ) has limit type, then $\tau_\infty \leq \delta$, Σ is δ-Suslin, and $\mathrm{cof}(\delta) = \mathrm{cof}(\tau_\infty) < \delta$.
(3) If P has a top block, then $\beta_\infty \leq \delta$, and Σ^{sh} is δ-Suslin; moreover

[25] One could look instead at a $<^*$-minimal (P, Σ) such that $\Sigma^{\mathrm{rl}} \notin \mathbf{\Delta}$, or equivalently, $o(\mathbf{\Delta}) \leq o(M_\infty(P, \Sigma))$. In the pure extender hierarchy, the pointclass generator for the projective sets in this alternative sense is (P, Σ), where P is the minimal ladder mouse (in the sense of [12]), and Σ is its unique iteration strategy.

(a) if P is not stable, then Σ is δ-Suslin, and $\mathbf{\Gamma}$ is not closed under $\exists^{\mathbb{R}}$,

(b) if P is stable and Σ is δ-Suslin, then $\mathbf{\Gamma}$ is not closed under $\exists^{\mathbb{R}}$, and

(c) if $\mathbf{\Gamma}$ is closed under $\exists^{\mathbb{R}}$, then $\beta_\infty = \delta$, and τ_∞ is the least Suslin cardinal $> \delta$, and Σ is τ_∞-Suslin.

Proof: For (1): If $(Q, \Psi) <^* (P, \Sigma)$, then $\text{Code}(\Psi) \in \mathbf{\Delta}$. The mouse order below (Q, Ψ) is represented by a wellfounded relation on $\mathbb{R}$ that is Δ^1_1 in $\text{Code}(\Psi)$ (namely, $N < M$ iff M and N are Ψ-iterates of Q, and N is a Ψ_M-iterate of M via a tree that drops on its main branch). So (Q, Ψ) has mouse rank $< \delta$. Thus (P, Σ) has mouse rank $\leq \delta$.

To see that (P, Σ) has rank at least δ, let $\xi < \delta$. We claim there is a $(Q, \Psi) \in \mathbf{\Delta}$ such that setting

$$\gamma = \begin{cases} \tau_\infty(Q, \Psi) & \text{if } Q \text{ has limit type, and} \\ \beta_\infty(Q, \Psi) & \text{otherwise,} \end{cases}$$

we have

$$\xi \leq \gamma.$$

For otherwise, by Corollary 3.16, every (Q, Ψ) in $\mathbf{\Delta}$ is μ-Suslin, where μ least Suslin cardinal $> \xi$. (In applying Corollary 3.16, note that $o(M_\infty(Q, \Psi)) < \gamma^+$ by Kunen-Martin.) Let

$$C = \{\eta < \gamma \mid \eta \text{ is a strong cutpoint of } Q\}.$$

It is easy to see that C has order type γ, and that C is mapped order preservingly into the order type of the mouse order by sending $\eta \in C$ to the mouse rank of all (R, Φ) such that $M_\infty(R, \Phi) = M_\infty(Q, \Psi)|\eta$. So the mouse order below (Q, Ψ) has order type at least ξ, as desired.

For (2): Suppose (P, Σ) has limit type. Then τ_∞ is a limit of successor cardinal cutpoints η in $M_\infty(P, \Sigma)$, and for each such η there is a $(Q, \Psi) <^* (P, \Sigma)$ such that $M_\infty(Q, \Psi) = M_\infty(P, \Sigma)|\eta$. It follows that $\tau_\infty \leq \delta$. It is easy to see that such pairs (Q, Ψ) are cofinal in the mouse order below (P, Σ), so $\text{cof}(\tau_\infty) = \text{cof}(\delta)$. Thus $\text{cof}(\tau_\infty) > \omega$, so $\pi_{P,\infty}$ is discontinuous at τ^P, so there is a $\mu < \tau^P$ so that $\mu < \rho_{k(P)}(P)$ and μ is measurable in P and some $r\Sigma_{k(P)}(P)$ map of μ cofinally into τ^P. But then $\text{cof}(\tau_\infty) = \text{cof}(\pi_{P,\infty}(\mu))$, so τ_∞ and δ are singular. Finally, for each $\beta < \delta$, $o(H_\beta) < \delta$, and thus by Corollary 3.16, Σ is δ-Suslin.

We turn to (3). Suppose P has a top block. The proof in (2) that $\tau_\infty \leq \delta$ shows that $\beta_\infty \leq \delta$ in this case. Since $o(H_\alpha) < \delta$ for all $\alpha < \delta$, we then have, by Corollary 3.40(c), that Σ^{sh} is δ-Suslin.

If P is not stable, then Σ is δ-Suslin by Lemma 3.33(c). In this case, $\mathbf{\Gamma}$ cannot be closed under $\exists^{\mathbb{R}}$, for otherwise, $\mathbf{\Gamma}$ is the class of all δ-Suslin sets, so $\text{Code}(\Sigma) \in \mathbf{\Gamma}$. But Σ is a complete strategy, so s is not by Σ iff there is a t diverging from s at some limit step such that t is by Σ. So $\neg\,\text{Code}(\Sigma)$ is positive-$\mathbf{\Sigma}^1_1$ in $\text{Code}(\Sigma)$, and hence $\neg\,\text{Code}(\Sigma)$ is δ-Suslin. This implies $\text{Code}(\Sigma) \in \mathbf{\Delta}$, contradiction. This proves (3)(a).

For (3)(b), suppose P is stable. Clearly, P is minimal. The argument of the last paragraph shows that if Σ is δ-Suslin, then $\mathbf{\Gamma}$ is not closed under $\exists^{\mathbb{R}}$.

For (3)(c), let $\mathbf{\Gamma}$ be closed under $\exists^{\mathbb{R}}$, so that $\mathbf{\Gamma}$ is the class of δ-Suslin sets.

Claim 1. $\Sigma^{\text{rl}} \notin \mathbf{\Gamma}$.

Proof: Otherwise Σ^{rl} is δ-Suslin, so $o(H_\delta) < \delta^+$ by Kunen-Martin. But $o(H_\alpha) < \delta$ for all $\alpha < \delta$, so $\text{Code}(\Sigma)$ is δ-Suslin by Corollary 3.16. But Σ is a complete strategy, so $\neg\,\text{Code}(\Sigma)$ is positive-$\mathbf{\Sigma}^1_1$ in $\text{Code}(\Sigma)$, and hence $\neg\,\text{Code}(\Sigma)$ is δ-Suslin. Thus $\text{Code}(\Sigma) \in \mathbf{\Delta}$, contradiction. □

Claim 2. $\beta_\infty = \delta$.

Proof: β_∞ is the sup of all $\tau_\infty(Q|\beta^Q, \Sigma_{Q|\beta^Q})$ such that (Q, Σ_Q) is a non-dropping iterate of (P, Σ), and thus $\beta_\infty \leq \delta$. δ is a limit of Suslin cardinals, so if $\beta_\infty < \delta$ then $\tau_\infty < \delta$ by Corollary 3.46. But that would imply $\Sigma^{\text{rl}} \in \mathbf{\Delta}$, contrary to claim 1. Thus $\beta_\infty = \delta$. □

The claims together with Theorem 3.47 complete the proof of (3)(c). □

Remark 4.5: We do not know whether the converse of (3)(b) holds. If P is top block stable and $\mathbf{\Gamma}$ is not inductive-like, must Σ be δ-Suslin? A related question is: what is the pointclass generator for the sets Kleene-recursive in the real quantifier, the "Kleene pointclass"?

It is worth noting that our results show that assuming HPC, the class of Suslin cardinals is closed strictly below its sup.

Corollary 4.6: *Assume HPC, and let δ be a limit of Suslin cardinals such that there are Suslin cardinals $> \delta$; then δ is a Suslin cardinal.*

Proof: Let $\mathbf{\Delta} = \{A \mid \exists\gamma < \delta(A \text{ is } \gamma\text{-Suslin}\}$, and let (P, Σ) be a pointclass generator for $\mathbf{\Delta}$. If Σ is δ-Suslin, then Σ witnesses that δ is a Suslin cardinal. If Σ is not δ-Suslin, then P has a top block and is stable, and the proofs of claims (1) and (2) just given show that $\beta_\infty(P, \Sigma) = \delta$. But then Σ^{sh} witnesses that δ is a Suslin cardinal. □

Of course, AD^+ implies the class of Suslin cardinals is closed, period. We have developed the theory of mouse pairs assuming AD^+, so it is a tacit hypothesis in Corollary 4.6. However, we would guess that $\mathsf{AD} + \mathsf{DC}_\mathbb{R}$ is sufficient to develop the theory, perhaps without much difficulty. Further, we conjecture

Conjecture 4.6.1: $\mathsf{AD} + \mathsf{HPC}$ implies AD^+.

This may be significantly easier to prove than the well-known conjecture that AD implies AD^+.

Finally, we sketch a proof of Theorem 1.10 from the introduction. We shall rely on the fact that the theory of mouse pairs under AD^+ can be extended so that it applies to pure extender pairs (P, Σ) such that P is active, with its last extender $\dot{F}^P$ being long, and all other extenders on the P-sequence being short. This seems to be a completely routine matter of putting together [19] and [10], but as of this writing, it has not yet been done. That is why this is a sketch, rather than a full proof.

Theorem 4.7: *Assume that there is a countable $\omega_1 + 1$-iterable pure extender mouse with a long extender on its sequence. Let $M \models \mathsf{AD}^+ + \mathsf{NLE}$ be such that $\mathbb{R} \cup OR \subseteq M$; then $M \models \mathsf{LEC}$.*

Proof: (*Sketch.*) Let (P, Σ) be a minimal pure extender pair such that $\dot{F}^P$ is long. Our hypothesis implies that there is such a pair. Because P is active, it has a top block, and because $\dot{F}^P$ is long, $\beta(P, \Sigma) < \text{crit}(\dot{F}^P)$. We have that $(Q, \Psi) <^* (P, \Sigma)$ iff (Q, Ψ) is a pure extender pair with only short extenders on its sequence.

Let $M \models \mathsf{AD}^+ + \mathsf{NLE}$ with $\mathbb{R} \cup \mathrm{OR} \subseteq M$, and let

$$\mathbf{\Delta} = \{A \mid \exists (Q, \Psi) \in M (A \leq_w \text{Code}(\Psi))\}.$$

Suppose toward contradiction that $M \not\models \mathsf{LEC}$, so that we have a Suslin cardinal μ such that $o(\mathbf{\Delta}) < \mu < \theta^M$. It is enough to show that there is a $(Q, \Psi) <^* (P, \Sigma)$ such that $(Q, \Psi) \notin \mathbf{\Delta}$, for then the mouse-least such (Q, Ψ) is μ-Suslin by Theorem 4.4, and hence in M, contradiction.

So suppose all $(Q, \Psi) <^* (P, \Sigma)$ are in $\mathbf{\Delta}$. Then

$$\beta_\infty(P, \Sigma) \leq \sup(\{o(M_\infty(Q, \Psi)) \mid (Q, \Psi) <^* (P, \Sigma)\}) = o(\mathbf{\Delta}).$$

But (P, Σ) is not top block stable, because it has a last extender, and so Σ is $o(\boldsymbol{\Delta})$-Suslin. This implies $\Sigma \in M$, contrary to $M \models \mathsf{NLE}$. □

5. Local HOD Computation

We assume $\mathsf{AD}_{\mathbb{R}} + \mathsf{HPC} + V = L(P(\mathbb{R}))$ throughout this section. Our goal is to show that $\mathrm{HOD}|\theta$ is a least branch premouse.

Some preliminary definitions:

Definition 5.1: If (Q, Ψ) and (R, Ω) are mouse pairs, then $(Q, \Psi))$ is a *strong initial segment* of (R, Ω) iff $Q = R|\eta$ for some cutpoint η of R such that η is not the critical point or length of any (even partial) extender on the R-sequence, and $\Psi = \Omega_Q$. We write $(Q, \Psi) \trianglelefteq^* (R, \Omega)$ in this case.

Note that if Q is passive and $(Q, \Psi) \trianglelefteq^* (R, \Omega)$, then $E^R_{o(Q)} = \emptyset$.

Definition 5.2: Let (P, Σ) and (R, Ω) be mouse pairs of the same type such that $(P, \Sigma) \leq^* (R, \Omega)$. We call $\langle \mathcal{T}, \mathcal{U} \rangle$ a *minimal comparison* of (P, Σ) with (R, Ω) iff

(a) $\mathcal{T}$ is a normal tree on (P, Σ) (by Σ) with last model (Q, Ψ), and $\mathcal{U}$ is a normal tree on (R, Ω) (by Ω) with last model (S, Φ),
(b) P-to-Q does not drop, and $(Q, \Psi) \trianglelefteq^* (S, \Phi)$, and
(c) for all α, $\mathrm{crit}(E^{\mathcal{U}}_\alpha) \leq o(Q)$.

It is clear that if $(P, \Sigma) \leq^* (R, \Omega)$, then there is a minimal comparison of the two.[26]

Now let

$$\bar{\theta} \in \{\theta_\xi \mid \xi = 0 \text{ or } \exists \alpha(\xi = \alpha + 1)\},$$

and recalling that $P_\gamma(\mathbb{R}) = \{A \mid |A|_w < \gamma\}$, let

$$\boldsymbol{\Gamma} = \{A \mid P_{\bar{\theta}}(\mathbb{R}) \models \ A \text{ is Suslin}\},$$

and

$$\boldsymbol{\Delta} = \boldsymbol{\Gamma} \cap \breve{\boldsymbol{\Gamma}}.$$

So $\boldsymbol{\Gamma}$ is closed under both real quantifiers and has the scale property, $o(\boldsymbol{\Delta})$ is the largest Suslin cardinal $< \bar{\theta}$, and $\bar{\theta}$ is itself a Suslin cardinal. Fix an

[26] This comparison may not proceed by iterating away least extender disagreements, however.

lbr hod pair (P, Σ) that is a pointclass generator for $\boldsymbol{\Delta}$. By Theorem 4.4, (P, Σ) has a top block, is top block stable and minimal, and

$$\beta_\infty(P, \Sigma) = o(\boldsymbol{\Delta}),$$

and

$$\tau_\infty(P, \Sigma) = \bar{\theta}.$$

We shall show that $M_\infty(P, \Sigma) = \mathrm{HOD}|\bar{\theta}$. The first step is to prove that (P, Σ) is mouse-fullness preserving below images of β^P.

Lemma 5.3: *Let (Q, Ψ) be an iterate of (P, Σ) via an iteration such that P-to-Q does not drop, let κ be a strong cutpoint cardinal of Q, and let $\eta = \kappa^{+,Q}$; then there is no mouse pair (S, Φ) such that $(Q|\eta, \Psi_{Q|\eta}) \trianglelefteq (S, \Phi)$ and $\rho(S) \leq \beta^Q$.*

Proof: Suppose $(Q|\eta, \Psi_{Q|\eta}) \trianglelefteq (S, \Phi)$ and $\rho(S) \leq \kappa$, and no proper initial segment of (S, Φ) has this property. Let $k = k(S)$, so that $\kappa < \rho_k(S)$. We may assume that κ is not measurable by the S-sequence, for otherwise we can replace S with $R = \mathrm{Ult}_k(S, D)$, where D is the order zero total measure of S on κ. By strategy coherence, $(Q|\eta, \Psi_{Q|\eta}) \trianglelefteq (R, \Phi_{\langle D}\,)$, so the replacement is valid. Similarly, we may assume that S is $\kappa + 1$-sound, in that

$$S = \mathrm{Hull}^S_{k+1}(\kappa + 1 \cup p_{k+1}(S)).$$

It follows from [19] that (S, Φ) is ordinal definable from $(Q|\eta, \Psi_{Q|\eta})$.[27] Since $(Q|\eta, \Psi_{Q|\eta}) <^* (P, \Sigma)$, $(Q|\eta, \Psi_{Q|\eta}) \in \boldsymbol{\Delta}$, so $(S, \Phi) \in P_{\bar{\theta}}(\mathbb{R})$, so $(S, \Phi) <^* (Q, \Psi)$. This leads to a contradiction, the argument being more awkward than usual because we do not have comparison by least extender disagreement available.

Let $\mathcal{T}$ on S and $\mathcal{U}$ on Q be a minimal comparison witnessing $(S, \Phi) <^* (Q, \Psi)$, with models $S_\xi = \mathcal{M}^{\mathcal{T}}_\xi$ and $Q_\xi = \mathcal{M}^{\mathcal{U}}_\xi$. Let $\mathcal{T} \restriction \gamma + 1$ be the initial segment of $\mathcal{T}$ that is on $S|\eta = Q|\eta$; then $\mathcal{T} \restriction \gamma + 1$ and $\mathcal{U} \restriction \gamma + 1$ use the same extenders and pick the same branches because $\Psi_{Q|\eta} = \Phi_{S|\eta}$. Let

$$i = i^{\mathcal{T}}_{0,\gamma} \restriction (S|\eta) = i^{\mathcal{U}}_{0,\gamma} \restriction (Q|\eta).$$

Letting S_θ and Q_δ be the last models, we have that $\gamma \leq_T \theta$ and $\gamma \leq_U \delta$ because $i(\kappa)$ is a strong cutpoint in S_γ and Q_γ.

[27] See Claim 3 in the proof of Theorem 11.3.2 in [19].

Suppose $S_\theta \in Q_\delta$; then $i(\eta)$ is not a cardinal in Q_δ because $S_\gamma = \mathrm{Hull}^{S_\theta}(i(\kappa)+1 \cup r)$ for some finite r, and $i(\eta) \subseteq S_\gamma$. But $i(\eta)$ is a cardinal of Q_γ, and $\mathrm{lh}(E^{\mathcal{U}}_\gamma) > i(\eta)$, so $i(\eta)$ is a cardinal of Q_δ, contradiction.

So S_θ and Q_δ are the same as bare premice. If $[0,\delta)_U$ does not drop, then $S_\theta \lhd Q_\delta$, and since $k(Q_\delta) = 0$, $S_\theta \in Q_\delta$, contradiction. Thus $[0,\delta)_U$ drops, and $k = k(Q_\delta)$ as well. But $[0,\gamma)_U$ does not drop, and $Q_\gamma | i(\eta) \trianglelefteq Q_\delta$, so $i(\eta) \leq \rho_{k+1}(Q_\delta)$, contrary to $\rho_{k+1}(S_\theta) \leq i(\kappa)$. □

Now we show (P,Σ) has a kind of mouse-fullness at images of $o(P)$.

Lemma 5.4: *Let (R,Ω) be a projectum stable lbr hod pair such that $(P,\Sigma) \leq^* (R,\Omega)$, and let $\langle \mathcal{T},\mathcal{U}\rangle$ be a minimal comparison of (P,Σ) with (R,Ω), with $(Q,\Psi) \trianglelefteq^* (S,\Phi)$ being the last models; then*

(a) R-to-S does not drop (in model or degree), and
(b) $o(Q) \leq \rho_{k(S)}(S)$, $o(Q)$ is $r\Sigma_{k(S)}(S)$-regular.

Proof: Fix $\mathcal{T}$ and $\mathcal{U}$ with last models (Q,Ψ) and (S,Φ) such that $(Q,\Psi) \trianglelefteq^* (S,\Phi)$.

Claim 1. There is no (N,Λ) such that $(Q,\Psi) \trianglelefteq (N,\Lambda) \trianglelefteq (S,\Phi)$, and for some $\eta < o(Q)$ and $r \in N$, $\mathrm{Hull}^N_{k(N)+1}(\eta \cup r)$ is cofinal in $o(Q)$.

Proof: Assume not, and let (N,Λ) be the first such initial segment of (S,Φ). It follows that $o(Q)$ is $r\Sigma^N_{k(N)}$-regular, and a cutpoint cardinal of N. But then

$$M_\infty(Q,\Psi) \trianglelefteq M_\infty(N,\Lambda),$$

and

$$\pi^\Lambda_{N,\infty} \restriction o(Q) = \pi^\Psi_{Q,\infty}.$$

Letting η_∞ and r_∞ be the images of η and r under $\pi^\Lambda_{N,\infty}$, we have that

$$\mathrm{Hull}^{M_\infty(N,\Lambda)}_{k+1}(\eta_\infty \cup r_\infty) \text{ is cofinal in } o(M_\infty(Q,\Psi)).$$

But $M_\infty(N,\Lambda) \in \mathrm{HOD}$, and $\bar{\theta} = o(M_\infty(P,\Sigma)) = o(M_\infty(Q,\Psi))$, so $\bar{\theta}$ is singular in HOD. However, successor points in the Solovay sequence are regular in HOD.[28] □

Claim 2. For all $\alpha + 1 < \mathrm{lh}(\mathcal{U})$, $\mathrm{crit}(E^{\mathcal{U}}_\alpha) < o(Q)$.

[28]In fact, they are Woodin in HOD, but regularity is much easier to prove.

Proof: $\text{crit}(E^{\mathcal{U}}_\alpha) \leq o(Q)$ by minimality of the comparison. Suppose $\text{crit}(E_\alpha)^{\mathcal{U}} = o(Q)$. Then $Q \trianglelefteq \mathcal{M}^{\mathcal{U}}_\alpha$ because $\mathcal{U}$ is normal, so β^Q is strong to $\text{crit}(E^{\mathcal{U}}_\alpha)$ in $\mathcal{M}^{\mathcal{U}}_\alpha$, so by coherence, $o(\beta^Q) > o(Q)$ in $\mathcal{M}^{\mathcal{U}}_\alpha$. So in fact, $E^{\mathcal{U}}_\alpha$, which by normality is the first extender F on the $\mathcal{M}^{\mathcal{U}}_\alpha$ sequence with $\text{lh}(F) \geq o(Q)$, must have critical point β^Q. □

Claim 3. The branch R-to-S in $\mathcal{U}$ does not drop.

Proof: Suppose toward contradiction that R-to-S drops.

Suppose first that there is a last extender F used on the branch R-to-S. If $\lambda(F) < o(Q)$, then because the branch dropped, $S = \text{Hull}^{S}_{k(S)+1}(\lambda(F) \cup r)$ for $r = p_{k+1}(S)$. This contradicts Claim 1, so $\lambda(F) \geq o(Q)$. The proof of Claim 2 then yields that $\kappa = \text{crit}(F)$ is a strong cutpoint of Q, and a cardinal of Q. (Thus $\kappa \leq \beta_Q$.) This implies that $\rho(S) \leq \kappa$, which contradicts Lemma 5.3. Thus there is no last extender used in R-to-S.

But then we can find some model W far enough out on R-to-S that there is no further dropping, and $i^{\mathcal{U}}_{W,S}(\delta) = o(Q)$ for some δ. Letting $k = k(W)$, we must have that $\text{Hull}^{W}_{k+1}(\eta \cup r)$ is cofinal in δ, for η the sup of the generators of R-to-W, and some r. But $\eta < \delta$, so using $i^{\mathcal{U}}_{W,S}$ to move this situation up to S, we have a contradiction to Claim 1. (Note here that $i^{\mathcal{U}}_{V,S}$ is continuous at δ, because δ is $r\Sigma_{k(W)}$-regular and not measurable in W, because $o(Q)$ has those properties in S.) □

So we have (a) of the lemma, and (b) follows at once from Claim 1. □

Corollary 5.5: *If (P, Σ) is a generator of $P_{\bar{\theta}}(\mathbb{R})$, where $\bar{\theta} = \theta_0$ or $\bar{\theta} = \theta_{\alpha+1}$ for some α, then whenever $(P, \Sigma) \leq^* (R, \Omega)$, then $M_\infty(P, \Sigma) \trianglelefteq^* M_\infty(R, \Omega)$.*

Proof: Let $(Q, \Psi) \trianglelefteq^* (S, \Phi)$ be the last pairs in a minimal comparison of (P, Σ) with (R, Ω). By (b) of the lemma, $M_\infty(Q, \Psi) \trianglelefteq^* M_\infty(S, \Phi)$. But $M_\infty(P, \Sigma) = M_\infty(Q, \Psi)$, and by part (a) of the lemma, $M_\infty(R, \Omega) = M_\infty(S, \Phi)$. □

We are assuming $\mathsf{AD}_{\mathbb{R}}$, so the successor points in the Solovay sequence are cofinal in θ.

Definition 5.6: $\mathcal{H}$ is the unique least branch premouse N such that $o(N) = \theta$, and whenever (P, Σ) is a pointclass generator for $P_{\theta_{\alpha+1}}(\mathbb{R})$ for some α, then $M_\infty(P, \Sigma) \trianglelefteq^* N$.

The following is Theorem 1.11 from the introduction.

Theorem 5.7: *($AD_{\mathbb{R}}+HPC+V=L(P(\mathbb{R}))$) The universe of $\mathcal{H}$ is V_θ^{HOD}, and $HOD=L[\mathcal{H}]$.*

Proof: It is clear that $\mathcal{H}\in$ HOD. It is known that under $\mathsf{AD}_{\mathbb{R}}+V=L(P(\mathbb{R}))$, $\text{HOD}=L(V_\theta^{\text{HOD}})$ (see [23]), and that Θ is a strong limit cardinal in HOD, so it is enough to show that every bounded subset of Θ that is ordinal definable belongs to $\mathcal{H}$. So fix A bounded in Θ such that A is ordinal definable.

Lemma 5.8: *There are $\xi<\gamma<\Theta$ and a formula $\varphi(u,v)$ of the language of set theory such that $A\subseteq\xi$, and whenever $X\subseteq\mathbb{R}$ has Wadge rank at least γ, then for all α,*

$$\alpha\in A\Leftrightarrow L(X,\mathbb{R})\models\varphi[\alpha,\xi].$$

Proof: We can define A from the stage at which it becomes ordinal definable. More precisely, let $A\subseteq\mu$, and let Y be a prewellorder of $\mathbb{R}$ of length μ. Woodin's results show that A is OD in $L(X,\mathbb{R})$ for some X of Wadge rank $<\theta(Y)$. We wellorder the OD subsets of μ: for any OD set $B\subseteq\mu$, let $f(B)$ be the lexicographically least triple $\langle\nu,\eta,\psi\rangle$ such that whenever Z has Wadge rank ν, then B is the unique t such that $L(Z,\mathbb{R})\models\psi[\eta,t]$. f is definable over $L(X,\mathbb{R})$ from μ whenever X has Wadge rank at least $\theta(Y)$, via a formula that does not depend on X. Let ξ be an ordinal that codes $(\xi)_0=\mu$ and $(\xi)_1=f(A)$ in some simple way, and let

$$\varphi(u,v)=\text{“}\exists B\subseteq(v)_0(f(B)=(v)_1\wedge u\in B)\text{”}.$$

It is easy to see that this works. □

Fix ξ, γ, and φ as in Lemma 5.8. Let $\gamma<\theta_{\alpha+1}$, and let (P,Σ) be a pointclass generator for $P_{\theta_{\alpha+1}}(\mathbb{R})$. We can take (P,Σ) so that ξ and γ are in $\text{ran}(\pi^\Sigma_{P,\infty})$, say

$$\pi_{P,\infty}(\langle\bar{\xi},\bar{\gamma}\rangle)=\langle\xi,\gamma\rangle.$$

For any iterate (Q,Ψ) of (P,Σ) put

$$\alpha\in A^Q\Leftrightarrow\pi_{Q,\infty}(\alpha)\in A.$$

Lemma 5.9: *There is an iterate (R,Λ) of (P,Σ) such that whenever (Q,Ψ) is an iterate of (R,Λ), then $A^Q\in Q$.*

Lemma 5.9 suffices for the theorem. For let (R, Λ) be as in Lemma 5.9. There is no infinite chain $(Q_n, \Psi_n) \prec^{\mathcal{F}(P,\Sigma)} (Q_{n+1}, \Psi_{n+1})$ such that $(Q_0, \Psi_0) = (R, \Lambda)$ and $\pi_{Q_n,Q_{n+1}}(A^{Q_n}) \neq A^{Q_{n+1}}$ for all n. (Consider the preimage of A^{Q_ω}.) So there is a (Q, Ψ) such that whenever (S, Ω) is an iterate of (Q, Ψ), then $\pi_{Q,S}(A^Q) = A^S$. This implies $A = \pi_{Q,\infty}(A^Q)$, so $A \in \mathcal{H}$.

So let us prove Lemma 5.9. This is precisely where we use that (P, Σ) is a hod pair, and not a pure extender pair, and that it is mouse full in the sense of Lemma 5.4. Suppose we could find an iterate (R, Λ) of (P, Σ) and an (S, Φ) such that

(i) $(R, \Lambda) \trianglelefteq^* (S, \Phi)$, and
(ii) $S \models$ "$\lambda > o(R)$ and λ is a limit of Woodin cardinals".

By the generic interpretability theorem of [19], Code(Λ) is in the derived model of S at λ, and this enables S to compute the theory of $L(\text{Code}(\Lambda), \mathbb{R})$, as well as the map $\pi_{R,\infty}\colon R \to M_\infty(R, \Lambda)$.[29] Thus $A^R \in S$, and hence $A^R \in R$ by Lemma 5.4. Moreover, if (Q, Ψ) is an iterate of (R, Λ), then the iteration is by Φ and carries S along, so $A^Q \in Q$ by the same proof.

The problem with this sketch is that S is a strategy mouse with Woodin cardinals, and such mice are difficult to construct. With more care, we could make do with an S that has a small finite number of Woodin cardinals above $o(R)$, but that doesn't help much at this stage. Once we have shown that $\mathcal{H} = \text{HOD}$, the fact that Θ is a limit of Woodins in HOD implies the existence of (R, Λ) and (S, Φ) as in (i) and (ii).[30] But at this stage we are trying to show $\mathcal{H} = \text{HOD}$, so this doesn't help.

We solve our problem by letting S be a hybrid, an extension of R that continues to insert the strategy information in Λ, but is a pure extender premouse otherwise. That is, $S = M_\omega^{\Lambda,\sharp}$, the minimal active such hybrid with ω Woodin cardinals. $M_\omega^{\Lambda,\sharp}$ can compute the theory of $L(\text{Code}(\Lambda), \mathbb{R})$ by consulting its derived model, and it can be constructed using AD^+ in $L(\text{Code}(\Lambda)^\sharp, \mathbb{R})$. Let us proceed to the details.

Proof: [Proof of Lemma 5.9] Let Γ be a good pointclass closed under $\forall^{\mathbb{R}}$ with the scale property, and such that $\text{Code}(\Sigma) \in \Gamma \cap \breve{\Gamma}$. Let

[29] Via an $\mathbb{R}$-genericity iteration of S above $o(R)$ we can arrange that the reals of the derived model are the reals of V.

[30] With S satisfying some reasonable fragment of ZFC. One could not ask for a measurable cardinal in S above λ, because we could be living in the minimal model of $\mathsf{AD}_{\mathbb{R}}$.

$\langle N^*, \delta, S, T, w, \Sigma^* \rangle$ be a coarse Γ-Woodin tuple, and let $\mathbb{C}$ be the associated maximal hod pair construction. By Theorem 2.3, we may fix $\nu < \delta$ such that $(M_{\nu,0}^{\mathbb{C}}, \Omega_{\nu,0}^{\mathbb{C}})$ is an iterate of (P, Σ), and (P, Σ) iterates strictly past all earlier levels of $\mathbb{C}$. Set

$$(R, \Lambda) = (M_{\nu,0}^{\mathbb{C}}, \Omega_{\nu,0}^{\mathbb{C}}).$$

Claim 1. For all $\langle \alpha, l \rangle$ such that $\langle \nu, 0 \rangle \leq_{\text{lex}} \langle \alpha, l \rangle \leq_{\text{lex}} \langle \delta, 0 \rangle$,

(a) $o(R)$ is a cutpoint of $M_{\alpha,l}^{\mathbb{C}}$, and $o(R)$ is not the critical point of an extender on the $M_{\alpha,l}^{\mathbb{C}}$ sequence,
(b) $o(R) \leq \rho(M_{\alpha,l}^{\mathbb{C}})$, and
(c) $o(R)$ is regular in $M_{\alpha,l}^{\mathbb{C}}$.

Proof: By induction on $\langle \alpha, l \rangle$. For (a), it is enough to see that if $F = \dot{F}^{M_{\alpha,0}}$ and $F \neq \emptyset$, then $o(R) < \text{crit}(F)$. But if not, then letting $F^* = B^{\mathbb{C}}(F)$ be the background extender for F,

$$(R, \Lambda) = (M_{\nu,0}, \Omega_{\nu,0})^{i_{F^*}(\mathbb{C})}$$

and

$$\nu < i_{F^*}(\nu).$$

$i_{F^*}(P) = P$ and $i_{F^*}(\Sigma) \subseteq \Sigma$, so (P, Σ) iterates to (R, Λ) in $\text{Ult}(N^*, F^*)$, but (P, Σ) iterates strictly past (R, Λ) in $\text{Ult}(N^*, F^*)$ because $\nu < i_{F^*}(\nu)$, contradiction.

For (b), suppose that

$$\rho = \rho_{l+1}(M_{\alpha,l}) < o(R).$$

Let

$$\begin{aligned} (M, \Omega) &= (M_{\alpha,l}, \Omega_{\alpha,l}), \\ (H, \Phi) &= (M_{\alpha,l+1}, \Omega_{\alpha,l+1}), \\ \pi &= \text{ anticore map from } H \text{ to } M. \end{aligned}$$

As in the proof of Lemma 5.4, we reach a contradiction by showing that $o(M_\infty(R, \Lambda))$ is singular in HOD. The new element here is that we have not reached $(M_{\alpha,l}, \Omega_{\alpha,l})$ in an iteration. Instead, let us consider the comparison of (M, H, ρ) with M, done in some larger Γ_1-Woodin universe N^{**} where M is countable and Code(Ω) has been captured, as in the proof in [19] that (M, Ω) is parameter solid.

Assume first that M is projectum stable. By [19, §10.6] there are (S, Ψ) and iteration maps

$$j\colon (M, \Omega) \to (S, \Psi)$$

and

$$i\colon (H, \Phi) \to (S, \Psi)$$

such that $\operatorname{crit}(i) > \rho$, $\operatorname{crit}(j) > \rho$, and $i = j \circ \pi$. Letting $p = p(M) = p_{l+1}(M)$, we claim that there is a $\tau < j(o(R))$ such that $\mathrm{Hull}_{l+1}(\tau \cup j(p))$ is cofinal in $j(o(R))$. For let $\mathcal{T}$ be the iteration tree on (M, H, ρ) giving rise to i, $S = \mathcal{M}^{\mathcal{T}}_{\gamma}$, and $\xi <_T \gamma$ be least such that $j(o(R)) \in \operatorname{ran}(i^{\mathcal{T}}_{\xi,\gamma})$. Let

$$i^{\mathcal{T}}_{\xi,\gamma}(\mu) = j(o(R)).$$

Either $\xi = 1$ and $\mathcal{M}^{\mathcal{T}}_{\xi} = H$, or $\xi = \tau + 1$ for some $\tau \geq 1$, and in either case there is a $\beta < \mu$ such that

$$\mathrm{Hull}^{\mathcal{M}^{\mathcal{T}}_{\beta}}_{l+1}(\beta \cup i_{1,\xi} \circ \pi^{-1}(p)) \text{ is cofinal in } \mu.$$

(If $\xi = 1$, then $\beta = \rho$, and otherwise $\beta = \lambda(E^{\mathcal{T}}_{\xi-1})$.) By induction, $j(o(R))$ is $r\Sigma_l$ regular and not measurable in S, so μ is $r\Sigma_l$ regular and not measurable in $\mathcal{M}^{\mathcal{T}}_{\xi}$, and hence $i^{\mathcal{T}}_{\xi,\gamma}$ is continuous at μ. Thus letting $\tau = i^{\mathcal{T}}_{\xi,\gamma}(\beta)$ and noting that $i_{1,\gamma} \circ \pi^{-1}(p) = j(p)$, we get

$$\tau < j(o(R))$$

and

$$\mathrm{Hull}^{S}_{l+1}(\tau \cup j(p)) \text{ is cofinal in } j(o(R)).$$

But then, since $o(R)$ is an $r\Sigma_l$ regular cutpoint in M, $(j(R), \Psi_{j(R)})$ is an iterate of (R, Λ) via an initial segment of the tree from M to S, so

$$M_\infty(R, \Lambda) = M_\infty(j(R), \Psi_{j(R)}) \trianglelefteq M_\infty(S, \Psi).$$

Since $M_\infty(S, \Psi) \in \mathrm{HOD}$, $o(M_\infty(R, \Lambda))$ is singular in HOD, a contradiction.

If M is not projectum stable, then let D be the order zero measure of M on η, where η is the $r\Sigma^M_l$ cofinality of $\rho_l(M)$. M is a pfs premouse, so $\eta < \rho_{l+1}(M)$. Let N be the strong core of M, $\pi\colon N \to M$ be the anticore map (cf. [19]). π is the identity on $\rho_l(M)$. Let

$$(M_1, \Omega_1) = (\mathrm{Ult}_l(N, D), \Omega^{\sigma}_{\langle D \rangle}),$$

where $\sigma\colon \mathrm{Ult}_l(N,D) \to \mathrm{Ult}_l(M,D)$ is the copy map. (M_1, D_1) is strongly stable.[31] Let

$$(R_1, \Lambda_1) = (\mathrm{Ult}_0((R,\Lambda), D), \Lambda_{\langle D\rangle}).$$

$R_1 \lhd M_1$ and $\sigma \restriction R_1 = \mathrm{id}$, so $\Lambda_1 = (\Omega_1)_{R_1}$. Moreover $\rho_{l+1}(M_1) \leq i_D(\rho_{l+1}(M) < i_D(o(R)) = o(R_1)$. Since $M_\infty(R,\Lambda) = M_\infty(R_1,\Lambda_1)$, we now have the bad situation from above, but with M_1 being projectum stable, a contradiction.

For (c), suppose that $o(R)$ is $r\Sigma_l^M$ regular and $r\Sigma_{l+1}^M$ singular, and not measurable by the M-sequence. Let $\pi\colon M \to M_\infty(M,\Omega)$; then the relevant ultrapower maps are all continuous at the current image of $o(R)$, so π is continuous at $o(R)$, so again $o(M_\infty(R,\Lambda))$ is singular in HOD, contradiction. □

Now let

$$(K, \Upsilon) = (M^{\mathbb{C}}_{\delta,0}, \Omega^{\mathbb{C}}_{\delta,0}).$$

$(R,\Lambda) \trianglelefteq^* (K,\Upsilon)$, and in K, $o(R)$ is a limit cardinal, a cutpoint, and not measurable. Working in K, we do the maximal pure extender construction $\mathbb{D}$ relative to (R,Λ). That is,

$$(M^{\mathbb{D}}_{0,0}, \Omega^{\mathbb{D}}_{0,0}) = (R,\Lambda),$$

and we continue by adding extenders F whenever $o(R) < \mathrm{crit}(F)$, F has a nice background extender F^* that is on the K-sequence, and we get a hybrid Λ-premouse by doing so. We add branch information about Λ at branch active stages just as in [19]. No branch information beyond that in Λ gets added. Because we are working in K, no level projects below $o(R)$, so

$$(R,\Lambda) \trianglelefteq (M^{\mathbb{D}}_{\nu,k}, \Omega^{\mathbb{D}}_{\nu,k})$$

for all $\langle \nu, k\rangle \leq_{\mathrm{lex}} \langle \delta, 0\rangle$. The well-known arguments for pure extender constructions show that $\mathbb{D}$ does not break down.

For any lbr hod pair (S,Φ), let $M_\omega^\sharp(S,\Phi)$ be the minimal active pure extender (S,Φ)-mouse N with ω Woodin cardinals such that $S \trianglelefteq N$. Let $\Psi(S,\Phi)$ be its canonical iteration strategy. Here $\Psi(S,\Phi)$ only acts on iteration trees that are above $o(S)$, and these iterations move the predicate for

---

[31] $\rho_l(M_1) = \sup i_D$"$\rho_l(M)$, and therefore has $r\Sigma_l$ cofinality η is M_1. But η is not measurable by the M_1-sequence.

Λ correctly. In this general case, some level of N may project across $o(S)$; nevertheless $S \trianglelefteq N$ because all cores include $S \cup \{S\}$ by fiat.

Claim 2. There is a $\nu < \delta$ such that $M_\omega^\sharp(R, \Lambda)$ is the core of $(M_{\nu,0}^{\mathbb{D}}, \Omega_{\nu,0}^{\mathbb{D}})$.

Proof: Let $Q = M_\omega^\sharp(R, \Lambda)$ and $\Psi = \Psi(R, \Lambda)$. In the comparison of Q with $M_{\delta,0}^{\mathbb{D}}$ only Q moves, so if the claim is false, then Q iterates past $M_{\delta,0}^{\mathbb{D}}$ by Ψ. Let $\mathcal{U}$ be the tree on Q of length δ whereby it does so, and $b = \Psi(\mathcal{U})$. Since Ψ is projective in $\text{Code}(\Sigma)^\sharp$ and $\text{Code}(\Sigma)^\sharp \in \Gamma \cap \check{\Gamma}$, $\mathcal{U}$ and b belong to $L[N^*, T_0]$, where T_0 is the tree of a Γ scale on a universal Γ set. But then δ is singular in $L[N^*, T_0]$, contrary to its being Woodin there. □

Now let

$$(Q, \Psi) = (M_{\nu,0}^{\mathbb{D}}, \Omega_{\nu,0}^{\mathbb{D}}),$$

where ν is as in Claim 2. It is enough to show that $A^R \in Q$. For this it will be enough to show that (Q, Ψ) can compute truth in $L(\mathbb{R}, \text{Code}(\Lambda))$ by consulting its derived model. The difficulty here is that Q has only been directly given $\Lambda \cap Q$, and not the action of Λ on trees in its derived model.

For $n < \omega$ and S a non-dropping iterate of (Q, Ψ), let δ_n^S be the nth Woodin cardinal of S above $o(R)$, and let $\lambda^S = \sup_{n<\omega} \delta_n^S$. Let $\mathcal{A}(n, S)$ be the set of normal trees $\mathcal{U}$ on S such that for some cutpoint cardinal $\eta < \delta_{n+1}^S$, $\mathcal{U}$ is based on $S|\eta$ and $\delta_n^S < \text{crit}(E_\alpha^{\mathcal{U}})$ for all $\alpha + 1 < \text{lh}(\mathcal{U})$.

Claim 3. Let (S, Φ) be a non-dropping iterate of (Q, Ψ) and $n < \omega$; then $\Phi(\mathcal{U}) \in S$ for all $\mathcal{U} \in \mathcal{A}(n, S) \cap S|\lambda^S$, and $\Phi \restriction \mathcal{A}(n, S) \cap S|\lambda^S$ is definable over S. Moreover, the definition is uniform in S and n.

Proof: We use Q-structures to compute Φ. Given $\mathcal{U} \in \mathcal{A}(n, S)$ of limit length that is by Φ, S uses its own extender sequence (restricted to critical points $> \delta(\mathcal{U})$) and $\Lambda \cap S$ to rebuild a Λ hybrid premouse extending $\mathcal{M}(\mathcal{U})$. It must reach a first structure $Q(\mathcal{U})$ projecting across or killing the Woodinness of $\delta(\mathcal{U})$. Then $\Phi(\mathcal{U})$ is the unique b such that $Q(\mathcal{U}) \trianglelefteq \mathcal{M}_b^{\mathcal{U}}$. □

Claim 4. Let (S, Φ) be a non-dropping iterate of (Q, Ψ) and $n < k < \omega$; then there is a term

$$\tau = \tau_{n,k}^S$$

in S such that whenever g is $\text{Col}(\omega, \delta_k^S)$-generic over S then $\Phi(\mathcal{U}) \in S[g]$ for all $\mathcal{U} \in \mathcal{A}(n, S) \cap (S|\lambda^S)[g]$, and

$$\Phi \restriction \mathcal{A}(n, S) \cap (S|\lambda^S)[g] = \tau_g.$$

Moreover, the definition of $\tau_{n,k}^S$ is uniform in S, n, and k.

Proof: (Sketch.) We use the Boolean-valued comparison method to compute Φ on $S[g]$. We work in $S[g]$. Let σ be a term such that $\mathcal{U} = \sigma_g$, and suppose that the empty condition forces that σ is a tree of limit length in $\mathcal{A}(n, S)$ on $S|\eta$, and σ is according to the procedure we are defining now. For each $p \in \mathrm{Col}(\omega, \delta_k^S)$ let $g_p(i) = p(i)$ if $i \in \mathrm{dom}(p)$ and $g_p(i) = g(i)$ otherwise. Let $\mathcal{U}_p = \sigma_{g_p}$, and

$$N_p = \mathcal{M}(\mathcal{U}_p).$$

We assume that our procedure is correct so far, that is, that each $\mathcal{U}_p$ is by Φ. We must find branches b_p such that $\mathcal{U}_p{}^\frown b_p$ is by Φ. For this, it is enough to find the Q-structures for them, that is to find Λ-hybrids

$$N_p \trianglelefteq Q_p$$

that are iterable above $\delta(\mathcal{U}_p)$ in a way that moves the predicate for Λ correctly. Such Q_p do exist, but we must show that they belong to $S[g]$, and identify them.

We do this by simultaneously comparing $S|\eta$ with all the N_p. This results in trees $\mathcal{T}$ on $S|\eta$ and $\mathcal{T}_p$ on N_p. By Claim 3 and the fact that we have symmetrized the situation by considering all N_p simultaneously, $\mathcal{T} \in S$. This allows us to identify the correct branches for $\mathcal{T}_p$; given $\mathcal{T}_p \restriction \gamma$ for γ a limit ordinal, the comparison will have produced some $\mathcal{M}_\xi^\mathcal{T}$ such that

$$\mathcal{M}(\mathcal{T}_p \restriction \gamma) \trianglelefteq \mathcal{M}_\xi^\mathcal{T}.$$

We then extend $\mathcal{T}_p \restriction \gamma$ by choosing the unique b such that $Q(b, \mathcal{T}_p \restriction \gamma) \trianglelefteq \mathcal{M}_\xi^\mathcal{T}$.

For any p we must reach a stage γ such that

$$\mathcal{M}_\gamma^{\mathcal{T}_p} \trianglelefteq \mathcal{M}_\xi^\mathcal{T}$$

for some ξ, and $[0, \gamma)_{T_p}$ does not drop, so that we have $i = i_{0,\gamma}^\mathcal{T} : N_p \to M$, where $M \trianglelefteq \mathcal{M}_\xi^\mathcal{T}$. Let $Q(M)$ be the first level of $\mathcal{M}_\xi^\mathcal{T}$ that projects across $o(M)$ or kills its Woodinness, and let r be the standard parameter of $Q(M)$. Since $\mathcal{T}_p$ is by Φ, i can be extended to a map from Q_p to $Q(M)$, and hence

$$Q_p = \text{ transitive collapse of } \mathrm{Hull}^{Q(M)}(\mathrm{ran}(i) \cup r).$$

This lets us identify Q_p, and hence b_p, in $S[g]$. □

Claim 5. Let (S, Φ) be a non-dropping iterate of (Q, Ψ), $k < \omega$, and g be $\mathrm{Col}(\omega, \delta_k^S)$-generic over S; then

(a) for all $\mathcal{U} \in S[g]$ such that $\mathrm{lh}(\mathcal{U}) < \lambda^S$ and $\mathcal{U}$ is by Λ, $\Lambda(\mathcal{U}) \in S[g]$, and

(b) there is a term $\tau = \tau_k^S$ in S such that

$$\Lambda \cap (S|\lambda^S)[g] = \tau_g.$$

Moreover, τ_k^S is independent of g, and definable over S, uniformly in S and k.

Proof: Let $\delta = \delta_0^S$. Working in S, let $\mathcal{F}$ be the collection of nice extenders E such that $\mathrm{crit}(E) > o(R)$ and $E = E_\xi^S \restriction \mathrm{lh}(E)$ for some $\xi < \delta$, and let w be the canonical wellorder of $S|\delta$. Let $\mathbb{G}$ be the maximal $(w, \mathcal{F})$ hod pair construction[32] of length δ. By Claim 3, S can compute Φ on all $\mathcal{F}$-trees in $S|\lambda^S$, so the construction $\mathbb{G}$ can be done in S. The arguments of [19] show that it does not break down.

Still in S, we can apply Theorem 2.3 to (R, Λ) and $\mathbb{G}$. Since δ is Woodin, (R, Λ) cannot iterate past $(M_{\delta,0}, \Omega_{\delta,0})^{\mathbb{G}}$. Thus we can fix $\xi < \delta$ and a normal tree $\mathcal{W}$ such that letting

$$(R_1, \Lambda_1) = (M_{\xi,0}, \Omega_{\xi,0})^{\mathbb{G}}$$

we have

$$S \models \ (R, \Lambda) \text{ iterates to } (R_1, \Lambda_1) \text{ via } \mathcal{W}.$$

Let $i = i^{\mathcal{W}} \colon R \to R_1$ be the iteration map.

The iteration strategy Σ^* acts on all trees in V, so the strategy $\Upsilon = \Omega(K, \mathbb{C}, \Sigma^*)$ that it induces can be extended so as to act on all trees in V. Thus the strategy $\Psi = \Omega(Q, \mathbb{D}, \Upsilon)$ for Q extends so as to act on all trees in V. Letting $j \colon (Q, \Psi) \to (S, \Phi)$ be the iteration map, this means that the $j(\mathbb{D})$-induced strategy $\Lambda_1 = \Omega(R_1, \mathbb{G}, \Phi)$ extends so as to act on all trees in V. Let

$$\Lambda_1^* = \ \text{extension of } \Lambda_1 \text{ to } V \text{ determined by } \mathbb{G}, \mathbb{D}, \mathbb{C}, \text{ and } \Sigma^*.$$

By Claim 4, $S[g]$ has a term for $\Lambda_1^* \cap S[g]$.

We claim that $\Lambda_1^* = \Lambda_{\mathcal{W}, R_1}$ holds in V, not just in S. This follows from the proof of Theorem 2.3. For by [19], if $\mathcal{U}$ is of limit length and is by both strategies, and $b = \Lambda_1^*(\mathcal{U})$, then

$$W(\mathcal{W}, \mathcal{U}^\frown b) \text{ is a pseudo-hull of } i_b^*(\mathcal{W}),$$

and

$$\mathcal{M}_b^{\mathcal{U}^*} \models i_b^*(\mathcal{W}) \text{ is by } i_b^*(\Lambda).$$

[32] Over nothing, like $\mathbb{C}$, not beginning with (R, Λ) as did $\mathbb{D}$.

So it is enough to see that Φ-iterations move the internal predicate of S for Λ correctly. But Σ^* iterations with critical point $> o(R)$ move $\Lambda \cap N^*$ correctly (because they move Σ^* correctly), so by strong hull condensation for the relevant lifting maps, Υ-iterations above $o(R)$ move the predicate of K for Λ correctly. Similarly, Ψ iterations by Q (which perforce are above $o(R)$) move the predicate of Q for Λ correctly. $\mathcal{M}_b^{\mathcal{U}^*}$ comes from such an iteration, so indeed its internal Λ-predicate is correct.

So S has a term for $\Lambda_{\mathcal{W},R^1}$. But S has i in it, and $\Lambda = (\Lambda_{\mathcal{W},R_1})^i$ by pullback consistency. So S has a term for $\Lambda \cap S[g]$, as desired. □

We can now finish the proof of Lemma 5.9. Let S be an $\mathbb{R}$-genericity iterate of Q by Ψ, with $i\colon Q \to S$ the iteration map. We have g $\text{Col}(\omega, < \lambda^S)$-generic over S such that $\mathbb{R} = \mathbb{R}^*_g$, where $\mathbb{R}^*_g$ are the reals of the symmetric collapse. Using $i(\langle \tau_k^Q \mid k < \omega \rangle)$ and Claim 5, we see that $\Lambda \in S(\mathbb{R}^*_g)$, and is definable in $S(\mathbb{R}^*_g)$ from parameters in S. But then $M_\infty(R, \Lambda)$ and in the direct limit map $\pi\colon R \to M_\infty(R, \Lambda)$ are also in $S(\mathbb{R}^*_g)$ and definable there from parameters in S. But let $\pi(\langle \xi_{,0}, \gamma_0 \rangle) = \langle \xi, \gamma \rangle$; then

$$\xi \in A^R \text{ iff } L(\text{Code}(\Lambda, \mathbb{R}) \models \varphi[\alpha, \pi(\xi_0), \pi(\gamma_0)].$$

So $A^R \in S$ by the homogeneity of $\text{Col}(\omega, < \lambda^S)$, and hence $A^R \in R$.

The proof just given shows that whenever (R_1, Λ_1) is an iterate of (R, Λ), then $A^{R_1} \in R_1$. This is what we need. □

The proof of Lemma 5.9 completes the proof of Theorem 5.7. □

Theorem 5.7 characterizes HOD in models of $\mathsf{AD}_{\mathbb{R}} + \mathsf{HPC}$. It leaves open

Question. Let $M \models \mathsf{AD}^+ + \mathsf{HPC}$, and suppose that M has a largest Suslin cardinal. What is HOD^M, as a mouse?

We believe that the arguments above show that $\text{HOD}|\theta$ is an lpm, but there are difficulties in analyzing the full HOD. Even when $M = L(\mathbb{R})$, we have no natural characterization of HOD^M in the least branch hierarchy. One seems to need to shift over to an extender-biased hierarchy at θ. We do not know how to do this properly for larger M.

6. Woodins in HOD and the Solovay Sequence

By pushing the proof of Theorem 5.7 further, we get our characterization of the successor points in the Solovay sequence as the cutpoint Woodins in HOD. We can also identify the successor Woodins in HOD as the successor points in a certain refinement of the Solovay sequence.

The following is Theorem 1.13 from the introduction.

Theorem 6.1: *Assume $AD_{\mathbb{R}}$ + HPC; then for any $\eta < \theta$, the following are equivalent:*

(a) $\eta = \theta_0$, or $\eta = \theta_{\alpha+1}$ for some α,
(b) η is a cutpoint Woodin cardinal of HOD.

Proof: Let us prove (a) $\Rightarrow$ (b). Let η be as in (a). Of course, Woodin showed ([9]) that η is Woodin in HOD. Let (P, Σ) be a pointclass generator of $P_\eta(\mathbb{R})$. By Corollary 5.5, $M_\infty(P, \Sigma)$ is a strong initial segment of $\mathcal{H}$. Since $\eta = o(M_\infty(P, \Sigma))$, this means that η is a cutpoint of $\mathcal{H}$.

Now let us prove (b) $\Rightarrow$ (a). Let η_0 be a cutpoint Woodin cardinal of HOD. Let

$$\eta_0 < \eta_1 < ... < \eta_5 < \bar{\theta}$$

where $\bar{\theta}$ and the η_i for $i \geq 1$ are successor points in the Solovay sequence, and thus cutpoint Woodins in HOD, as we have just shown. Let (P, Σ) be a pointclass generator for $P_{\bar{\theta}}(\mathbb{R})$,

$$\pi^{\Sigma}_{P,\infty}(\delta_i) = \eta_i,$$

and

$$N = P|\delta_0.$$

Thus $M_\infty(N, \Sigma_N) \trianglelefteq^* M_\infty(P, \Sigma) \trianglelefteq^* \mathcal{H}$, where $\mathcal{H}$ is HOD viewed as an lpm, and the initial segments are cardinal cutpoint initial segments in both cases. N is top block stable, and

$$\eta_0 = \tau_\infty(N, \Sigma_N).$$

Let

$$\Psi = \Sigma_N^{\text{rl}}$$

and

$$\Psi_0 = \Sigma_N^{\text{sh,rl}}.$$

We write

$$\beta_\infty = \beta_\infty(N, \Sigma_N).$$

By our results on optimal Suslin representations, Ψ is η_0-Suslin, but not α-Suslin for any $\alpha < |\tau_\infty|$, while Ψ_0 is β_∞-Suslin, but not α-Suslin for any $\alpha < |\beta_\infty|$.

Recall that for $A \subseteq \mathbb{R}$, $\theta(A)$ is the least ordinal not the image of $\mathbb{R}$ under a map f that is OD(A). Equivalently, $\theta(A) = \theta_{\alpha+1}$, where α is least such that $A \in P_{\theta_{\alpha+1}}(\mathbb{R})$. The main step in our proof is:

Claim 1. $\beta_\infty < \theta(\text{Code}(\Psi_0)) \leq \eta_0$.

Proof: The direct limit system $\mathcal{F}^0(N, \Sigma_N)$ is definable from Ψ_0, and it yields a prewellorder of order type $o(M^0_\infty(N, \Sigma))$. So $\beta_\infty < \theta(\text{Code}(\Psi_0))$.

Our proof that $\theta(\text{Code}(\Psi_0)) \leq \eta_0$ follows the outline of an argument due to Hjorth and/or Woodin. (See [2], and the proof of Lemma 3.28 in [30].) Let $f\colon \mathbb{R} \to \eta_0$ be ordinal definable from Code(Ψ_0). We shall show that ran(f) is bounded in η_0.

By AD^+, f is OD from Code(Ψ_0) in $L(X, \mathbb{R})$, whenever X has Wadge rank $\geq \theta(\text{Code}(\Psi_0))$. Code($\Sigma_{P|\delta_1}$) is one such X, so we have a formula $\varphi(u, v, w, z)$ and an ordinal $\nu_0 < \eta_1$ such that for all reals x and $\alpha < \eta_0$

$$f(x) = \alpha \Leftrightarrow L(\Sigma_{P|\delta_1}, \mathbb{R}) \models \varphi[x, \alpha, \nu_0, \text{Code}(\Psi_0)].$$

We may assume

$$\nu_0 = \pi^{\Sigma}_{P,\infty}(\bar{\nu}).$$

Let (Q, Σ_Q) be a nondropping iterate of (P, Σ), with iteration map i. We write

$$\delta^Q_k = i(\delta_k), \text{ and } \Sigma^Q_k = \Sigma_{Q|i(\delta_k)}.$$

We want to look at what happens when we iterate P below δ_0 in such a way that some real x is extender-algebra-generic over Q at δ^Q_0. In particular, we want to see that if the generic g adds enough information, then $Q[g]$ can identify some $\bar{\alpha}$ such that $\pi^{\Sigma}_{Q,\infty}(\bar{\alpha}) = f(x)$, from x and the image of $\bar{\nu}$. (Then we use the fact that the extender algebra of Q is δ^Q_0-c.c. to find a bound $\xi < \delta^Q_0$ on all the possible $\bar{\alpha}$, and we get that $\pi^{\Sigma}_{Q,\infty}(\xi)$ is a bound on ran(f).)

The generic interpretability theorem of [19] says that each iterate Q of P has terms $\dot{\Sigma}_i$ such that for all h set generic over Q, $(\dot{\Sigma}_i)_h = \Sigma^Q_i \cap Q[h]$. Going one step further, we have

Subclaim A. There is a term $\tau \in P$ such that whenever $i\colon (P, \Sigma) \to (Q, \Sigma_Q)$ is an iteration map, and g is is $\text{Col}(\omega, \delta^Q_2)$-generic over Q, and $\mathbb{R}^*_g = \mathbb{R} \cap Q[g]$, then

$$\begin{aligned} i(\tau)_g &= \text{Code}(\Sigma^Q_1)^\sharp \cap Q[g] \\ &= (\text{Code}(\Sigma^Q_1) \cap Q[g])^\sharp. \end{aligned}$$

Proof: The proofs of Claims 3–5 in the proof of Theorem 5.7 show that $\text{Code}(\Sigma_1)^\sharp$ is Wadge reducible to $\text{Code}(\Sigma_2)$ via a recursive function.[33] By Lemma 7.3 of [19], there is a term $\sigma \in P$ such that $\sigma_h = \text{Code}(\Sigma_2) \cap P[h]$ for all h set generic over P. This gives us τ such that $\tau_h = \text{Code}(\Sigma_1)^\sharp \cap P[h]$ for all h set generic over P.

We get the second equality from the fact that τ^g satisfies the witness condition for sharps of sets of reals. That is because the witness condition for interpretations of τ is a Π^1_2 fact about the interpretation, and we have Woodin cardinals δ_3 and δ_4 in P, and in $P[g]$, we have a UB code for $\text{Code}(\Sigma_1)^\sharp$.[34]

The same proof shows that if $i\colon (P, \Sigma) \to (Q, \Sigma_Q)$ is an iteration map, then $i(\tau)$ captures $\text{Code}(\Sigma_1^Q)^\sharp$. □

Note that if (Q, Σ_Q) is as in the subclaim, then $L(\Sigma_1^Q, \mathbb{R}) = L(\Sigma_1^P, \mathbb{R})$, so $i(\tau)$ can be used in the evaluation of $f(x)$. The problem lies with the parameter $\text{Code}(\Psi_0)$ that occurs in the definition of $f(x)$. We cannot replace Ψ_0 with $(\Sigma_0^Q)^{\text{sh}}$ when we move from P to Q, for then our formula φ will define a different function. We solve this problem by showing that for (Q, Σ_Q) arbitrarily far out in $\mathcal{F}(P, \Sigma)$, we can add a UB code for $\text{Code}(\Psi_0)$ to Q via the δ_0^Q-generator version of the extender algebra at δ_0^Q. It is crucial here that we are only trying to add a UB code for the short tree component of $\Sigma_{P|\delta_0}$. The full $\Sigma_{P|\delta_0}$ lets us compute the iteration map $i\colon P|\delta_0 \to Q|i(\delta_0)$, so it will in general collapse δ_0^Q.

Given our UB code for Ψ_0 in $Q[g]$, where g has not collapsed δ_0^Q, we then consider homogeneous collapse extensions $Q[g][h]$ in our definition of $f(x)$ over $Q[g]$. But it is important that the parameters in the definition were added by δ_0^Q-c.c. forcing. The further collapse extensions are just a tool whereby $Q[g]$ computes what is true of these parameters in $L(\Sigma_1^Q, \mathbb{R})$.

Remark 6.2: The argument below can be simplified if $\theta(\Psi_0) = \theta(\Sigma_{P|\alpha})$ for some $\alpha < \delta_0$. In that case, we can make $\text{Code}(\Sigma_{P|\alpha})$ our parameter in the definition of f, and add a real coding it relative to $\text{Code}(\Sigma_1^Q)$ via the ω-generator extender algebra at δ_0^Q. This simpler case is exactly the case when $L(P_{\theta(\Psi_0)}(\mathbb{R}))$ does not satisfy LSA. See the next section.

[33]We reconstruct an iterate Q of $M_\omega^{(P|\delta_1, \Sigma_1)}$, together with the iteration strategy Ψ for Q, inside $P|\delta_2$. Ψ acts on all trees in V, and Q has a term for Σ_1 on its derived model that is moved correctly by Ψ. So from $\text{Code}(\Psi)$ we can recover $\text{Code}(\Sigma_1)^\sharp$. We can recover Ψ from Σ_2.

[34]Note $\pi^{-1}(\sigma)_k = \text{Code}(\Sigma_2) \cap \pi^{-1}(P|\eta)[g][k]$ for $\pi\colon N \to P|\eta$ with $\text{crit}(\pi) > \delta_2$, and genericity iterations of N inside $P[h]$ also move $\pi^{-1}(\sigma)$ correctly.

The next two subclaims execute our plan. For any premouse Q and any $\delta < o(Q)$, let $\mathbb{B}^Q_\delta$ be the δ-generator extender algebra of Q at δ.[35]

Subclaim B. There is a $\mathbb{B}^P_{\delta_0}$ term ρ such that whenever $(S, \Sigma_S) \in \mathcal{F}(P, \Sigma)$, and $x \in \mathbb{R}$; then there is a $(Q, \Sigma_Q) \in \mathcal{F}(S, \Sigma_S)$ such that letting $\mathcal{U}$ be the unique normal tree on P by Σ that has last model Q, $i = i^{\mathcal{U}}$, and $\delta = i(\delta_0)$, we have

(1) $i^{\mathcal{U}}_{0,\delta}(\delta_0) = \delta$, and
(2) $\langle P, \mathcal{U} \upharpoonright \delta, x\rangle = i(\rho)_g$, for some $\mathbb{B}^Q_\delta$-generic g.

Proof: Let (S, Σ_S) and x be given. We show how to iterate it to a (Q, Σ_Q) such that for the corresponding $\mathcal{U}$, the $A \subset \delta$ naturally coding $\langle P, \mathcal{U}, x\rangle$ is $\mathbb{B}^Q_\delta$ generic. The desired term ρ is then just the decoding method.

Let Γ be a good pointclass closed under $\forall^{\mathbb{R}}$ with the scale property, and such that $\text{Code}(\Sigma) \in \Gamma \cap \breve{\Gamma}$. Let $\langle N^*, \delta^*, S, T, w, \Sigma^*\rangle$ be a coarse Γ-Woodin tuple such that $x, S, P \in N^*$, and let $\mathbb{C}$ be the associated maximal hod pair construction. By Theorem 2.3, we may fix $\nu < \delta^*$ such that $(M^{\mathbb{C}}_{\nu,0}, \Omega^{\mathbb{C}}_{\nu,0})$ is an iterate of (P, Σ), and (P, Σ) iterates strictly past all earlier levels of $\mathbb{C}$. Set

$$(Q, \Sigma_Q) = (M^{\mathbb{C}}_{\nu,0}, \Omega^{\mathbb{C}}_{\nu,0}),$$

and let $\mathcal{U}$ be the normal tree by Σ from P to Q.

By Theorem 2.3, there is also a level of $\mathbb{C}$ that is a normal iterate of (S, Σ_S), and such that (S, Σ_S) iterates strictly past all earlier levels. By Dodd-Jensen, this level must be $(M_{\nu,0}, \Omega_{\nu,0})$ as well. So $(S, \Sigma_S) \prec (Q, \Sigma_Q)$ in $\mathcal{F}(P, \Sigma)$. Letting $\delta = \delta^Q_0 = i^{\mathcal{U}}_{P,Q}(\delta_0)$, we have that δ is Woodin in Q, and hence a limit of measurable cardinals in N^*. But P is countable in N^*. This implies $\mathcal{U} \upharpoonright \delta$ is a tree on $P|\delta_0$, and $i^{\mathcal{U}}_{0,\delta}(\delta_0) = \delta$.

We are left to verify part (2). Let $A(P, \mathcal{U}, x)$ be a subset of δ that codes $\langle P, \mathcal{U} \upharpoonright \delta, x\rangle$ in some fixed, natural way. We must see that $A(P, \mathcal{U}, x) \models$

---

[35] For the reader's convenience: letting $\mathcal{L}$ be the propositional language with sentence symbols $\dot{A}_\alpha$ for $\alpha < \delta$, $\mathbb{B}^Q_\delta$ is the Lindenbaum algebra of an $\mathcal{L}_{\delta,0}$ theory $T(Q, \delta)$. The axioms of T are induced by E on the $Q||\delta$ sequence. If $\kappa = \text{crit}(E)$, then

$$\bigvee_{\alpha<\kappa} \varphi_\alpha \leftrightarrow i_E\left(\bigvee_{\alpha_\kappa} \varphi_\alpha\right) \upharpoonright \nu$$

is an axiom of $T(Q, \delta)$, whenever $i_E(\bigvee_{\alpha_\kappa} \varphi_\alpha) \upharpoonright \nu \in Q|\eta$, for some cardinal η of Q such that $\eta < i_E(\kappa)$. If δ is Woodin in Q, then $\mathbb{B}^Q_\delta$ is δ-c.c. in Q. Moreover, if $A \subset \delta$ and $A \models T(Q, \delta)$ (where $A \models \dot{A}_\alpha$ iff $\alpha \in A$), then $G_A = \{[\varphi] \mid A \models \varphi\}$ is $\mathbb{B}^Q_\delta$ generic over Q.

$T(Q, \delta)$.[36] In order to suppress the coding a little, let us write $\langle P, \mathcal{U}, x\rangle \models \varphi$ instead of $A(P, \mathcal{U}, x) \models \varphi$ below. Let

$$\bigvee_{\alpha<\kappa} \varphi_\alpha \leftrightarrow i_E\left(\bigvee_{\alpha_\kappa} \varphi_\alpha\right) \restriction \gamma$$

be an axiom induced by E. This implies that $i_E(\bigvee_{\alpha<\kappa} \varphi_\alpha) \restriction \gamma \in Q|\mu$, where $\mu < \lambda(E)$ and μ is a cardinal of Q. Our natural coding is such that $A(P, \mathcal{U}, x) \cap \mu$ is determined by P, x, and $\mathcal{U} \restriction \mu$, and thus only $\langle P, \mathcal{U} \restriction \mu, x\rangle$ is relevant to whether $\langle P, \mathcal{U}, x\rangle \models i_E(\bigvee_{\alpha<\kappa} \varphi_\alpha) \restriction \gamma$.

E occurs on the $M^{\mathbb{C}}_{\nu,k}$ sequence. Let $\sigma\colon M_{\nu,k}|\,\mathrm{lh}(E) \to M_{\eta,0}$ be the resurrection map of $\mathbb{C}$, and E^* the background extender for $\sigma(E) = \dot{F}^{M_{\eta,0}}$ given by $\mathbb{C}$. Note $\sigma \restriction (M_{\nu,k}|\mu) =$ identity. It follows that

$$\begin{aligned} i_E\left(\bigvee_{\alpha<\kappa} \varphi_\alpha\right) \restriction \gamma &= i_{\sigma(E)}\left(\bigvee_{\alpha<\kappa} \varphi_\alpha\right) \restriction \gamma \\ &= i_{E^*}\left(\bigvee_{\alpha_\kappa} \varphi_\alpha\right) \restriction \gamma. \end{aligned}$$

Now assume that

$$\langle P, \mathcal{U}, x\rangle \models i_E\left(\bigvee_{\alpha<\kappa} \varphi_\alpha\right) \restriction \gamma,$$

where again $\langle P, \mathcal{U}, x\rangle$ should be replaced by its code $A(P, \mathcal{U}, x)$ as a subset of δ. We must show that $\langle P, \mathcal{U}, x\rangle \models \bigvee_{\alpha<\kappa} \varphi_\alpha$. We have that

$$\langle P, \mathcal{U}, x\rangle \models i_{E^*}\left(\bigvee_{\alpha<\kappa} \varphi_\alpha\right) \restriction \gamma.$$

Moreover,

$$i_{E^*}(\mathcal{U}) \restriction \mu = \mathcal{U} \restriction \mu.$$

This is because $i_{E^*}(\Sigma) \subseteq \Sigma$, since N^* has captured Σ, and $i_{E^*}(M^{\mathbb{C}}_{\nu,0})|\mu = M^{\mathbb{C}}_{\nu,0}|\mu$ by coherence. So $A(P, \mathcal{U}, x) \cap \mu = i_{E^*}(A(P, \mathcal{U}, x)) \cap \mu$, so

$$i_{E^*}(\langle P, \mathcal{U}, x\rangle) \models i_{E^*}\left(\bigvee_{\alpha<\kappa} \varphi_\alpha\right),$$

[36] $T(Q, \delta)$ is defined in a previous footnote.

and thus

$$\langle P, \mathcal{U}, x\rangle \models \bigvee_{\alpha<\kappa} \varphi_\alpha,$$

as desired. This proves Subclaim B. □

Now we show that if Q and g are as in Subclaim B, then $Q[g]$ has a UB code for $\mathrm{Code}(\Psi_0)$. Fix terms $\tau, \rho \in P$ as above.

Subclaim C. There is a term $\sigma \in P$ such that whenever $Q, \mathcal{U}, x, i, \delta$, and g are such that

(a) $(Q, \Sigma_Q) \in \mathcal{F}(P, \Sigma)$, $x \in \mathbb{R}$, $\mathcal{U}$ is the unique normal tree on P by Σ with last model Q, and $i = i^{\mathcal{U}}$,
(b) $\delta = i(\delta_0) = i^{\mathcal{U}}_{0,\delta}(\delta_0)$, and
(c) $\langle P, \mathcal{U} \restriction \delta, x\rangle = i(\rho)_g$,

then $i(\sigma)_g$ is a $\mathrm{Col}(\omega, \delta_2^Q)$ term such that for any $Q[g]$-generic h on $\mathrm{Col}(\omega, \delta_2^Q)$,

$$\mathrm{Code}(\Psi_0) \cap Q[g][h] = (i(\sigma)_g)_h.$$

Proof: We show how to define $\Psi_0 \cap Q[g][h]$ inside $Q[g][h]$ using parameters from Q in $\mathrm{ran}(i)$, together with $\langle P, \mathcal{U} \restriction \delta, x\rangle$. The procedure is uniform, so we get the fixed term $\sigma \in P$.

Let $S = Q|\delta$. If k is $Q[g][h]$-generic over $\mathrm{Col}(\omega, \delta_3^Q)$, then inside $Q[g][h][k]$ we can form $M_\infty(S, \Sigma_S)$ and the normal tree $\mathcal{U}(S, \Sigma_S)$ from S to $M_\infty(S, \Sigma_S)$. Here the trees entering into the direct limit are those that are countable in $Q[g][h][k]$. Σ_S has a UB code in Q, and the forcing is homogeneous, so $M_\infty(S, \Sigma_S), \mathcal{U}(S, \Sigma_S) \in Q$. Let us write

$$\mathcal{T} = \mathcal{U}(S, \Sigma_S)^{Q[g][h][k]}$$

for this tree. From $\mathcal{T}$ and $\mathcal{U} \restriction \delta$, we can construct

$$\mathcal{X} = X(\mathcal{U} \restriction \delta, \mathcal{T}).$$

Here we are thinking of $\mathcal{U} \restriction \delta$ and $\mathcal{X}$ as trees on $N = P|\delta_0$. Some care is needed, because $\mathcal{U} \restriction \delta$ is a tree of limit length, and its good branch $b = [0, \delta]_U$ is not available in $Q[g]$. However, $\delta(\mathcal{U} \restriction \delta) = \delta$ is a cutpoint in $\mathcal{M}_b^{\mathcal{U}}$, and $\mathcal{T}$ is a tree on the common part model $M(\mathcal{U} \restriction \delta)$, so $X((\mathcal{U} \restriction \delta)^\frown b), \mathcal{T})$ has the form $\mathcal{X}^\frown c$, where only c, and not $\mathcal{X}$, depends on b. So we may write $X(\mathcal{U} \restriction \delta, \mathcal{T}) = \mathcal{X}$. We have

$$\mathcal{X} \in Q[g],$$

although b and c are not in $Q[g]$, since they collapse δ.

All weak hulls of $\mathcal{X}$ are by Ψ, so it is enough to show that whenever $\mathcal{W} \in Q[g][h]$ is a short, normal tree of countable limit length by Ψ_0 (hence relevant), then in $Q[g][h]$ there is a proper extension of $\mathcal{W}$ that is a weak hull of $\mathcal{X}$. By a simple absoluteness argument, it is enough to show that

$$Q[g][h] \models \mathrm{Col}(\omega, \delta_3^Q) \Vdash \exists \mathcal{V}(\mathcal{V} \text{ properly extends } \mathcal{W} \text{ and } \mathcal{V} \text{ is a weak hull of } \mathcal{X}).$$

For this, let $\mathcal{V}$ be a proper normal extension of $\mathcal{W}^\frown c$ with last model M such that the branch $P|\delta_0$-to-M does not drop, and is generated by the image of $\beta(P|\delta_0)$. Let us iterate Q above δ_2^Q to make $\mathcal{V}$ generic at the image of δ_3^Q, at the same time making $[0, \delta]_U$ generic. This gives us

$$j \colon Q[g][h] \to R[g][h]$$

with $\mathrm{crit}(j) > \delta_2^Q$, and k on $\mathrm{Col}(\omega, \delta_3^R)$ such that

$$\mathcal{V}, [0, \delta]_U \in R[g][h][k].$$

Since j is elementary, it is enough to see that in $R[g][h][k]$ there is a weak hull embedding of $\mathcal{V}$ into $j(\mathcal{X})$. In $R[g][h][k]$ we have $\pi = i^{\mathcal{U}}_{0,\delta} \restriction P|\delta_0$, and we have Σ_S on all countable trees, so we can compute $\Sigma_{P|\delta_0} = \Sigma_S^\pi$ on all countable trees. Thus we can compute $\Sigma_M = (\Sigma_{P|\delta_0})_{\mathcal{V},M}$, and we have the normal tree

$$\mathcal{S} = \mathcal{U}(M, \Sigma_M)^{R[g][h][k]}$$

from M to

$$M_\infty(M, \Sigma_M)^{R[g][h][k]} = j(M_\infty(S, \Sigma_S)^{Q[g][h]}).$$

From this we obtain

$$\begin{aligned} X(\mathcal{V}, \mathcal{S}) &= \mathcal{U}(P|\delta_0, \Sigma_{P|\delta_0})^{R[g][h][k]} \\ &= X(\mathcal{U} \restriction (\delta+1), \mathcal{U}(Q|\delta, \Sigma_{Q|\delta})^{R[g][h][k]}). \end{aligned}$$

Thus there is a weak hull embedding Φ of $\mathcal{V}$ into $X(\mathcal{U} \restriction (\delta+1), \mathcal{U}(Q|\delta, \Sigma_{Q|\delta})^{R[g][h][k]})$. But this tree is just $j(\mathcal{X})^\frown d$, for some d. Since the generators of $P|\delta_0$-to-M are contained in the image of $\beta(P|\delta_0)$, Φ is actually a weak hull embedding into $j(\mathcal{X})$, as desired. □

Let $\sigma \in P$ be a term witnessing Subclaim C.

Subclaim D. There is a term $\dot{z} \in P$ and a Σ_2^1 formula γ such that whenever $Q, \mathcal{U}, x, i, \delta$, and g are as in Subclaim C, and h is $\mathrm{Col}(\omega, \delta_2^Q)$-generic over $Q[g]$, then $\dot{z}_{g,h} = z$ is a real such that

$$\mathrm{Code}(\Psi_0) \cap Q[g][h] = \{y \mid Q[g][h] \models \gamma[y, \mathrm{Code}(\Sigma_1^Q), z]\}.$$

Proof: It is enough to see that $Q[g] \models \gamma_0[i(\sigma)_g, i(\tau)]$, where $\gamma_0(u, v)$ is the formula saying that for collapse generics h, $(i(\sigma)g)_h$ is Σ^1_2 in $(\dot{\Sigma}^Q_1)_{g,h}$ and some real. (Then we can just take $\dot{z}$ to be a name such that $i(\dot{z})$ interprets as such a real whenever one exists.) But let

$$j\colon Q[g] \to R[g]$$

come from a genericity iteration with all critical points $> \delta$ such that there is a collapse δ^R_2 generic h with

$$[0, \delta]_U \in R[g][h].$$

In $R[g][h]$ we have $i^{\mathcal{U}}_{0,\delta}\colon P|\delta_0 \to Q|\delta = R|\delta$, and we have $\Sigma_{R|\delta}$, so we have the full $\Psi = \Sigma^{i_{0,\delta}}_{R|\delta}$. Moreover, Code($\Psi$) is Σ^1_2 in Code(Σ_1) and a real coding $i_{0,\delta}$. Thus

$$R[g][h] \models j(i(\sigma)_g)_h \in L(\Sigma^R_1, \mathbb{R}).$$

So $R[g] \models \gamma[j(i(\sigma)_g), j(i(\tau))]$, so $Q[g] \models \gamma_0[i(\sigma)_g, i(\tau)]$. □

Now let (Q, Σ_Q) be any nondropping iterate of (P, Σ), and $i\colon P \to Q$ the iteration map. We define an ordinal $\xi < \delta = \delta^Q_0$ by considering generics g on $\mathbb{B}^Q_\delta$. Let such a g be given. We define an ordinal $\xi(g) < \delta$ in $Q[g]$ from g as follows:

(i) if $i(\rho)_g$ is not a triple whose third coordinate is a real y, let $\xi(g) = 0$. Otherwise, let y be that real, and go on.
(ii) Let h be $\mathrm{Col}(\omega, \delta^Q_2)$ generic, and $z = \dot{z}_{g,h}$. If it is not the case that z is a real, and in $Q[g][h]$, $(i(\sigma_g)_h)$ is defined by γ from z and Code(Σ^Q_1), then set $\xi(g) = 0$.
(iii) Otherwise, working in $Q[g][h]$, consider $L(\Sigma^Q_1, \mathbb{R})$. Let $\pi\colon Q|\delta \to M_\infty(Q|\delta, \Sigma^Q_0)$ be the direct limit map of the model $L(\Sigma^Q_1, \mathbb{R})$.
(iv) For each x recursive in y, using z and $i(\tau)_{g,h} = (\Sigma^Q_1)^\sharp$ as oracles, find the least $\mu < \delta$ such that $L(\Sigma^Q_1, \mathbb{R}) \models \varphi[\pi(\mu), \pi(\bar{\nu}), x, Z]$, where $Z = \{u \mid \gamma(u, z, \mathrm{Code}(\Sigma^Q_1))\}$, if such a μ exists. Let $\bar{\alpha}(x) = \mu$ if μ exists, and $\bar{\alpha}(x) = 0$ otherwise.
(v) Finally, let $\xi(g)$ be the sup of all $\bar{\alpha}(x)$ for x recursive in y.

Since $\mathbb{B}$ is δ-cc, $\xi(g) < \delta$ for all g, and in fact, there is a $\xi < \delta$ such that for all g, $\xi(g) < \xi$. We let ξ^Q be the least such ξ.

It is clear from the definition of ξ^Q that whenever (R, Σ_R) is an iterate of (Q, Σ_Q), then

$$\pi^\Sigma_{Q,R}(\xi^Q) = \xi^R.$$

But our Subclaims show that for any real x, there is an iterate (Q, Σ_Q) of (P, Σ) and a g that is $\mathbb{B}_\delta^Q$ generic such that $i(\rho)_g$ has the form $\langle P, \mathcal{U}, x\rangle$, and $\pi_{Q,\infty}^{\Sigma}(\bar{\alpha}(x)) = f(x)$. So in fact

$$\mathrm{ran}(f) \subseteq \pi_{P,\infty}^{\Sigma}(\xi^P) < \eta_0.$$

This proves Claim 1. □

Claim 2. $\eta_0 = \theta(\Psi_0)$.

Proof: We must show $\eta_0 \leq \theta(\Psi_0)$. We have that β_∞ is strong to η_0 in HOD, so if $\theta(\Psi_0) < \eta_0$, then $\theta(\Psi_0)$ is not a cutpoint in HOD. However, $\theta(\Psi_0)$ is in the Solovay sequence, so by the first part of the theorem, it is a cutpoint in HOD. □

Claim 2 clearly finishes the proof of the theorem. □

We now characterize the successor Woodins in HOD in terms of a modified Solovay sequence. The characterization was suggested to the author by Grigor Sargsyan.

Definition 6.3: For any ordinal κ, $\theta({}^\omega\kappa)$ is the least ordinal α such that there is no ordinal definable map of ${}^\omega\kappa$ onto α.

Notice that there is no $f\colon {}^\omega\kappa \overset{\text{onto}}{\longrightarrow} \theta({}^\omega\kappa)$ such that for some $s\colon \omega \to \kappa$ f is $\mathrm{OD}(s)$.

Theorem 6.4: *Assume $AD_{\mathbb{R}}$ + HPC, let HOD $\models$ "η is a Woodin cardinal but not a limit of Woodin cardinals". Let $\kappa = \sup(\{\gamma < \eta \mid HOD \models$ "γ is Woodin"$\})^{+,HOD}$; then $\eta = \theta({}^\omega\kappa)$.*

Proof: We show first that $\eta \leq \theta({}^\omega\kappa)$. Let $\alpha < \eta$; we want to define a surjection f from ${}^\omega\kappa$ onto α. Let $\gamma_0 > \eta$ be a cardinal cutpoint of HOD.

Claim 1. Let (P, Σ) and (P, Φ) be lbr hod pairs, $\pi = \pi_{P,\infty}^{\Sigma}$, and $\sigma = \pi_{P,\infty}^{\Phi}$. Suppose

(a) $M_\infty(P, \Sigma) = M_\infty(P, \Phi) = \mathrm{HOD}|\gamma_0$,
(b) $\pi(\langle\bar{\kappa}, \bar{\alpha}, \bar{\eta}\rangle) = \sigma(\langle\bar{\kappa}, \bar{\alpha}, \bar{\eta}\rangle) = \langle\kappa, \alpha, \eta\rangle$, and
(c) $\pi \restriction \bar{\kappa} = \sigma \restriction \bar{\kappa}$;

then $\pi \restriction \bar{\alpha} = \sigma \restriction \bar{\alpha}$.

Proof: Let $\pi = \pi^{\Sigma}_{P,\infty}$ and $\sigma = \pi^{\Phi}_{P,\infty}$. Let $\pi(\bar{\kappa}) = \kappa$. We are given that $\pi \restriction \bar{\kappa}+1 = \sigma \restriction \bar{\kappa}+1$.

Let

$$\mathcal{T} = \mathcal{U}(P, \Sigma)$$

and

$$\mathcal{U} = \mathcal{U}(P, \Phi)$$

be the normal trees from P to $\mathrm{HOD}|\gamma$ by Σ^+ and Φ^+ respectively. The main branch maps of $\mathcal{T}$ and $\mathcal{U}$ are π and σ respectively. Let ξ be largest on the main branch of $\mathcal{T}$ such that whenever E is used in $[0,\xi)_T$, then $\lambda(E) \subseteq \kappa$, and let β be largest on the main branch of $\mathcal{U}$ such that whenever E is used in $[0,\beta)_U$, then $\lambda(E) \subseteq \kappa$. (Equivalently, ξ is least on the main branch such that $\kappa < \mathrm{crit}(i^{\mathcal{T}}_{\xi,\infty})$, and least anywhere in $\mathcal{T}$ such that $\kappa < \lambda(E^{\mathcal{T}}_{\xi})$. Similarly for β and $\mathcal{U}$.) The branch extender of $[0,\xi)_T$ is $E_\pi \restriction \kappa$, and that of $[0,\beta)_U$ is $E_\sigma \restriction \kappa$, so the two branch extenders are the same.[37] The base model is P in each case, so

$$\mathcal{M}^{\mathcal{T}}_{\xi} = \mathcal{M}^{\mathcal{U}}_{\beta}.$$

(This does not imply $\xi = \beta$, however.)

Now let ξ_1 be largest on the main branch of $\mathcal{T}$ such that whenever E is used in $[0,\xi_1)_T$ then $\lambda(E) \subseteq \alpha$, and β_1 be largest on the main branch of $\mathcal{U}$ such that whenever E is used in $[0,\beta_1)_T$, then $\lambda(E) \subseteq \alpha$. We claim that the part of $\mathcal{T}$ between ξ and ξ_1 is a tree on $\mathcal{M}^{\mathcal{T}}_{\xi}$, and the same as the part of $\mathcal{U}$ between β and β_1. More precisely, we show by induction on $\xi \leq \gamma \leq \xi_1$ that there is a tree $\mathcal{W}_\gamma$ on $\mathcal{M}^{\mathcal{T}}_{\xi} = \mathcal{M}^{\mathcal{U}}_{\beta}$ such that, setting $\gamma^* = \beta + (\gamma - \xi)$,

$$\mathcal{T} \restriction \gamma + 1 = \mathcal{T} \restriction (\xi+1)^\frown \mathcal{W}_\gamma$$

and

$$\mathcal{U} \restriction \gamma^* + 1 = \mathcal{U} \restriction (\beta+1)^\frown \mathcal{W}_\gamma.$$

The main thing is that in this interval, we are comparing $\mathcal{M}^{\mathcal{T}}_{\xi} = \mathcal{M}^{\mathcal{U}}_{\beta}$ with $\mathrm{HOD}|\alpha$, and HOD has no Woodin cardinals between κ and η. So if we have $\mathcal{W}_\gamma$ for γ a limit ordinal, then $\mathcal{W}_{\gamma+1}$ is determined by the Q-structure for $M(\mathcal{W}_\gamma)$ that HOD provides. At successor steps, granted $\mathcal{M}^{\mathcal{T}}_{\gamma} = \mathcal{M}^{\mathcal{U}}_{\gamma^*}$ by induction, $E^{\mathcal{T}}_{\gamma} = E^{\mathcal{U}}_{\gamma^*}$ because we are iterating away the least extender

[37] At this point we use that κ is a successor cardinal of HOD. Since $\bar{\kappa}$ is a successor cardinal of P, and π is continuous at κ, $E_\pi \restriction \kappa$ is determined by $\pi \restriction (P|\bar{\kappa})$.

disagreement with HOD in both cases. Further, $\kappa < \text{crit}(E_\gamma^{\mathcal{T}})$, because otherwise $\text{crit}(E_\gamma)$ is a limit of Woodin cardinals of HOD, and hence in $\mathcal{M}_\xi^{\mathcal{T}}$, which implies there are arbitrarily large Woodins in $\mathcal{M}_\gamma^{\mathcal{T}}|\lambda(E_\gamma^{\mathcal{T}})$, and since $\lambda(E_\gamma^{\mathcal{T}})$ is a cardinal of HOD, that there are Woodins of HOD between κ and α.[38] So $E_\gamma^{\mathcal{T}}$ and $E_{\gamma^*}^{\mathcal{T}}$ are applied to the same model in $\mathcal{T}$ and $\mathcal{U}$ (though it is indexed differently), and we have $\mathcal{M}_{\gamma+1}^{\mathcal{T}} = \mathcal{M}_{\gamma^*+1}^{\mathcal{U}}$, and we have $\mathcal{W}_{\gamma+1}$.

We thus have $\mathcal{M}_{\xi_1}^{\mathcal{T}} = \mathcal{M}_{\xi_1^*}^{\mathcal{U}}$. $\xi_1^* \leq \beta_1$, because the extenders E in $\mathcal{W}_{\xi_1}$ have $\lambda(E) \subseteq \alpha$. If $\xi_1^* < \beta_1$, then $\mathcal{M}_\xi^{\mathcal{T}} = \mathcal{M}_{\xi_1^*}^{\mathcal{U}}$ disagrees with HOD below α, contradiction. Thus $\xi_1^* = \beta_1$, so $\mathcal{M}_{\xi_1}^{\mathcal{T}} = \mathcal{M}_{\beta_1}^{\mathcal{U}}$. Moreover, $[\xi, \xi_1)_T$ and $[\beta, \beta_1)_U$ have the same branch extender, namely the extender of the main branch of $\mathcal{W}_{\xi_1}$.

It follows that $[0, \xi_1)_T$ and $[0, \beta_1)_U$ have the same branch extender, and thus $\pi \restriction \bar{\alpha} = i_{0,\xi_1}^{\mathcal{T}} \restriction \bar{\alpha} = i_{0,\beta_1}^{\mathcal{U}} \restriction \bar{\alpha} = \sigma \restriction \bar{\alpha}$, as desired. □

We can now define the desired surjection of ${}^\omega\kappa$ onto α. For $s \in {}^\omega\kappa$, we let $f(s) = 0$ unless s decodes as a tuple $\langle t, P, \bar{\kappa}, \bar{\alpha}, \bar{\eta}, \xi \rangle$ such that

(i) P is a countable lpm, $\bar{\kappa} < \xi < \bar{\alpha} < \bar{\eta}$ are ordinals of P,
(ii) $t \colon \bar{\kappa} \to \kappa$,
(iii) there is a Σ such that (P, Σ) is an lbr hod pair, $M_\infty(P, \Sigma) = \text{HOD}|\gamma_0$, $\pi_{P,\infty}^{\Sigma}(\langle \bar{\kappa}, \bar{\alpha}, \bar{\eta} \rangle) = \langle \kappa, \alpha, \eta \rangle$, and $\pi_{P,\infty}^{\Sigma} \restriction \bar{\kappa} = t$.

If s does decode to a tuple $\langle t, P, \bar{\kappa}, \bar{\alpha}, \bar{\eta}, \xi \rangle$ as in (i)–(iii), then we let $f(s)$ be the common value of $\pi_{P,\infty}^{\Sigma}(\xi)$, for all Σ as in (iii). By the claim, there is a common value. Clearly, f is surjective.

This shows $\eta \leq \theta({}^\omega\kappa)$. The proof that $\theta({}^\omega\kappa) \leq \eta$ follows the outline of the proof of Claim 1 in the proof of Theorem 6.1, but it is easier. We just give a sketch.

Let $\eta_0 = \eta$, and let

$$\eta_0 < \eta_1 < \ldots < \eta_5 < \bar{\theta}$$

where $\bar{\theta}$ and the η_i for $i \geq 1$ are successor points in the Solovay sequence, and thus cutpoint Woodins in HOD. Let

$$\text{HOD}|\bar{\theta} = M_\infty(P, \Sigma),$$

---

[38] Let $\sigma = \sup(\{\xi < \eta \mid \text{HOD} \models \xi \text{ is Woodin}\})$. If $\text{crit}(E_\gamma) = \sigma$, then κ is a limit of Woodins in HOD. If $\text{crit}(E_\gamma) < \sigma$, then E_γ overlaps a Woodin of HOD, hence a Woodin of $\mathcal{M}_\xi^{\mathcal{T}}$, so again $\text{crit}(E_\gamma)$ is a limit of Woodins in HOD.

and $\pi_{P,\infty}(\bar{\kappa}) = \kappa$, and

$$\pi_{P,\infty}(\delta_i) = \eta_i, \text{ for } 0 \le i \le 5.$$

We omit the superscript Σ in these formulae, because from now on all strategies will be tails of Σ. Suppose that $f\colon {}^\omega\kappa \to \eta_0$ is ordinal definable; we want to show that its range is bounded. We may assume that we have a formula $\varphi(u, v, w)$ and an ordinal $\nu_0 < \eta_1$ such that for all $s \in^\omega \kappa$, $\alpha < \eta_0$,

$$f(s) = \alpha \Leftrightarrow L(\Sigma_{P|\delta_1}, \mathbb{R}) \models \varphi[s, \alpha, \nu_0].$$

We may also assume

$$\nu_0 = \pi^{\Sigma}_{P,\infty}(\bar{\nu}).$$

We show that there is a $\xi < \delta_0$ such that $\text{ran}(f) \subseteq \pi_{P,\infty}(\xi)$, so $\text{ran}(f)$ is bounded in η_0, as desired.

ξ is defined as follows. Let $\mathbb{B}$ be the $\bar{\kappa}$-generator extender algebra of P at δ_0, where the axioms are obtained using only extenders with critical points $> \bar{\kappa}$. Via some simple decoding term ρ, a $\mathbb{B}$-generic g over P yields $t = \rho_g$ such that $t \in^\omega \bar{\kappa}$. Let

$$s(n) = \pi_{P,\infty}(t(n)).$$

$P[g]$ now consults its term for $(\Sigma_{P|\delta_1})^\sharp$ (cf. Subclaim A in the proof of Theorem 6.1) to determine whether there is a $\bar{\alpha} < \delta_0$ such that

$$L(\Sigma_{P|\delta_1}, \mathbb{R}) \models \varphi[s, \pi_{P|\delta_1,\infty}(\bar{\alpha}), \pi_{P|\delta_1,\infty}(\bar{\nu})].$$

If so, then $\bar{\alpha}(g)$ is the least such $\bar{\alpha}$, otherwise $\bar{\alpha}(g) = 0$. By the δ_0-cc, we have $\xi < \delta_0$ such that for all such g, $\bar{\alpha}(g) < \xi$. We write $\xi = \xi^P$.

If $i\colon (P, \Sigma) \to (Q, \Sigma_Q)$ is an iteration map, then $\xi^Q = i(\xi^P)$ is defined from $i(\mathbb{B})$ and $i(\bar{\nu})$ over Q in the same way. To see that $\text{ran}(f)$ is bounded by $\pi_{P,\infty}(\xi^P)$, let $s \in^\omega \kappa$ be given. We can iterate (P, Σ) to (R, Σ_R) so that

$$s \cup \{f(s)\} \subseteq \text{ran}(\pi_{R,\infty}).$$

Let $j\colon P \to R$ be the iteration map. Let

$$\pi_{R,\infty}(t(n)) = s(n).$$

Now we can genericity iterate (R, Σ_R) to (Q, Σ_Q) with all critical points $> j(\bar{\kappa})$ so that, letting $i\colon P \to Q$ be the iteration map, there is a g on $i(\mathbb{B})$ generic over Q such that

$$t = \rho_g.$$

But then $\pi_{Q,\infty}(t(n)) = \pi_{R,\infty}(t(n)) = s(n)$, so $\pi^{-1}_{Q,\infty}(f(s)) < \bar{\alpha}(g)^Q < \xi^Q$, as desired. □

The following definition is due to Sargsyan.[39]

Definition 6.5: Assume AD^+. We set

$$\eta_0 = \theta({}^\omega\omega) = \theta_0,$$
$$\eta_{\alpha+1} = \theta({}^\omega\kappa), \text{ where } \kappa = (\eta_\alpha)^{+,\mathrm{HOD}},$$
$$\eta_\lambda = \bigcup_{\alpha<\lambda} \eta_\alpha.$$

We have shown

Corollary 6.6: *Assume $\mathsf{AD}_{\mathbb{R}}$ + HPC; then for any $\delta < \theta$, δ is a successor Woodin cardinal of HOD iff $\delta = \eta_0$ or $\delta = \eta_{\alpha+1}$ for some α.*

7. LSA from Least Branch HOD Pairs

We go back to AD^+ as our background theory.

Recall that LSA is the theory AD^+ + "there is an α such that θ_α is the largest Suslin cardinal". That is, the largest Suslin cardinal exists, and belongs to the Solovay sequence. LSA stands for "Largest Suslin Axiom". The members of the Solovay sequence are all Suslin cardinals, so if θ_α is the largest Suslin, then $\theta = \theta_{\alpha+1}$.

LSA is stronger than many familiar determinacy theories. For example, if θ_α is the largest Suslin, then $L(P_{\theta_\alpha}(\mathbb{R})) \models \mathsf{AD}_{\mathbb{R}}$ + "θ is regular". A lower bound for LSA in terms of hod pair existence follows from the fact that if θ_α is the largest Suslin, then in HOD, θ_α is a limit of Woodin cardinals, and θ_α is strong to θ, which is itself a Woodin cardinal. In [15], Sargsyan showed that a hypothesis on the existence of hod pairs in the rigidly layered hierarchy that is close to this lower bound implies the existence of models of LSA. Here we shall obtain something close to Sargsyan's upper bound, but in the least branch hierarchy.

Here is a theory still stronger than LSA.

Definition 7.1: LSA^+ is the theory: $\mathsf{AD}^+ + \exists A \subseteq \mathbb{R} \forall \alpha < \theta \forall s \colon \omega \to \alpha (A \notin \mathrm{OD}(s))$.

Proposition 7.2: *LSA^+ implies LSA.*

[39] One might call this the *Sargsyan sequence.*

Proof: If B is Suslin and co-Suslin, then B is homogeneously Suslin (cf. [11]), so B is definable from a homogeneity system $\bar{\mu}$ such that $B = \{x \mid \bar{\mu}_x \text{ is wellfounded}\}$. Since any measure on an ordinal $< \theta$ is ordinal definable, B is OD(s) for some $s \in \bigcup_{\alpha<\theta} {}^{\omega}\alpha$.

So LSA^+ implies that not all sets are Suslin, and thus there is a largest Suslin cardinal κ. Let A be a complete κ-Suslin set. It is enough to see that A is not OD(B), for all Suslin-co-Suslin sets B. But if it were, A would be OD(s) for some $s\colon \omega \to \kappa$. By results of Woodin, if $\theta(A) < \theta$, then $\theta(A)$ is a Suslin cardinal. So $\theta(A) = \theta$; that is, every set of reals is OD from A and a real. So every set of reals is OD from s and a real, contrary to LSA^+. □

It seems very likely that LSA^+ is strictly stronger than LSA, but we do not have a full proof. The most natural proof would proceed by showing that HOD in the minimal model of LSA is a weaker mouse than HOD in the minimal model of LSA^+. This involves proving HPC in those models, which is work in progress.[40]

The next theorem gives hod mouse upper bounds on LSA and LSA^+.

Theorem 7.3: *Suppose that there is an lbr hod pair (P, Σ) such that for some δ, λ, the following hold in P:*

(a) ZFC + λ is a limit of Woodin cardinals,
(b) $\delta < \lambda$, and δ is Woodin, and
(c) letting κ be the least $< \delta$-strong cardinal, κ is a limit of Woodins.

Then there is a pointclass Γ such that $L(\Gamma, \mathbb{R}) \models \mathsf{LSA}$. If in addition

(d) δ is a limit of Woodin cardinals in P,

then there is a pointclass Γ such that $L(\Gamma, \mathbb{R}) \models \mathsf{LSA}^+$.

Remark 7.4: One could obtain what is essentially equivalent to Sargsyan's mouse by taking P as in the theorem, letting $Q = P|\delta$, letting $\Psi = \Sigma_Q^{\mathrm{sh}}$ be the short-tree component of Σ_Q, and considering the mouse $N = M_\omega^\Psi(Q)$. Our proof of Theorem 7.3 should work assuming only the existence of this weaker mouse. Unfortunately, N is not in the least branch hierarchy, because above δ we are only inserting Σ_Q^{sh} and not the full Σ. Objects like N show up in an important way in the theory of [15], but not at all in the theory of [19]. We therefore do not have at this moment a reasonable

[40][15] proves that the analog of HPC in a rigidly layered hierarchy holds in the minimal model of LSA.

theory of N, and cannot use its existence as our hypothesis in Theorem 7.3. Developing such a theory is closely related to the problem of characterizing HOD assuming only $\mathsf{AD}^+ + \mathsf{HPC}$; "short-tree-strategy mice" like N should be part of the characterization.

Proof: Let $(P, \Sigma), \delta$, and λ be as in the hypotheses. We may assume $\lambda = \sup_{i<\omega} \delta_i$, where δ_i is the ith Woodin cardinal above $\delta = \delta_0$. Let g be $\mathrm{Col}(\omega, <\lambda)$ generic over P, and

$$L(\mathbb{R}^*_g, \mathrm{Hom}^*_g) = D(P, <\lambda)$$

be the associated derived model. By [19, §7], $L(\mathbb{R}^*_g, \mathrm{Hom}^*_g) \models \mathsf{AD}_{\mathbb{R}} + P(\mathbb{R}) = \mathrm{Hom}^*_g$, and the $\mathrm{Code}(\Sigma_{P|\delta_i}) \cap \mathbb{R}^*_g$, for $i < \omega$, are Wadge cofinal in Hom^*_g. Thus $L(\mathbb{R}^*_g, \mathrm{Hom}^*_g) \models \mathsf{HPC}$. It is also shown in [19] that the lbr hod pairs $(P|\delta_i, \Sigma_{P|\delta_i})$ are fullness preserving in $L(\mathbb{R}^*_g, \mathrm{Hom}^*_g)$, so that

$$\mathcal{H} = (\mathrm{HOD}|\theta)^{L(\mathbb{R}^*_g, \mathrm{Hom}^*_g)} = \bigcup_{i<\omega} M_\infty(P|\delta_i, \Sigma_{P|\delta_i}).$$

We will be working in $L(\mathbb{R}^*_g, \mathrm{Hom}^*_g)$ now, so let us write $\mathbb{R} = \mathbb{R}^*_g$ and $P(\mathbb{R}) = \mathrm{Hom}^*_g$. Let

$$\eta_i = o(M_\infty(P|\delta_i, \Sigma_{P|\delta_i})) = \tau_\infty(P|\delta_i, \Sigma_{P|\delta_i}).$$

The η_i are cutpoint Woodins of HOD, and thus successor points in the Solovay sequence. Moreover,

$$\beta_\infty = \beta_\infty(P|\delta_0, \Sigma_{P|\delta_0})$$

is a limit of cutpoint Woodins of HOD, and hence a limit point in the Solovay sequence. Since there are no Suslin cardinals strictly between $\beta_\infty(P|\delta_0, \Sigma_{P|\delta_0})$ and $|\tau_\infty(P|\delta_0, \Sigma_{P|\delta_0})| = \eta_0$,

$$L(P_{\eta_0}(\mathbb{R})) \models \beta_\infty \text{ is the largest Suslin cardinal.}$$

Thus $L(P_{\eta_0}(\mathbb{R})) \models \mathsf{LSA}$.

Now suppose that δ_0 is a limit of Woodins in P. We shall show that $L(P_{\eta_0}(\mathbb{R})) \models \mathsf{LSA}^+$. Let us write

$$\Psi_0 = (\Sigma_{P|\delta_0})^{\mathrm{sh}}$$

as before. $\mathrm{Code}(\Psi_0)$ is β_∞-Suslin, and hence belongs to $L(P_{\eta_0}(\mathbb{R}))$. Let $s\colon \omega \to \alpha$, where $\alpha < \eta_0$. It will be enough to show that Ψ_0 is not ordinal definable from s in $L(P_{\eta_0}(\mathbb{R}))$. The ordinal parameter in such a definition could be taken $< \eta_0$, and then absorbed in s. Moreover, η_0 is definable in

$L(P(\mathbb{R}))$. So assume toward contradiction that we have a formula φ such that

$$\Psi_0(\mathcal{T}) = b \Leftrightarrow L(P(\mathbb{R})) \models \varphi[s, \mathcal{T}, b].$$

$L(P(\mathbb{R}))$ is a derived model of P, and also of any Σ-iterate Q of P such that the iteration is given by a tree coded in $\mathbb{R}$. Thus P has a $\mathrm{Col}(\omega, < \lambda)$-term $\dot{X}$ such that whenever $k \colon P \to N$ is an iteration map via a tree coded in $\mathbb{R}$, and h is $\mathrm{Col}(\omega, < k(\lambda^N))$ generic over N with $\mathbb{R}^*_h = \mathbb{R}$, then

$$k(\dot{X})_h = \{\langle s, \mathcal{T}, b\rangle \mid L(P(\mathbb{R})) \models \varphi[s, \mathcal{T}, b]\}.$$

Let (Q, Σ_Q) be an iterate of (P, Σ) via a tree coded in $\mathbb{R}$ such that $\{\alpha\} \cup s \subseteq \mathrm{ran}(\pi_{Q,\infty})$, and $i \colon P \to Q$ the iteration map. (All strategies now are tails of Σ, so we have dropped the superscript.) Let t be such that for all n,

$$\pi_{Q,\infty}(t(n)) = s(n),$$

and let $\pi_{Q,\infty}(\bar{\alpha}) = \alpha$. Let γ be a Woodin cardinal of Q such that $\bar{\alpha} < \gamma < i(\delta_0)$. (This is precisely where we use that δ_0 is a limit of Woodin cardinals in P!) Now let

$$j \colon Q \to R$$

come from a genericity iteration with critical points above $\bar{\alpha}$ and be such that

$$t \in R[g]$$

for some g that is $\mathrm{Col}(\omega, j(\gamma))$-generic over R. In $R[g]$ we can compute Ψ_0 on all trees of size $< j(i(\lambda))$, for

$$\Psi_0(\mathcal{T}) = b \Leftrightarrow \mathrm{Col}(\omega, < j(i(\lambda))) \Vdash \langle s, \mathcal{T}, b\rangle \in j(i(\dot{X})).$$

In particular, we can compute $\mathcal{U} \restriction j(i(\delta_0))$, where $\mathcal{U}$ is the normal tree from P to R. But $\mathcal{U} \restriction j(i(\delta_0))$ collapses all cardinals of $R[g]$ in the interval $(j(\gamma), j(i(\delta_0)))$, contradiction. □

References

1. L.A. Harrington and A.S. Kechris, On the determinacy of games on ordinals. Annals of Math. Logic, vol. 20 (1981), pp. 109–154.
2. G. Hjorth, A boundedness lemma for iterations. J. Symbolic Logic 66 (2001), no. 3, 1058–1072.

3. S. Jackson, Structural consequences of AD. In Handbook of Set Theory, M. Foreman and A. Kanamori eds., Springer-Verlag, v. 3, pp. 1754–1873 (2010).
4. R. B. Jensen, Smooth iterations. Preprint, 2018, available at `https://www.mathematik.hu-berlin.de/~raesch/org/jensen.html`.
5. A.S. Kechris, The theory of countable analytical sets. Trans. Amer. Math. Soc. **202** (1975), pp. 259–297.
6. A.S. Kechris, Countable ordinals and the analytical hierarchy I. Pacific Journal of Math., vol. 60 (1975), pp. 223–227.
7. A.S. Kechris, Countable ordinals and the analytical hierarchy II. Annals of Math. Logic, vol. 15 (1978), pp. 193–223.
8. A. S. Kechris, R. M. Solovay and J. R. Steel, The Axiom of Determinacy and the Prewellordering Property. In A. S. Kechris, D. A. Martin and Y. N. Moschovakis, Cabal Seminar 77–79, vol. 839 of Lecture Notes in Logic, pp. 101–125, Springer, Berlin, 1981.
9. P. Koellner and W. H. Woodin, Large cardinals from determinacy. in Handbook of Set Theory, M. Foreman and A. Kanamori eds., Springer-Verlag, v. 3, pp. 1951–2120 (2010).
10. I. Neeman and J. R. Steel, Plus one premice. Available at `www.math.berkeley.edu/~steel`.
11. D. A. Martin and J. R. Steel, The tree of a Moschovakis scale is homogeneous. In Games, scales, and Suslin cardinals, The Cabal seminar, volume I, Kechris et al. eds., Cambridge University Press, 2008, pp. 404–420.
12. M. Rudominer, Mouse sets. Annals of Pure and Applied Logic 87, 1–100 (1997).
13. G. Sargsyan, A tale of hybrid mice. ProQuest LLC, Ann Arbor, MI, 2009. Thesis (Ph.D.)–University of California, Berkeley.
14. G. Sargsyan, Hod mice and the mouse set conjecture. Memoirs of the Amer. Math. Soc., vol. 236, no. 1111 (2015).
15. G. Sargsyan and N. Trang, The Largest Suslin Axiom. Lecture Notes in Logic, Vol. 56 (ASL), Cambridge Univ. Press (2024).
16. F. Schlutzenberg, Iterability for (transfinite) stacks. Journal of Math. Logic, vol. 21, no. 2 (2021).
17. S. Jackson, G. Sargsyan and J. Steel, Suslin cardinals and cutpoints in mouse limits. Math. ArXiv `http://arxiv.org/abs/2207.04042` (2022).
18. B. Siskind and J. R. Steel, Full normalization for mouse pairs. Math. ArXiv `http://arxiv.org/abs/2207.11065` (2022).
19. J. R. Steel, A comparison process for mouse pairs. Lecture Notes in Logic, vol. 51 (ASL), Cambridge Univ. Press (2022).
20. J. R. Steel, Ordinal definability on models of determinacy. In Ordinal definability and recursion theory: the Cabal Seminar, volume III, A.S. Kechris et al. eds., volume 43 of Lecture Notes in Logic, ASL, CUP (2016) pp. 3–48.
21. J. R. Steel, Local HOD computation. Handwritten notes July 2016, available at `www.math.berkeley.edu/~steel`.
22. J. R. Steel, A theorem of Woodin on mouse sets. In Ordinal definability and recursion theory: the Cabal Seminar, volume III, A.S. Kechris et al. eds., volume 43 of Lecture Notes in Logic, ASL, CUP (2016) pp. 243–256.

23. J. R. Steel, Notes on V as a derived model. Notes distributed at the Conference on hod mice and the core model induction, Münster 2010, available at `https://ivv5hpp.uni-muenster.de/u/rds`.
24. J. R. Steel, Derived models associated to mice. In Computational prospects of infinity. Part I. Tutorials, volume 14 of Lect. Notes Ser. Inst. Math. Sci. Natl. Univ. Singap., pp. 105–193. World Sci. Publ., Hackensack, NJ, 2008.
25. J. R. Steel, Hod pair capturing and short tree strategies. Handwritten notes October 2016, available at `www.math.berkeley.edu/~steel`.
26. J. R. Steel, LSA from least branch hod pairs. Handwritten notes July 2016, available at `www.math.berkeley.edu/~steel`.
27. J. R. Steel, Closure properties of pointclasses. In A. S. Kechris, D. A. Martin and Y. N. Moschovakis, Cabal Seminar 77–79, vol. 839 of Lecture Notes in Logic, pp. 147–163, Springer, Berlin, 1981.
28. J. R. Steel and N. Trang, Condensation for mouse pairs. Math. ArXiv `http://arxiv.org/abs/2207.03559` (2022).
29. J. R. Steel and N. Trang, Square in least branch hod mice. In preparation.
30. J. R. Steel and W. H. Woodin, HOD as a core model. In Ordinal definability and recursion theory: the Cabal Seminar, volume III, A.S. Kechris et al. eds., volume 43 of Lecture Notes in Logic, ASL (2016) pp. 257–343.
31. T. M. Wilson, The envelope of an inductive-like pointclass under a local determinacy hypothesis, Annals of Pure and Applied Logic, 166 (10), pp. 991–1018, 2015.

https://doi.org/10.1142/9789819806584_0014

A Worked Example of the Functional Interpretation

Henry Towsner
Department of Mathematics, University of Pennsylvania,
209 South 33rd Street, Philadelphia, PA 19104-6395, USA
http://www.math.upenn.edu/~htowsner
htowsner@math.upenn.edu

Proof mining is the branch of proof theory concerned with examining actual proofs, as they appear in the mathematical literature, and extracting additional from them. In the most common case, we begin with a suitable non-effective proof and convert this proof, by purely syntactic means, into an effective one. In this chapter, we illustrate the main tool, the functional interpretation, by working through a concrete, fairly simple example before describing the underlying proof-theoretic ideas.

1. Introduction

Suppose we have a *non-effective* proof: a proof that some quantity exists but which does not actually tell us what the value is. Such a thing might happen because the proof proceeds by contradiction, because it uses the axiom of choice, or because the proof depends on an abstract technique which takes us far afield from the quantity we original set out to prove the existence of.

We might then ask whether there is any effective proof of the statement—a proof which actually calculates the quantities in the statement.[1] The answer

Partially supported by NSF grant DMS-1340666.

[1]We might—and, in a draft, did—use the word "constructive" instead of "effective", which is perhaps the more common term in mathematics. However, in logic, "constructive" often has a more technical meaning, namely a proof carried out in a formal system of constructive logic. We will *not* not be concerned with this; for instance, we will freely use the law of excluded middle when convenient, so long as we can still obtain explicit bounds. We use "effective" and "non-effective" to emphasize that we are interested in proofs which calculate bounds, irrespective of the logical means those proofs use.

may be no. Even if the answer is yes, it may require finding a completely different proof, which could be as hard as—or harder than—proving the theorem in the first place.

Our topic, however, is circumstances which guarantee that the answer is "yes": when knowing that there is a non-effective proof guarantees the existence of an effective proof, and further, tells us how to find it. This is the subject of *proof mining*. Strikingly, at least to those unfamiliar with proof theory, both the circumstances and the method for finding an effective proof are *syntactic*: they depend on the written form of the theorem and its proof. While there exists more than one such method, our focus in this chapter will be the *functional interpretation* (also called the "Dialectica" interpretation, after the journal where it first appeared [19]).

If we hold a strongly formal view of mathematics, we can ask for proper "meta-theorems": we could take the view that a proof in mathematics is always, at heart, a formal deduction in our favorite system of axioms—say, ZFC. Then we could hope for a formal proof that, given any deduction of a theorem with the necessary syntactic property, there exists an effective deduction of the same theorem. For many choices of axioms (not currently including ZFC, but including systems which suffice to formalize most of mathematical practice), such meta-theorems actually exist.

These theorems are of limited applicability by themselves, however. Even if one believes that actual proofs, as written in textbooks and articles, are intended as descriptions of formal deductions, obtaining those formal deductions is an arduous process. The usefulness of the functional interpretation comes from the fact that it can be applied, with a feasible amount of effort, to actual proofs as written by and for humans. Of course, we cannot have a theorem about "all proofs accepted by mathematicians"; in the broader setting we have only a heuristic: the functional interpretation is a practical method for producing ordinary effective proofs from ordinary non-effective ones as already written up in journals.[2]

As the title promises, our goal in this chapter is to introduce this method

[2]Because proof theory is historically tied to intuitionist and formalist philosophical views, its dependence on these philosophies is sometimes overstated. One need not believe that formal deductions, effective proofs, or syntax play any special role in mathematics for proof theoretic methods to be useful. Even someone who believes that axiomatic proofs are artificial constructs of no intrinsic importance should recognize that large swaths of mathematics just happen to be formalizable, and therefore that methods derived from their study just happen to be useful in practice. Our attitude can be the one attributed to Niels Bohr regarding the lucky horseshoe above his door: "I am told it works even if you don't believe in it".

by focusing on a concrete example. In Sec. 2, we will present a non-effective proof of a theorem on approximations of L_1 functions. In Sec. 3, we will present a corresponding quantitative theorem with an effective proof. These two sections will be completely elementary—all that is needed is some very basic real analysis. (There is nothing new in our analysis of this theorem, which derives from Kohlenbach and Oliva's work [36]. We have chosen this example because the underlying theorem and proof are simple enough that they can be discussed in detail in a reasonable space.)

With a motivating example in place, in Sec. 4 we will finally introduce the functional interpretation, illustrating that the example in Sec. 3 is an instance of a general method. This section will necessarily involve some actual formal logic.

We will conclude with some references to further applications in the literature.

The author is grateful to Jeremy Avigad and Ulrich Kohlenbach for helpful suggestions on a draft of this chapter.

2. A Theorem of Jackson's

Throughout, we will only be concerned with real-valued functions on $[0,1]$, which will refer to as just "functions". When we write $\int f\,dx$ without bounds, we always mean the integral $\int_0^1 f\,dx$.

Definition 2.1: We say f is an L_1-*function* if $\int |f|\,dx$ exists. The L_1-*norm* is defined on such functions by

$$||f||_1 = \int |f|\,dx.$$

Many bizarre, complicated, or difficult-to-work-with functions are nonetheless L_1, so it is natural to ask about approximating such functions by nicer classes of functions; when the distance is measured using the L_1-norm, this is known as *mean approximation*. In particular:

Definition 2.2: Let f be an L_1-function, and let $\mathcal{P}$ be a collection of L_1-functions. A *best (mean) approximation to f in $\mathcal{P}$* is a function $p \in \mathcal{P}$ such that for every $q \in \mathcal{P}$,

$$||f - p||_1 \leq ||f - q||_1.$$

We will mostly be interested in approximation by polynomials of low degree.

Definition 2.3: We write $\mathcal{P}_n$ for the collection of polynomials of degree at most n.

When $\mathcal{P}$ is finite dimensional, the existence of a best approximation follows immediately from a theorem of Riesz [45]. Our focus will be on the question of whether the best approximation is *unique*. For an arbitrary set $\mathcal{P}$, it need not be:

Example 2.4: Let f be the function which is constantly equal to 2, and let $\mathcal{P}$ consist of all piecewise continuous L_1-functions p such that $||p||_1 \leq 1$. Clearly the function p which is constantly equal to 1 is a best approximation to f in $\mathcal{P}$, but the function

$$q(x) = \begin{cases} 2 \text{ if } 0 \leq x \leq 1/2 \\ 0 \text{ if } 1/2 < x \leq 1 \end{cases}$$

is *also* a best approximation.

Uniqueness of approximations is related to a property known as *strict convexity* of the norm (see, for instance, [13] for more general theory on the subject of approximations); since the L_1-norm is not strictly convex, in general there is not a unique best mean approximation. However for particular classes of functions, ad hoc arguments can still give uniqueness, and an example of such a result is due to Jackson [22]:

Theorem 2.5: *Let f be a continuous function. Then for any n, there is a unique best approximation to f in $\mathcal{P}_n$.*

There are a number of proofs of this theorem [39, 44, 51], but we will examine a direct proof (in particular, avoiding the use of measure theory) due to Cheney [12], which we break into several lemmas to make the subsequent analysis easier. (Certain equations are labeled because we will want to refer to them later.)

We need a few definitions.

Definition 2.6: If f is a continuous function, $||f||_\infty = \sup_{x\in[0,1]} |f(x)|$.

(This is a simplification of the usual L_∞ norm; we do not need the L_∞ norm for discontinuous functions, and so can get away with this simplified definition.) Note that, since f is assumed to be continuous, $||f||_\infty$ is defined and finite.

Definition 2.7: If g is a function, we write $\operatorname{sgn} g$ for the *sign* function, given by

$$\operatorname{sgn} g(x) = \begin{cases} 1 & \text{if } g(x) > 0 \\ 0 & \text{if } g(x) = 0 \\ -1 & \text{if } g(x) < 0 \end{cases} .$$

Lemma 2.8: *Let g and h be continuous functions such that g has finitely many zeros and $\int h \operatorname{sgn} g\, dx \neq 0$. Then for some λ,*

$$||g - \lambda h||_1 < ||g||_1.$$

Proof: Let $x_1, \ldots, x_k$ be the zeros of g. We partition the interval $[0,1]$ into two sets, A and B. B will consist of a small open interval around each x_i:

$$B = \bigcup_{i \leq k} (x_i - \epsilon, x_i + \epsilon) \cap [0,1]$$

while $A = [0,1] \setminus B$. We will have to choose ϵ small enough. Observe that

$$\int_B |h(x)|\, dx \leq ||h||_\infty \cdot k \cdot 2\epsilon, \tag{2.1}$$

so by choosing ϵ very small, we may arrange for $\int_B |h(x)|\, dx$ to be as small as we need.

On the other hand, we know that $|\int h \operatorname{sgn} g\, dx| > 0$, and we have

$$\begin{aligned} \left|\int_A h \operatorname{sgn} g\, dx\right| &\geq \left|\int h \operatorname{sgn} g\, dx\right| - \left|\int_B h \operatorname{sgn} g\, dx\right| \\ &\geq \left|\int h \operatorname{sgn} g\, dx\right| - \int_B |h|\, dx. \end{aligned} \tag{2.2}$$

In particular, by choosing ϵ small enough that $2\int_B |h(x)|\, dx < \left|\int h \operatorname{sgn} g\, dx\right|$, we may arrange to have

$$\left|\int_A h \operatorname{sgn} g\, dx\right| > \int_B |h(x)|\, dx. \tag{2.3}$$

$|g|$ is continuous, and therefore the restriction of $|g|$ to A has a minimum somewhere on the closed set A, and since g has no zeros in A, this infimum $m = \inf\{|g(x)| \mid x \in A\}$ must be positive. Choose λ so that $0 < |\lambda| \cdot ||h||_\infty < m$ and $\operatorname{sgn} \lambda = \operatorname{sgn}\left(\int_A h \operatorname{sgn} g\, dx\right)$. Then we have $|\lambda h(x)| < m \leq |g(x)|$ and

$\operatorname{sgn} g = \operatorname{sgn}(g - \lambda h)$ for every $x \in A$. Now we can compute:

$$\begin{aligned}
||g - \lambda h||_1 &= \int |g - \lambda h|\, dx \\
&= \int_A |g - \lambda h|\, dx + \int_B |g - \lambda h|\, dx \\
&= \int_A (g - \lambda h) \operatorname{sgn} g\, dx + \int_B |g - \lambda h|\, dx \\
&= \int_A |g|\, dx - \lambda \int_A h \operatorname{sgn} g\, dx + \int_B |g - \lambda h|\, dx \\
&= \int |g|\, dx - \int_B |g|\, dx - \lambda \int_A h \operatorname{sgn} g\, dx + \int_B |g - \lambda h|\, dx \\
&\leq \int |g|\, dx - \lambda \int_A h \operatorname{sgn} g\, dx + \int_B |g| + |\lambda|\, |h| dx - \int_B |g|\, dx \\
&= \int |g|\, dx - |\lambda| \left| \int_A h \operatorname{sgn} g\, dx \right| + |\lambda| \int_B |h|\, dx \\
&= \int |g|\, dx - |\lambda| \left(\left| \int_A h \operatorname{sgn} g\, dx \right| - \int_B |h|\, dx \right) \qquad (2.4) \\
&< \int |g|\, dx \\
&= ||g||_1.
\end{aligned}$$

□

Lemma 2.9: *If f is continuous and p is a best approximation to f in $\mathcal{P}_n$ then $f - p$ has at least $n + 1$ zeros.*

Proof: Let $g(x) = f(x) - p(x)$. Suppose the statement is false, so g has at most n zeros. Since f, and therefore g, is continuous, there are at most n points where g changes sign. For some $m \leq n$, we may choose points

$$0 < r_1 < r_2 < \cdots < r_m < 1$$

so that the r_i are exactly the interior points where g changes sign. Consider the polynomial

$$h(x) = \prod_{i=1}^{m} (x - r_i).$$

Since $h(x)$ also changes sign exactly at the r_i, it follows that $h \operatorname{sgn} g$ is either always non-negative or always non-positive, and is not constantly 0, and therefore

$$\int h \operatorname{sgn} g\, dx \neq 0. \qquad (2.5)$$

By Lemma 2.8, there is a λ such that $||g-\lambda h||_1 < ||g||_1$, and equivalently,

$$||f-(p+\lambda h)||_1 = ||g-\lambda h||_1 < ||g||_1.$$

But $p+\lambda h$ is a polynomial of degree at most n, contradicting the assumption that p was a best approximation. □

In order to derive (2.5), we used the following seemingly trivial fact, which we will need again:

Lemma 2.10: *Let q be a continuous function with $\int |q(x)|\,dx = 0$. Then $||q||_\infty = 0$.*

Theorem 2.11: *Let f be a continuous function. Then for any n, there is a unique best approximation to f in $\mathcal{P}_n$.*

Proof: Suppose p_1, p_2 are both best approximations of f by polynomials of degree at most n. Let p be the average, $p(x) = \frac{1}{2}(p_1(x)+p_2(x))$. Clearly p is also a polynomial of degree at most n. Also

$$\begin{aligned} ||f-p||_1 &= \int |f(x)-p(x)|\,dx \\ &= \int |f(x)-\frac{1}{2}(p_1(x)+p_2(x))|\,dx \\ &\leq \int \frac{1}{2}|f(x)-p_1(x)| + \frac{1}{2}|f(x)-p_2(x)|\,dx \\ &= \frac{1}{2}||f-p_1||_1 + \frac{1}{2}||f-p_2||_1. \end{aligned}$$

This means that p is also a best approximation of f.

Since p_1 and p_2 are both best approximations, we have $||f-p||_1 = ||f-p_1||_1 = ||f-p_2||_1$, and therefore

$$\begin{aligned} 0 &= \frac{1}{2}||f-p_1||_1 + \frac{1}{2}||f-p_2||_1 - ||f-p||_1 \\ &= \int \frac{1}{2}|f(x)-p_1(x)| + \frac{1}{2}|f(x)-p_2(x)| - |f(x)-p(x)|\,dx. \end{aligned} \tag{2.6}$$

On the other hand, by the triangle inequality, for every x,

$$\begin{aligned} |f(x)-p(x)| &= |f(x)-\frac{1}{2}(p_1(x)+p_2(x))| \\ &= |\frac{1}{2}(f(x)-p_1(x)) + \frac{1}{2}(f(x)-p_2(x))| \\ &\leq \frac{1}{2}|f(x)-p_1(x)| + \frac{1}{2}|f(x)-p_2(x)|. \end{aligned}$$

In particular,

$$\frac{1}{2}|f(x) - p_1(x)| + \frac{1}{2}|f(x) - p_2(x)| - |f(x) - p(x)| \geq 0 \tag{2.7}$$

for every x.

Combining (2.6) with (2.7) using Lemma 2.10, for all x

$$\frac{1}{2}|f(x) - p_1(x)| + \frac{1}{2}|f(x) - p_2(x)| - |f(x) - p(x)| = 0.$$

Therefore for every x,

$$|f(x) - p(x)| = \frac{1}{2}|f(x) - p_1(x)| + \frac{1}{2}|f(x) - p_2(x)|.$$

Since p is a best approximation to f, by Lemma 2.9, $f - p$ must have at least $n+1$ zeros.

Suppose r is a zero of $f - p$. Then

$$0 = |f(r) - p(r)| = \frac{1}{2}|f(r) - p_1(r)| + \frac{1}{2}|f(r) - p_2(r)|$$

and therefore

$$\frac{1}{2}|f(r) - p_1(r)| = \frac{1}{2}|f(r) - p_2(r)| = 0.$$

Then $p_1(r) = f(r) = p_2(r)$, and therefore $p_1(r) - p_2(r) = 0$. $p_1 - p_2$ is a polynomial of degree at most n which is 0 at every zero of $f - p$. The only polynomial of degree at most n which has $n+1$ zeros is 0, so $p_1 - p_2$ must be constantly 0, and therefore $p_1 = p_2$. □

3. Rates of Unicity

Not all uniqueness results are created equal. Once we know there is a unique approximation, we can ask for more detailed quantitative information about the approximation.

Definition 3.1: $p \in \mathcal{P}_n$ is a *ϵ-nearly best approximation to f in $\mathcal{P}_n$* if for every $q \in \mathcal{P}_n$,

$$||f - p||_1 < ||f - q||_1 + \epsilon.$$

We know that two best approximations in $\mathcal{P}_n$ must be equal, so the analogous question to ask is whether ϵ-nearly best approximations must be near each other. More precisely, we look for a function Φ_f such that if p_1 and p_2 are $\Phi_f(\delta)$-nearly best approximations then $||p_1 - p_2||_1 < \delta$. Such a function Φ_f is known as a *modulus of uniqueness* for f [30, 34].

It is clear from the definition that if there is any modulus of uniqueness then there are many functions meeting the definition (for instance, if Φ_f is a modulus of uniqueness for f and $\Psi(\delta) \geq \Phi_f(\delta)$ for every δ then Ψ is also a modulus of uniqueness for f). However if there is a unique best approximation to f, we would expect that there is a "nice" modulus of uniqueness—one that is continuous, at least from one side, and where $\lim_{\delta\to 0} \Phi_f(\delta) = 0$. The optimal modulus has these properties under suitable conditions, and is sometimes called "the" modulus of uniqueness. An arbitrary modulus of uniqueness, however, need not be so nice. The particular case where $c\delta$ is a modulus of uniqueness for some constant c is called *strong unicity*, and is well-studied in approximation theory (again, see [13]). (In fact, it is known that the L^1-approximation does not have strong unicity [6].)

A second type of quantitative information is the stability of the mean approximation under small changes to the function. Let us write $\mathcal{A}$ for the functional mapping f to its best approximation in $\mathcal{P}$. When $||f - f'||_1 \leq \epsilon$, what can we say about $||\mathcal{A}(f) - \mathcal{A}(f')||_1$? A modulus of uniqueness immediately answers this question.

Theorem 3.2: *Suppose f, f' have unique mean approximations in $\mathcal{P}$, Φ_f is the modulus of uniqueness for f, and $||f - f'||_1 < \frac{1}{2}\Phi_f(\delta)$. Then $||\mathcal{A}(f) - \mathcal{A}(f')||_1 < \delta$.*

Proof: Suppose $||f - f'||_1 < \frac{1}{2}\Phi_f(\delta)$. For any $q \in \mathcal{P}$,

$$\begin{aligned} ||f - \mathcal{A}(f')||_1 &\leq ||f - f'||_1 + ||f' - \mathcal{A}(f')||_1 \\ &\leq ||f - f'||_1 + ||f' - q||_1 \\ &\leq 2||f - f'||_1 + ||f - q||_1 \\ &< ||f - q||_1 + \Phi_f(\delta). \end{aligned}$$

Therefore $\mathcal{A}(f')$ is a $\Phi_f(\delta)$-nearly best approximation to f, and since Φ_f is a modulus of uniqueness,

$$||\mathcal{A}(f) - \mathcal{A}(f')||_1 < \delta.$$

□

So, given a theorem like Jackson's theorem showing that some function is unique, a modulus of uniqueness is the quantitative information which tells us *how* unique. Our goal in the remainder of this section is to find a modulus of uniqueness corresponding to Jackson's Theorem, closely following [36]. More precisely, we want to find a modulus corresponding to the *particular proof* given in the previous section—we think of this as the process of making our arguments more quantitative: instead of working with best approximations, we will work with nearly best approximations; instead of working with non-zero values, we will include numerical bounds on how large the values must be; and so on.

We begin with a simple (but, as it turns out, essential) observation: if $||p||_1 > 2||f||_1$ then $||f - 0||_1 < ||f - p||_1$, and therefore when considering approximations, we can consider only those polynomials which satisfy $||p||_1 \leq 2||f||_1$. When considering nearly best approximations, we have to deal with the possibility that a nearly best approximation might have norm slightly more than $2||f||_1$, so we add a bit of leeway.

Definition 3.3: $\mathcal{Q}_n \subseteq \mathcal{P}_n$ is the set of polynomials p of degree $\leq n$ such that $||p||_1 \leq \frac{5}{2}||f||_1$.

In particular, the best approximation to f must belong to $\mathcal{Q}_n$. From here on, we will restrict our attention to $\mathcal{Q}_n$.

3.1. *A quantitative Lemma 2.8*

We can expect that in order to obtain a quantitative version of the main theorem, we will need quantitative versions of the intermediate lemmas as well. We will start, naturally, with Lemma 2.8; we need to strengthen the conclusion so that $||g - \lambda h||_1 + \delta < ||g||_1$ for some fixed value δ.

Of course, the stronger conclusion is not true unless we make stronger assumptions as well—in order to get a quantitative conclusion, we will need quantitative assumptions. We look through the proof to find out which steps contribute to the size of the gap between $||g||_1$ and $||g - \lambda h||_1$, and then take steps to bound those values.

The actual guarantee that the gap exists is given by (2.4), where we can see that the size of the gap is $\lambda(\left|\int_A h \operatorname{sgn} g \, dx\right| - \int_B |h| \, dx)$. This gives us two values we need to bound away from 0: λ and $\left|\int_A h \operatorname{sgn} g \, dx\right| - \int_B |h| \, dx$. The latter is shown to be positive at (2.3); from this and (2.2), we can see that there are two factors contributing to the size of $\left|\int_A h \operatorname{sgn} g \, dx\right| - \int_B |h| \, dx$: the size of $\left|\int h \operatorname{sgn} g \, dx\right|$ and the size of $\int_B |h(x)| \, dx$. The first of these is simply one of our assumptions: instead of making the qualitative assumption that $\int h \operatorname{sgn} g \, dx \neq 0$, we will make a quantitative assumption that $\left|\int h \operatorname{sgn} g \, dx\right| \geq \theta$ for some appropriate θ.

We turn to the bound on $\int_B |h(x)| \, dx$. The bound on this value is given by (2.1), where it is established that

$$\int_B |h(x)| \, dx \leq ||h||_\infty \cdot k \cdot 2\epsilon.$$

A bound on $||h||_\infty$ is just an additional assumption, as is a particular value for k, but ϵ is some value we have chosen. As long as we are willing to make ϵ small, we can make $\int_B |h(x)|\,dx$ as small as we want. So as we make ϵ smaller, we make $\int_B |h(x)|\,dx$ smaller, and therefore seem to make $||g||_1 - ||g - \lambda h||_1$ larger. However when we turn to λ, we will see that making ϵ small may force λ to be small, which makes $||g||_1 - ||g - \lambda h||_1$ small.

λ was given by the rule that $\lambda||h||_\infty < m = \inf\{|g(x)| \mid x \in A\}$, so to keep λ bounded away from 0, we need to bound m away from 0. But here we run into a problem: what if g has a point which dips very close to 0, but is not near any of the zeros? Then this point is included in the set A, but it forces m to be very small, which in turn forces λ to be small, and therefore causes $||g - \lambda h||_1$ to be very close to $||g||_1$.

Suppose g_ζ is a function which "almost" has a zero—say, $g_\zeta(x) = (x - 1/2)^2 + \zeta$ where ζ is very small. If we take the proof given in the previous section at face value, ζ being small just forces us to choose λ_ζ very small. Indeed, following this argument, as ζ approaches 0, the gap $||g_\zeta||_1 - ||g_\zeta - \lambda_\zeta h||_1$ approaches 0 as well.

When ζ reaches 0, however, the situation changes. Now $1/2$ is a zero of $g_0(x)$, and λ only has to be concerned with the value of $g_0(x)$ when x is outside the interval $(1/2 - \epsilon, 1/2 + \epsilon)$. In particular, we can still choose a value λ_0 so that $||g_0||_1 - ||g_0 - \lambda_0 h||_1 > 0$. It is a general principle that quantitative results ought not exhibit such discontinuities—all parameters in our proof should vary continuously in the function g. It is therefore not surprising that we can easily modify our proof to eliminate this discontinuity: when ζ is very small, we could make $||g_\zeta||_1 - ||g_\zeta - \lambda_\zeta h||_1$ larger by treating $1/2$ as if it were a zero and removing the interval around it. This has a price (it makes k larger, because we have a new zero, which in turn makes $\int_B |h(x)|\,dx$ larger, and therefore $\left|\int_A h \operatorname{sgn} g\,dx\right| - \int_B |h|\,dx$ smaller), but when ζ is very small, we still obtain better bounds this way.

Instead of assuming that g has finitely many zeros, we could assume there are finitely many points where the value is below some parameter ζ (with the number of such points and the size of ζ contributing to our ultimate bound on $||g||_1 - ||g - \lambda h||_1$). Stated like this, we still do not have the formulation quite right, since when g is continuous, having any zeros at all will lead to having uncountably many points below ζ; clearly having many small points "close together" does not count. We will remove a small interval around each almost-zero, so what matters—what forces the set B to be large—is if there are many points close to 0 which are separated by large gaps. It is more convenient to jump to our ultimate purpose, and simply require that there is a set B of measure at most ϵ containing all points r with $|g(r)| < \zeta$.

We can now prove this quantitative formulation using essentially the same argument as in the proof of Lemma 2.8, but filling in explicit calculations where appropriate.

Lemma 3.4: *Let g, h, ϵ, ζ be given such that:*

- *g and h are L_1-functions,*
- *$\epsilon > 0$ and $\zeta > 0$,*
- *$|\int h \operatorname{sgn} g \, dx| \geq 3K\epsilon$,*
- *$||h||_\infty \leq K$,*
- *There is a set B of measure at most ϵ such that if $|g(x)| < \zeta$ then $x \in B$.*

Then

$$||g - \frac{\zeta}{2K} h||_1 + \epsilon\frac{\zeta}{2} \leq ||g||_1.$$

Proof: Let $A = [0,1] \setminus B$. We have

$$\int_B |h| \, dx \leq K\epsilon.$$

We have

$$\left|\int_A h \operatorname{sgn} g \, dx\right| \geq \left|\int h \operatorname{sgn} g \, dx\right| - \int_B |h| \, dx \geq 2K\epsilon$$

and therefore

$$\left|\int_A h \operatorname{sgn} g \, dx\right| - \int_B |h| \, dx \geq K\epsilon.$$

Note that if $x \in A$ then $|g(x)| \geq \zeta$. Let $\lambda = \frac{\zeta}{2K}$. Then for any $x \in A$, $|\lambda h(x)| \leq \frac{\zeta}{2} < \zeta \leq |g(x)|$. In particular, $\operatorname{sgn} g(x) = \operatorname{sgn}(g(x) - \lambda h(x))$ for every $x \in A$. Then, by the same calculations as in the proof of Lemma 2.8,

$$||g - \lambda h||_1 \leq ||g||_1 - \lambda\left(\left|\int_A h \operatorname{sgn} g \, dx\right| - \int_B |h(x)| \, dx\right) \leq ||g||_1 - \epsilon\frac{\zeta}{2}. \quad \square$$

3.2. *A quantitative Lemma 2.9*

We now turn to formulating a quantitative version of Lemma 2.9. We would expect to need a quantitative version of Lemma 2.10 to derive (2.5). In fact, a particularly nice version which suffices for our purposes is available in the existing literature:

Theorem 3.5: (Markov brothers' inequality[3]). *If p has degree $\leq n$, $||p'||_\infty \leq 2n^2||p||_\infty$.*

An immediate consequence of this is

Lemma 3.6: *If p has degree $\leq n$, $||p||_\infty \leq 2(n+1)^2||p||_1$.*

Proof: Let $q(x) = \int_0^x p(y)\,dy$ be the integral of p. Applying Markov's inequality to q gives

$$\begin{aligned}||p||_\infty &\leq 2(n+1)^2||q||_\infty \\ &= 2(n+1)^2 \sup_x \left|\int_0^x p(y)\,dy\right| \\ &\leq 2(n+1)^2 \sup_x \int_0^x |p(y)|\,dy \\ &= 2(n+1)^2 \int_0^1 |p(y)|\,dy \\ &= 2(n+1)^2||p||_1.\end{aligned}$$ □

As for stating a quantitative version of Lemma 2.9, our work in the previous subsection gives us a good guess what will need to happen: we will need to replace the assumption that p is a best approximation with the assumption that p is a nearly best approximation, and the conclusion that $f-p$ has n zeros with the conclusion that $f-p$ has a collection of n "almost" zeros which are well spread out.

Lemma 3.7: *Suppose f is continuous and p is a $\frac{\zeta}{20(n+2)^2}$-nearly best approximation to f in $\mathcal{Q}_n$. Then there are $n+1$ points $r_1 < \cdots < r_{n+1}$ such that $r_{i+1} - r_i \geq \frac{1}{20(n+2)^2 n}$ for $i < n+1$ and $|f(r_i) - p(r_i)| \leq \zeta$ for each $i \leq n+1$.*

Proof: Let $\theta = \frac{1}{20(n+2)^2 n}$, $\epsilon = \frac{1}{10(n+2)^2}$, and $g(x) = f(x) - p(x)$. Suppose the conclusion is false; then there must be some $k \leq n$ and some $r'_1 < \cdots < r'_k$ such that $|g(r'_i)| \leq \zeta$ for each $i \leq k$, but whenever $|g(r)| \leq \zeta$ for any

[3]This inequality is sometimes called just "Markov's inequality", but as that name is also used for an inequality from probability theory, this name is perhaps less confusing. The inequality given here was proven by Andrey Markov, the same one who gave his name to the other Markov's inequality. His younger brother Vladimir, also a mathematician, proved a generalization to multiple derivatives, leading to the name given here, which, properly, applies only to the generalization.

$r \in [0,1]$, there is some $i \leq k$ such that $|r - r'_i| < \theta$. In particular, the set $B = \bigcup_{i\leq k}(r'_i - \theta, r'_i + \theta)$ contains all points r with $|g(r)| \leq \zeta$. Observe that $\mu(B) \leq 2n\theta = \frac{1}{10(n+2)^2} = \epsilon$.

If $r'_i + \theta < r'_{i+1} - \theta$ then, since $g(x)$ is continuous, g does not change sign on the interval $[r'_i + \theta, r'_{i+1} - \theta]$. So all sign changes take place in some interval $(r'_i - \theta, r'_i + \theta)$. We choose a subsequence $r_1, \ldots, r_m$ and consider the polynomial

$$h'(x) = \prod_{i=1}^{m}(x - r_i).$$

By choosing the subsequence r_i appropriately, we can ensure that $h(x) = (-1)^b h'(x)$ has the same sign as $g(x)$ whenever $x \notin B$. (Saying this precisely is tricky: roughly, we want to include r'_i in our list if $\operatorname{sgn} g(r'_i - \theta) \neq \operatorname{sgn} g(r'_i + \theta)$, but the situation is complicated by the fact that we could have $r'_{i+1} - \theta < r'_{i+1} + \theta$ and have the sign change occur somewhere in that interval; in that case we would want to include either of r'_i and r'_{i+1}, but not both.) Let $K = ||h||_\infty$; by Lemma 3.6, $||h||_1 \geq \frac{K}{2(n+1)^2}$.

On the other hand, we have

$$\left|\int_B h \operatorname{sgn} g \, dx\right| \leq \int_B |h(x)| \, dx \leq K\mu(B) = K\epsilon,$$

and, taking $A = [0,1] \setminus B$,

$$||h||_1 = \int |h(x)| \, dx = \int_A |h(x)| \, dx + \int_B |h(x)| \, dx.$$

Putting these together gives

$$\int_A |h(x)| \, dx \geq \frac{K}{2(n+1)^2} - \frac{K}{10(n+1)^2} = \frac{2K}{5(n+1)^2}.$$

Therefore

$$\begin{aligned}
\left|\int h \operatorname{sgn} g \, dx\right| &\geq \left|\int_A h \operatorname{sgn} g \, dx\right| - \int_B |h(x)| \, dx \\
&= \int_A |h(x)| \, dx - \int_B |h(x)| \, dx \\
&\geq \frac{2K}{5(n+1)^2} - \frac{K}{10(n+1)^2} \\
&= \frac{3K}{10(n+1)^2} \\
&= 3K\epsilon.
\end{aligned}$$

Therefore we may apply Lemma 3.4 and conclude that there is a λ so that $||g - \lambda h||_1 + \epsilon\frac{\zeta}{2} \leq ||g||_1$, contradicting the assumption that p was a $\frac{\zeta}{20(n+1)^2}$-nearly best approximation. □

3.3. *A quantitative Lemma 2.10*

We now turn to Lemma 2.10. We replaced one use of this lemma with the Markov brothers' inequality, but there is a second use, in the proof of the main theorem, which cannot be replaced by the Markov brothers' inequality (since that only applies to polynomials). In its qualitative form, this lemma seems like an obvious fact about integrals, but we now need a quantitative version. A quantitative formulation should weaken the assumption to merely $\int |q(x)|\, dx \leq \epsilon$ for some bound ϵ. Of course, we can no longer hope to have $q(x) = 0$ for every x, since we can easily think of continuous functions which have small but non-zero integrals but are not always 0. Therefore we expect to weaken the conclusion as well—instead of having $||q||_\infty = 0$, we might hope to show that $||q||_\infty < \delta$ for some value δ.

We can easily imagine continuous functions q with $||q||_1$ small but $||q||_\infty$ quite large: q could be 0 except on a small "bump" where it gets very large. By making the bump very narrow, we can let the bump be very tall, causing $||q||_\infty$ to be large even though $||q||_1$ is staying small. These bumps do not create discontinuity, but it's not unreasonable to say that a function with a tall, thin bump is "less" continuous than functions without such features. We can quantify this effect:

Definition 3.8: Let q be a continuous function. A *modulus of uniform continuity* of q is a function $\omega_q(\epsilon)$ such that for any $\epsilon > 0$ and any $x, y \in [0, 1]$ such that $|y - x| < \omega_q(\epsilon)$, $|q(y) - q(x)| < \epsilon$.

It's not hard to see that the modulus of uniform continuity is sufficient to give a quantitative version of the lemma: if $|q(x)| \geq \epsilon$ for some x, the modulus of uniform continuity ensures that $|q(y)| \geq \epsilon/2$ when y is near x, and this ensure that $\int |q(x)|\, dx$ cannot be too small. To make this precise:

Lemma 3.9: *Let q be a function with modulus of continuity ω_q and*

$$\int |q|\, dx < \frac{\epsilon}{2} \cdot \min\left\{\frac{1}{2}, \omega_q(\epsilon/2)\right\}.$$

Then $||q||_\infty < \epsilon$.

Proof: We prove the contrapositive. If $|q(x)| \geq \epsilon$ then for every y with $|y - x| < \omega_q(\epsilon/2)$, $|q(y)| > \epsilon/2$. We cannot be sure the whole interval

$(x-\omega_q(\epsilon/2), x+\omega_q(\epsilon/2))$ belongs to $[0,1]$, but setting $\eta = \min\{\frac{1}{2}, \omega_q(\epsilon/2)\}$, we can be sure that if $x < 1/2$ then $(x, x+\eta) \subseteq [0,1]$, while if $x \geq 1/2$ then $(x-\eta, x) \subseteq [0,1]$. The two cases are symmetric, so consider the case where $x < 1/2$, and therefore

$$\begin{aligned}\int |q(x)|\,dx &= \int_0^x |q(x)|\,dx + \int_x^{x+\eta} |q(x)|\,dx + \int_{x+\eta}^1 |q(x)|\,dx \\ &> 0 + (\epsilon/2)\eta + 0 \\ &\geq (\epsilon/2)\eta.\end{aligned}$$

□

Now when we apply Lemma 3.9, we need to know the modulus of continuity for the function we apply the lemma to. The only application in our proof not covered by the Markov brothers' inequality is to the function

$$\frac{1}{2}|f(x) - p_1(x)| + \frac{1}{2}|f(x) - p_2(x)| - |f(x) - p(x)|.$$

f is a function given to us; it is no surprise that our bounds will depend on modulus of continuity of f, so we will include as an assumption that we are given particular bounds on the modulus of continuity of f.

For the polynomials p_1, p_2, we can use a combination of the Markov brothers' inequality and the fact that we restricted our polynomials to those in $\mathcal{Q}_n$ to obtain a modulus of continuity. (This is where we use the fact that we are optimizing over $\mathcal{Q}_n$ instead of $\mathcal{P}_n$.)

Lemma 3.10: *Let $p \in \mathcal{Q}_n$. Then $\omega_p(\epsilon) = \frac{\epsilon}{4n^2(n+1)^2||f||_1}$ is a modulus of uniform continuity for p.*

Proof: p is differentiable, so for any $x < y$ in $[0,1]$,

$$|p(y) - p(x)| \leq \left|\int_x^y p'(z)\,dz\right| \leq (y-x)||p'||_\infty.$$

Applying the Markov brothers' inequality to p', we have

$$||p'||_\infty \leq 2n^2||p||_\infty \leq 2n^2(n+1)^2||p||_1 \leq 4n^2(n+1)^2||f||_1.$$

Therefore for any $x < y$,

$$|p(y) - p(x)| \leq 4n^2(n+1)^2||f||_1(y-x)$$

and so the function $\omega_p(\epsilon) = \frac{\epsilon}{4n^2(n+1)^2||f||_1}$ is a modulus of uniform continuity for p.

□

The function we ultimately need is a linear combination of absolute values of functions. It suffices to observe the following:

Lemma 3.11: *Let f, g be functions, let c be a constant, and let ω_f, ω_g be corresponding moduli of uniform continuity. Then:*

(1) ω_f is a modulus of uniform continuity for $|f(x)|$,
(2) $\omega_{cf}(\epsilon) = \omega_f(\epsilon/|c|)$ is a modulus of uniform continuity for $cf(x)$,
(3) $\omega_{f+g}(\epsilon) = \min\{\omega_f(\epsilon/2), \omega_g(\epsilon/2)\}$ is a modulus of uniform continuity for $f(x) + g(x)$.

Proof:

(1) Suppose $|y - x| < \omega_f(\epsilon)$. Then $||f(y)| - |f(x)|| \leq |f(y) - f(x)| < \epsilon$.
(2) Suppose $|y - x| < \omega_f(\epsilon/c)$. Then $|cf(y) - cf(x)| = |c| \cdot |f(y) - f(x)| < |c|\epsilon/|c| = \epsilon$
(3) Suppose $|x - y| < \min\{\omega_f(\epsilon/2), \omega_g(\epsilon/2)\}$. Then

$$\begin{aligned} |(f+g)(x) - (f+g)(y)| &\leq |f(x) - f(y)| + |g(x) - g(y)| \\ &< \epsilon/2 + \epsilon/2 = \epsilon. \end{aligned}$$ □

By the same argument, we may replace the last item in the lemma with $\omega_{f+g+h}(\epsilon) = \min\{\omega_f(\epsilon/3), \omega_g(\epsilon/3), \omega_h(\epsilon/3)\}$ for the sum of three functions.

3.4. *A quantitative version of Jackson's Theorem*

Finally, we turn to the main theorem, Theorem 2.11. We know that we need to strengthen the assumption by stipulating a modulus of continuity for the function f, and that in exchange we expect to calculate a modulus of uniqueness for the approximation of f in $\mathcal{Q}_n$.

Once again, we go through the proof systematically adding quantitative information to statements. Suppose that instead of being best approximations, p_1 and p_2 are ϵ-nearly best approximations. Our goal is to obtain some kind of bound on $||p_1 - p_2||_1$. The function $p(x) = \frac{1}{2}(p_1(x) + p_2(x))$ is no longer a best approximation, but it is still an ϵ-nearly best approximation. As a result, the function

$$q(x) = \frac{1}{2}|f(x) - p_1(x)| + \frac{1}{2}|f(x) - p_2(x)| - |f(x) - p(x)|$$

is still non-negative, and has a small (but not necessarily 0) integral.

The main difference is that we will ultimately obtain points $r_1, \ldots, r_{n+1}$ with the property that $|p_1(r) - p_2(r)|$ is small, but not necessarily 0. In the qualitative version, we used the presence of $n + 1$ zeros, and the fact that $p_1 - p_2$ was a polynomial of degree at most n, to conclude that $p_1 - p_2$ was constantly 0. A quantitative version should allow us to conclude from the presence of $n+1$ "well-separated almost-zeros" that $p_1 - p_2$ is close to the constantly 0 polynomial.

We know that $n+1$ points are enough to specify a polynomial of degree at most n, so if we write down *any* polynomial of degree at most n going through all the points $(r_i, p_1(r_i) - p_2(r_i))$, we will have written down the polynomial $p_1 - p_2$. We just need to express this polynomial in a way that makes explicit that the polynomial is always small. As it happens, one of the most common ways of writing down a polynomial from its zeros has precisely this property: the *Lagrange interpolant* is given by

$$L(x) = \sum_{i=1}^{n+1} (p_1(r_i) - p_2(r_i)) \prod_{j \neq i} \frac{x - r_j}{r_i - r_j}.$$

Observe that for any k, for each $i \neq k$ the term $\prod_{j \neq i} \frac{r_k - r_j}{r_i - r_j}$ is 0, and therefore

$$L(r_k) = (p_1(r_k) - p_2(r_k)) \prod_{j \neq k} \frac{r_k - r_j}{r_k - r_j} = p_1(r_k) - p_2(r_k),$$

so the Lagrange interpolant does goes through the desired points. It is also clear that we have retained exactly enough information about the points $(r_i, p_1(r_i) - p_2(r_i))$ to place a bound on $||L||_\infty$: we know that each $p_1(r_i) - p_2(r_i)$ is small and that as long as $j \neq i$, $r_j - r_i$ is not too small.

Stated formally, we have the following theorem:

Theorem 3.12: *Let f be a function with modulus of continuity ω_f. Then for any n,*

$$\Phi_f(\epsilon) = \frac{\zeta}{2} \min \left\{ \frac{1}{10(n+2)^2}, \omega_f(\zeta/12), \frac{\zeta}{24n^2(n+1)^2||f||_1} \right\}$$

where $\zeta = \frac{\epsilon}{4(n+1)20^n(n+2)^{2n}n^n}$, is a modulus of uniqueness for the approximation by $\mathcal{Q}_n$.

Proof: We fix some values: $\gamma = \frac{1}{20(n+2)^2 n}$, $\upsilon = \frac{\epsilon\gamma^n}{n+1}$, $\zeta = \upsilon/4$, and $\rho = \frac{\zeta}{2}\min\{\frac{1}{10(n+2)^2}, \omega_f(\zeta/12), \frac{\zeta}{24n^2(n+1)^2||f||_1}\}$.

Let p_1, p_2 be ρ-nearly best approximations to f in $\mathcal{Q}_n$. Let $p(x) = \frac{1}{2}(p_1(x) + p_2(x))$. p is also an ρ-nearly best approximation to f since

$$||f - p||_1 \leq \frac{1}{2}||f - p_1||_1 + \frac{1}{2}||f - p_2||_1.$$

This means the equation (2.6) gets weakened to

$$\int \frac{1}{2}|f(x) - p_1(x)| + \frac{1}{2}|f(x) - p_2(x)| - |f(x) - p(x)|\, dx < \rho.$$

But (2.7),

$$\frac{1}{2}|f(x) - p_1(x)| + \frac{1}{2}|f(x) - p_2(x)| - |f(x) - p(x)| \geq 0,$$

is still valid.

Let

$$q(x) = \frac{1}{2}|f(x) - p_1(x)| + \frac{1}{2}|f(x) - p_2(x)| - |f(x) - p(x)|.$$

We wish to apply Lemma 3.9, which requires a modulus of continuity. We have the modulus ω_f for f and the modulus $\omega_p(\delta) = \frac{\delta}{4n^2(n+1)^2||f||_1}$ for p, p_1, and p_2, so we may apply Lemma 3.11 to obtain the modulus of continuity

$$\omega_q(\delta) = \min\left\{\omega_f(\delta/6), \frac{\delta}{12n^2(n+1)^2||f||_1}\right\}.$$

In particular, $\rho \leq \frac{\zeta}{2}\min\{1/2, \omega_q(\zeta/2)\}$, so by Lemma 3.9, we must have $||q||_\infty < \zeta$.

$\rho \leq \frac{\zeta}{20(n+2)^2}$, so by Lemma 3.7, we have $n+1$ points $r_1 < \cdots < r_{n+1}$ such that $r_{i+1} - r_i \geq \gamma$ for $i < n+1$ and $|f(r_i) - p(r_i)| \leq \zeta$ for each $i \leq n+1$. Therefore for each $i \leq n+1$,

$$\begin{aligned}
|p_1(r_i) - p_2(r_i)| &\leq |f(r_i) - p_1(r_i)| + |f(r_i) - p_2(r_i)| \\
&\leq 2\left[q(r_i) + |f(r_i) - p(r_i)|\right] \\
&< 2[\zeta + \zeta] \\
&= \upsilon.
\end{aligned}$$

The Lagrange interpolant gives us an expression for the polynomial $p_1(x) - p_2(x)$:

$$p_1(x) - p_2(x) = \sum_{i=1}^{n+1}(p_1(r_i) - p_2(r_i))\prod_{j \neq i}\frac{x - r_j}{r_i - r_j}.$$

We may bound this:

$$
\begin{aligned}
|p_1(x) - p_2(x)| &= \left| \sum_{i=1}^{n+1} (p_1(r_i) - p_2(r_i)) \prod_{j \neq i} \frac{x - r_j}{r_i - r_j} \right| \\
&\leq \sum_{i=1}^{n+1} |p_1(r_i) - p_2(r_i)| \prod_{j \neq i} \frac{|x - r_j|}{|r_i - r_j|} \\
&< \sum_{i=1}^{n+1} \upsilon \gamma^{-n} \\
&= (n+1)\upsilon\gamma^{-n} \\
&= \epsilon,
\end{aligned}
$$

and therefore

$$||p_1 - p_2||_1 = \int |p_1 - p_2|\, dx < \epsilon. \qquad \square$$

An additional merit of this proof is that we have obtained information on the *uniformity* of the bounds: we now know that if someone gives us only ω_f and $||f||_1$, we can compute a modulus Φ which will work for *any* function g for which ω_f is a modulus of continuity and with $||g||_1 \leq ||f||_1$. (One can also obtain this uniformity from a suitable use of compactness on the statement of the theorem [34, 40], without having to rewrite the proof.)

As discussed in [36], with just a little more work, one can also eliminate the dependence on $||f||_1$. Suppose we are given f and a modulus of continuity ω_f.

Consider the modified function $\tilde{f}(x) = f(x) - f(0)$. Then $\tilde{f}$ has the same modulus of continuity, and we can use it to find $||\tilde{f}||_1$: choosing n so that $1/n < \omega_f(1)$, it must be that if $r \leq i/n$ then $|\tilde{f}(r)| < i$, so

$$||\tilde{f}||_1 = \int |\tilde{f}| dx \leq \sum_{i \leq n} \int_{[i/n, i+1/n]} i + 1\, dx = \sum_{i=0}^{n} i.$$

So we can apply the theorem to $\tilde{f}$. Since the best approximations to $\tilde{f}$ are simply the best approximations to f but shifted over by $f(0)$, the same modulus of uniqueness works for f.

4. The Functional Interpretation

We have now seen that it was possible not only to prove an explicit quantitative version of Jackson's Theorem, but to find such a proof closely related to the qualitative proof. Still, the arguments we used appeared to be somewhat ad hoc, and we seem to have gotten lucky in various places. We made a string of fortunate guesses, for instance replacing $\mathcal{P}_n$ with $\mathcal{Q}_n$, or strengthening the assumption of Lemma 3.4, and therefore weakening the conclusion of Lemma 3.7, in a way which just so happened to allow us to finish the proof anyway. We also depended on existing arguments, like the Markov brothers' inequality and the properties of the Lagrange interpolant, to complete our quantitative proof.

Our goal in this section is to describe a formalization of the procedure used in the previous section which applies systematically to certain kinds of proofs. (Indeed, this method is precisely the way the arguments in the previous section were found by Kohlenbach and Oliva.)

In order to make this precise, we need to pin down what counts as a quantitative calculation. For our purposes, we will identify "quantitative" with "computable". Then our goal in this section is to describe how to systematically take a proof of a statement and:

(1) Find a related statement for which it is appropriate to investigate the existence of computable bounds,
(2) Convert the proof into a calculation of those computable bounds.

As mentioned in the introduction, one of the things that makes the functional interpretation useful in practice is that we can take two different perspectives on it. In the first perspective, the functional interpretation is a completely formal idea; it refers to a family of theorems of the following form:

Theorem 4.1: *Suppose that there is a proof of ϕ in the formal theory T. Then there is a proof of ϕ^{ND} in the formal theorem T'.*

Here T is some particular formal theory, T' is a theory related to T but with the additional property that proofs in T' are effective,[4] and $\cdot^{ND}$ is an operation mapping formulas of (many-sorted) first-order logic to formulas in some particular nice form. The main reason we view these theorems as instances of a uniform idea is that the transformation $\phi \mapsto \phi^{ND}$ is very similar across all these theorems.[5] Such theorems have been proven for a variety of theories T.

[4]In fact, the theory T' is typically a constructive theory based on intuitionistic logic.
[5]The name ϕ^{ND} comes from [1], where the ND translation is actually the composition of two translations: the *N*egative translation from classical to intuitionistic logic and the *D*ialectica translation from first-order intuitionistic logic to a higher-order system.

The second perspective is that the functional interpretation is a heuristic: across many theories and situations, there is an operation which takes any formula ϕ (in particular, where there may be no computable bounds at all for ϕ) and transforms ϕ to ϕ^{ND}, a statement for which computable bounds do exist, which is reliably *implication preserving*—if we can conclude ψ from ϕ then we can also conclude ψ^{ND} from ϕ^{ND}. Since we can convert individual mathematical statements into formulas ϕ without too much difficulty, we can also convert mathematical statements into the quantitative formula ϕ^{ND}.

Taking the first perspective would require choosing a suitable formal theory T, often the theory of Peano arithmetic or a variant, and carefully formalizing our statement in this theory. This tends to involve a great deal of tedious coding—one must interpret statements about the real numbers as statements about sequences of integers, statements about integration as formulas involving quantification over partitions, and so on. Nonetheless, the formal approach is often useful, and can give insights that the informal approach cannot.

Here, however, we will take the second approach, and work only semi-formally. We will use the notation of first-order logic, but deal somewhat informally with our exact choice of formal language and theory.

4.1. *Formalizing statements*

The first thing we need to do is translate ordinary mathematical language into the more formal language of first-order logic. As promised, we will work semi-formally, writing formulas using quantifiers $\forall, \exists$ and the connectives $\wedge$ (and), $\vee$ (or), and $\rightarrow$ (implies), but without pinning down an exact language. In particular, we will freely write things like $\forall q \in \mathbb{Q}\, \exists n \in \mathbb{N} \ldots$ to indicate quantification over various particular domains.

In order to get meaningful results, we do need some restrictions on the formulas we use. (These restrictions stand in for actually formalizing a statement in a particular fixed language—essentially, the restrictions we are choosing are the ones which ensure that a proper formalization is possible.) The core restriction we need is on our atomic formula: our atomic formula should represent computable operations. This means our atomic operations ultimately need to be concerned with countable sets and fixed computable

Both components have several variants, leading to many possible translations. Here we abuse notation and use ϕ^{ND} as a general name for the whole family of translations, to be interpreted as whichever is suitable for a given application.

encodings in order for this to even make sense. For our purposes, it suffices to take our atomic formulas to be the comparisons $x < y$ and $x \leq y$ between rational numbers.

In order to talk about other things, like real numbers, we have to represent them using approximations. For example, we can represent a real number r by a Cauchy sequence (r_i); usually we want to restrict ourselves to Cauchy sequences with a known rate of convergence—say, $|r_i - r| < 1/i$ for all i. Then basic questions about reals become existential or universal statements; for instance, $r < s$ if and only if there is some i such that $2/i < s_i - r_i$—clearly if there is such an i then $r < s$, while if $r < s$ then there is a $\delta > 0$ so that $\delta < s - r$, and choosing $i \geq 4/\delta$ suffices.

Similarly, continuous functions $f : \mathbb{R} \to \mathbb{R}$ are determined by their value on rationals, so we could represent f by a function $\hat{f} : \mathbb{Q} \times \mathbb{N} \to \mathbb{Q}$ so that $|\hat{f}(x, n) - f(x)| < 1/n$ for all x, n.

To illustrate the idea, let us consider how we translate the uniqueness part of Jackson's theorem into a formula of the specified form. In English, the uniqueness part of Jackson's theorem says

> Let f be a continuous function on $[0, 1]$. Then for any n, there is at most one best approximation to f in $\mathcal{P}_n$.

Not all functions from $\mathbb{Q} \times \mathbb{N}$ to $\mathbb{R}$ correspond to continuous functions, so we first need a formula $\text{cont}(f)$ which should hold exactly when f is continuous. The usual ϵ-δ formulation of continuity would be appropriate, but since $[0, 1]$ is compact, continuity is equivalent to uniform continuity, which will slightly simplify things later.[6] So we take

$$\text{cont}(f) \equiv \forall \epsilon \in \mathbb{Q}^+ \exists \delta \in \mathbb{Q}^+ \forall x, y \in \mathbb{Q} \cap [0, 1](|x - y| < \delta \to |f(x) - f(y)| < \epsilon) .$$

More precisely, we can make this a statement about the representation $\hat{f}$, such as

$$\text{cont}(\hat{f}) \equiv \forall \epsilon \in \mathbb{Q}^+ \, \exists \delta \in \mathbb{Q}^+ \, \forall x, y \in \mathbb{Q} \cap [0, 1] \\ \left(|x - y| < \delta \to |\hat{f}(x, \lceil 3/\epsilon \rceil) - \hat{f}(y, \lceil 3/\epsilon \rceil)| < \epsilon\right) .$$

[6]The more common approach in the literature is to build the assumption that f is continuous into the domain of quantification, and further require that f come equipped with a modulus of continuity. This approach avoids some messy issues about extensionality—the question of whether two different representations of a function need, for instance, have the same modulus of continuity. The description here, while informal, emphasizes the way the modulus of continuity falls out of the definitions.

In order to talk about best approximations, we need to talk about L_1 norms. The L_1 norm of a function is a real number, so what we need to do is explain how we calculate the ith approximation—that is, we want $||f||_1 = r$, where r is represented by (r_i) and $|r - r_i| < 1/i$, so we need to explain how we calculate each r_i.

When f is represented by a polynomial, this is easily done: if $f = \sum_d s_d x^d$ with s_d represented by $(s_{d,i})$ where $|s_d - s_{d,i}| < 1/i$, we have

$$\left| ||\sum_d s_d x^d||_1 - ||\sum_d s_{d,j} x^d||_1 \right| \leq \int \sum_d |s_d - s_{d,j}| x^d dx \leq \frac{d}{j},$$

so by choosing $j > id$, we approximate the L_1 norm to within $1/i$.

The situation with general L_1 functions is messier; in fact, in order to be sure that the L_1 norm is even meaningful, we need to know that f is continuous (or have some other guarantee that the norm exists). It is therefore not surprising that in order to actually approximate the L_1 norm of f computably, we need to use the quantitative properties of continuity, the modulus of continuity. Since we will ultimately assume f is continuous, we take for granted that $||f||_1$ is a real number—that is, a sequence of rationals—that we can work with in our formulas. We will write $||f||_{1,i}$ for a sequence of rational numbers with the property that, for all i, $|||f||_{1,i} - ||f||| < 1/i$.

To say that there is a unique best approximation is to say that, if p and p' are distinct best approximations then they must be equal. That is,

$$\begin{aligned} \forall p_0 \in \mathcal{P}_n \; \forall p_1 \in \mathcal{P}_n \qquad & [\exists p' \in \mathcal{P}_n \; ||f - p'||_1 < ||f - p_0||_1] \\ & \vee [\exists p' \in \mathcal{P}_n \; ||f - p'||_1 < ||f - p_1||_1] \\ & \vee ||p_0 - p_1||_1 = 0. \end{aligned}$$

Rephrasing this to account for the fact that the L_1 norm is real valued, this becomes

$$\begin{aligned} \forall p_0 \in \mathcal{P}_n \; \forall p_1 \in \mathcal{P}_n \qquad & [\exists p' \in \mathcal{P}_n \exists i \; ||f - p'||_{1,i} + 2/i < ||f - p_0||_{1,i}] \\ & \vee [\exists p' \in \mathcal{P}_n \exists i \; ||f - p'||_{1,i} + 2/i < ||f - p_1||_{1,i}] \\ & \vee \forall j \; ||p_0 - p_1||_{1,j} < 1/j. \end{aligned}$$

The quantifiers over i and j, and the approximations like $||f - p'||_{1,i}$ are a bit ugly, and they distract from the actual meaning. Here we can introduce a trick which eliminates this issue, at the cost of making our notation slightly more treacherous—we will hide these issues behind a formalism.

When r is a real number and we express $0 < r$ as an existential formula, we would prefer to focus on the gap δ bounding r away from 0 than on the details of the approximation to r—that is, we would like to represent $0 < r$ as the formula $\exists\delta \in \mathbb{Q}^+ \, \delta < r$. The problem is that $\delta < r$ is not computable—if $r = \delta$ then the sequence r_i converges to δ, but at no finite stage i can we be sure that the sequence will not ultimately converge to slightly more or slightly less than δ. If we cared specifically about converging to exactly δ this would be important, but as part of the formula $\exists\delta \in \mathbb{Q}^+ \, \delta < r$, this issue is a distraction: we only care about δ as an approximate bound to how far r is from 0, so if r is very close to δ, we've already established that r is strictly positive, and bounded away from 0 by *almost* δ.

The right comparison is to allow a bit of error in our comparison. When δ is rational and r is real, we define

$$\delta \lessdot r \quad \Leftrightarrow \quad \delta < r_{\lceil 2/\delta \rceil}.$$

Then we have that if $\delta \lessdot r$ then $\delta/2 < r$, while if $\neg(\delta \lessdot r)$ then we must have $r \leq \frac{3}{2}\delta$. When $r \in [\delta/2, \frac{3}{2}\delta)$, whether or not $\delta \lessdot r$ holds depends on the exact nature of how the sequence r_i converges to r. However $\exists\delta \in \mathbb{Q}^+ \, \delta \lessdot r$ is equivalent to $\exists\delta' \in \mathbb{Q}^+ \, \delta' < r$, and the bound δ which witnesses the first statement is within a small constant of the correct bound for the second statement.[7]

We similarly define

$$r \lessdot \delta \quad \Leftrightarrow \quad r_{\lceil 2/\delta \rceil} < \delta.$$

Rewriting the formula above, we get

$$\begin{aligned} \mathrm{approx}(f) \equiv \forall p_0 \in \mathcal{P}_n \, \forall p_1 \in \mathcal{P}_n \quad & [\exists p' \in \mathcal{P}_n \exists\delta \; \delta \lessdot ||f - p_0||_1 - ||f - p'||_1] \\ & \vee [\exists p' \in \mathcal{P}_n \exists\delta \; \delta \lessdot ||f - p_1||_1 - ||f - p'||_1] \\ & \vee \forall\epsilon \; ||p_0 - p_1||_{1,i} \lessdot \epsilon. \end{aligned}$$

Putting this all together, the uniqueness part of Jackson's theorem is:

$$\forall f \in \mathbb{R}^{\mathbb{Q}\cap[0,1]} \left(\mathrm{cont}(f) \to \mathrm{approx}(f)\right). \tag{4.1}$$

4.2. *Extracting quantitative statements*

Once we have placed a formula in the form

$$\forall x \exists y \, \phi(x, y)$$

[7]When one is concerned with finding approximate bounds, it generally causes no difficulties to ignore this point and pretend that $\delta < r$ is a reasonable formula, so it may be easier to ignore the distinction between $\lessdot$ and $<$ on a first read through this chapter.

where x and y may be tuples of multiple variables, there is a natural way to identify a potential quantitative analog of this formula: replace the existence of y with some calculated value—$\forall x\, \phi(x, g(x))$ for some reasonable function g. In practice, computing exact values is often messy, so it usually enough to settle for some kind of bound: $\forall x \exists y \in G(x)\, \phi(x, y)$, where $G(x)$ is always some sort of bounded (really, compact) set. In the simplest, but representative, case, y ranges over the natural numbers and $G(x)$ always has the form $[0, n]$, so $G(x)$ really gives a bound on the size of y.

The functional interpretation depends on the fact that whether or not there is a computable g is closely related to the syntactic properties of ϕ: if every quantifier in ϕ is over a compact domain (and our formalization was appropriate) then there is guaranteed to be a computable g. Conversely, if there is a computable bound g then there must be some formula equivalent to g which contains only quantifiers over compact domains. (The one complication to this equivalence is that when ϕ has quantifiers over non-compact domains, there is no easy way to tell whether it is equivalent to a simpler formula).

To see why this should be the case, consider the simplest situation: x and y are natural numbers and ϕ is a formula of *arithmetic*, where the only domain being quantified over is the natural numbers. If all quantifiers in ϕ are over compact subsets of the natural numbers then all quantifiers in ϕ are really over finite sets. The atomic formulas should themselves be computable, and therefore there is a computer program which, given values n, m, checks in finite time whether or not $\phi(n, m)$ is true. If $\forall x \exists y \phi(x, y)$ is true then g is a computable function: given the input n, $g(n)$ first checks whether $\phi(n, 0)$ holds; if so, $g(n)$ returns 0, and otherwise, $g(n)$ checks whether $\phi(n, 1)$ holds, if so returns 1, and if not continues similarly. (The standard encoding of computability in arithmetic gives the other half of a correspondence—any computable function can be converted into a formula of the right form.)

A formula in the form $\forall x \exists y \phi(x, y)$ where ϕ only has quantifiers over compact domains is called a Π_2 *formula*. (The Π indicates that the outermost quantifier is $\forall$ and the 2 indicates that there are two blocks of quantifiers over non-compact domains.) In general, quantifiers over finite, or more generally, compact, domains will be simpler for us than quantifiers over non-compact domains, so we usually only count the non-compact quantifiers when we discuss how complex a formula is.

Let's consider what this means for Jackson's Theorem.

Recall that in the previous section we replaced the space $\mathcal{P}_n$ with the space $\mathcal{Q}_n$. Now we see what motivated this change: unlike $\mathcal{P}_n$, the space $\mathcal{Q}_n$

is compact. Standard equivalences let us bring the non-compact quantifiers out to the front and rearrange to an implication

$$\begin{aligned}\text{approx}(f) \equiv \forall \epsilon \in \mathbb{Q}^+ \exists \delta \in \mathbb{Q}^+ \quad & \forall p_0 \in \mathcal{Q}_n \ \forall p_1 \in \mathcal{Q}_n \exists p' \in \mathcal{Q}_n \\ & \epsilon \lessdot ||p_0 - p_1||_1 \to \\ & [\delta \lessdot ||f - p_0||_1 - ||f - p'||_1 \\ & \vee \ \delta \lessdot ||f - p_1||_1 - ||f - p'||_1] \,.\end{aligned}$$

To emphasize the way this is Π_2, we write

$$\begin{aligned}\text{approxQ}(f, \epsilon, \delta) \equiv \ & \forall p_0 \in \mathcal{Q}_n \ \forall p_1 \in \mathcal{Q}_n \exists p' \in \mathcal{Q}_n \\ & \epsilon \lessdot ||p_0 - p_1||_1 \to \\ & [\delta \lessdot ||f - p_0||_1 - ||f - p'||_1 \\ & \vee \ \delta \lessdot ||f - p_1||_1 - ||f - p'||_1] \,.\end{aligned}$$

To deal with the assumption $\text{cont}(f)$, we *Skolemize*. The idea is that we can replace statements like

$$\forall x \in \mathbb{Q} \, \exists y \in \mathbb{Q} \ \phi(x, y)$$

with the equivalent statement

$$\exists Y \in \mathbb{Q}^{\mathbb{Q}} \, \forall x \in \mathbb{Q} \ \phi(x, Y(x)).$$

In other words, we can move existential quantifiers outwards by replacing them with functions. We can rewrite $\text{cont}(f)$ as:

$$\exists \omega_f \in (\mathbb{Q}^+)^{\mathbb{Q}^+} \forall \epsilon \in \mathbb{Q}^+ \forall x, y \in \mathbb{Q} \cap [0,1](|x - y| \lessdot \omega_f(\epsilon) \to |f(x) - f(y)| \lessdot \epsilon).$$

ω_f should look familiar: this is just the statement that ω_f is a modulus of uniform continuity for f. We write

$$\text{ucont}(f, \omega_f) \equiv \forall \epsilon \in \mathbb{Q}^+ \forall x, y \in \mathbb{Q} \cap [0,1](|x - y| \lessdot \omega_f(\epsilon) \to |f(x) - f(y)| \lessdot \epsilon).$$

Both $\exists \delta \in \mathbb{Q}^+ \text{approxQ}(f, \epsilon, \delta)$ and $\neg \text{ucont}(f, \omega_f)$ have the form $\exists y \, \phi(y)$ where $\phi(y)$ has only quantifiers over compact domains. Then we can rewrite Jackson's theorem as

$$\begin{aligned}\forall f \in \mathbb{R}^{\mathbb{Q} \cap [0,1]} \forall \omega_f \in (\mathbb{Q}^+)^{\mathbb{Q}^+} \forall \epsilon \in \mathbb{Q}^+ [\neg \text{ucont}(f, \omega_f) \vee \\ \exists \delta \in \mathbb{Q}^+ \text{approxQ}(f, \epsilon, \delta)].\end{aligned}$$

So we have converted Jackson's theorem into a Π_2 form.

Without going further, we have already obtained the uniformity of the theorem: we expect there to be a computable function which, given f, ω_f, and ϵ, computes a δ such that $\text{approxQ}(f, \epsilon, \delta)$. Actually, since we do not use anything about f other than ucont(f), we do not expect the bound to depend on f: in other words, we expect the modulus of uniqueness to depend only on ω_f.

4.3. *The significance of syntax*

It is worth stating explicitly what the previous subsection implied: if a theorem can be put into the Π_2 form $\forall x \exists y \phi(x, y)$, we expect there to be a computable bound G so that $G(x)$ is always compact and such that we can prove that $\forall x \exists y \in G(x)\, \phi(x, y)$. Furthermore, as we will describe in the next subsection, we expect to be able to take a proof of the original statement and systematically convert it into a particular choice of G and a proof of the bounded statement.

If a theorem *cannot* be put into the Π_2 form then this might not happen. It is possible that we can both prove $\forall x \exists y \forall z \phi(x, y, z)$ and also prove that the value of x does not suffice to give a computable bound on the value of y. Further, given a proof of $\forall x \exists y \forall z \phi(x, y, z)$, it may be very hard to determine whether there is a computable bound on y.

If we want to think about which proofs have effective bounds and which do not, we have to begin with the fact that there is a qualitative difference between Π_2 statements and other statements.

4.4. *Proof extraction*

So suppose we have a proof of a Π_2 statement, as in Jackson's Theorem. We have said that we expect there to be a computable bound, and we now turn to the question of extracting such a bound from a proof of the original statement.

Let us begin by considering what we could do with completely formal proofs—that is, proofs which consist of a sequence of formulas, and with each step justified by some formal axiom or inference rule. (We will not worry too much about the exact rules allowed; any reasonable choice will do.) If every formula in the proof were in the Π_2 form then we would expect the proof to directly provide an explicit algorithm: typically any Π_2 axiom will either have an obvious computable bound or one that can be taken as an input to a calculation, and standard inference rules all preserve the property of having explicit computable bounds.

However many proofs of Π_2 statements have intermediate steps which are *not* Π_2. Cheney's proof of Jackson's theorem is an example: the statement of Lemma 2.8 is not Π_2. (Of course, this statement requires substantial massaging for it to even be meaningful to ask whether the statement is Π_2, and showing that a statement cannot, in any way, be rearranged into a Π_2 form is much harder than showing that it can be. But the germ of the idea is given in our discussion before Lemma 3.4: the bounds are discontinuous.) Such statements break the flow of explicit computations through our proof, and must usually be replaced by Π_2 statements if we want to recover explicit bounds.

We will focus on Lemma 2.8, since it was in the process of quantifying that lemma that we had to do the most work. We face some difficulty translating the assumption that g has at most n zeros into our format. A first attempt at a translation would look something like this:

$$\forall g, h\, [(\exists n \exists x_1, \ldots, x_n \forall y\, [g(y) = 0 \rightarrow \exists i \leq n\; y = x_i]) \rightarrow \text{rest}(g, h)]$$

where $\text{rest}(g, h)$ is a formalization of "if g and h are continuous and $\int h \operatorname{sgn} g\, dx \neq 0$ then for some λ, $||g - \lambda h||_1 < ||g||_1$". Standard manipulations on first-order logic allow us to pull some of the quantifiers to the front:

$$\forall g, h, n, x_1, \ldots, x_n \exists y\, [(g(y) = 0 \rightarrow \exists i \leq n\; y = x_i) \rightarrow \text{rest}(g, h)]\,.$$

We can't directly compare real numbers, because our only atomic formulas are comparisons on rationals, so we have to work with the sequences of approximations. The approximations to y and x_i might not be equal, but they will approach 0. So we can rephrase this as:

$$\begin{aligned}\forall g, h, n, x_1, \ldots, x_n \exists y\, [([\forall \zeta > 0\; |g(y)| \lessdot \zeta] \rightarrow \exists i \leq n \forall \delta > 0\; |y - x_i| \lessdot \delta)\\ \rightarrow \text{rest}(g, h)]\,.\end{aligned}$$

We can now pull the quantifiers over ζ and δ out to the front; we actually get to choose which quantifier will be the outermost, and we should make this choice with the goal of minimizing the alternations of quantifiers, and therefore the ultimate complexity of the statement:

$$\begin{aligned}\forall g, h, n, x_1, \ldots, x_n \forall \zeta > 0 \exists \delta > 0\, [(\forall y\; |g(y)| \lessdot \zeta \rightarrow \exists i \leq n\; |y - x_i| \lessdot \delta)\\ \rightarrow \text{rest}(g, h)]\,.\end{aligned}$$

To complete the process of finding the form of Lemma 3.4, we expand rest(g, h). Initially, we might have:

$$\forall g, h, \omega_g, \omega_h, n, x_1, \ldots, x_n \forall \zeta > 0 \forall \gamma > 0 \Big[\exists \delta > 0 \neg (\forall y\ |g(y)| \lessdot \zeta \to \exists i \leq n\ |y - x_i| \lessdot \delta) \vee \neg \mathrm{ucont}(g, \omega_g) \vee \neg \mathrm{ucont}(h, \omega_h) \vee \Big|\int h \,\mathrm{sgn}\, g\, dx\Big| \lessdot \gamma \vee \exists \lambda \exists \delta\ \delta \lessdot ||g||_1 - ||g - \lambda h||_1\Big].$$

As noted above, $\neg$ucont(f, ω) has the $\exists y \phi(y)$ form, so this is a Π_2 formula.

The actual Lemma 3.4 stated above includes some ad hoc simplification—replacing the particular witnesses $x_1, \ldots, x_n$ and δ with the measurable set $B = \bigcup_{i \leq n}(x_i - \delta, x_i + \delta)$, and replacing the continuity of g and h with just an L_1 bound on g, h, and the integrability of $h \,\mathrm{sgn}\, g$. These all weaken the assumptions, and therefore strengthen the overall theorem; they are discovered by observing what properties actually get used in the proof. Note that while the proof of the resulting statement goes through with the weakened assumptions—in particular, without the continuity of g—we actually used the continuity of g to derive the equivalence of the modified statement with the original one.

Most importantly, the syntactic manipulations have guided us into discovering that we should replace counting zeros with a bound on those y with $|g(y)| < \zeta$. What the proof-theoretic methods tells us is that this is guaranteed to be the right thing to look at: both that we will be able to obtain computable bounds for this lemma by using such a restriction, and that the rest of our proof will ultimately match up with this choice. In particular, this is a *local* transformation: while proving a quantitative version of Lemma 2.8, we do not need to look at the quantitative part of the other lemmas to know whether the proof will still work after this modification.

4.5. *More complicated sentences*

In the case of Cheney's proof of Jackson's Theorem, the statements of all our lemmas naturally unravel to Π_2 sentences, even if they did not start that way. In a more general proof (see below for references to some examples), that might not happen. For instance, an intermediate step of the proof might be to show a statement of the form

$$\forall x \exists y \forall z\, \phi(x, y, z)$$

(if all quantifiers in ϕ are over compact domains then this is called a Π_3 formula) and use this to prove

$$\forall u \exists v\, \psi(u, v).$$

The proof might proceed something like this:

> Given u, calculate suitable values of x. Then there is a y such that $\forall z\, \phi(x, y, z)$, and we can use this y to calculate a value v such that $\psi(u, v)$ holds.

However, since there might be no way to compute y from x, this might not actually give us a computable algorithm.

The essential idea of the functional interpretation is to replace the non-quantitative formula $\forall x \exists y \forall z \phi(x, y, z)$ with a new quantitative analog which suffices to carry out the proof. This is best illustrated by an example at the purely syntactic level: we will replace

$$\forall x \in \mathbb{X}\, \exists y \in \mathbb{Y}\, \forall z \in \mathbb{Z}\, \phi(x, y, z)$$

with

$$\forall x \in \mathbb{X}\, \forall Z \in \mathbb{Z}^{\mathbb{Y}}\, \exists y \in \mathbb{Y}\, \phi(x, y, Z(y)).$$

In other words, given x and a *function* Z, we can find a y which works, not necessarily for every z, but at least for the particular value $Z(y)$ returned by the function. Ultimately we will want to restrict Z to be a computable function, though Π_3 statements are simple enough that it does not matter. For statements this simple, we can easily see the equivalence with the original version:

Lemma 4.2: *$\forall x \exists y \forall z\, \phi(x, y, z)$ holds iff $\forall x \forall Z \exists y\, \phi(x, y, Z(y))$ holds.*

Proof: The left to right direction is obvious: if $\forall x \exists y \forall z\, \phi(x, y, z)$ then, given x, we let y be the corresponding witness, and then $\forall z\, \phi(x, y, z)$ holds, so in particular, $\phi(x, y, Z(y))$.

For the other direction, we show the contrapositive. Suppose $\exists x \forall y \exists z\, \neg\phi(x, y, z)$. Take some value x such that $\forall y \exists z\, \neg\phi(x, y, z)$, and consider the function Z which, given y, choose a value $Z(y)$ such that $\neg\phi(x, y, Z(y))$. Then x, Z form a counterexample to $\forall x \forall Z \exists y\, \phi(x, y, Z(y))$. □

Notice that in the left to right direction, our argument was effective, in the sense that, given a value of y which worked on the left, we obtained a value which worked on the right. The right to left direction, however, was not effective—it was a proof by contradiction. Knowing a particular value of y satisfying the formula of right, or even knowing a general method for finding, from each x and Z, a value of y is not enough to find the single value of y which works on the left.

The statement

$$\forall x \forall Z \exists y\, \phi(x, y, Z(y))$$

is Π_2, so we expect to be able to concretely calculate y from x and Z. More surprising is that when we have a proof of a Π_2 statement $\forall u \exists v\, \psi(u, v)$ from $\forall x \exists y \forall z\, \phi(x, y, z)$, we also have a proof of $\forall u \exists v\, \psi(u, v)$ from $\forall x \forall Z \exists y\, \phi(x, y, Z(y))$, and the bounds on v (as a function of u) depend only on the bounds on y as a function of x and Z. In other words, the statement $\forall x \forall Z \exists y\, \phi(x, y, Z(y))$ captures all the computable information present in the original statement.

To handle even more complicated sentences, with yet more alternations of quantifiers, we need not only functions, but *functionals*—operations which map functions to functions, and then operations which map functionals to functionals, and so on. In order to meet our goal of working with quantitative data, in general we need to restrict ourselves to computable functions. If the domains in the original statement are all countable (and coded appropriately) then it makes sense to talk about computable functions from a countable domain to another countable domain, and there are only countably many such functions.

To each formula ϕ we will assign a new formula, ϕ^{ND} which will always have the form

$$\exists y \forall x\, \phi_D(x, y),$$

where ϕ_D will only have quantifiers over compact domains. The intention is that we will have a systematic method for converting proofs of ϕ into a particular choice of y together with a proof that $\forall x\, \phi_D(x, y)$ holds. The definitions of ϕ_D and ϕ^{ND} are given by induction on the form of ϕ.

Remember that we are now restricting ourselves to computable functions, so when we write $\mathbb{X}^{\mathbb{Y}}$ in the following definition, we mean the domain of computable functions from $\mathbb{Y}$ to $\mathbb{X}$.

When ϕ is atomic, $\phi_D = \phi$. For the inductive case, suppose we already have $\phi_D(x, y)$ and $\psi_D(u, v)$ where x has type $\mathbb{X}$, y has type $\mathbb{Y}$, u has type $\mathbb{U}$, and v has type $\mathbb{V}$. Then

(1) $(\phi \wedge \psi)_D(x, u, y, v)$ is $\phi_D(x, y) \wedge \psi_D(u, v)$ and $(\phi \wedge \psi)^{ND} = \exists y, v \forall x, u\, \phi_D(x, y) \wedge \psi_D(u, v)$,
(2) $(\neg\phi)_D(X, y)$ is $\neg\phi_D(X(y), y)$ and $(\neg\phi)^{ND}$ is $\exists X \in \mathbb{X}^{\mathbb{Y}} \forall y \neg\phi_D(X(y), y)$,
(3) $(\phi \to \psi)_D(X, V, y, u)$ is $\phi_D(X(y, u), y) \to \psi_D(u, V(y))$ and $(\phi \to \psi)^{ND}$ is

$$\exists X \in \mathbb{X}^{\mathbb{Y}\times\mathbb{U}} \exists V \in \mathbb{V}^{\mathbb{Y}} \forall y, u(\phi_D(X(y, u), y) \to \psi_D(u, V(y))),$$

(4) $(\forall z \in S\, \phi)_D(x, Y, z)$ is $\phi_D(x, Y(z), z)$ and $(\forall z \in S\, \phi)^{ND}$ is $\exists Y \forall x, z \phi_D(x, Y(z), z)$.

Instead of defining cases for $\vee$ and $\exists$, we derive them using the de Morgan laws: $(\exists z\phi)^{ND}$ is $(\neg\forall z\neg\phi)^{ND}$ and $(\phi \vee \psi)^{ND}$ is $(\neg(\neg\phi \wedge \neg\psi))^{ND}$.[8]

It is generally useful to think of ϕ^{ND} as saying "y is a demonstration that ϕ is true" where a purely mechanical verification that a statement is true would require checking that $\phi_D(x, y)$ holds for all possible values of x. Then some of the inductive cases are easy to interpret: for instance, if y demonstrates ϕ holds and v demonstrates that ψ holds then the pair v, y demonstrates $\phi \wedge \psi$.

A more interesting case is implication; what does it mean to have an explicit demonstration for $\phi \to \psi$? It means we should have an algorithm which converts demonstrations of ϕ into demonstrations of ψ. This is the function V: given a y which demonstrates ϕ, $V(y)$ is a demonstration of ψ. The complication is that when y fails to be a demonstration of ϕ—there exists some x with $\neg\phi_D(x, y)$—we do not want $V(y)$ to be arbitrary. Instead we want to have the property that not only can we determine $V(y)$ from y, but when u is a counter-example to $V(y)$, we can find a counter-example $X(y, u)$ to y. Because it is the most important case, it is worth dwelling on this point: the interpretation of $\to$ here requires that we have an algorithm which converts *any* y into a value $V(y)$, without needing to know whether y actually works or not—the commitment is that *if* y works for ϕ then $V(y)$ works for ψ as well, and furthermore that we know how to translate a counter-example to $V(y)$ into a counter-example to y itself. This imposes something like a continuity requirement on V: if y is "almost right", in the sense that counter-examples are rare (for instance, the only counter-examples are very large values of x) then $V(y)$ should be "almost right" as well.

[8]This last step is the N in the ND-translation: the D translation gives a direct, constructive interpretation of $\vee$ and $\exists$ which would not be able to interpret some classically provable theorems. Instead we interpret $\vee$ and $\exists$ by first inserting some negations, leading to a more complicated translation which does apply to classical logic.

We would like to add the following clause which would cause compact domains to behave more like finite sets:

> When S is a compact domain, $(\forall z \in S\,\phi)_D(x, Y, z)$ is $\exists y \leq Y(x)\,\phi_D(x, y, z)$ and $(\forall z \in S\,\phi)^{ND}$ is $\exists Y \forall x \forall z \in S \exists y \leq Y(x)\,\phi_D(x, y, z)$.

This says, roughly, that when we quantify over a compact domain, we obtain a bound on all witnesses needed for all elements of that domain. Making this idea work turns out to be a bit difficult—this is the subject of the *monotone* [29] and *bounded* [15] functional interpretations. However for most practical applications, the additional effort is justified.

There are two properties that are needed to make the ND translation useful. The first is that if we have a proof of ϕ in a reasonable theory (for instance, Peano arithmetic) then we have a particular value of y together with a proof that $\forall x \phi_D(x, y)$ actually holds. To prove this, we could pin down a particular formal system of axioms and inference rules, and then prove that the translation of each axiom is justified, and also that each inference rule preserves being justified. It is essential that the proof itself is completely explicit—given a formal proof ϕ, there is an explicit procedure for converting it into a proof of ϕ^{ND} together with the associated algorithm. It is convenient that the algorithm is *short*: a single line in the original proof generally translates to a fixed, small number of lines in the new proof (the exact values, of course, depending on the particular formal system).

The second important property of the ND translation is that it does not alter Π_2 formulas:

Lemma 4.3: *When ϕ is quantifier-free, $(\forall x \exists y \phi(x, y))^{ND}$ is equivalent to the existence of a computable function Y such that $\forall x \phi(x, Y(x))$.*

Proof: Since ϕ has no quantifiers, ϕ^{ND} is actually the formula ϕ except that $\vee$ has been replaced by $\neg(\neg \cdots \wedge \neg \cdots)$, which is equivalent by the de Morgan law. Then simply plugging in the definitions above, $(\forall x \exists y \phi(x, y))^{ND}$ is equivalent to

$$\exists Y \forall x \phi(x, Y(x))$$

where Y is a computable function. □

More generally, we expect this to hold when ϕ has only quantifiers over compact domains, subject to suitable computability conditions on atomic formulas.

This, of course, is exactly what we want: given a proof of $\forall x \exists y \phi(x, y)$, we translate it, step by step, to an explicit proof of $(\forall x \exists y \phi(x, y))^{ND}$, which means we have an actual function Y such that $\forall x \phi(x, Y(x))$. However the intermediate steps of the proof have been changed to bring out explicit information which may have been hidden.

If we used the monotone functional interpretation in place of the ND translation described above, we would get a stronger result; essentially

Lemma 4.4: *When ϕ only has quantifiers over compact domains, $(\forall x \exists y \phi(x, y))^{ND}$ is equivalent to the existence of a computable function Y such that $\forall x \exists y \leq Y(x)\, \phi(x, y)$.*

Making precise what $y \leq Y(x)$ means when y comes from a domain of complicated functions, however, is a bit tricky.

We have described this as a process applied to completely formal proofs. This is not a very practical approach; real proofs, as written in journals, are far from being strings of logical formulas. If we had to first translate those proofs into formal strings of logical inferences, that alone would be a huge process, and empirical experience suggests that the proof becomes several times as long when reduced to a completely formal proof.[9] What makes the functional interpretation useful for actual substantial proofs is that it can be applied relatively directly to journal proofs.

The functional interpretation is a local transformation: it tells us how to translate each individual statement. So we can translate particular statements—say, the statements of individual lemmas—and then fill in the proofs by hand, knowing that there is a proof (and one close to the original). If this proves too difficult, we can simply break the proof in half at some convenient point, translate the statement of the halfway point, and prove two shorter lemmas.

We need the functional interpretation to complete our goal that extraction of bounds depends on the syntactic form of the conclusion. No matter what the intermediate steps look like, we can use the ND translation to convert every step of the proof into an argument with explicit quantitative bounds.

[9]The ratio of length of the formalized proof to the informal one is known as the *de Bruijn factor*, and a value of 4 seems to be common [53].

4.6. *Logical issues*

We have tried to minimize the discussion of technical logical issues in this presentation, but as the fuller discussion in [36] shows, these issues cannot be ignored when attempting to extract effective bounds from actual mathematical proofs.

In a more formal approach, one has to actually formalize the proof in a particular formal system, and the complexity of that system can make a substantial difference in what techniques are needed to find an effective proof. We have discussed here the simplest set of techniques, which would suffice for theories like Peano arithmetic or the higher-order (i.e. containing functions and functions on functions and so on) analog PA^ω [1].

In fact, these methods by themselves do not suffice for the most interesting results in the literature, including Cheney's proof. Cheney's proof has two features which push it out of this range. First, it makes use of the sign function, which in general does not exist in PA^ω. Actually, because it is not continuous—more precisely, not *majorizable*—the sign function is incompatible with the monotone functional interpretation, and its presence represents a substantial further complication to the process of finding effective bounds. In this case, we were able to eliminate the use by breaking the interval $[0,1]$ into finitely many intervals with the sign function constant on each one.

Cheney's proof also makes use of the compactness of the interval $[0,1]$ in the proof of Lemma 2.8, when we note that the infimum of a continuous function with no zeros on a closed set exists and is positive. Compactness (usually called WKL, "weak König's lemma", in this context) is a further complication which appears in many, and the development of techniques for eliminating it in some situations [34], including Cheney's proof, was one of the major steps towards making proof mining viable in practice.

5. Some Applications of the Functional Interpretation

5.1. *Fixed point theorems*

One place where non-effective proofs occur naturally is metric spaces (often special kinds of metric spaces, like Banach spaces or C^*-algebras), where compactness is a powerful, frequently used tool. The functional interpretation has been used extensively to extract quantitative information from such proofs [7–11, 18, 27, 30, 31, 33, 35, 41–43].

We will consider just one family of examples, fixed point theorems, with an eye towards the importance of the syntactic form of statements. A typical fixed point theorem is Edelstein's fixed point theorem [14]:

Theorem 5.1: *Let (K, d) be a compact metric space and $f : K \to K$ be* contractive*—for any $x, y \in K$, $d(f(x), f(y)) < d(x, y)$. Then for any x, the sequence given by $x_0 = x$, $x_{k+1} = f(x_k)$ converges to a unique c such that $f(c) = c$.*

In general, the statement that a sequence converges is not Π_2—it has the form

$$\forall\epsilon\exists m\forall n(n \geq m \to d(x_n, x_{n+1}) < \epsilon).$$

For Edelstein's theorem, contractivity lets us make the following observation: once we have $d(x_m, c) < \epsilon/2$, we also have $d(x_n, c) < \epsilon/2$ for all $n \geq m$, and in particular, $d(x_n, x_{n+1}) < \epsilon$ for $n \geq m$. If we know what c is, we see that any m with $d(x_m, c) < \epsilon/2$ is an m we are looking for. The statement

$$\forall\epsilon\exists m\ d(x_m, c) < \epsilon$$

is Π_2, so we can expect to find explicit bounds for this. While we cannot expect to actually use c when finding bounds, the choice of c is unique, and it has an explicit modulus of uniqueness. By methods similar to the ones in the effective proof of Jackson's theorem, Kohlenbach showed [30] that from the modulus of uniqueness, one can construct a modulus of continuity—that is, one can figure out how c varies as x_0 varies. Putting these facts together—the rate of convergence in the original statement depends on the convergence of x_m to c, where c is continuous in x_0—is enough to find a computable function $N(\epsilon)$ so that for each x, each ϵ, and each $n \geq N(\epsilon)$, $d(x_n, c) < \epsilon$ [28]. The function N depends on the diameter of K (that is, $\sup_{x,y\in K} d(x, y)$) and a *modulus of contractivity* of f—a function $\eta(\epsilon)$ such that $d(x, y) > \epsilon$ implies $d(f(x), f(y)) + \eta(\epsilon) < d(x, y)$.

To see that these considerations are really necessary, one can consider the Krasnoselski fixed point theorem:

Theorem 5.2: ([38]). *Let K be a convex, closed and bounded set in a uniformly convex Banach space $(X, ||\cdot||)$, f a mapping of K into a compact subset of K such that f is* non-expansive*—that is, $||f(x) - f(y)|| \leq ||x - y||$ for all x, y. Then for every $x_0 \in K$, the sequence given by*

$$x_{k+1} = \frac{x_k + f(x_k)}{2}$$

converges to a p such that $f(p) = p$.

Again this is a convergence statement, so not in a Π_2 form. Unlike Edelstein's fixed point theorem, this is really unavoidable—Kohlenbach gives an example [32] showing that there can be no algorithm finding bounds from x_0 and f. However a similar idea is to observe that the quantity $||x_n - f(x_n)||$ is decreasing, and therefore given ϵ one can hope to find an n such that $||x_n - f(x_n)|| \leq \epsilon$. (This is a bound on the *asymptotic regularity* of the sequence (x_n).) Asymptotic regularity *is* Π_2, and it is therefore unsurprising that many explicit bounds are known, both by analytic methods [24, 37] and by use of the functional interpretation [32].

5.2. *Ultraproducts and similar constructions*

Another common source of non-effective proofs is the compactness of first-order logic—that is, nonstandard analysis, or, more generally, the use of ultraproducts.[10] We begin with a sequence of models $\mathfrak{M}_N$ of a language $\mathcal{L}$ and combine them into a single model $\mathfrak{M}$ with the property that $\mathfrak{M} \models \phi$ for a formula ϕ in $\mathcal{L}$ if and only if for "most" N, $\mathfrak{M}_N \models \phi$. If we prove that ϕ must hold in $\mathfrak{M}$, we can conclude that it holds in $\mathfrak{M}_N$ for "most", and certainly for infinitely many N. (See [3] for discussion of the reverse argument, that ultraproducts can be used to show the existence of uniform bounds.)

A typical example is the ergodic-theoretic proof of Szemerédi's Theorem [16, 17]:

Theorem 5.3: *For every $\epsilon > 0$ and every k, there is an N such that if $A \subseteq [1, N]$ with $|A| \geq \epsilon N$ then there are a, d with*

$$\{a, a + d, a + 2d, \ldots, a + (k - 1)d\} \subseteq A.$$

Note that this is Π_2—for every ϵ and k, there is an N; the quantifiers over A and over a, d are over finite sets.

If the statement is not true, we may fix $\epsilon > 0$ and k, and for each N find an $A_N \subseteq [1, N]$ with $|A_N| \geq \epsilon N$ so that a contains no arithmetic progression of length k. We take the language $\mathcal{L}$ containing a unary relation symbol $\underline{A}$, a unary function $\underline{T}$, and for each formula $\phi(x)$ with only the displayed free variables and each $q \in \mathbb{Q}^{>0}$, a 0-ary relation symbol $m_{\phi,q}$. We interpret $([1, 2N], A_N)$ as a model of this language by taking A_N as the interpretation

[10]This general idea has been rediscovered a number of times, especially various special cases that do not depend as heavily on the general logical framework, and is therefore known by a number of names: Banach limits, the Furstenberg correspondence, vague convergence, and graphons. These are not all exactly the same notion, but when our concern is the extraction of computable bounds, the differences are not significant.

of $\underline{A}$, $x \mapsto x+1 \mod 2N$ as the interpretation of $\underline{T}$, and

$$([1,N],A_N) \vDash m_{\phi,q} \Leftrightarrow |\{x \mid ([1,N],A_n) \vDash \phi(x)\}| \geq 2qN.$$

(The last clause is technical; it gives us some ability to talk about the measure of sets using formulas in the language. This is a special case of more general approaches for considering measures in the context of first-order logic [4, 20, 23].)

We then work in the ultraproduct of these models, which we call (X, A). Observe that if we can prove, for some d, that the ultraproduct satisfies the formula

$$\exists x\, x \in \underline{A} \wedge T^d x \in \underline{A} \wedge \cdots \wedge T^{(k-1)d} x \in \underline{A}$$

then infinitely many of the finite models $([1,2N],A_N)$ satisfy this formula. Take N much larger than d and let $a \in [1,2N]$ witness this fact. Since $A_N \subseteq [1,N]$ and each $\underline{T}^{id}a \in A_N$, we must have $\underline{T}^{id}a = a+id$ for each $i<k$. (We are going to a small amount of trouble here to prevent the case where the progression involves "wrapping around", since T is interpreted as addition mod $2N$.) Therefore $a, a+d, \ldots, a+(k-1)d$ is an arithmetic progression in A_N. This gives the desired contradiction, since the A_N were chosen to be sets with no arithmetic progressions.

Furstenberg gives a proof [17] that the ultraproduct satisfies this formula by way of interpreting the ultraproduct as dynamical system. (He phrases his construction quite differently, but the essential idea is the same.)

If we want to give an effective version of this proof, we face the following obstacle. The ergodic theoretic argument involves statements about the measure—say, $\mu(A \cap TA) > 0$. Translated into a formula, this is

$$\exists q \in \mathbb{Q}^{>0} m_{Ax \wedge A(Tx),q}.$$

We need to distinguish between formulas in the precise, formally defined language $\mathcal{L}$ and formulas in the informal sense we used them in the previous section. This formula, with its quantifier over $\mathbb{Q}^{>0}$, is a formula in the broader sense, but it is not actually a formula in $\mathcal{L}$. In particular, we do not automatically have that (X, A) satisfies this formula exactly when most $([1,2N],A_N)$ satisfy this formula.

There is an obvious attempt at a solution: we could extend $\mathcal{L}$ to a bigger language $\mathcal{L}'$, with two sorts, where the second sort is intended to represent $\mathbb{Q}^{>0}$. We would replace $m_{\phi,q}$ with $m_\phi(q)$, where q will obviously range over the second sort. Then we start with two-sorted models $([1,2N],A_N,\mathbb{Q}^{>0})$. The problem is that when we take the ultraproduct, we also have to take the ultraproduct in the second sort, so we get a model $(X, A, (\mathbb{Q}^{>0})^*)$, where $(\mathbb{Q}^{>0})^*$ is the (positive) *nonstandard* rational numbers—which includes infinitesimal rationals.

But the formula we want, $\exists q \in \mathbb{Q}^{>0} m_{Ax \wedge A(Tx)}(q)$, is not the same as the formula $\exists q \in (\mathbb{Q}^{>0})^* m_{Ax \wedge A(Tx)}(q)$; the latter allows for the possibility that the measure of $A \cap TA$ is "positive" but infinitesimal.

The issue is that when we talk about the distinction between sets of positive measure and sets of measure 0, we want to work with the non-compact domain $\mathbb{Q}^{>0}$. In an ultraproduct, all the domains in the language of the model are compact, so we can only discuss the distinction between positive and 0 measure by including quantifiers outside the language of the ultraproduct.

However the *transfer principle* tells us that we have the following correspondence:

Lemma 5.4: *Consider a statement in the form $\forall x \in D \exists y \in S \phi(x, y)$ where D is arbitrary, S is countable, and ϕ is a statement in the language $\mathcal{L}$. Then $\mathfrak{M} \models \forall x \exists y \phi(x, y)$ iff for every $x \in D$ there is a $y \in S$ such that for most N, $\mathfrak{M}_N \models \phi(x, y)$.*

In other words, we have a correspondence between Π_2 sentences in $\mathfrak{M}$ and those in the $\mathfrak{M}_N$, namely that Π_2 statements which are *uniformly* true in the finite models are true in the infinite one.

For statements which are not Π_2, we do not have such a correspondence. For example, the ergodic proof of Szemerédi's Theorem uses the mean ergodic theorem, the relevant case of which is the following statement:

> Let $\alpha_n(x) = \frac{1}{n} \sum_{i \leq n} \chi_A(x+i)$ and let $||\alpha(x)|| = \sqrt{\frac{1}{|X|} \sum_{x \in X} \alpha^2(x)}$. For every $\epsilon > 0$, there is an n such that for all $m \geq n$
>
> $$||\alpha_n(x) - \alpha_m(x)|| < \epsilon.$$

We may treat the inner part of the statement—$||\alpha_n(x) - \alpha_m(x)|| < \epsilon$—as being a formula in $\mathcal{L}$ (working this out in detail requires expanding the language and getting into technicalities about representing the measure as part of $\mathcal{L}$). Even so, this is not Π_2, since we still have the three alternating quantifiers "for every ϵ there is an n such that for all m".

In fact, this statement is true in the finite models, for utterly trivial reasons—in $([1, 2N], A_N)$, take n to be much, much larger than N, so that for any $m \geq n$, $\alpha_m(x)$ is the function which is very close to $\frac{1}{N} \sum_{i \leq N} \chi_A(i)$ at every point. This is a vacuous argument, and it does not reflect the real mathematical content of the mean ergodic theorem. We cannot expect to express the full content of saying that an infinite sequence converges in a finite model.

The functional interpretation tells us that every statement implies a Π_2 statement which captures its computational content. The right statement here is

> For every $\epsilon > 0$ and every function F, there is an n such that if $F(n) \geq n$ then $||\alpha_n - \alpha_{F(n)}|| < \epsilon$.

This is also true in the finite models, with a bound on n that depends on ϵ and F, but *not* on the size of the finite model. This is the statement about finite models which matches up correctly with the mean ergodic theorem in the ultraproduct.

We can make this statement slightly nicer with the following observation: given a function F, define F' by

$$F'(n) = \begin{cases} \text{The least } m \in [n, F(n)] \text{ such that } ||\alpha_n - \alpha_m|| \geq \epsilon & \text{if there is one} \\ F(n) & \text{otherwise} \end{cases}.$$

Then F' is computable from F, and by applying the statement above to F' instead of F, we obtain

> For every $\epsilon > 0$ and every function F, there is an n such that if $F(n) \geq n$ then for all $m \in [n, F(n)]$, $||\alpha_n - \alpha_m|| < \epsilon$.

The usual mean ergodic theorem tells us that the averages α_n stabilize. This version tells us that we can find arbitrarily long intervals on which the average remains stable. This is known as the *metastability* of the average [2,26,46,52].

A recent generalization of this idea is in [50], where the corresponding notion of metastability for a double limit is used to give an effective version of a theorem about L^1 functions; this is applied in [48] to give explicit bounds for a theorem about Banach spaces—the "failure of local unconditionality of the James space"—whose usual proof involves an ultraproduct.

Note that in this case it's incidental that what the functional interpretations extracts from proofs which used ultraproducts is computable—the same method would work with any language and any replacement for our finite models. What's really important is that the functional interpretation tells us how to correspond statements about the ultraproduct model to statements in the original models. In almost every situation where we care about this, it makes sense to view us as extracting information computable from the original models, but there has been some recent investigation of this idea more abstractly, where one has the right syntactic features to extract information using the functional interpretation, but where one need not be working with computable information [5, 21, 47, 49].

6. Further Reading

There are two well-established and complementary introductions to the functional interpretation. Avigad and Feferman [1] give a thorough introduction to the formal theory of the functional interpretation in the most important settings, including a number of variants and many applications within proof theory. Kohlenbach [25] gives a detailed guide to the applications of the functional interpretation outside logic, especially in analysis.

References

1. J. Avigad and S. Feferman. "Gödel's functional ("Dialectica") interpretation". In: *Handbook of proof theory*. Vol. 137. Stud. Logic Found. Math. Edited by Jon Barwise, With the cooperation of H. J. Keisler, K. Kunen, Y. N. Moschovakis and A. S. Troelstra, Studies in Logic and the Foundations of Mathematics, Vol. 90. Amsterdam: North-Holland, 1998, pp. 337–405. ISBN: 0-7204-2285-X.
2. J. Avigad, P. Gerhardy and H. Towsner. "Local stability of ergodic averages". In: *Trans. Amer. Math. Soc.* 362.1 (2010), pp. 261–288. ISSN: 0002-9947.
3. J. Avigad and J. Iovino. "Ultraproducts and metastability". In: *New York J. Math.* 19 (2013), pp. 713–727. ISSN: 1076-9803.
4. S.-M. Bagheri and M. Pourmahdian. "The logic of integration". In: *Arch. Math. Logic* 48.5 (2009), pp. 465–492. ISSN: 0933-5846.
5. B. van den Berg, E. Briseid and P. Safarik. "A functional interpretation for nonstandard arithmetic". In: *Ann. Pure Appl. Logic* 163.12 (2012), pp. 1962–1994. ISSN: 0168-0072.
6. B. Björnestal. "Continuity of the metric projection operator I-III". preprint. 1974–1975.
7. E. M. Briseid. "A rate of convergence for asymptotic contractions". In: *J. Math. Anal. Appl.* 330.1 (2007), pp. 364–376. ISSN: 0022-247X.
8. E. M. Briseid. "Fixed points of generalized contractive mappings". In: *J. Nonlinear Convex Anal.* 9.2 (2008), pp. 181–204. ISSN: 1345-4773.
9. E. M. Briseid. "Some results on Kirk's asymptotic contractions". In: *Fixed Point Theory* 8.1 (2007), pp. 17–27. ISSN: 1583-5022.
10. E. M. Briseid. "A new uniformity for asymptotic contractions in the sense of Kirk". In: *Int. J. Math. Stat.* 6.S10 (2010), pp. 2–13. ISSN: 0974-7117.
11. E. M. Briseid. "Logical aspects of rates of convergence in metric spaces". In: *J. Symbolic Logic* 74.4 (2009), pp. 1401–1428. ISSN: 0022-4812.
12. E. W. Cheney. "An elementary proof of Jackson's theorem on meanapproximations". In: *Math. Mag.* 38 (1965), pp. 189–191. ISSN: 0011-801x.
13. E. W. Cheney. *Introduction to approximation theory*. Reprint of the second

(1982) edition. AMS Chelsea Publishing, Providence, RI, 1998, pp. xii+259. ISBN: 0-8218-1374-9.
14. M. Edelstein. "On fixed and periodic points under contractive mappings". In: *J. London Math. Soc.* 37 (1962), pp. 74–79. ISSN: 0024-6107.
15. F. Ferreira and P. Oliva. "Bounded functional interpretation". In: *Ann. Pure Appl. Logic* 135.1-3 (2005), pp. 73–112. ISSN: 0168-0072.
16. H. Furstenberg, Y. Katznelson and D. Ornstein. "The ergodic theoretical proof of Szemerédi's theorem". In: *Bull. Amer. Math. Soc. (N.S.)* 7.3 (1982), pp. 527–552. ISSN: 0273-0979.
17. H. Furstenberg. "Ergodic behavior of diagonal measures and a theorem of Szemerédi on arithmetic progressions". In: *J. Analyse Math.* 31 (1 1977). 10.1007/BF02813304, pp. 204–256. ISSN: 0021-7670.
18. P. Gerhardy and U. Kohlenbach. "General logical metatheorems for functional analysis". In: *Trans. Amer. Math. Soc.* 360.5 (2008), pp. 2615–2660. ISSN: 0002-9947.
19. K. Gödel. "Über eine bisher noch nicht benützte Erweiterung des finiten Standpunktes". In: *Dialectica* 12 (1958), pp. 280–287. ISSN: 0012-2017.
20. I. Goldbring and H. Towsner. "An approximate logic for measures". English. In: *Israel Journal of Mathematics* 199.2 (2014), pp. 867–913. ISSN: 0021-2172.
21. M.-D. Hernest. "Light functional interpretation: an optimization of Gödel's technique towards the extraction of (more) efficient programs from (classical) proofs". In: *Computer science logic*. Ed. by L. Ong. Vol. 3634. Lecture Notes in Comput. Sci. Berlin: Springer, 2005, pp. 477–492. ISBN: 978-3-540-28231-0; 3-540-28231-9.
22. D. Jackson. "Note on a class of polynomials of approximation". In: *Trans. Amer. Math. Soc.* 22.3 (1921), pp. 320–326. ISSN: 0002-9947.
23. H. J. Keisler. "Probability quantifiers". In: *Model-theoretic logics*. Ed. by J. Barwise and S. Feferman. Perspect. Math. Logic. Springer, New York, 1985, pp. 509–556. ISBN: 0-387-90936-2.
24. W. A. Kirk and C. Martínez-Yañez. "Approximate fixed points for nonexpansive mappings in uniformly convex spaces". In: *Ann. Polon. Math.* 51 (1990), pp. 189–193. ISSN: 0066-2216.
25. U. Kohlenbach. *Applied proof theory: proof interpretations and their use in mathematics*. Springer Monographs in Mathematics. Berlin: Springer-Verlag, 2008, pp. xx+532. ISBN: 978-3-540-77532-4.
26. U. Kohlenbach and L. Leuştean. "A quantitative mean ergodic theorem for uniformly convex Banach spaces". In: *Ergodic Theory Dynam. Systems* 29.6 (2009), pp. 1907–1915. ISSN: 0143-3857.
27. U. Kohlenbach and L. Leuştean. "Asymptotically nonexpansive mappings in uniformly convex hyperbolic spaces". In: *J. Eur. Math. Soc. (JEMS)* 12.1 (2010), pp. 71–92. ISSN: 1435-9855.
28. U. Kohlenbach and P. Oliva. "Proof mining: a systematic way of analyzing proofs in mathematics". In: *Tr. Mat. Inst. Steklova* 242.Mat. Logika i Algebra (2003), pp. 147–175. ISSN: 0371-9685.
29. U. Kohlenbach. "Effective bounds from ineffective proofs in analysis: an application of functional interpretation and majorization". In: *J. Symbolic Logic* 57.4 (1992), pp. 1239–1273. ISSN: 0022-4812.

30. U. Kohlenbach. "Effective moduli from ineffective uniqueness proofs. An unwinding of de la Vallée Poussin's proof for Chebycheff approximation". In: *Ann. Pure Appl. Logic* 64.1 (1993), pp. 27–94. ISSN: 0168-0072.
31. U. Kohlenbach. "New effective moduli of uniqueness and uniform a priori estimates for constants of strong unicity by logical analysis of known proofs in best approximation theory". In: *Numer. Funct. Anal. Optim.* 14.5-6 (1993), pp. 581–606. ISSN: 0163-0563.
32. U. Kohlenbach. "On the computational content of the Krasnoselski and Ishikawa fixed point theorems". In: *Computability and complexity in analysis (Swansea, 2000).* Ed. by J. Blanck, V. Brattka and P. Hertling. Vol. 2064. Lecture Notes in Comput. Sci. Selected papers from the 4th International Workshop (CCA 2000) held at the University of Wales, Swansea, September 17–19, 2000. Berlin: Springer, 2001, pp. 119–145. ISBN: 3-540-42197-1.
33. U. Kohlenbach. "Some logical metatheorems with applications in functional analysis". In: *Trans. Amer. Math. Soc.* 357.1 (2005), 89–128 (electronic). ISSN: 0002-9947.
34. U. Kohlenbach. "Theorie der majorisierbaren und stetigen Funktionale und ihre Anwen- dung bei der Extraktion von Schranken aus inkonstruktiven Beweisen: Effektive Ein- deutigkeitsmodule bei besten Approximationen aus ineffektiven Beweisen". PhD thesis. Frankfurt am Main, 1990.
35. U. Kohlenbach and B. Lambov. "Bounds on iterations of asymptotically quasi-nonexpansive mappings". In: *International Conference on Fixed Point Theory and Applications.* Ed. by J. G. Falset, E. Llorens Fuster and B. Sims. Yokohama Publ., Yokohama, 2004, pp. 143–172. ISBN: 4-946552-13-8.
36. U. Kohlenbach and P. Oliva. "Proof mining in L1-approximation". In: *Ann. Pure Appl. Logic* 121.1 (2003), pp. 1–38. ISSN: 0168-0072.
37. M. A. Krasnoselski. "Two remarks on the method of successive approximations". In: *Acad. R. P. Romî ne An. Romî no-Soviet. Ser. Mat. Fiz. (3)* 10.2(17) (1956), pp. 55–59.
38. M. A. Krasnosel′skiĭ. "Two remarks on the method of successive approximations". In: *Uspehi Mat. Nauk (N.S.)* 10.1(63) (1955), pp. 123–127. ISSN: 0042-1316.
39. B. R. Kripke and T. J. Rivlin. "Approximation in the metric of $L^1(X, \mu)$". In: *Trans. Amer. Math. Soc.* 119 (1965), pp. 101–122. ISSN: 0002-9947.
40. A. Kroó. "On the continuity of best approximations in the space of integrable functions". In: *Acta Math. Acad. Sci. Hungar.* 32.3-4 (1978), pp. 331–348. ISSN: 0001-5954.
41. B. Lambov. "Rates of convergence of recursively defined sequences". In: *Proceedings of the 6th Workshop on Computability and Complexity in Analysis (CCA 2004).* Ed. by V. Brattka, L. Staiger and K. Weihrauch. Vol. 120. Electron. Notes Theor. Comput. Sci. Held in Lutherstadt Wittenberg, August 16–20, 2004. Amsterdam: Elsevier, 2005, pp. 125–133.
42. L. Leuştean. "A quadratic rate of asymptotic regularity for CAT(0)-spaces". In: *J. Math. Anal. Appl.* 325.1 (2007), pp. 386–399. ISSN: 0022-247X.
43. L. Leuştean. "Proof mining in ℝ-trees and hyperbolic spaces". In: *Proceedings of the 13th Workshop on Logic, Language, Information and Computation*

(WoLLIC 2006). Ed. by G. Mints and R. de Queiroz. Vol. 165. Electron. Notes Theor. Comput. Sci. Held at Stanford University, Stanford, CA, July 18–21, 2006. Amsterdam: Elsevier, 2006, 95–106 (electronic).

44. V. Pták. "On approximation of continuous functions in the metric $\int_a^b |x(t)|dt$". In: *Czechoslovak Math. J.* 8(83) (1958), 267–273, supplement, 464. ISSN: 0011-4642.
45. F. Riesz. "Über lineare Funktionalgleichungen". In: *Acta Math.* 41.1 (1916), pp. 71–98. ISSN: 0001-5962.
46. T. Tao. "Norm convergence of multiple ergodic averages for commuting transformations". In: *Ergodic Theory Dynam. Systems* 28.2 (2008), pp. 657–688. ISSN: 0143-3857.
47. H. Towsner. *What do ultraproducts remember about the original structures?* draft. Apr. 2018.
48. H. Towsner. "An Inverse Ackermannian Lower Bound on the Local Unconditionality Constant of the James Space". draft. Apr. 2015.
49. H. Towsner. "Partial impredicativity in reverse mathematics". In: *J. Symbolic Logic* 78.2 (2013), pp. 459–488. ISSN: 0022-4812.
50. H. Towsner. "Towards an effective theory of absolutely continuous measures". In: *Studies in weak arithmetics. Vol. 3.* Ed. by P. Cégielski, A. Enayat and R. Kossak. Vol. 217. CSLI Lecture Notes. Papers based on the conference "Journées sur les Arithmétiques Faibles (Weak Arithmetics Days)" (JAF33) held at the University of Gothenburg, Gothenburg, June 16–18, 2014 and the conference (JAF34) held at CUNY, New York, July 7–9, 2015. CSLI Publ., Stanford, CA, 2016, pp. 171–229. ISBN: 978-1-57586-953-7; 1-57586-953-5.
51. J. L. Walsh and T. S. Motzkin. "Polynomials of best approximation on an interval". In: *Proc. Nat. Acad. Sci. U.S.A.* 45 (1959), pp. 1523–1528. ISSN: 0027-8424.
52. M. N. Walsh. "Norm convergence of nilpotent ergodic averages". In: *Ann. of Math. (2)* 175.3 (2012), pp. 1667–1688. ISSN: 0003-486X.
53. F. Wiedijk. "Formal proof-getting started". In: *Notices Amer. Math. Soc.* 55.11 (2008), pp. 1408–1417. ISSN: 0002-9920.

https://doi.org/10.1142/9789819806584_0015

Completely Determined Borel Sets and Measurability

Linda Westrick
Department of Mathematics, Penn State University
University Park, Pennsylvania USA
westrick@psu.edu

We consider the reverse math strength of the statement CD-M: "Every completely determined Borel set is measurable." Over WWKL, we obtain the following results analogous to the previously studied category case:

(1) CD-M lies strictly between ATR_0 and $\mathsf{L}_{\omega_1,\omega}$-CA.

(2) Whenever $M \subseteq 2^\omega$ is the second-order part of an ω-model of CD-M, then for every $Z \in M$, there is a $R \in M$ such that R is Δ^1_1-random relative to Z.

On the other hand, without WWKL, all sets have measure zero (as measured according to CD-M), and it follows vacuously that ¬WWKL implies CD-M over RCA_0.

1. Introduction

The notion of a *completely determined Borel set* was introduced in [ADM+19] to permit the reverse mathematics analysis of weak principles involving Borel sets. In the standard treatment of Borel sets in reverse mathematics [Sim09], a Borel set is any well-founded tree T whose leaves are labeled with clopen sets and whose interior nodes are labeled with intersections or unions. A real $X \in 2^\omega$ is then said to belong to the set coded by T if and only if there is an *evaluation map*, a function $f : T \to \{0,1\}$ such that $f(\sigma) = 1$ if and only if X is in the set coded by $T_\sigma := \{\tau : \sigma^\frown\tau \in T\}$.

The author was supported by grant DMS-1854107 from the National Science Foundation of the United States and by the Cada R. and Susan Wynn Grove Early Career Professorship in Mathematics. Part of the work was done during the author's IMS-supported visit to the Institute for Mathematical Sciences, National University of Singapore in 2019.

While *arithmetic transfinite recursion* (ATR_0) suffices to construct evaluation maps for each X, in general it is also required. As a result, most principles concerning an arbitrary Borel set reverse to ATR_0 simply because most such principles have a conclusion that presupposes an element X in the Borel set.

An exception was encountered by [DFSW17] in their analysis of the Borel dual Ramsey theorem. The hypothesis of this theorem posits ℓ-many Borel sets whose union is the entire space. In order to say the union is the entire space, the existence of evaluation maps for each X must be a part of the hypothesis. That is, an instance of the Borel dual Ramsey theorem is not well-defined unless the given Borel sets are *completely determined*, meaning that each X has an evaluation map.

This example fueled the idea that the lack of interesting reversals for weak principles involving Borel sets could be remedied by restricting attention to completely determined Borel sets. This was borne out in [ADM$^+$19], in which the following was proven about the principle $\mathsf{CD\text{-}PB}$: "Every completely determined Borel set has the property of Baire."

Theorem 1.1: ([ADM$^+$19]). *The principle* $\mathsf{CD\text{-}PB}$ *is strictly weaker than* ATR_0. *Every ω-model $\mathcal{M}$ of* $\mathsf{CD\text{-}PB}$ *is closed under hyperarithmetic reduction, and for every $Z \in M$, there is some $G \in \mathcal{M}$ that is $\Delta^1_1(Z)$-generic.*

In this chapter, we do the same for the principle "every Borel set is measurable." Similar results are obtained by similar methods. The only new twist is the need to work with an appropriate meaning of "measurable" for a Borel set; there are several candidates. This delicate task has already been undertaken by Simpson, X. Yu, Brown, Giusto and others (see for example [Sim09, Chapter X], [Yu93], [Yu94], and [BGS02]). We summarize their work and give the sometimes more detailed versions of the results needed for our application.

We then define the principle $\mathsf{CD\text{-}M}$: "Every completely determined Borel set is measurable." We show that $\mathsf{CD\text{-}M}$ follows from $\neg\mathsf{WWKL}$ (for the simple reason that $\neg\mathsf{WWKL}$ implies the Cantor space has measure 0, and thus every subset of it is also measure 0). On the other hand, working over WWKL, we obtain results similar to the category case.

In [ADM$^+$19], a model was constructed in which a Baire approximation to a given completely determined Borel set B was obtained without ATR_0 by polling $\Sigma^1_1(B)$-generics about their membership in B. We do essentially the same to construct a proof of measurability of a given completely determined set B, but using $\Pi^1_1(B)$-randoms. The result of this polling is exactly an element $f \in L^1(2^\omega)$, so no translation is required to obtain a code for a measurable set as defined in [Sim09, Chapter X]. The main results of this chapter are as follows.

Theorem 1.2: *The principle* $\mathsf{WWKL}+\mathsf{CD\text{-}M}$ *is strictly weaker than* ATR_0. *Every* ω*-model* $\mathcal{M}$ *of* $\mathsf{WWKL}+\mathsf{CD\text{-}M}$ *is closed under hyperarithmetic reduction, and for every* $Z \in \mathcal{M}$*, there is some* $R \in \mathcal{M}$ *that is* $\Delta^1_1(Z)$*-random.*

These results were first presented by the author at the Institute for Mathematical Sciences workshop Higher Recursion Theory and Set Theory in 2019, using a version of Proposition 4.5 to quickly move the base theory to ACA_0, and using an ad hoc notion of a "function measuring a set" which was later found to essentially coincide with the notion of a measurable characteristic function previously proposed by Simpson and several of his collaborators. The author would like to thank Steve Simpson for his suggestion to lower the base theory and for bringing that connection to light. Thanks go also to the anonymous referee who provided further helpful suggestions. Finally, the author would like to thank Ted Slaman, her PhD advisor, for his support and mentorship, his good humor and sound principles, and his excellent body of research which this volume celebrates.

2. Notation and Preliminaries

We use the notation and conventions of [ADM+19]. In that paper, much more background and context can be found in the introduction. The eth Turing functional is denoted Φ_e. Elements of $\omega^{<\omega}$ are denoted by σ, τ and elements of $2^{<\omega}$ by p, q. We write $\sigma \preceq \tau$ to indicate that σ is an initial segment of τ, with $\prec$ if $\sigma \neq \tau$. For $p \in 2^{<\omega}$, the notation $[p]$ refers to the cylinder $\{X \in 2^\omega : p \prec X\}$. The empty string is denoted by λ. A string with a single component of value $n \in \omega$ is denoted by $\langle n \rangle$. String concatenation is denoted by $\sigma\tau$. Usually we write σn instead of the more technically correct but uglier $\sigma\langle n \rangle$.

If U is a set of strings (for example, a tree, or a coded open subset of 2^ω), and σ is any string, we write $\sigma^\frown U$ to mean $\{\sigma\tau : \tau \in U\}$. If T is a tree and $\sigma \in T$, we write T_σ to mean $\{\tau : \sigma\tau \in T\}$, and if $\langle n \rangle \in T$, we write T_n to mean $\{\tau : n\tau \in T\}$.

We assume familiarity with reverse mathematics, in particular the systems RCA_0, WWKL, ACA_0 and ATR_0. We note that effective transfinite recursion and arithmetic transfinite induction can be carried out in ACA_0. We identify an ω-model $\mathcal{M}$ of second order arithmetic with its second-order part, writing $X \in \mathcal{M}$ to mean that X is an element of the second-order part of $\mathcal{M}$.

We assume familiarity with ordinal notations and pseudo-ordinals. Kleene's O is denoted by $\mathcal{O}$. The relation $<_*$ is the transitive closure of the relation defined by $1 <_* x$ if $x \neq 1$, $x <_* 2^x$, and $\Phi_e(n) <_* 3 \cdot 5^e$. We will not distinguish between ordinals and their notations. Additionally, if $b \in \mathcal{O}$, we write $b+1$ for the successor of b (rather than the more technically correct but cumbersome 2^b) and $b + O(1)$ for the outcome of taking some fixed constant number of successors of b. If $b \in \mathcal{O}$ the unique jump hierarchy on b is denoted H_b. All these concepts can be relativized to an oracle Z. Kleene's $\mathcal{O}$ also has a Σ^1_1 superset $\mathcal{O}^*$, defined as the intersection of all $X \in HYP$ such that $1 \in X$, $a \in X \implies 2^a \in X$, and

$$\forall n[\Phi_e(n) \in X \text{ and } \Phi_e(n) <_* \Phi_e(n+1)] \implies 3 \cdot 5^e \in X.$$

Observe also that $\mathcal{O}$ is contained in $\mathcal{O}^*$. The elements of $\mathcal{O}^* \setminus \mathcal{O}$ are called pseudo-ordinals. For more details, see the introduction of [ADM$^+$19].

A $T \subseteq \omega^{<\omega}$ is well-founded if it has no infinite path. If T is any tree, and $\rho : T \to \mathcal{O}^*$, we say that ρ *ranks* T if for all σ and n such that $\sigma^\frown n \in T$, we have $\rho(\sigma^\frown n) <_* \rho(\sigma)$, and for each leaf $\sigma \in T$, $\rho(\sigma) = 1$. If T is ranked by ρ and $\rho(\lambda) = a$, we say that T is *a-ranked* by ρ. If $a \in \mathcal{O}$ and T is a-ranked then T is well-founded, but it is possible and useful for an ill-founded tree to be ranked by a pseudo-ordinal. A tree T is *alternating* if whenever $\sigma \in T$ is a $\bigcap$, then each $\sigma n \in T$ is either a $\bigcup$ or a leaf, and similarly if $\sigma \in T$ is a $\bigcup$, then each $\sigma n \in T$ is either a $\bigcap$ or a leaf.

A labeled *Borel code* is a well-founded tree $T \subseteq \omega^{<\omega}$ whose leaves are labeled by basic open sets or their complements, and whose inner nodes are labeled by $\bigcup$ or $\bigcap$. The Borel set associated to a Borel code is defined by induction, interpreting the labels in the obvious way. Any Borel set can be represented this way, by applying DeMorgan's laws to push complementation out to the leaves. A formula of $L_{\omega_1,\omega}$ is a well-founded tree whose interior nodes are labeled with $\bigwedge$ (conjunction) and $\bigvee$ (disjunction) and whose leaves are labeled with the symbols `true` or `false`.

There is a computable procedure which, for any $b \in \mathcal{O}$ and any $n \in \omega$, outputs a $b + O(1)$-ranked alternating formula of $L_{\omega_1,\omega}$ which holds true if and only if $n \in H_b$.

If T is a labeled Borel code and $X \in 2^\omega$, an *evaluation map* for $X \in T$ is a function $f : T \to \{0, 1\}$ such that

- If σ is a leaf, $f(\sigma) = 1$ if and only if X is in the clopen set coded by $\ell(\sigma)$.

- If σ is a union node, $f(\sigma) = 1$ if and only if $f(\sigma^\frown n) = 1$ for some $n \in \omega$.
- If σ is an intersection node, $f(\sigma) = 1$ if and only if $f(\sigma^\frown n) = 1$ for all $n \in \omega$.

We say that X is in the set coded by T, denoted $X \in |T|$, if there is an evaluation map f for X in T such that $f(\lambda) = 1$. Note that $X \in |T|$ is a Σ^1_1 statement. In ACA_0, evaluation maps are unique when they exist. If T is ill-founded, the notation $|T|$ may not have meaning outside of a given model. If T is a truly well-founded Borel code, we do use $|T|$ outside of the context of a model to denote the elements of the set that T codes.

A Borel code T is *completely determined* if every $X \in 2^\omega$ has an evaluation map in T. A formula ϕ of $L_{\omega_1,\omega}$ is *completely determined* if there is map $f : \phi \to \{\mathtt{true}, \mathtt{false}\}$ that agrees with ϕ on the leaves and satisfies the logic of ϕ at interior nodes. The principle $\mathsf{L}_{\omega_1,\omega}$-$\mathsf{CA}$ states that whenever $\langle \phi_n \rangle_{n\in\omega}$ is a sequence of completely determined formulas of $L_{\omega_1,\omega}$, then $\{n : \phi_n \text{ is true}\}$ exists.

We assume familiarity with higher randomness. The key theorems we need are:

Theorem 2.1: ([Ste73, Ste75]). *A real $R \in 2^\omega$ is Π^1_1-random if and only if it is Δ^1_1-random and $\omega_1^R = \omega_1^{ck}$.*

Theorem 2.2: ([HN07]). *For $R_0, R_1 \in 2^\omega$, we have $R_0 \oplus R_1$ is Π^1_1-random if and only if R_0 and R_1 are relatively Π^1_1-random.*

Theorem 2.3: ([CNY08]). *If $R_0 \oplus R_1$ is Π^1_1-random, then $\Delta^1_1(R_0) \cap \Delta^1_1(R_1) = \Delta^1_1$.*

3. Measure Theory in Reverse Mathematics

Historically, measure theory developed as a third-order theory. Classically, a measure is a set function from a σ-algebra of subsets of a space to the non-negative reals. Therefore, although much of measure theory can be developed within second-order arithmetic, this development has required some care and some non-trivial choices. We now summarize work of Simpson, X. Yu, Brown, and Giusto [Yu90, YS90, Yu93, Yu94, BGS02, Sim09], in which this development took place.

In the context of second-order arithmetic, all the relevant information about a measure space $(X, \mu, \mathcal{S})$ is already contained in the values that μ takes on an algebra which generates $\mathcal{S}$ as a σ-algebra. When X is a separable complete metric space space and $\mathcal{S}$ is the Borel sets, a countable generating algebra is naturally obtained by taking all finite Boolean combinations of basic open sets. In the case of Cantor space 2^ω, this approach works out very cleanly because the basic open sets (and thus all elements of the generating algebra) are clopen. However, for an arbitrary separable complete metric space, a problem arises. What if there is an atom on the boundary of a basic open set U? Is it fair to ask that our encoding of a measure μ be able to precisely compute $\mu(U)$ and $\mu(U^c)$? (Because a typical open set V can only be represented as an infinite enumeration of its basic open subsets, its measure $\mu(V)$ would be at best c.e., not computable, in a description of μ and V.) Another way of asking the same question is: for the purposes of constructive mathematics, what is a suitable topology to put on the space of Borel measures on X?

When X is Cantor space, a popular representation choice has been to name a measure μ with a function from $2^{<\omega}$ to $\mathbb{R}$ which records the measure of each basic clopen set (see for example [DM13]). This representation induces the so-called weak topology on the space of probability measures on X (see for example [Bog07, Definition 8.2.1]). This is the same topology induced by the Prohorov metric (see for example [Bog07, Theorem 8.3.2]), and also coincides with the weak-$*$ topology on $C(X)^*$ (see the discussion following Definition 8.2.1 in [Bog07]). Restricting attention to probability measures on compact complete separable metric spaces, Yu also settled on the same topology in [Yu93], and made the following definition.

Definition 3.1: Let X be a compact complete separable metric space. A *Borel probability measure* μ on X is a bounded positive linear functional $\mu : C(X) \to \mathbb{R}$ with $\mu(1) = 1$.

Here $C(X)$ denotes the Banach space of continuous real-valued functions on X with the supremum norm, and $1 \in C(X)$ denotes the constant function. Care is required in the definition of $C(X)$. It is not simply the collection of continuous function on X equipped with the supremum norm, because in weak subsystems of second-order arithmetic, a continuous function on a compact space X need not have a supremum. Instead, $C(X)$ is defined as a complete separable metric space by choosing a particularly well-behaved collection of continuous functions to be the dense subset. The details are given in [Sim09, Exercise 4.2.13], in which it is also established that $C(X)$ consists of precisely those continuous functions from X to $\mathbb{R}$ which also possess a modulus of uniform continuity. Therefore, while a measure μ on X is defined by specifying how to integrate elements of $C(X)$

with respect to μ, it does not follow that every continuous function on X is μ-integrable; only those with a modulus of uniform continuity come with this guarantee.

An unavoidable drawback to Definition 3.1 is that it puts a small distance between the definition of a measure and its basic function of assigning sizes to sets. Therefore, it is necessary to make a further definition for "the measure of an open set" (and subsequently a further definition for the measure of an arithmetic set, etc. leading up to the notion of a measurable set). At each point of definition, a choice arises: should the measure assignment be *intensional* (depending only on the *description* of the set in question) or *extensional* (depending on only on the *membership* of the set in question)?

To understand the tension here, consider that if U is any component of a universal Martin-Löf test in Cantor space with its usual fair-coin measure, then statement $U = 2^\omega$ holds in REC. Thus in REC, we cannot simultaneously have both of these two desirable properties:

(1) If $S \subseteq 2^{<\omega}$ is prefix-free, then $\mu\left(\bigcup_{\sigma\in S}[\sigma]\right) = \sum_{\sigma\in S} 2^{-|\sigma|}$.
(2) If $A = B$ then $\mu(A) = \mu(B)$.

Note that the first is an intensional property and the second is an extensional property. Although both are clearly wanted, the second seems more essential. Thus the extensional definition for the measure of an open set is the one which appears in [Sim09].

Definition 3.2: (RCA_0). Let μ be a Borel probability measure on X. Let U be an open subset of X. The *μ-measure* of U is defined as

$$\mu(U) = \sup\{\mu(f) : f \in C(X), 0 \leq f \leq 1, f(x) = 0 \text{ for } x \in X \setminus U\}.$$

In the absence of ACA_0, this supremum may not exist as a number, but statements about $\mu(U)$ may still be made in weaker systems by simply substituting the above definition of $\mu(U)$ in any sentence which makes a claim about this quantity. For example, it holds in RCA_0 that $U \subseteq V$ implies that $\mu(U) \leq \mu(V)$. Such statements are said to hold in a "virtual" or "comparative" sense.

Observe that this extensional definition also gives the "right" values on Cantor space with the fair coin measure when U is a finite union of non-intersecting cylinders $U = \cup_{i<n}[p_i]$. That is, $\mu(\cup_{i<n}[p_i]) = \sum_{i<n} 2^{-|p_i|}$.

On the other hand, in RCA_0 we can always assume that open subsets of Cantor space are given by prefix-free enumerations of elements of $2^{<\omega}$, so we can also give the following intensional definition of measure of an open set in Cantor space:

Definition 3.3: (RCA_0). If U is an open subset of 2^ω given by $U = \bigcup_{i<\omega}[p_i]$, where each $p_i \in 2^{<\omega}$ and where $\{p_i : i \in \omega\}$ is prefix-free, then define the *intensional measure* of U by $\mu_I(U) = \sum_i 2^{-|p_i|}$.

The intensional and extensional definitions fully coincide under WWKL.

Theorem 3.4: ([YS90]; see also [BGS02]). *Over* RCA_0, WWKL *is equivalent to the statement that for every compact separable metric space X and every measure μ on X, μ is countably additive. That is, for every sequence of open sets U_n,*

$$\lim_N \mu(\cup_{n<N} U_n) = \mu(\bigcup_n U_n).$$

Corollary 3.5: (WWKL). *For all open sets* $U \subseteq 2^\omega$, $\mu(U) = \mu_I(U)$.

One final intensional notion of a measurable set is needed for the development of measure theory.

Definition 3.6: A *rapidly null G_δ set* is a G_δ set $\bigcap_n U_n$ such that for each n, $\mu_I(U_n) < 2^{-n}$.

Note: a Martin-Löf test is just a computably presented rapidly null G_δ set.

Theorem 3.7: ([ADR12]). *Over* RCA_0, WWKL *is equivalent to the statement that if A is a rapidly null G_δ subset of 2^ω, then $A \neq 2^\omega$.*

Thus in WWKL, a μ-measurable set may be non-vacuously defined as follows. Let $\mu : C(X) \to \mathbb{R}$ be a positive Borel probability measure. Let $L^1(X,\mu)$ denote the completion of $C(X)$ with respect to the L^1 norm defined by $||f-g||_1 = \int |f-g|$. Recall that a sequence $\langle x_n \rangle$ of points of a metric space is called *rapidly Cauchy* if for all n, we have $d(x_n, x_{n+1}) < 2^{-n}$. Each element of $L^1(X,\mu)$ is represented by many *names*, where a name is a sequence $\langle f_n \rangle_{n\in\omega}$ of functions from $C(X)$ that is rapidly Cauchy for the L^1 norm.

Definition 3.8: ([BGS02]). A *measurable characteristic function* is a function $f \in L^1(X,\mu)$ such that $f(x) \in \{0,1\}$ for all x outside a rapidly null G_δ set. A set E is *measurable* if there is some $f \in L^1(X,\mu)$ such that $f = \chi_E$ outside a rapidly null G_δ set.

Here χ_E denotes the characteristic function of E. The measure of E is then defined as $\mu(E) = \mu(f)$, where $f = \chi_E$ almost everywhere as above.

This is well-defined and locally well-behaved by the following results of X. Yu [Yu94].

Theorem 3.9: (WWKL). *For $f, f' \in L^1(X, \mu)$, $||f - f'||_1 = 0$ if and only if $f = f'$ outside of a rapidly null G_δ set. If $f \leq f'$ outside of a rapidly null G_δ set, then $\mu(f) \leq \mu(f')$.*

For the rest of this paragraph, WWKL is assumed. Observe now that if $U \subseteq 2^\omega$ is open and if U is measurable in the above sense (that is, $\chi_U \in L^1(X, \mu)$), then we have $\mu(U) = \mu_I(U) = \mu(\chi_U)$. The last equality follows because if $U = \bigcup_{i<\omega}[p_i]$, the functions $\chi_{\bigcup_{i<n}[p_i]}$ are continuous and converge to χ_U in the L^1 norm. Finally, if A is a rapidly null G_δ set, then $\mu(\chi_A) = \mu_I(A) = 0$ because $\chi_A = 0$ outside of A itself. Therefore, when measurable characteristic functions for open or rapidly null G_δ sets exist, all our ways of defining measures for these sets coincide. The existence of a measurable characteristic function for an open set also guarantees that the measure of that open set exists in the model (and thus can be discussed directly, not just comparatively).

Finally, we will need to make use of some more explicit versions of known results from the literature. For example, we want to use Theorem 3.9, but as stated it does not give any bounds on the complexity of the rapidly null G_δ set. However, those bounds do exist and we need the uniformity that comes with them. So below we reprove several results in order to clarify the complexity of the null set of points that are being discarded. From here forward, we also restrict our attention to Cantor space with the fair coin measure, which is denoted by λ.

First, recall that if A_n is a sequence of rapidly null G_δ sets $A_n = \bigcap_i A_{n,i}$, the same trick used for producing a universal Martin-Löf test can also produce a rapidly null G_δ set $A \supseteq \bigcup_n A_n$. Just let $U_j = \bigcup_n A_{n,n+j+1}$, and let $A = \bigcap_j U_j$.

Much but not all of the rest of this section has been presented in [BGS02].

Proposition 3.10: (WWKL). *Suppose that $\langle f_i \rangle$ is a sequence of ideal continuous functions of $C(X)$ which is rapidly Cauchy for the L^1 norm. Let*

$$A_n = \left\{ x : \exists N \sum_{i=2n+1}^{N} |f_i(x) - f_{i+1}(x)| > 2^{-n} \right\}.$$

Then $\mu(A_n) \leq 2^{-n}$.

Proof: Formally, A_n is a union of basic open sets $\bigcup_j [p_j]$ satisfying the condition. We can assume the $[p_j]$ are disjoint. By countable additivity, it suffices to show that $\mu(B) < 2^{-n}$ for all sets $B = \cup_{j<k}[p_j]$. Let N be large enough to witness that $[p_j] \subseteq A_n$ for all $j < k$. We have

$$2^{-n}\mu(B) = \int 2^{-n}\chi_B \leq \int \sum_{i=2n+1}^{N} |f_i - f_{i+1}| = \sum_{i=2n+1}^{N} \int |f_i - f_{i+1}| < 2^{-2n}.$$

Thus $\mu(B) < 2^{-n}$, as needed. □

The corollaries use ACA_0 only to guarantee that a Cauchy sequence converges.

Corollary 3.11: (ACA_0). *A name $\langle f_i \rangle$ for an element of $L^1(2^\omega)$ converges pointwise a.e. Furthermore, this pointwise convergence is achieved outside of the rapidly null G_δ set*

$$\bigcap_k \bigcup_{n>k} A_n$$

where A_n are defined as above.

Corollary 3.12: (WWKL). *A name $\langle f_i \rangle$ for an element of $L^1(2^\omega)$ converges uniformly on each closed set*

$$B_k = 2^\omega \setminus \bigcup_{n \geq k} A_n,$$

where A_n are defined as above. Furthermore, the modulus of uniform convergence of f_i on B_k is primitive recursive: if $m > 2\max\{\ell, k\}$, then $|f_m(x) - f(x)| \leq 2^{-\ell}$.

Proof: Let $n = \max\{\ell, k\}$ and $x \in B_k$. Then $B_k \cap A_n = \emptyset$, and thus the series $f_m(x) + \sum_{i=m}^{\infty}(f_{i+1}(x) - f_i(x))$ converges absolutely, with

$$\sum_{i=m}^{\infty} |f_{i+1}(x) - f_i(x)| \leq \sum_{i=2n+1}^{\infty} |f_{i+1}(x) - f_i(x)| \leq 2^{-n} \leq 2^{-\ell}.$$

□

Corollary 3.13: (ACA_0). *If $\langle f_i \rangle$ and $\langle g_i \rangle$ are two names for the same element of $L^1(2^\omega)$, then*

$$\lim_i f_i(x) = \lim_i g_i(x)$$

for almost all x. Furthermore, this pointwise convergence is achieved outside of a rapidly null G_δ set given by an explicit formula.

Proof: Let $A_n(f), A_n(g)$, and $A_n(f,g)$ be defined as in Proposition 3.10 applied to the rapidly Cauchy sequences $\langle f_i\rangle$, $\langle g_i\rangle$, and $\langle f_2, g_3, f_4, g_5, \dots\rangle$ respectively. Then the limits of $f_i(x)$ and $g_i(x)$ exist and agree for any x outside of three rapidly null G_δ sets. Combine these rapidly null G_δ sets into a single rapidly null G_δ set. □

Proposition 3.14: (ACA_0). *If $\langle h_j\rangle$ is a sequence of functions of $L^1(2^\omega)$ rapidly converging to a function $g \in L^1(2^\omega)$, then*

$$\lim_{j\to\infty} h_j(x) = g(x)$$

for almost all x. Furthermore, this pointwise convergence is achieved outside of a rapidly null G_δ set given by an explicit formula.

Proof: Define $\langle f^i\rangle_{i\in\omega}$ by $f^i = h_i^{2i+1}$, where $\langle h_j^i\rangle_{i<\omega}$ is the given name for h_j. Then $\langle f^{i+2}\rangle_{i\in\omega}$ is rapidly Cauchy and is another name for g, which we can see because

$$\int |f^i - f^{i+1}| \le \int |f^i - h_i| + \int |h_i - h_{i+1}| + \int |h_{i+1} - f^{i+1}| \le 2^{-2i} + 2^{-i} + 2^{-2i}$$

and

$$\int |f^i - g| \le \int |f^i - h_i| + \int |h_i - g| \le 2^{-2i} + 2^{-i+1}.$$

Let $A_n(g)$ and $A_n(h_j)$ be the building blocks of infinitely many rapidly null G_δ sets as in Corollary 3.11, so that outside of these sets the notations $g(x)$ and $h_j(x)$ are well-defined as the pointwise limits of the given names for g and each h_j. Additionally, letting

$$C_k = \bigcup_{\substack{j>k\\ n>j}} A_n(h_j),$$

by Proposition 3.10, we have $\lambda(\bigcup_{n>j} A_n(h_j)) < 2^{-j}$ and thus $\lambda(C_k) < 2^{-k}$ and $\bigcap_k C_k$ is a rapidly null G_δ set. Combine into a single test

(1) the infinitely many rapidly null G_δ sets which result from applying Corollary 3.11 to the given names for g and each h_j

(2) the rapidly null G_δ set guaranteed by Corollary 3.13, so that for x outside of B, $\lim_i f^i(x) = g(x)$.

(3) $\bigcap_k C_k$.

By (1), if x avoids this test, then $h_j(x)$ and $g(x)$ are well-defined as the pointwise limit of the given names of g and h_j. By (2), if x avoids this test, then $\lim_i f^i(x) = g(x)$. Finally, we claim that if x avoids this test, then $\lim_j h_j(x) = \lim_i f^i(x)$. The limit on the right hand side exists, so it suffices to show that $\lim_j |f^j(x) - h_j(x)| = 0$. This follows by (3) because if $x \notin C_k$ for some k, then for all $j > k$ we have

$$|f^j(x) - h_j(x)| \leq \sum_{i=2j+1}^{\infty} |h_j^i(x) - h_j^{i+1}(x)| \leq 2^{-j}.$$

□

We have the following relationship between higher randomness and measure theory. This is surely known (and one could surely do better than Δ^1_1-random) but it is enough for our purposes.

Lemma 3.15: *Suppose that $f \in L^1(2^\omega)$, with name $\langle f^i \rangle_{i<\omega}$. Suppose that R is Δ^1_1-random relative to $\langle f^i \rangle_{i<\omega}$. Then*

$$\lim_{N\to\infty} \frac{1}{N} \sum_{j<N} f(R^{[j]}) = \int_{2^\omega} f.$$

Proof: Note that the randomness of R ensures that $f(R^{[j]})$ is well-defined as $\lim_i f^i(R^{[j]})$. For any ε, we can find a measurable function $f_\varepsilon = \sum_{k=-\infty}^{\infty} k\varepsilon\chi_{A_k}$ where A_k are measurable sets which have Borel definitions uniformly in the name $\langle f^i \rangle$, and such that $|f(x) - f_\varepsilon(x)| < \varepsilon$ for all x outside of a G_δ set which also has a Borel definition relative to $\langle f^i \rangle$. Then the randomness of R ensures that the $R^{[j]}$ visit each A_k with the right limiting frequency, and that $|f(R^{[j]}) - k\varepsilon| < \varepsilon$ whenever $R^{[j]} \in A_k$. Thus $\frac{1}{N}\sum_{j<N} f(R^{[j]})$ is within ε of $\frac{1}{N}\sum_{j<N} f_\varepsilon(R^{[j]})$, and the latter tends to $\int_{2^\omega} f_\varepsilon$ as N increases. Letting ε go to zero completes the proof. □

4. Regularity Approximations and Measure Approximations

The following version of measurability for a set was implicit in [Yu93].

Definition 4.1: A set B is *regularity-measurable* if there are G_δ sets A and C such that $A^c \subseteq B \subseteq C$ and $A \cap C$ is rapidly null.

We bring up this definition because such a pair (A, C), which we could call a *regularity approximation* to B, would seem an obvious analog to the *Baire approximation* to a set B defined in [ADM$^+$19]. We can use this notion of measurability to define the principle CD-M as follows.

Definition 4.2: Let CD-M be the principle "Every completely determined Borel set is regularity-measurable".

A difference between measure and category now arises. The Baire Category Theorem holds in RCA_0, so RCA_0 knows that the whole space is not meager. However, WWKL is needed in order to know that the whole space is not null.

Proposition 4.3: *Over* RCA_0, $\neg\mathsf{WWKL}$ *implies* CD-M.

Proof: By Theorem 3.7, let A be an rapidly null G_δ set with empty complement. Let $C = 2^\omega$. Then for any set B, we have $A^c = \emptyset \subseteq B \subseteq C$, but $A \cap C$ is rapidly null because A is rapidly null. □

In the presence of WWKL, however, regularity-measurable coincides with the same notion of measurability given in Definition 3.8.

Proposition 4.4: (WWKL). *Let $B \subseteq 2^\omega$ be any set. (Formally, the membership of B can be given by any formula in the language of second order arithmetic). Then B is regularity-measurable if and only if it is measurable in the sense of Definition 3.8.*

Proof: Suppose B is regularity-measurable. It follows that $A \cup C = 2^\omega$. Therefore, if $A = \bigcap_n A_n$ and $C = \bigcap_n C_n$, we have for each n that $A_n \cup C_n = 2^\omega$. Using WWKL, it follows that $\mu(A_n \cup C_n) = 1$, while $\mu(A_n \cap C_n) < 2^{-n}$ because $A \cap C$ is rapidly null. Define a sequence of functions $f_n : 2^\omega \to \{0, 1\}$ and open sets B_n as follows. Given n, let s be large enough that $\mu(D_{n+1,s}) < 2^{-(n+1)}$, where we define

$$D_{n,s} = 2^\omega \setminus (A_{n,s} \cup C_{n,s}).$$

Let f_n be the characteristic function of $C_{n+1,s}$, and let

$$B_n = (A_{n+1} \cap C_{n+1}) \cup D_{n+1,s}.$$

Then $\mu(B_n) < 2^{-n}$. We have

$$||f_n - f_m||_1 = \mu(A_{n+1,s} \Delta A_{m+1,t})$$

where s and t are chosen as in the definition. Since $A_{n+1,s} \Delta A_{m+1,t} \subseteq B_n \cup B_m$, the sequence $\langle f_n \rangle$ is rapidly Cauchy and $f_n(x)$ converges to $\chi_B(x)$ for all x outside of $\bigcap_n (\bigcup_{k>n} B_k)$.

On the other hand, if B is measurable in the sense of Definition 3.8, then if $\langle f_n \rangle_{n\in\omega}$ is an L^1-name for χ_B, the sets $A_n = \{x : f_n(x) < 2/3\}$ and $C_n = \{x : f_n(x) > 1/3\}$ demonstrate that B is regularity-measurable. This follows because, letting $D = A_n \cap C_n$, we have

$$\frac{1}{3}\mu(D) = \int_D \frac{1}{3} \leq \int_D |f_n - \chi_B| \leq ||f_n - \chi_B||_1 \leq 2^{-n+1}.$$ □

The first step in evaluating the strength of CD-M+WWKL is immediate.

Proposition 4.5: *Over* WWKL, *the statement "Every open subset of* 2^ω *is measurable" is equivalent to* ACA_0.

Proof: It is clear that ACA_0 proves the given statement. In the other direction, given an increasing sequence of real numbers $\langle a_n \rangle$ with each $a_n < 1$, let U be an open set designed so that $\mu_I(U) = \sup_n a_n$. For example, let U be the set which contains exactly those cylinders $[p^\frown 0]$ such that for some n, we have $.p^\frown 1 < a_n$, where $.p^\frown 1$ denotes the rational number with binary decimal expansion given by $p^\frown 1$. By WWKL, $\mu_I(U) = \mu(U)$. But $\mu(U)$ exists as a number, thus $\sup_n a_n$ exists. □

Combining Propositions 4.3 and 4.5, we arrive at the following curiosity. Let OSM be the statement "Every open set is regularity-measurable". Then by Proposition 4.5, we have that ACA_0 is equivalent to WWKL+OSM, while Proposition 4.3 shows that RCA_0 proves WWKL $\vee$ OSM (here $\vee$ denotes a disjunction of two principles, not a a join operator on those principles). Thus we have a diamond formed of reasonably natural principles, though it must be admitted that OSM does not mean much outside of WWKL. We are not aware of any other diamond in reverse mathematics. By a diamond here we just mean informally an incomparable pair of principles A and B such that $A + B$ is equivalent to some principle of interest, while $A \vee B$ follows from RCA_0.

We return now to our main discussion of the principle CD-M. One direction of Proposition 4.5 can be extended to the Borel case as follows.

Proposition 4.6: *Over* WWKL, CD-M *implies* $\mathsf{L}_{\omega_1,\omega}$-CA.

Proof: Given $\langle \phi_n \rangle_{n\in\omega}$ a sequence of completely determined formulas of $L_{\omega_1,\omega}$, turn them into Borel codes by change $\bigcap$ to $\bigwedge$, $\bigcup$ to $\bigvee$, and changing their leaves as follows. If ϕ_n has `true` at a leaf, replace it with $[0^n 1]$. If ϕ_n has `false` at a leaf, replace it with $\emptyset$. Now take the union of all of these codes. The resulting code is completely determined because each ϕ_n was completely determined and each $X \in 2^\omega$ belongs to at most one cylinder $[0^n 1]$. If f is a measurable characteristic function, then f is almost surely 1 on $[0^n 1]$ whenever ϕ_n is true, and almost surely 0 on $[0^n 1]$ whenever ϕ_n is false. Thus the sequence $\langle 2^n \int_{[0^n 1]} f \rangle_{n\in\omega}$ witnesses the satisfaction of $\mathsf{L}_{\omega_1,\omega}$-CA; this sequence assigns 1 to the true formulas and 0 to the false ones. □

The classical way of showing that every Borel set is measurable is to use arithmetic transfinite recursion to define a regularity approximation to $|T_\sigma|$ for each $\sigma \in T$. We present an effectivization of the classical proof which is particularly well-suited to our subsequent analysis.

Definition 4.7: Let T be a code for a Borel set. A *measure decomposition* for T is a collection $\langle f_\sigma : \sigma \in T\rangle$, where each $f_\sigma \in L^1(2^\omega)$, such that

(1) If σ is a leaf, then f_σ is the characteristic function of $|T_\sigma|$.
(2) If σ is a union, then $f_\sigma = \sup_n f_{\sigma n}$.
(3) If σ is an intersection, then $f_\sigma = \inf_n f_{\sigma n}$.

All three equalities above refer to equality in the sense of the metric space $L^1(2^\omega)$. For example, the equation $f_\sigma = \sup_n f_{\sigma n}$ is shorthand for

$$\lim_{N\to\infty}\left(\sup_{n<N} f_{\sigma n}\right) = f_\sigma$$

and similarly for the other equation. In all cases, n ranges only over those numbers for which $\sigma n \in T$.

Proposition 4.8: (ACA_0). *Suppose T is a code for a completely determined Borel set. If T has a measure decomposition, then $|T|$ is measurable.*

Proof: We need to show that $f_\emptyset$ is a.e. equal to the characteristic function of $|T|$. This is proved by arithmetic transfinite induction on T.

Observe that if we were willing to use Σ^1_2 transfinite induction and Σ^1_1-AC, the proof which inducts on the following statement would be very short: there is a rapidly null G_δ such that for all X outside of it, $f_\sigma(X) = 1$ if and only if $X \in |T_\sigma|$. Since we want to get away with arithmetic transfinite induction only, we need to identify the rapidly null G_δ in advance, then fix some X outside it, and then prove $f_\emptyset(X)$ is correct by transfinite induction on T.

We claim the following collection of rapidly null G_δ sets exists:

(1) For all σ, a rapidly null G_δ such that for all x outside of it, the name of f converges at x.
(2) For all leaf σ, a rapidly null G_δ set such that on its complement, f_σ is the characteristic function of $|T_\sigma|$.
(3) For all union σ, a rapidly null G_δ set such that for all x in its complement, $f_\sigma(x) = \sup_n f_{\sigma n}(x)$.
(4) For all intersection σ, same as the above except using $\inf_n f_{\sigma n}$.

The sets in (1) are obtained by uniform application of Corollary 3.11 to the given names for the functions f_σ. The sets in (2) are obtained by uniform application of Corollary 3.13 to f_σ and a standard name for the characteristic function of the clopen set $|T_\sigma|$. To obtain (3), use the fact that

$$\lim_{N\to\infty}\left(\sup_{n<N} f_{\sigma n}\right) = f_\sigma,$$

define $h_N = \sup_{n<N} f_{\sigma n}$, and find a sequence N_i such that $\langle h_{N_i}\rangle_{i\in\omega}$ is rapidly convergent to f_σ. Then apply Proposition 3.14 to $\langle h_{N_i}\rangle_{i\in\omega}$ together with the given name for f_σ. Although we have passed to a subsequence, because $h_N(x) \leq h_{N+1}(x)$ for all x, it follows that $h_N(x)$ converges if and only if $h_{N_i}(x)$ converges. (It will happen in our situation that $h_N(x)$ converges for all x, though we do not need this.) The procedure for (4) is similar.

Let A be a rapidly null G_δ set which contains all the bad-behavior sets above. Fix $X \notin A$. We claim that the map which sends σ to $f_\sigma(X)$ is an evaluation map for X in T. That is, we claim $f_\sigma(X) = 1$ if and only if $X \in |T_\sigma|$. The claim is proved by arithmetic transfinite induction on T. Observe that A contains all the points at which the proposed evaluation map fails to be right at the leaves or fails to satisfy the logic of the tree.

In particular, $f_\emptyset(X) = 1$ if and only if $X \in |T|$. □

Uniformly arithmetic in a sequence $\langle f_{\sigma n}\rangle_{n\in\omega}$, we may produce the functions $\sup_n f_{\sigma n}$ and $\inf_n f_{\sigma n}$. Therefore, ATR_0 suffices to create measure decompositions for all Borel sets. However, ACA_0 is enough to guarantee their uniqueness.

Proposition 4.9: (ACA_0). *Suppose that T is a Borel code and $\langle f_\sigma\rangle_{\sigma\in T}$ and $\langle g_\sigma\rangle_{\sigma\in T}$ are two measure decompositions for T. Then for all $\sigma \in T$, $f_\sigma = g_\sigma$ as L^1 functions.*

Proof: By arithmetic transfinite induction. If for all n, $f_{\sigma n} = g_{\sigma n}$, then for all N, $\sup_{n<N} f_{\sigma n} = \sup_{n<N} g_{\sigma n}$. Therefore, these sequences have the same limit in the sense of L^1. □

Although we will show in the next section that $\mathsf{WWKL} + \mathsf{CD\text{-}M}$ is strictly weaker than ATR_0, the existence of measure decompositions is still necessary for $\mathsf{WWKL} + \mathsf{CD\text{-}M}$ to hold. Therefore, any model of $\mathsf{WWKL} + \mathsf{CD\text{-}M} + \neg\mathsf{ATR}_0$ will need some other way of producing measure decompositions.

Proposition 4.10: (ACA_0). *If* CD-M *holds, then every completely determined Borel set has a measure decomposition.*

Proof: For any Borel code S, define an operation $S[n]$ as follows. Whenever a leaf of S is labeled by the clopen set $[p_0] \cup \cdots \cup [p_k]$, replace it with the clopen set $[0^n 1 p_0] \cup \cdots \cup [0^n 1 p_k]$. This has the effect of shrinking the set coded by S and relocating it to live completely inside the cone $[0^n 1]$.

Let $h : \omega \to T$ be a computable surjection. If T is completely determined, so is $\tilde{T}$, where

$$\tilde{T} = \bigcup_{n \in \omega} T_{h(n)}[n].$$

Colloquially, $\tilde{T}$ has been formed by taking each subtree T_σ of T and giving it its own dedicated part of the Cantor space. Now, if $\tilde{T}$ is measurable via the function $f \in L^1$, then the functions

$$f_\sigma(X) = f(0^{\min h^{-1}(\sigma)} 1^\frown X)$$

are a measure decomposition for T. $\square$

5. Results

In this section, we construct an ω-model $\mathcal{M}$ which satisfies CD-M but not ATR_0. Let R be a Π^1_1-random. Let $\mathcal{M}$ be the ω-model whose second-order part is $\bigcup_{i<\omega} \Delta^1_1(\bigoplus_{k<i} R^{[k]})$, where $R^{[k]}$ denotes the kth column of R.

Since the strings of $2^{<\omega}$ are in one-to-one correspondence with ω, we can assume such a correspondence is fixed and abuse notation to also let $G^{[p]}$ denote a column of G whenever $p \in 2^{<\omega}$ and $G \in 2^\omega$.

Proposition 5.1: *The model $\mathcal{M}$ does not satisfy* ATR_0.

Proof: Let a^* be a computable pseudo-ordinal. Then $a^* \in \mathcal{M}$. We claim that a^* has neither a descending sequence, nor a jump hierarchy, in $\mathcal{M}$. If $\Delta^1_1(R_0)$ had one, where $R_0 = \bigoplus_{k<i} R^{[k]}$, then by Theorem 2.1, $\omega_1^{R_0} = \omega_1^{ck}$. Thus there is an ordinal $b \in \mathcal{O}$ such that $H_b^{R_0}$ computes either a jump hierarchy on or a descending sequence in a^*. But recognizing a jump hierarchy or a descending sequence is arithmetic. So

"H_b^X computes a jump hierarchy or descending sequence for a^*"

is a $\Sigma^0_{b+O(1)}$ statement, and it has measure either 0 or 1 because it describes a property of the tail of X. Because R_0 is sufficiently random, and satisfies the statement, the set has measure 1. But then any $b + O(1)$-generic also satisfies the statement. This is a contradiction because there are $b + O(1)$-generics in HYP, but a^* has no hyperarithmetic descending sequence nor any hyperarithmetic jump hierarchy. □

Proposition 5.2: *The model $\mathcal{M}$ satisfies $\mathsf{L}_{\omega_1,\omega}$-$\mathsf{CA}$. Furthermore, whenever $R_0 \in M$ and $\langle \phi_i \rangle \in \Delta^1_1(R_0)$, if $\langle \phi_i \rangle$ is completely determined in $\mathcal{M}$, then it is completely determined in $\Delta^1_1(R_0)$.*

Proof: Suppose that $\langle \phi_j \rangle \in \Delta^1_1(\bigoplus_{i<k} R^{[i]})$ is a sequence of formulas of $\mathsf{L}_{\omega_1,\omega}$-$\mathsf{CA}$ which is completely determined in $\mathcal{M}$. Since $\mathsf{L}_{\omega_1,\omega}$-$\mathsf{CA}$ is a theory of hyperarithmetic analysis, it suffices to show that the sequence is determined in $\Delta^1_1(R_0)$, where $R_0 = \bigoplus_{i<k} R^{[i]}$. Fixing j, there is an $m > k$ such that $\Delta^1_1(\bigoplus_{i<m} R^{[i]})$ contains an evaluation map for ϕ_j. Let $R_1 = \bigoplus_{k\leq i<m} R^{[i]}$. By Van Lambalgen's Theorem for Π^1_1-randoms, R_0 and R_1 are relatively Π^1_1-random. Since $\omega_1^{R_0\oplus R_1} = \omega_1^{ck}$, there is some $a \in \mathcal{O}$ such that this evaluation map is computable from $H_a^{R_0\oplus R_1}$. Then

$$C_j := \{X : H_a^{R_0\oplus X} \text{ computes an evaluation map for } \phi_j\}$$

is a $\Delta^1_1(R_0)$ set which contains the $\Pi^1_1(R_0)$-random R_1. Therefore, C_j has measure 1, so any sufficiently random element computes an evaluation map for ϕ_j. Here, sufficiently random just means more random (relative to R_0) than the descriptive complexity of C_j. So there are elements of $\Delta^1_1(R_0)$ that are sufficiently random. Thus ϕ_j is determined in $\Delta^1_1(R_0)$. □

To show that $\mathcal{M}$ models CD-M, the following classical fact will be useful. It says roughly that if you approximate a bounded function f by using its average values on smaller and smaller partitions of the domain, the resulting sequence converges to f in the L^1 sense.

Lemma 5.3: (WWKL). *If $f \in L^1(2^\omega)$ is bounded and $h_i = \sum_{p\in 2^i}(2^i \int_{[p]} f)\chi_{[p]}$, then $h_i \to f$ in the L^1 norm.*

Proof: Given ε, use Corollary 3.12 to find a closed set B such that the restriction of f to B is continuous, and $\mu(2^\omega) - B < \varepsilon/M$, where M is a bound on f. Let i be large enough that on B, if $x \upharpoonright i = y \upharpoonright i$, then

$|f(x) - f(y)| < \varepsilon$. Then for all strings $p \in 2^i$ and all $x_0 \in [p]$,

$$\begin{aligned}
|h_i(x) - f(x)| &= |(2^i \textstyle\int_{[p]} f) - f(x_0)| \\
&= |(2^i \textstyle\int_{[p]\cap B} f) + (2^i \textstyle\int_{[p]\setminus B} f) - 2^i \textstyle\int_{[p]} f(x_0)| \\
&\qquad (f(x_0) \text{ is a constant.}) \\
&\leq |2^i \textstyle\int_{[p]\cap B}(f - f(x_0))| + |2^i \textstyle\int_{[p]\setminus B} f| + |2^i \textstyle\int_{[p]\setminus B} f(x_0)| \\
&\leq 2^i \textstyle\int_{[p]\cap B} \varepsilon + 2(2^i \textstyle\int_{[p]\setminus B} M).
\end{aligned}$$

Therefore,

$$\begin{aligned}
\int |h_i - f| &= \sum_{p\in 2^i} \int_{[p]} |h_i - f| \\
&\leq \sum_{p\in 2^i} \int_{[p]} 2^i (\textstyle\int_{[p\cap B]} \varepsilon + 2 \textstyle\int_{[p]\setminus B} M) \\
&\leq \sum_{p\in 2^i} (\textstyle\int_{[p\cap B]} \varepsilon + 2 \textstyle\int_{[p]\setminus B} M) \\
&= \textstyle\int_B \varepsilon + 2 \textstyle\int_{2^\omega\setminus B} M \leq \varepsilon + 2\varepsilon.
\end{aligned}$$

□

Lemma 5.4: *Suppose that* $f \in L^1(2^\omega)$, *with name* $\langle f^i\rangle_{i<\omega}$. *Suppose that* R *is* Δ^1_1*-random relative to* $\langle f^i\rangle_{i<\omega}$. *Define a sequence of functions* g^i *by*

$$g^i(X) = \lim_{N\to\infty} \frac{1}{N} \sum_{j<N} f((X \upharpoonright i)^\frown R^{[j]}).$$

Then the functions are well-defined and $g^i \to f$ *in the* L^1 *norm.*

Proof: By 2^i-many applications of Lemma 3.15 to the functions $f_p(R) := f(p^\frown R)$, and since $2^i \int_{[p]} f = \int_{2^\omega} f_p$, we have $g^i(X) = \sum_{p\in 2^i}(2^i \int_{[p]} f)\chi_{[p]}$. Then $g^i \to f$ by Lemma 5.3. □

Theorem 5.5: *Over* WWKL, $\mathsf{CD\text{-}M}$ *is strictly weaker than* ATR_0. *In particular,* $\mathcal{M}$ *satisfies* $\mathsf{WWKL} + \mathsf{CD\text{-}M}$ *but not* ATR_0.

Proof: Suppose that we are given T, a completely determined Borel code. To simplify notation, we assume that $T \in \Delta^1_1$; the result for arbitrary $T \in \mathcal{M}$ follows by relativization. Let $R_0 = R^{[0]}$. Then abusing the column notation further, consider R_0 as being made out of infinitely many distinct and computably identifiable columns, one column for each pair (σ, j), where

$\sigma \in \omega^{<\omega}, j \in \omega$, and let $R^{[\sigma,j]}$ denote the column allocated to that pair. Then letting

$$U := \{(p, \sigma, j) : p^\frown R_0^{[\sigma,j]} \in |T_\sigma|\}$$

we have $U \in \Delta_1^1(R_0)$ by Proposition 5.2. By the same reasoning, we also have that $U_\sigma := \{(p, j) : (\sigma, p, j) \in U\}$ satisfies $U_\sigma \in \Delta_1^1(R_0^{[\sigma]})$, where $R_0^{[\sigma]} = \bigoplus_{j<\omega} R_0^{[\sigma,j]}$.

Therefore, in $\Delta_1^1(R_0)$ we can also find the array of functions $\langle f_\sigma^i \rangle_{\sigma\in T, i\in\omega}$ defined as follows.

$$f_\sigma^i(X) := \limsup_{N\to\infty} \frac{1}{N} \sum_{j<N} U_\sigma(X \restriction i, j).$$

Then define $f_\sigma = \limsup_i f_\sigma^i$. Since the functions $X \mapsto U_\sigma(X \restriction i, j)$ are continuous and bounded above by 1, $f_\sigma \in L^1(2^\omega)$ by the monotone convergence theorem. Of course, the intention is to show that all limsups above can be replaced by limits a.e., and that f_σ represents $|T_\sigma|$ as a measurable set. We prove that $\langle f_\sigma \rangle$ is a measure decomposition by arithmetic transfinite induction within $\mathcal{M}$.

If σ is a leaf then the sequence of functions f_σ^i is eventually constant and equal to the characteristic function of the clopen set coded by T_σ, as desired.

So to complete the proof that $\langle f_\sigma \rangle_{\sigma\in T}$ is measure decomposition, it suffices to show that $\mathcal{M}$ models the following statement for each non-leaf $\sigma \in T$:

"If for all n, $\langle f_{\sigma n \tau} \rangle_{\tau\in T_{\sigma n}}$ is a measure decomposition, then $\langle f_{\sigma\tau} \rangle_{\tau\in T_\sigma}$ is a measure decomposition." That is, assuming $\mathcal{M}$ models the hypothesis, we need to show that $\mathcal{M}$ models:

(1) If σ is a union, then $f_\sigma = \sup_n f_{\sigma n}$.
(2) If σ is an intersection, then $f_\sigma = \inf_n f_{\sigma n}$.

We show the union case; the intersection case is completely symmetric. By Proposition 4.8, for each n, there is a rapidly null G_δ set such that on its complement, $f_{\sigma n}$ is the characteristic function of $|T_{\sigma n}|$. Inspecting the proof of Proposition 4.8, we see that the rapidly null G_δ sets guaranteed there have a uniform Δ_1^0 definition relative to the data $\langle f_{\sigma n \tau} : \tau \in T_{\sigma n}, n \in \omega \rangle$. Let A denote the rapidly null G_δ set obtained by combining these infinitely many tests into a single test. Define

$$R_0^{[<\sigma]} = \bigoplus_{\substack{\tau\in T_{\sigma n} \\ n\in\omega}} R_0^{[\sigma n \tau]}.$$

Since

$$A \leq_T \langle f_{\sigma n \tau} : \tau \in T_{\sigma n}, n \in \omega \rangle \leq_T \bigoplus_{\substack{\tau \in T_{\sigma n} \\ n \in \omega}} U_{\sigma n \tau} \in \Delta^1_1(R^{[<\sigma]})$$

and $R^{[\sigma]}$ is Δ^1_1-random relative to $R^{[<\sigma]}$, each column $R_0^{[\sigma,j]}$ avoids A. Therefore, for each $p \in 2^{<\omega}$ and each j and n, we have

$$p^\frown R_0^{[\sigma,j]} \in |T_{\sigma n}| \iff f_{\sigma n}(p^\frown R_0^{[\sigma,j]}) = 1.$$

Therefore,

$$\begin{aligned} (p,j) \in U_\sigma &\iff p^\frown R_0^{[\sigma,j]} \in |T_\sigma| \\ &\iff \exists n p^\frown R_0^{[\sigma,j]} \in |T_{\sigma n}| \\ &\iff \exists n f_{\sigma n}(p^\frown R_0^{[\sigma,j]}) = 1 \\ &\iff \sup_n f_{\sigma n}(p^\frown R_0^{[\sigma,j]}) = 1. \end{aligned}$$

Here $\sup_n f_{\sigma n}$ has a canonical L^1 name arithmetic in $\langle f_{\sigma n} : n \in \omega \rangle$, and the last bi-implication is justified by Proposition 3.14, since $p^\frown R_0^{[\sigma,j]}$ also avoids the rapidly null G_δ guaranteed there. Thus by Lemma 5.4,

$$\lim_{N \to \infty} \frac{1}{N} \sum_{j<N} \sup_n f_{\sigma n}(p^\frown R_0^{[\sigma,j]})$$

exists for all σ, and $\langle f^i_\sigma \rangle_{i \in \omega}$ is actually a name for $\sup_n f_{\sigma n}$. Therefore, by Proposition 3.14 and Corollary 3.13, for almost all x we have $f_\sigma(x) = \lim_i f^i_\sigma(x) = \sup_n f_{\sigma n}(x)$. Theorem 3.9 then implies that $f_\sigma = \sup_n f_{\sigma n}$, which is what we wanted to prove. □

6. ω-models of CD-M are Closed Under Δ^1_1-randoms

In this section, we show that any ω-model $\mathcal{M}$ of CD-M must be closed under Δ^1_1-randoms, in the sense that for every $Z \in \mathcal{M}$, there is an $R \in \mathcal{M}$ that is Δ^1_1-random relative to Z. We first review the machinery of decorating trees from [ADM+19]. All results summarized here relativize and they will be used in a relativized form, but we state them in unrelativized form to reduce clutter.

The purpose of the operation Decorate is to take a code for a Borel set which may not be completely determined, and force it to become determined for some "small" set of inputs, while not changing its membership facts for other inputs. In our case, "small" will mean measure 0. Roughly speaking, we are going to make a code T and add decorations to ensure that all non Δ^1_1-randoms are determined in T. We will also make sure any measure decomposition is complicated enough to compute a Δ^1_1-random. That way, if there are no Δ^1_1-randoms then the tree is completely determined, at which point the existence of a computationally powerful measure decomposition leads to a contradiction.

Definition 6.1: ([ADM+19]). A *nice decoration generator* is a partial computable function which maps any $b \in \mathcal{O}^*$ to alternating, b-ranked trees (P_b, N_b), where each P_b and N_b have an intersection or a leaf at their root.

For example (and this is what we will use), there is a finite number k such that the following almost defines a nice decoration generator.

$$\begin{aligned} P_{b+k} &= \{X : X \text{ is not } MLR^{H_b}, \text{ but for all } c <_* b,\, X \text{ is } MLR^{H_c}\} \\ &\qquad \cap \{X : X <_{\text{lex}} H_b\} \\ N_{b+k} &= \{X : X \text{ is not } MLR^{H_b}, \text{ but for all } c <_* b,\, X \text{ is } MLR^{H_c}\} \\ &\qquad \cap \{X : X \geq_{\text{lex}} H_b\}. \end{aligned}$$

All that remains is to define P_b and N_b when b is within k successors of a limit ordinal; in that case we set both P_b and N_b to be b-ranked alternating codes for the empty set.

The operation Decorate is defined below using effective transfinite recursion (with parameter $<_*$ which is computable from $\emptyset'$), and therefore is well-defined on a-ranked trees T for all $a \in \mathcal{O}^{*,T}$.

Definition 6.2: ([ADM+19]). The operation Decorate is defined as follows. The inputs are an a-ranked labeled tree T and a nice decoration generator h.

$$\begin{aligned} \text{Decorate}(T, h) = \{\lambda\} &\cup \bigcup_{\langle n \rangle \in T} \langle 2n \rangle^\frown \text{Decorate}(T_{\langle n \rangle}, h) \\ &\cup \bigcup_{b <_* \rho_T(\lambda)} \langle 2b+1 \rangle^\frown \text{Decorate}(Q_b, h) \end{aligned}$$

where $Q_b = P_b$ if λ is a $\bigcup$ in T, and $Q_b = N_b^c$ if λ is a $\bigcap$ in T.

The rank and label of λ in Decorate(T, h) are defined to coincide with the rank and label of λ in T. The ranks and labels of the other nodes in Decorate(T, h) are inherited from Decorate$(T_{\langle n \rangle}, h)$ or Decorate(Q_b, h) as appropriate.

If T is a-ranked, so is Decorate(T, h). Similarly, if T and each P_b and N_b are alternating, then Decorate(T, h) will also be alternating. (Note that in this case, N_b^c has a union at its root).

Lemma 6.3: ([ADM$^+$19]). *Let h be a nice decoration generator. Suppose $b \in \mathcal{O}$, and suppose that $X \notin |P_d| \cup |N_d|$ for any $d <_* b$. Then for any b-ranked tree T, $X \in |\operatorname{Decorate}(T,h)|$ if and only if $X \in |T|$.*

Lemma 6.4: ([ADM$^+$19]). *Let $a \in \mathcal{O}^*$ and $b \in \mathcal{O}$ with $b <_* a$. Let T be an alternating, a-ranked tree and let h be a nice decoration generator. Suppose $X \in |P_b| \cup |N_b|$. Then*

(1) X has a unique evaluation map in $\operatorname{Decorate}(T,h)$.
(2) This evaluation map is $H^{X\oplus T}_{b+O(1)}$-computable.

Theorem 6.5: *Suppose that $\mathcal{M}$ is an ω-model of* WWKL + CD-M. *Then for any $Z \in \mathcal{M}$, there is an $R \in \mathcal{M}$ such that R is Δ^1_1-random relative to Z.*

Proof: If $\mathcal{M}$ is a β-model, then $\mathcal{M}$ is already closed under Δ^1_1-randoms in the sense described above, because the statement $\exists R(R$ is $\Delta^1_1(Z)$-random$)$ is a true $\Sigma^1_1(Z)$ statement, and any witness to its truth computes such an R.

On the other hand, if $\mathcal{M}$ is not a β-model, then there is a tree $S \in M$ such that $\mathcal{M}$ believes S to be well-founded, but in fact S is ill-founded. Without loss of generality, assume that $Z \geq_T S$; otherwise we end up with a Δ^1_1-random relative to $Z \oplus S$. There is a Z-computable procedure which, given any truly well-founded tree as input, produces an element of $\mathcal{O}^Z$ which bounds its rank. Apply this procedure to S to produce a pseudo-ordinal $a^* \in (\mathcal{O}^*)^Z$. Then $\mathcal{M}$ thinks that a^* is an ordinal. Let T be any Z-computable, alternating, (a^*+1)-ranked tree such that each level-one subtree T_n is a^*-ranked. We can assume T has a union at the root, though the symmetric choice would also work. Let h be the nice decoration generator which produces codes for P^Z_b and N^Z_b as follows (this is just the relativized form of what was defined above).

$$\begin{aligned} P^Z_{b+k} &= \{X : X \text{ is not } MLR^{H^Z_b}, \text{ but for all } c <^Z_* b, X \text{ is } MLR^{H^Z_c}\} \\ &\quad \cap \{X : X <_{\text{lex}} H^Z_b\} \\ N^Z_{b+k} &= \{X : X \text{ is not } MLR^{H^Z_b}, \text{ but for all } c <^Z_* b, X \text{ is } MLR^{H^Z_c}\} \\ &\quad \cap \{X : X \geq_{\text{lex}} H^Z_b\}. \end{aligned}$$

As above, we also define P^Z_b and N^Z_b to be b-ranked codes for the empty set in case b is within k successors of a limit ordinal. Now consider the tree $\operatorname{Decorate}^Z(T,h)$. Is it completely determined?

Suppose it is not completely determined; let X be an element that does not have an evaluation map. Since CD-M + WWKL implies $\mathsf{L}_{\omega_1,\omega}$-CA, every element of $HYP(X \oplus Z)$ is in $\mathcal{M}$. So by Lemma 6.4, for any $b \in \mathcal{O}^Z$, $X \notin |P_b^Z| \cup |N_b^Z|$ (if it were in this set, it would have a $HYP(X \oplus Z)$ evaluation map). But this means that X is Δ_1^1-random relative to Z, since each non-random belongs to some $|P_b^Z| \cup |N_b^Z|$.

So suppose that $\text{Decorate}^Z(T, h)$ is completely determined. Then by CD-M, it has a measure decomposition. We claim that any element R that is 1-random relative to the measure decomposition is in fact Δ_1^1-random relative to Z. It suffices to show that the measure decomposition computes H_b^Z for all $b \in \mathcal{O}^Z$. Fix $b \in \mathcal{O}^Z$ with $b <_*^Z a^*$ and observe that $\text{Decorate}^Z(P_{b+k}, h)$ appears as a level-one subtree of $\text{Decorate}^Z(T, h)$. Thus, by examining the definition of P_{b+k}, which has an intersection at the root and $\{X : X <_{\text{lex}} H_b^Z\}$ as a level-one subtree, we see that $\text{Decorate}^Z(\{X : X <_{\text{lex}} H_b^Z\}, h)$ appears as a level-two subtree of $\text{Decorate}^Z(T, h)$. (Here of course, $\{X : X <_{\text{lex}} H_b^Z\}$ is represented using an approximately b-ranked formula of $L_{\omega_1,\omega}$, but this formula contributes computational, not topological, complexity.) Therefore, there is an L^1 function f included in the measure decomposition which is equal to the characteristic function of $\text{Decorate}^Z(\{X : X <_{\text{lex}} H_b^Z\}, h)$ almost everywhere. We claim that $\int f = H_b^Z$, where here we regard H_b^Z as a number in $[0, 1]$ given by its binary expansion. Using WWKL, it suffices to provide another L^1 function g which has $\int g = H_b^Z$ and such that g is equal to the characteristic function of $\text{Decorate}^Z(\{X : X <_{\text{lex}} H_b^Z\}, h)$ almost everywhere. Let g be the canonical measurable characteristic function of the open set $\bigcup_{p <_{\text{lex}} H_b^Z} [p]$. Then by Lemma 6.3, for any X that is $MLR^{H_b^Z}$, since $X \notin |P_d^Z| \cup |N_d^Z|$ for any $d <_*^Z b + k$, we have $X \in \text{Decorate}^Z(\{X : X <_{\text{lex}} H_b^Z\}, h)$ if and only if $X <_{\text{lex}} H_b^Z$, which is true if and only if $g(X) = 1$. This completes the proof. □

References

ADM+19. Eric P. Astor, Damir Dzhafarov, Antonio Montalbán, Reed Solomon and Linda Brown Westrick. The determined property of Baire in reverse math. *J. Symbolic Logic*, Accepted 2019. Available arXiv:1809.03940.

ADR12. Jeremy Avigad, Edward T. Dean and Jason Rute. Algorithmic randomness, reverse mathematics, and the dominated convergence theorem. *Ann. Pure Appl. Logic*, 163(12):1854–1864, 2012.

BGS02. Douglas K. Brown, Mariagnese Giusto and Stephen G. Simpson. Vitali's theorem and WWKL. *Arch. Math. Logic*, 41(2):191–206, 2002.

Bog07. V. I. Bogachev. *Measure theory. Vol. I, II*. Springer-Verlag, Berlin, 2007.

CNY08. C. T. Chong, Andre Nies and Liang Yu. Lowness of higher randomness notions. *Israel J. Math.*, 166:39–60, 2008.

DFSW17. Damir Dzhafarov, Stephen Flood, Reed Solomon and Linda Brown Westrick. Effectiveness for the Dual Ramsey Theorem. Submitted 2017. Available arXiv:1710.00070.

DM13. Adam R. Day and Joseph S. Miller. Randomness for non-computable measures. *Trans. Amer. Math. Soc.*, 365(7):3575–3591, 2013.

HN07. Greg Hjorth and André Nies. Randomness via effective descriptive set theory. *J. Lond. Math. Soc. (2)*, 75(2):495–508, 2007.

Sim09. Stephen G. Simpson. *Subsystems of Second Order Arithmetic.* Perspectives in Logic. Cambridge University Press, Cambridge; Association for Symbolic Logic, Poughkeepsie, NY, second edition, 2009.

Ste73. Jacques Stern. Réels aléatoires et ensembles de mesure nulle en théorie descriptive des ensembles. *C. R. Acad. Sci. Paris Sér. A-B*, 276:A1249–A1252, 1973.

Ste75. Jacques Stern. Some measure theoretic results in effective descriptive set theory. *Israel J. Math.*, 20(2):97–110, 1975.

YS90. Xiaokang Yu and Stephen G. Simpson. Measure theory and weak König's lemma. *Arch. Math. Logic*, 30(3):171–180, 1990.

Yu90. Xiaokang Yu. Radon-Nikodým theorem is equivalent to arithmetical comprehension. In *Logic and computation (Pittsburgh, PA, 1987)*, volume 106 of *Contemp. Math.*, pp. 289–297. Amer. Math. Soc., Providence, RI, 1990.

Yu93. Xiaokang Yu. Riesz representation theorem, Borel measures and subsystems of second-order arithmetic. *Ann. Pure Appl. Logic*, 59(1):65–78, 1993.

Yu94. Xiaokang Yu. Lebesgue convergence theorems and reverse mathematics. *Math. Logic Quart.*, 40(1):1–13, 1994.

www.ingramcontent.com/pod-product-compliance
Lightning Source LLC
LaVergne TN
LVHW020024220325
806384LV00002B/43
* 9 7 8 9 8 1 9 8 0 6 5 7 7 *